SENTIENT ARCHAEOLOGIES

SENTIENT ARCHAEOLOGIES

GLOBAL PERSPECTIVES ON PLACES, OBJECTS, AND PRACTICE

Essays in honour of Professor Chris Gosden

Edited by

COURTNEY NIMURA, REBECCA O'SULLIVAN, AND
RICHARD BRADLEY

Oxford & Philadelphia

Published in the United Kingdom in 2023 by
OXBOW BOOKS
The Old Music Hall, 106–108 Cowley Road, Oxford, OX4 1JE

and in the United States by
OXBOW BOOKS
1950 Lawrence Road, Havertown, PA 19083

© Oxbow Books and the individual authors 2023

Hardback Edition: ISBN 978-1-78925-932-2
Digital Edition: ISBN 978-1-78925-933-9 (ePub)

A CIP record for this book is available from the British Library

Library of Congress Control Number: 2023933482

All rights reserved. No part of this book may be reproduced or transmitted in any form or by any means, electronic or mechanical including photocopying, recording or by any information storage and retrieval system, without permission from the publisher in writing.

Printed in Malta by Melita Press
Typeset in India by DiTech Publishing Services

For a complete list of Oxbow titles, please contact:

UNITED KINGDOM
Oxbow Books
Telephone (0)1226 734350
Email: oxbow@oxbowbooks.com
www.oxbowbooks.com

UNITED STATES OF AMERICA
Oxbow Books
Telephone (610) 853-9131, Fax (610) 853-9146
Email: queries@casemateacademic.com
www.casemateacademic.com/oxbow

Oxbow Books is part of the Casemate Group

Front and back cover: 'Waterland' by Miranda Creswell (oil on canvas, 40 × 51 cm, 2022)

Contents

Contributors	vii
List of Figures and Tables	xi

1. Living Archaeology — 1
   Rebecca O'Sullivan, Courtney Nimura, and Richard Bradley

2. Reflections on Populating the Western Pacific — 9
   Glenn Summerhayes

3. Diversity and Difference in New Britain, Papua New Guinea: Seeking Indigenous Communities in the Archaeological Record — 19
   Jim Specht and Robin Torrence

4. Why the Concept of Near and Remote Oceania Fails Island Melanesian Prehistory — 27
   Christophe Sand and Jim Allen

5. Storied Landscapes in the Palaeolithic? The View from the Cave — 39
   Graeme Barker and Chris O. Hunt

6. A Circular Tomb with 'Stones' of Clay: The Tomb of Lord Bai of Zhongli, Anhui Province, Central China, Early 6th Century BC — 51
   Jessica Rawson

7. Agricultural Places as Processes — 59
   Amy Bogaard

8. A *Viereckschanze* in Oxfordshire, England? Enclosure and Memory at Marcham — 65
   Gary Lock and Sheila Raven

9. A Landscape's Memory: The Long-term Impact of Proto-industrial Salt Extraction in the Seille Valley in France — 73
   Laurent Olivier

10. Taking, Using, and Giving Back Again: The Deposition of Living Matter in Ancient Europe — 81
    Richard Bradley

11. Rock Art: A Marker of Concepts and Practices — 87
    Courtney Nimura, Rebecca O'Sullivan, and Peter Hommel

12. Celtic Art Beyond Metal: Material Matters in Iron Age and Early Roman Southern England — 95
    Sarah Downum and Duncan Garrow

13. Jet and Gender in Late Roman Britain — 103
    Cameron Moffett

14. Using Coinage and Die-Studies to Obtain Evidence about Society in the Late Iron Age — 109
    John Talbot

15. 'Keep on Truckin' – Thoughts from the Back of a Bus — 117
    A. M. Pollard

16. Biography and Technology: A Bronze *Ding* Vessel of the Iron Age in China   123
    Xiuzhen Li
17. Rewriting Global Histories of Human–Material Relations in Different Cultural Contexts   131
    Shadreck Chirikure
18. Collections of Aboriginal Ground Stone Tools from the Murray Darling Basin: Function, Temporality, and Social Context   137
    Richard Fullagar, Elspeth Hayes, and Colin Pardoe
19. Cultural and Landscape Change in Australia's World Heritage Wet Tropics Bioregion, Northeast Queensland   145
    Richard Cosgrove
20. What's Involved in Technological Change? Aboriginal Marine Hunting in Tropical North Australia   155
    Harry Allen
21. The Yolŋu System as a Regional Polity   163
    Howard Morphy with Frances Morphy
22. Anthropology and Archaeology: A Necessary Unity   173
    Lambros Malafouris
23. On Ontological Impurity: Conceptualising Time in Archaeology   183
    John Robb
24. Archaeology, Heritage, and the Heritage of Archaeology   191
    Ian Lilley
25. Selling Photographs: Collecting Archaeology   199
    Elizabeth Edwards
26. On the Origins of Khami: Evidence from the Henry Balfour Collection, Pitt Rivers Museum, Oxford   209
    Innocent Pikirayi
27. *In Dreams the Heart*: Impermanence at the Museum   223
    Chantal Knowles
28. A Civil Servant Walks onto a Neolithic Barrow…: Sir Lindsay Scott and the Whiteleaf Oval Barrow   231
    Gill Hey
29. Redirecting the Field – Total Archaeologies, Flagships, and Sample Design   241
    Christopher Evans
30. Oxford Intelligence   257
    Lynn Meskell

Contributors

HARRY ALLEN, ASSOCIATE PROFESSOR OF ANTHROPOLOGY
School of Social Sciences, University of Auckland
32 Campbell Road, Maraetai Beach
Auckland, 2018
New Zealand
h.allen@auckland.ac.nz

JIM ALLEN, EMERITUS PROFESSOR
Department of Archaeology, School of Archaeology and History
La Trobe University
PO Box 5020
Broulee, NSW 2537
Australia
jjallen8@optusnet.com.au

GRAEME BARKER, DISNEY PROFESSOR OF ARCHAEOLOGY EMERITUS
McDonald Institute for Archaeological Research,
University of Cambridge
Downing Street
Cambridge, CB2 3ER
United Kingdom
gb314@cam.ac.uk

AMY BOGAARD, PROFESSOR OF NEOLITHIC AND BRONZE AGE ARCHAEOLOGY
School of Archaeology, University of Oxford
1 South Parks Rd
Oxford, OX1 3TG
United Kingdom
amy.bogaard@arch.ox.ac.uk

RICHARD BRADLEY, EMERITUS PROFESSOR
Department of Archaeology, University of Reading
Whiteknights Box 227
Reading, RG6 6AB
r.j.bradley@reading.ac.uk

SHADRECK CHIRIKURE, EDWARD HALL PROFESSOR OF ARCHAEOLOGICAL SCIENCE
School of Archaeology, University of Oxford
1 South Parks Rd
Oxford, OX1 3TG
United Kingdom
shadreck.chirikure@arch.ox.ac.uk

RICHARD COSGROVE, ADJUNCT PROFESSOR
Department of Archaeology and History
La Trobe University
Bundoora
Melbourne, Australia 3086
r.cosgrove@latrobe.edu.au

MIRANDA CRESWELL, ARTIST-IN-RESIDENCE
School of Archaeology, University of Oxford
1 S Parks Rd
Oxford, OX1 3TG
United Kingdom
miranda.creswell@arch.ox.ac.uk

SARAH DOWNUM, INDEPENDENT RESEARCHER
Department of Archaeology, University of Reading
Whiteknights Box 227
Reading, RG6 6AB
United Kingdom
s.m.downum@pgr.reading.ac.uk

ELIZABETH EDWARDS, EMERITUS PROFESSOR OF PHOTOGRAPHIC HISTORY
Photographic History Research Centre
School of Humanities, De Montfort University
The Gateway
Leicester, LE1 9BH
United Kingdom
ejmedwards10@gmail.com

CHRISTOPHER EVANS, SENIOR FELLOW
McDonald Institute for Archaeological Research,
Emeritus Director, Cambridge Archaeological Unit
University of Cambridge
Downing Street
Cambridge, CB2 3ER
cje30@cam.ac.uk

RICHARD FULLAGAR, HONORARY PROFESSOR
Flinders University, University of Western Australia, and
University of Wollongong
769 Park Street
Brunswick, VIC 3056
Australia
richard.fullagar@coreartefactresearch.com.au

DUNCAN GARROW, PROFESSOR OF PREHISTORIC ARCHAEOLOGY
Department of Archaeology, University of Reading
Whiteknights Box 227
Reading, RG6 6AB
United Kingdom
d.j.garrow@reading.ac.uk

ELSPETH HAYES, DIRECTOR OF MICROTRACE ARCHAEOLOGY
Honorary Research Fellow in the Centre for
Archaeological Science
University of Wollongong
PO Box 102
Wollongong, NSW 2520
Australia
ehayes@mtrace.com.au

GILL HEY, RETIRED CEO, OXFORD ARCHAEOLOGY
7 Kendal Green
Kendal
Cumbria, LA9 5PN
United Kingdom
gill.hey@oxfordarchaeology.com

PETER HOMMEL, LECTURER IN ARCHAEOLOGICAL MATERIALS
Department of Archaeology, Classics and Egyptology,
University of Liverpool
12–14 Abercromby Square
Liverpool, L69 7WZ
United Kingdom
peter.hommel@liverpool.ac.uk

CHRIS O. HUNT, EMERITUS PROFESSOR
School of Biological and Environmental Sciences,
Liverpool John Moores University
Byrom Street
Liverpool, L3 3AF
United Kingdom
C.O.Hunt@ljmu.ac.uk

CHANTAL KNOWLES, HEAD OF HUMAN HISTORY
Tāmaki Paenga Hira Auckland War Memorial Museum
Auckland Domain, Private Bag 92018
Victoria Street West, Auckland 1142
Aotearoa New Zealand
cknowles@aucklandmuseum.com

XIUZHEN LI, SENIOR RESEARCH FELLOW
Institute of Archaeology, University College London
31–34 Gordon Square
London, WC1H 0PY
United Kingdom
xiuzhen.li@ucl.ac.uk

IAN LILLEY, EMERITUS PROFESSOR
School of Social Science, University of Queensland
St Lucia, QLD 4072
Australia
i.lilley@uq.edu.au

GARY LOCK, EMERITUS PROFESSOR
Institute of Archaeology, University of Oxford
34–36 Beaumont Street
Oxford, OX1 2PG
United Kingdom
gary.lock@arch.ox.ac.uk

LAMBROS MALAFOURIS, PROFESSOR OF COGNITIVE AND
ANTHROPOLOGICAL ARCHAEOLOGY
Institute of Archaeology, University of Oxford
34–36 Beaumont Street
Oxford, OX1 2PG
United Kingdom
lambros.malafouris@arch.ox.ac.uk

LYNN MESKELL, PENN INTEGRATES KNOWLEDGE (PIK)
PROFESSOR OF ANTHROPOLOGY
Department of Anthropology, School of Arts & Sciences,
Penn Museum
Weitzman School of Design, University of Pennsylvania
3260 South Street
Philadelphia, PA 19104
USA
lmeskell@upenn.edu

CAMERON MOFFETT, CURATOR OF COLLECTIONS (WEST)
English Heritage
Boscobel House, Brewood
Shropshire, ST19 9AR
United Kingdom
cameron.moffett@english-heritage.org.uk

FRANCES MORPHY, HONORARY ASSOCIATE PROFESSOR
Centre for Aboriginal Economic Policy Research,
Australian National University
Copland Building, 24 Kingsley Pl
Acton, ACT 2601
Australia
frances.morphy@anu.edu.au

HOWARD MORPHY, EMERITUS PROFESSOR AND HEAD OF THE
CENTRE FOR DIGITAL HUMANITIES RESEARCH
Research School of Humanities and the Arts,
Australian National University
Sir Roland Wilson Building, 120 McCoy Cct
Acton, ACT 2601
Australia
howard.morphy@anu.edu.au

COURTNEY NIMURA, CURATOR FOR LATER EUROPEAN
PREHISTORY
Ashmolean Museum of Art and Archaeology
University of Oxford
Beaumont Street
Oxford, OX1 2PH
United Kingdom
courtney.nimura@ashmus.ox.ac.uk

LAURENT OLIVIER, CURATOR IN CHIEF OF CELTIC AND GALLIC
ARCHAEOLOGY
Musée d'Archéologie Nationale
Place Charles de Gaulle, 78100
Saint-Germain-en-Laye
France
laurent.olivier@culture.gouv.fr

REBECCA O'SULLIVAN, HUMBOLDT RESEARCH FELLOW
Department for Pre- and Early Historical Archaeology,
University of Bonn
Brühler Straße 7
53119 Bonn
Germany
rebecca.osullivan@uni-bonn.de

COLIN PARDOE
Department of Archaeology and Natural History
Australian National University
16 Hackett Gardens,
Turner, ACT 2612
Australia
colin.pardoe@ozemail.com.au

INNOCENT PIKIRAYI, PROFESSOR IN ARCHAEOLOGY
Department of Anthropology, Archaeology and
Development Studies,
Faculty of Humanities, University of Pretoria
Hatfield, 0028
South Africa
innocent.pikirayi@up.ac.za

MARK POLLARD, EMERITUS PROFESSOR
School of Archaeology, University of Oxford
1 South Parks Rd
Oxford, OX1 3TG
United Kingdom
mark.pollard@arch.ox.ac.uk

SHEILA RAVEN, FINDS SPECIALIST
24 Newland Street
Eynsham, OX29 4JZ
United Kingdom
sheila@wildbury.com

DAME JESSICA RAWSON, EMERITUS PROFESSOR
Merton College
Merton Street
Oxford, OX1 4JD
United Kingdom
jessica.rawson@merton.ox.ac.uk

JOHN ROBB, PROFESSOR OF EUROPEAN PREHISTORY
McDonald Institute for Archaeological Research,
University of Cambridge
Downing Street
Cambridge, CB2 3DZ UK
United Kingdom
jer39@cam.ac.uk

CHRISTOPHE SAND, DIRECTOR
Institute of Archaeology of New Caledonia and the
Pacific (IANCP)
65 rue Teyssandier de Laubarède
Montravel BP 11423
98802 NOUMEA
New Caledonia
christophe.sand@ird.fr

JIM SPECHT, SENIOR FELLOW AND HONORARY ASSOCIATE
Australian Museum and University of Sydney
1 William Street
Sydney NSW 2010
Australia
jspecht@bigpond.com

GLENN SUMMERHAYES, PROFESSOR OF ARCHAEOLOGY
School of Social Sciences, University of Otago
PO Box 56
Dunedin
New Zealand
glenn.summerhayes@otago.ac.nz

JOHN TALBOT, HONORARY RESEARCH ASSOCIATE
Institute of Archaeology, University of Oxford
34–36 Beaumont Street
Oxford, OX1 2PG
United Kingdom
john@jtalbot.co.uk

ROBIN TORRENCE, SENIOR FELLOW AND HONORARY ASSOCIATE
Australian Museum and University of Sydney
1 William Street
Sydney, NSW 2010
Australia
robin.torrence@australian.museum

List of Figures and Tables

List of Figures

Fig. 1.1.	Chris Gosden	1
Fig. 1.2.	Chris Gosden montage	2
Fig. 2.1.	Spread of Lapita over the western Pacific	10
Fig. 2.2.	Lapita sites in the Bismarck Archipelago	11
Fig. 2.3.	(a) Early Lapita pottery; (b) Later Lapita pottery	14
Fig. 2.4.	Later Lapita pottery (from the Arawe Islands)	15
Fig. 3.1.	The Bismarck Archipelago in relation to other western Pacific Islands, the main obsidian source regions, and other locations cited in the text	20
Fig. 3.2.	Stemmed tools from surface collections in New Britain	22
Fig. 3.3.	Distribution of Lapita sites and site clusters on New Britain and neighbouring islands	23
Fig. 4.1.	Map of Melanesia showing boundaries and localities mentioned in the text	28
Fig. 4.2.	Enveloped crosses on petroglyphs from Poro, New Caledonia	31
Fig. 4.3.	Megalithic ceremonial platform at Nasara Peterhil on Vao Island, Vanuatu	32
Fig. 5.1.	The locations of the Haua Fteah cave (Libya), Shanidar Cave (Iraqi Kurdistan), and the Niah Caves (Sarawak) with images of the caves	40
Fig. 5.2.	The sentient archaeologist in the field	41
Fig. 5.3.	The entrance to Shanidar Cave in spring 2022, showing dark green vegetation growing on the talus below the entrance arch and the wealth of wildflowers at this season	45
Fig. 6.1.	The tomb of Lord Bai, early 6th century BC	52
Fig. 6.2.	Drawings of the terrace wall around the tomb of Lord Bai	54
Fig. 6.3.	Plan of Arzhan 2	55
Fig. 8.1.	Geophysics of the western half of the Marcham site showing the three enclosures	67
Fig. 8.2.	Section drawn across Trench 2 showing cut features	68
Fig. 8.3.	Trench 3 showing the barrel-shaped enclosure entrance with internal cut features	68
Fig. 9.1.	Excavation at Marsal (Moselle) 'la Digue' in 2012	74
Fig. 9.2.	A reconstruction of the Iron Age salt industry in the upper Seille valley during the 6th century BC, around Marsal (Moselle)	75
Fig. 9.3.	The city of Marsal in the late 17th century AD	77
Fig. 11.1.	Map of Eurasia showing the three regions examined in the text: (a) Northern Europe; (b) Central Eurasia; (c) Eastern Asia	88
Fig. 11.2.	An eastern Eurasian vehicle motif	89
Fig. 11.3.	Examples of motifs found in southern Scandinavian rock art, from the site of Hästholmen, Sweden	91
Fig. 11.4.	Animal-headed dagger/knife depicted on a Late Bronze Age deer stone in northern Mongolia, with comparanda in the metalwork of the Karasuk Culture	91
Fig. 12.1.	Celtic Art on 'other' materials	96
Fig. 12.2.	Decoration by style zones	98
Fig. 12.3.	Materials and their decoration	99

Fig. 12.4.	The relationship between motifs and materials, ordered by prevalence in ceramics and metal	100
Fig. 13.1.	Jet pendant amulet in the form of a breast	105
Fig. 13.2.	Piece of jet inlay depicting a satyr(?)	106
Fig. 14.1.	Obverses of Gallo-Belgic A, Icenian, and Durotrigan staters	111
Fig. 14.2.	The reverse of Durotrigan staters	111
Fig. 14.3.	Pelleted sub-type (die-group 49)	111
Fig. 14.4.	Examples of staters from Durotrigan die-groups 1, 2, and 3	112
Fig. 14.5.	Distributions of Durotrigan (a) die-group 1, (b) die-group 2, (c) die-group 3, and (d) pelleted sub-type with rivers, counties (dashed lines), and major hillforts	113
Fig. 14.6.	The ageing of die YZ paired with reverse dies 4, 8, and 12, with the least worn on the left to the most worn on the right	114
Fig. 14.7.	The presence of the 15 die-group 1 reverse dies in the Brighstone hoard; the numbers of die-group 1 reverse dies in the Shorwell hoard; and die-group 1 reverse dies from neither Brighstone nor Shorwell	114
Fig. 16.1.	The bronze *ding* vessel. Emperor Qinshihuang's Mausoleum Site Museum collection, Shaanxi Province	124
Fig. 16.2.	Chiselled inscriptions of the Qin period, including the phrase *yi dou si sheng* and characters for Xian and Yiyang; magnification of the character *yi* showing the use of a combination of chisel marks; SEM micrograph showing detail of overlapping chisel marks used for the character *yi*	127
Fig. 16.3.	Wheel-incised inscription created in the Han Dynasty; SEM micrograph showing the curvature of lines used in the Han inscription	128
Fig. 18.1.	Location of Murray Darling Basin (blue rivers), locations mentioned in the text, and the distribution of potential grinding stone materials	138
Fig. 18.2.	A farmer's shed with grinding stones collected from paddocks	138
Fig. 18.3.	MDB axe heads; excavated quandong stone from Exford, Victoria; quandong stone from the MDB; family heirloom; MDB basket stones; MDB sandstone travelling plate pair; using travelling plates to grind seeds	142
Fig. 19.1.	Map of the study area showing the extent of current rainforest, archaeological sites mentioned in the text, ooyurka artefact localities and number found, directions of trade, and exchange items	146
Fig. 19.2.	The ooyurka artefact and its flat polished working edge	149
Fig. 19.3.	Eight ooyurkas demonstrate the variety of shapes and sizes of these tools	150
Fig. 20.1.	Distribution of Aboriginal watercraft types across northern Australia	157
Fig. 21.1.	Location of Yolŋu-matha languages and of known -*tjpi* toponyms	165
Fig. 21.2.	Location of the mapped -*tjpi* toponyms	166
Fig. 24.1.	Google Ngram comparing use in English of the terms 'heritage' and 'archaeology' since 1800	196
Fig. 25.1.	*Our Ancient Monuments*. Arthur's Round Table, Penrith, Cumbria	201
Fig. 25.2.	*Our Ancient Monuments*. Stonehenge	201
Fig. 25.3.	*Our Ancient Monuments*. Sueno's Stone, Forres	202
Fig. 25.4.	*Our Ancient Monuments*. Photographs of Avebury, Silbury Hill, Stonehenge, and Old Sarum	202
Fig. 25.5.	Loose albumen print. Cromlech, Trellys [Ffyst Samson], Pembrokeshire	203
Fig. 25.6.	Loose albumen print. Anglo Saxon chapel of St Laurence, Bradford-on-Avon	204
Fig. 26.1.	The location of the site of Khami on the southwestern Zimbabwe plateau	210
Fig. 26.2.	The Hill Complex at Khami	211
Fig. 26.3.	Lip morphology	214
Fig. 26.4.	Distribution of vessel thickness in the assemblage	214
Fig. 26.5.	Decorated bowls from Khami	215
Fig. 26.6.	Sherds from constricted bowls or globular vessels with in-sloping necks/rims	215
Fig. 26.7.	Vessels with tall necks decorated with complex incised triangular designs painted with alternative zones of red ochre and graphite burnish	216
Fig. 26.8.	Vessels with tall necks decorated with complex incised triangular designs painted with alternative zones of red ochre and graphite burnish	216
Fig. 26.9.	Vessels decorated with incised patterns of herringbone on the neck-shoulder region	217
Fig. 26.10.	Vessels with tall necks and fluted rims	218
Fig. 27.1.	*In Dreams the Heart* by Danie Mellor	224
Fig. 27.2.	A grouped display of shields in 'Danie Mellor: Exotic Lies Sacred Ties', a University of Queensland Art Museum 10-year touring survey curated by Maudie Palmer AO, 2014	225

Fig. 27.3.	Granny May Kelly photographed by Alfred Atkinson *ca.* 1908, Gimuy/Cairns, Queensland	227
Fig. 28.1.	Whiteleaf Hill from the air, taken by Major Allen in August 1934 just as work was beginning	232
Fig. 28.2.	The Neolithic 'barrow' as work began in 2003; cleaning the northwest quadrant of Scott's excavation area, showing his spoil heap in section	234
Fig. 28.3.	Lindsay and Winifred Scott sailing in the 1920s	237
Fig. 28.4.	Chris Gosden with staff and students on tea break at Dorchester	238
Fig. 29.1.	Mucking (I)	243
Fig. 29.2.	Mucking (II)	244
Fig. 29.3.	Great Wilbraham	245
Fig. 29.4.	Maxey	247
Fig. 29.5.	Surface deposit investigations	248
Fig. 29.6.	'School' affinities	251
Fig. 30.1.	D. G. Hogarth in the Middle East	259
Fig. 30.2.	T. E. Lawrence, D. G. Hogarth, and Lt. Col. Dawnay	261

List of Tables

Table 5.1.	Langley's proposed potential archaeological differences between Palaeolithic populations that participate in landscape socialisation and those that use symbolic material culture solely for the mediation of interpersonal social interaction (adapted from Langley 2013, table 1)	41
Table 5.2.	The major cultural phases of the Haua Fteah, Shanidar Cave, and the West Mouth of Niah Great Cave and their approximate chronologies in dates ka (thousands of years before the present)	43
Table 13.1.	Jet objects from civilian sites compared to jet objects from Richborough Fort	107
Table 14.1.	The relative presence of BHB Stater dies in Dallinghoo hoard	110
Table 14.2.	Indicative phases of Durotrigan coinage	112
Table 14.3.	Key statistics for three Durotrigan die-groups	112
Table 14.4.	Analysis of the contents of the Brighstone hoard	113
Table 14.5.	Analysis of the Durotrigan staters in the Shorwell hoard	113
Table 14.6.	Finds of die-group 23 coins with a provenance in chronological order	115
Table 26.1.	Vessel shapes from the Balfour collection compared with other assemblages excavated at Khami	213

1

Living Archaeology

Rebecca O'Sullivan, Courtney Nimura, and Richard Bradley

A major trend in archaeology over the last century has been the increasingly rapid departure from theoretical frameworks that treat the remains of past societies as static snapshots of particular moments in time. The move towards interpretations that prioritise change and variability as factors influencing the formation of the archaeological record has been notable, due in no small part to the work of Chris Gosden (Fig. 1.1).[1] Chris has at no point limited himself to one or two perspectives, engaging at various times with artefact biographies, cognition, creativity, social identity, and technology. Though seemingly abstract, Chris emphasises ideas and the best ways to use them, and these theories have underpinned his highly influential work and collaborations on various subjects, such as colonialism, magic, monumental architecture, and prehistoric art (see Selected Works appended). This approach has served to position the living world at the fore of archaeological investigation, a stark contrast from focusing on disembodied relics, casting archaeology in the role of creative process[2] as opposed to a set of fixed methods and approaches.

Though established analytical concepts, such as typology, remain key parts of the archaeologist's investigative toolkit, data-gathering strategies and interpretative frameworks have become infused progressively with the concept that archaeology – both in the sense of its objects of study and the discipline as a whole – is living, thus it is continually sensitive to developments in the immediate environment. The title of this volume, *Sentient Archaeologies*, reflects this concept and acknowledges the significant number of researchers across the world who are integrating ideas informed by relational epistemologies and mutually constructive ontologies into their work, from the initial stage of project design all the way down to post-excavation interpretation. The volume showcases examples of such work, highlighting the utility of these ideas to exploring material both old and new. The theories and methods used are as wide-ranging as Chris's own writings and share the same concern with illuminating the experience of past people.

The chapters are organised into three broad themes: 1) People & Places; 2) Form & Flow; and 3) History & Heritage. The themes are intentionally broad to capture the

Figure 1.1 Professor Chris Gosden (Photo: Ian R. Cartwright).

Figure 1.2 (a) Chris on Malai Island, Siassi District, Papua New Guinea 1984 (Photo: Ian Lilley); (b) Chris in the regalia of Peter Pasio, the Big Man of Iangpun village, Apugi Island, New Britain, Papua New Guinea, 2003 (Photo: Jim Specht © The Australian National Museum); (c) Chris in lower right-hand corner of a water-logged trench at the Makekur Lapita site on Adwe Island, Arawe Islands, New Britain, Papua New Guinea (Photo: Jim Specht © The Australian National Museum); (d) Chris explaining the stratigraphy while standing in a waterlogged test pit of the Makekur Lapita site on Adwe Island, Arawe Islands, New Britain, Papua New Guinea, 1992 (Photo: Jim Specht © The Australian National Museum); (e) Chris breaks ground at a Peking University-excavation in the Zhouyuan, Shaanxi, PR China, 2014 (Photo: Rebecca O'Sullivan); (f) Chris admiring the great tapestry from Pazyryk, Hermitage Museum, St Petersburg (Photo: Courtney Nimura).

diversity of theories and approaches being used across the world, a choice that was made to reflect the breadth of Chris's own research, which has ranged from Oceania to Europe and covered vast regions, such as northern Eurasia. Refusing to confine himself to a single specialism has contributed much to archaeology, and the discipline is better for it. The 29 chapters that follow share Chris's intellectual ambition and attempt to follow his aspiration – so often successfully achieved – of revitalising the discipline of archaeology.

Section 1: People & Place (Chapters 2–10) is concerned with the relationship between past societies and the places they have inhabited, with a focus on the mutually constructive nature of interactions. Fieldwork has always played a part in Chris's research, and many of the authors in this section have been with him on the sites presented in these chapters. Chapters 2 (Summerhayes), 3 (Specht & Torrence), and 4 (Sand & Allen) explore concepts of distance and diversity, social and technological change, and cultural connections in Oceania. Borneo, Libya, and Iraqi Kurdistan are the focus of Chapter 5 (Barker & Hunt), in which the very question of what it is to be human and how this is expressed in the landscape is addressed. Cultural connections are also the subject of Chapter 6 (Rawson), which presents a study of several centuries of burial practices in China and Siberia in the 1st millennium BC. Moving from East to western Asia and the Mediterranean, Chapter 7 (Bogaard) considers the impact of process on the development of agriculture and the creation of agricultural places. Persistent places in the landscape are also the subject of Chapters 8 (Lock & Raven), 9 (Olivier), and 10 (Bradley), which present case studies from Europe. Chapter 8 focuses on long-term activity at a single site, while Chapters 9 and 10 exemplify the power of landscapes to shape human behaviour.

Section 2: Form & Flow (Chapters 11–21) concentrates on the role of intangible socio-cultural processes in shaping/producing the material objects visible today, and the theoretical implications of this for archaeological study. Chris has often focused on the big picture, drawing on the material culture evidence to understand how people and ideas moved around the world. Methods for how this type of research can be undertaken is the subject of Chapter 11 (Nimura, O'Sullivan & Hommel), which centres on the rock art of Eurasia. Chapters 12 (Downum & Garrow), 13 (Moffett), and 14 (Talbot) tell detailed stories of artefacts from Iron Age and Roman Britain, from the regionality of style and material in so-called 'Celtic Art' to the magical and religious nature of jet objects to the production and circulation of Iron Age coins. Chapter 15 (Pollard) focuses on the form and flow of archaeological thought through a travelogue of conversations had 'on the road' in China. Technological change, object biographies, and materials and materialities are explored in Chapters 16 (Li) in China, 17 (Chirikure) on the African continent, and 18 (Fullagar, Hayes & Pardoe) in Oceania. This section continues in Oceania, where issues of cultural change and continuity are explored in Chapters 19 (Cosgrove), 20 (Allen), and 21 (Morphy & Morphy).

Section 3: History & Heritage (Chapters 22–30) turns to current themes concerning archaeology as practice, delving into the discipline's origins and history to evaluate its structure, as well as its applicability to global heritage. The relationship of archaeology to anthropology, the focus of Chris's 1999 book, is revisited in Chapter 22 (Malafouris). This sets the stage for this section, in which many chapters grapple with big ideas, such as the nature of ontologies discussed in Chapter 23 (Robb). Archaeology's critical relationships with anthropology and philosophy also extend to the politics of heritage, which is the difficult topic addressed in Chapter 24 (Lilley). Relatedly, Chapter 25 (Edwards) presents a study on how commercial markets for heritage affected archaeological practice in the early phase of the discipline's development. From field archaeologist to museum curator, Chris's research engaged with many disciplines and a diverse suite of material culture. Chapters 26 (Pikirayi) and 27 (Knowles) both focus on the museum space and the importance of critical practices of collecting and curating. The section ends with three reflections on British archaeology in Chapters 28 (Hey), 29 (Evans), and 30 (Meskell).

Taken together, *Sentient Archaeologies*, seeks not only to pay homage to a scholar who has greatly influenced the direction of global archaeology, it aims to challenge established theories concerning past societies on a global scale. The illuminating research and novel explanations presented in this volume contribute to resolving long-standing problems in regional archaeologies and reinvigorate approaches taken towards older material.

Acknowledgements

The editors would like to wholeheartedly thank: the anonymous peer reviewers, who gave their time and thoughtful criticism to the chapters in this book; Molly Masterson, who cast her eagle eye over the entire manuscript; Miranda Creswell, who generously provided permission to reproduce her painting 'Waterland' on the cover; Anwen Cooper, who helped shape the author list; an anonymous donor, who subsidised the colour images in this book; and the editors and staff at Oxbow for their guidance throughout this process. We are indebted to Jane Kaye for her advice on all things Gosden and her support for this project from its inception. Though the final list of authors included in this book reflects merely a fraction of the masses of current and former students, colleagues, and friends whose careers and lives have been positively impacted by Chris, we hope it is at least partly representative.

Notes

1. (1979–1983) PhD, University of Sheffield; (1984–1985) Visiting Fellow, Australian National University; (1986–1992) Lecturer, La Trobe University; (1992–1993) Senior Lecturer, La Trobe University; (1994–2006) University Lecturer in World Archaeology/Curator of Archaeology, Pitt Rivers Museum/Fellow, St Cross College, University of Oxford; (2004–2006) Professor of Archaeology, Institute of Archaeology/Dean, St Cross College, University of Oxford; (2006–2023) Professor of European Archaeology, Institute of Archaeology/Fellow, Keble College, University of Oxford; (2008–2023) Chairman of Trustees, Oxford Archaeology; (2009–2014) Member, Arts & Humanities Research Council Academic Advisory Board; (2010–2020) Trustee, The Art Fund; (2014–2019) Chair, Visitors of the Pitt Rivers Museum; (2018–2021) Chair, Section H7, British Academy; (2018–2022) Trustee, British Museum. (2004) Elected Fellow, Society of Antiquaries, London; (2005) Elected Fellow, The British Academy; (2016) Elected Corresponding Fellow, Australian Academy of Humanities.
2. This creative process is introduced through the volume cover, which features a painting by Miranda Creswell, whom Chris secured as the first Artist-in-Residence at the School of Archaeology at Oxford.

Selected Works by Chris Gosden

2025

Humans: The First Seven Million Years. Harmondsworth: Penguin Viking.

2023

Pollard, A. M. & Gosden, C. (2023) *An Archaeological Perspective on the History of Technology*. CUP Elements. Cambridge, Cambridge University Press.

2021

Gosden, C. (2021) *The History of Magic: From Alchemy to Witchcraft, from the Ice Age to the Present*. London, Penguin Books.

Gosden, C. & Pollard, M. (2021) Is the universe sentient? What implications might this have for archaeology? In M. J. Boyd & R. C. P. Doonan (eds), *Far from Equilibrium: An Archaeology of Energy, Life and Humanity*, 313–24. Oxford, Oxbow Books.

Gosden, C., Green, C., Cooper, A., Creswell, M., Donnelly, V., Franconi, T., Glyde, R., Kamash, Z., Mallet, S., Morley, L., Stansbie, D. & ten Harkel, L. (2021) *English Landscapes and Identities: Investigating Landscape Change from 1500 BC to AD 1086*. Oxford, Oxford University Press.

Iorga, A., Gosden, C., Lock, G. & Schulting, R. (2021) Stable carbon and nitrogen isotope analysis and Romano-British animal management along the Ridgeway, Oxfordshire. *Journal of Archaeological Science: Reports* 40(a), 103254.

Malafouris, L., Gosden, C. & Bogaard, A. (2021) Process archaeology. *World Archaeology* 53(1), 1–14.

Sainsbury, V. A., Bray, P., Gosden, C. & Pollard, A. M. (2021) Mutable objects, places and chronologies. *Antiquity* 95(379), 215–27.

2020

Gosden, C. & Knowles, C. (2020) *Collecting Colonialism: Material Culture and Colonial Change*. London, Routledge.

Gosden, C. (2020) Art, ambiguity and transformation. In Nimura *et al*. (2020a), 9–22. Oxford, Oxbow Books.

Gosden, C., Chittock, H., Hommel, P. & Nimura, C. (2020) Introduction: Context, connections and scale. In Nimura *et al*. (2020a), 1–8.

Grainger, A., Summerhayes, G. R. & Gosden, C. (2020) Investigating the nature of mobility patterns and interaction: Ceramic production at the Late Lapita site of Amalut, Papua New Guinea. *Australian Archaeology* 87(1), 93–104.

Hommel, P., Kovaleva, O., Whitlam, J., Amzarakov, P., Pouncett, J., Lim, J., Petrova, N., Gosden, C. & Esin, Y. (2020) 'Monumental myopia': Bringing the later prehistoric settlements of southern Siberia into focus. *Antiquity* 94(373), E2.

Malafouris, L. & Gosden, C. (2020) Mind, time, and material engagement. In I. Gaskell & S. A. Carter (eds), *The Oxford Handbook of History and Material Culture*, 105–122. Oxford, Oxford University Press.

Monteith, F., Liu, R., Pollard, A. M. & Gosden, C. (2020) Chaîne opératoire and the construction of Buddhist cave temples in Northwestern China. *Archaeological Review from Cambridge* 35(1), 44–56.

Nimura, C., Gosden, C., Hommel, P. & Chittock, H. (eds) (2020a) *Art in the Eurasian Iron Age: Context, Connections and Scale*. Oxford, Oxbow Books.

Nimura, C., Hommel, P., Chittock, H. & Gosden, C. (2020b) Collecting Iron Age art. In Nimura *et al*. (2020a), 23–36. Oxford, Oxbow Books.

2019

Gosden, C., Hommel, P. & Nimura, C. (2019) Making mounds: Monuments in Eurasian prehistory. In T. Romankiewicz, M. Fernández Götz, G. Lock & O. Büchsenschütz (eds), *Enclosing Space, Opening New Ground: Iron Age Studies from Scotland to Mainland Europe*, 141–52. Oxford, Oxbow Books.

Schulting, R. J., le Roux, P., Gan, Y. M., Pouncett, J., Hamilton, J., Snoeck, C., Ditchfield, P., Henderson, R., Lange, P., Lee-Thorp, J., Gosden, C. & Lock, G. (2019) The ups & downs of Iron Age animal management on the Oxfordshire Ridgeway, south-central England: A multi-isotope approach. *Journal of Archaeological Science* 101, 199–212.

Specht, J. & Gosden, C. (2019) New dates for the Makekur (FOH) Lapita pottery site, Arawe Islands, New Britain, Papua New Guinea. In S. Bedford & M. Spriggs (eds), *Debating Lapita: Distribution, Chronology, Society and Subsistence*. Terra Australis 52, 169–202. Canberra, Australian National University Press.

2018

Gosden, C. (2018) *Prehistory: A Very Short Introduction*. 2nd ed. Oxford, Oxford University Press.

Gosden, C. (2018) Trade and exchange. In C. Haselgrove, K. Rebay-Salisbury & P. S. Wells (eds), *The Oxford Handbook of the European Iron Age* [online]. Oxford, Oxford University Press.

2017

Ten Harkel, L., Franconi, T. & Gosden, C. (2017) Fields, ritual and religion: Holistic approaches to the rural landscape in long-term perspective (*c*. 1500 BC–AD 1086). *Oxford Journal of Archaeology* 36(4), 413–37.

2016

Barker, G., Hunt, C., Barton, H., Gosden, C., Jones, S., Lloyd-Smith, L., Farr, L., Nyiri, B. & O'Donnell, S. (2016) The 'cultured rainforests' of Borneo. *Quaternary International* 448, 44–61.

Green, C., Gosden, C., Cooper, A., Franconi, T., ten Harkel, L., Kamash, Z. & Lowerre, A. (2016) Understanding the spatial patterning of English archaeology: Modelling mass data, 1500 BC to AD 1086. *Archaeological Journal* 174(1), 244–80.

Specht, J., Gosden, C., Lentfer, C., Jacobsen, G., Matthew, P. J. & Lindsay, S. (2016) A pre-Lapita structure at Apalo, Arawe Islands, Papua New Guinea. *The Journal of Island and Coastal Archaeology* 12(2), 151–72.

Specht, J., Gosden, C., Pavlides, C., Richards, Z. & Summerhayes, G. R. (2016) Exploring Lapita diversity on New Britain's south coast, Papua New Guinea. *Journal of Pacific Archaeology* 7(1), 20–29.

2015

Bray, P., Cuénod, A., Gosden, C., Hommel, P., Liu, R. & Pollard, A. M. (2015) Form and flow: The 'karmic cycle' of copper. *Journal of Archaeological Science* 56, 202–9.

Cunliffe, B., Renfrew, C., Gosden, C. & Geake, H. (2015) The British Museum at 250. *Antiquity* 77(298), 828–33.

Gosden, C. (2015) Possession, property or ownership? In A. Klevnäs & C. Hedenstierna-Jonson (eds), *Own and be Owned: Archaeological Approaches to the Concept of Possession*, 215–21. Stockholm, Department of Archaeology and Classical Studies, Stockholm University.

Gosden, C. (2015) What use is the Palaeolithic in promoting new prehistoric narratives? In F. Coward, R. Hosfield, M. Pope & F. Wenban-Smith (eds), *Settlement, Society and Cognition in Human Evolution: Landscapes in Mind*, 1–14. New York, Cambridge University Press.

Gosden, C. & Malafouris, L. (2015) Process archaeology (p-arch). *World Archaeology* 47(5), 701–17.

Hamilton, D. W., Haselgrove, C. & Gosden, C. (2015) The impact of Bayesian chronologies on the British Iron Age. *World Archaeology* 47(4), 642–60.

Pollard, A. M., Bray, P., Gosden, C., Wilson, A. & Hamerow, H. (2015) Characterising copper-based metals in Britain in the first millennium AD: A preliminary quantification of metal flow and recycling. *Antiquity* 89(345), 697–713.

Specht, J., Lentfer, C., Gosden, C., Jacobsen, G. & Lindsay, S. (2015) Pre-Lapita decorated wood from Apalo, West New Britain, Papua New Guinea. *Archaeology in Oceania* 50(2), 105–10.

2014

Chippindale, C., Gosden, C., James, N., Pitts, M. & Scarre, C. (2014) New era for Stonehenge. *Antiquity* 88, 644–57.

Gosden, C. (2014) Cognitive landscapes: The origins of the English village. *Pragmatics and Cognition* 22(1), 93–108.

Gosden, C. (2014) Commentary: The archaeology of the colonized and global archaeological theory. In N. Ferris, R. Harrison & M. V. Wilcox (eds), *Rethinking Colonial Pasts Through Archaeology*, 476–82. Oxford, Oxford University Press.

Gosden, C., Crawford, S. & Ulmschneider, K. (2014) *Celtic Art in Europe: Making Connections*. Oxford, Oxbow Books.

Gosden, C., Crawford, S. & Ulmschneider, K. (2014) Introduction to *Celtic Art in Europe: Making Connections*. In Gosden *et al.* 2014, 1–5.

Malafouris, L., Gosden, C. & Overmann, K. A. (2014) Creativity, cognition, and material culture: An introduction. *Pragmatics and Cognition* 22(1), 1–4.

Morrison, W., Thomas, R. & Gosden, C. (2014) Laying bare the landscape: Commercial archaeology and the potential of digital spatial data. *Internet Archaeology* 36. doi:10.11141/ia.36.9.

Pollard, A. M., Bray, P. & Gosden, C. (2014) Is there something missing in scientific provenance studies of prehistoric artefacts? *Antiquity* 88(340), 625–31.

2013

Gosden, C. (2013) Extended and condensed relations: Bringing together landscapes and artefacts. In A. Meirion Jones, J. Pollard, J. Gardiner & M. J. Allen (eds), *Image, Memory and Monumentality: Archaeological Engagements with the Material World*, 127–35. Oxford, Oxbow Books.

Gosden, C. (2013) Fields. In S. Bergerbrandt & S. Sabatini (eds), *Counterpoint: Essays in Archaeology and Heritage Studies in Honour of Professor Kristian Kristiansen*, 111–17. Oxford, BAR International Series 2508.

Gosden, C. (2013) Humanized Environments. In M. I. J. Davies & F. N. M'Mbogori (eds), *Humans and the Environment: New Archaeological Perspectives for the Twenty-First Century*, 277–84. Oxford, Oxford University Press.

Gosden, C. (2013) Landscapes and scale: Some introductory thoughts. *Landscapes* 14(1), 3–6.

Gosden, C. (2013) Technologies of routine and enchantment. In L. Chua & M. Elliott (eds), *Distributed Objects: Meaning and Mattering After Alfred Gell*, 39–57. Oxford, Berghahn Books.

Lentfer, C., Matthews, P. J., Gosden, C., Lindsay, S. & Specht, J. (2013) Prehistory in a nutshell: A Lapita-age nut-cracking stone from the Arawe Islands, Papua New Guinea. *Archaeology in Oceania* 48(3), 121–9.

2012

Garrow, D. & Gosden, C. (2012) *Technologies of Enchantment? Exploring Celtic Art: 400 BC to AD 100*. Oxford, Oxford University Press.

Gosden, C. (2012) Magic, materials and matter: Understanding different ontologies. In J. Maran & P. W. Stockhammer (eds), *Materiality and Social Practice: Transformative Capacities of Intercultural Encounters*, 13–19. Oxford, Oxbow Books.

Gosden, C. (2012) On being more-than-one and doubts about mind. In N. Johannsen, M. Jessen & H. J. Jensen (eds), *Excavating the Mind: Cross-sections Through Culture, Cognition and Materiality*, 57–68. Aarhus, Aarhus University Press.

Gosden, C., Cooper, A., Creswell, M., Green, C., ten Harkel, L., Kamash, Z., Morley, L., Pybus, J. & Xiong, X. (2012) The English Landscape and Identities Project. *Antiquity* 86(332), http://www.antiquity.ac.uk/projgall/gosden332/.

ten Harkel, L., Gosden, C., Cooper, A., Creswell, M., Green, C. & Morley, L. (2012) Understanding the Relationship between Landscape and Identity: A Case Study from Dartmoor and the Tamar Valley, Devon, c. 1500 BC – AD 1086. *Journal for Ancient Studies* Special Volume 3, 181–88.

Wingfield, C. & Gosden, C. (2012) An Imperialist folklore? Establishing the folk-lore society in London. In T. Baycroft & D. Hopkin (eds), *Folklore and Nationalism in Europe During the Long Nineteenth Century*, 255–74. Leiden, Brill.

2011

Gosden, C. (2011) And your point is…? In J. Bate (ed.), *The Public Value of the Humanities*, 295–302. London, Bloomsbury Academic.

Gosden, C. (2011) The small worlds of the (pre-)Neolithic Mediterranean. In N. Phoca-Cosmetatou (ed.), *The First Mediterranean Islanders: Initial Occupation and Survival Strategies*, 173–6. Oxford, University of Oxford School of Archaeology Monograph 74.

Gosden, C. & ten Harkel, L. (2011) English landscapes and identities. The early medieval landscape: A perspective from the past. *Medieval Settlement Research* 26, 1–10.

2010

Gosden, C. (2010) The death of the mind. In L. Malafouris & C. Renfrew (eds), *The Cognitive Life of Things: Recasting the Boundaries of the Mind*, 39–46. Cambridge, McDonald Institute for Archaeological Research.

Gosden, C. (2010) When humans arrived in the New Guinea highlands. *Science* 330(6000), 41–2.

Gosden, C. (2010) Words and things: Thick description in archaeology and anthropology. In D. Garrow & T. Yarrow (eds), *Archaeology and Anthropology*, 110–16. Oxford, Oxbow Books.

Kamash, Z., Gosden, C. & Lock, G. (2010) Continuity and religious practices in Roman Britain: The case of the rural religious complex at Marcham/Frilford, Oxfordshire. *Britannia* 41, 95–125.

2009

Barker, G., Barton, H., Boutsikas, E., Britton, D., Davenport, B., Ewart, I., Farr, L., Ferraby, R., Gosden, C., Hunt, C. O., Janowski, M., Jones, S. E., Langub, J., Lloyd-Smith, L., Nyíri, B., Pearce, K. G. & Uppex, B. (2009) The Cultured Rainforest Project: The second (2008) field season. *Sarawak Museum Journal* 86, 119–84.

Díaz-Andreu, M., Price, M. & Gosden, C. (2009) Christopher Hawkes: His archive and networks in British and European archaeology. *The Antiquaries Journal* 89, 405–26.

Garrow, D., Gosden, C., Hill, J. D. & Bronk Ramsey, C. (2009) Dating Celtic art: A major radiocarbon dating programme of Iron Age and Early Roman metalwork in Britain. *Archaeological Journal* 166(1), 79–123.

Gosden, C. (2009) The Pacific Islands. In Gosden *et al.* 2009, 898–925.

Gosden, C., Cunliffe, B. & Joyce, R. A. (2009) Introduction. In Gosden *et al.* 2009, xiii–xviii.

Gosden, C., Cunliffe, B. & Joyce, R. A. (2009) *The Oxford Handbook of Archaeology*. Oxford, Oxford University Press.

Gosden, C., Kamash, Z., Kirkham, R. & Pybus, J. (2009) Joining the dots: Exploring technical and social issues in e-Science approaches to linking landscape and artefactual data in British archaeology. In *Proceedings of the 2009 5th IEEE, International Conference of E-Science Workshops*, Oxford, UK, 171–4. Piscataway NJ, IEEE. doi:10.1109/ESCIW.2009.5407968.

Gosden, C., Larson, F. & Petch, A. (2009) Origins and survivals: Tylor, Balfour and the Pitt Rivers Museum and their role within anthropology in Oxford 1883–1905. In P. Rivière (ed.), *A History of Oxford Anthropology*, 21–42. Oxford, Berghahn Books.

2008

Garrow, D., Gosden, C. & Hill, J. D. (eds) (2008) *Rethinking Celtic Art*. Oxford, Oxbow Books.

Gosden, C. & Hill, J. D. (2008) Introduction: Re-integrating 'Celtic' art. In Garrow *et al.* 2008, 1–14.

Gosden, C. (2008) Social ontologies. *Philosophical Transactions of The Royal Society B* 363(1499), 2003–10.

Gosden, C. (2008) The past and foreign countries: Colonial and post-colonial archaeology and anthropology. In L. Meskell & R. W. Preucel (eds), *A Companion to Social Archaeology*, 161–78. Oxford, Blackwell.

2007

Gosden, C. (2007) Holism, intelligence and time. In D. Parkin & S. Ulijaszek (eds), *Holistic Anthropology: Emergence and Convergence*, 182–93. Oxford, Berghahn Books.

Gosden, C., Hamerow, H., de Jersey, P. & Lock, G. (eds) (2007) *Communities and Connections: Essays in Honour of Barry Cunliffe*. Oxford, Oxford University Press.

Gosden, C. & Larson, F. (2007) *Knowing Things: Exploring the Collections at the Pitt Rivers Museum, 1884-1945*. Oxford, Oxford University Press.

2006

Edwards, E., Gosden, C. & Phillips, R. (eds) (2006) *Sensible Objects: Colonialism, Museums and Material Culture*. London, Routledge.

Edwards, E., Gosden, C. & Phillips, R. (2006) Introduction. In Edwards *et al.* 2006, 1–34. London, Routledge.

Gosden, C. (2006) Material culture and long-term change. In C. Tilley, W. Keane, S. Küchler, M. Rowlands & P. Spyer (eds), *Handbook of Material Culture*, 425–42. London, Sage.

Gosden, C. (2006) Warfare and colonialism in the Bismarck Archipelago, Papua New Guinea. In T. Otto, H. Thrane & H. Vandkilde (eds), *Warfare and Society: Archaeological and Social Anthropological Perspectives*, 201–10. Aarhus, Aarhus University Press.

Gosden, C. (2006) Race and racism in archaeology: Introduction. *World Archaeology* 38(1), 1–7.

Gosden, C. & Kirsanow, K. (2006) Timescales. In G. Lock & B. L. Molyneaux (eds), *Confronting Scale in Archaeology: Issues of Theory and Practice*, 27–37. Springer.

Gosden, C. & Lock, G. (2006) The aesthetics of landscape on the Berkshire Downs. In C. Haselgrove (ed.), *The Earlier Iron Age in Britain and the Near Continent*, 279–92. Oxford, Oxbow Books.

Robson, E., Treadwell, L. & Gosden, C. (eds) (2006) *Who Owns Objects? The Ethics and Politics of Collecting Cultural Artefacts*. Oxford, Oxbow Books.

2005

Gosden, C. (2005) Comments III: Is science a foreign country? *Archaeometry* 47(1), 182–5.

Gosden, C. (2005) What do objects want? *Journal of Archaeological Method and Theory* 12, 193–211.

Lock, G., Gosden, C. & Daly, P. (2005) *Segsbury Camp: Excavations in 1996 and 1997 at an Iron Age Hillfort on the Oxfordshire Ridgeway*. Oxford, University of Oxford School of Archaeology Monograph 61.

2004

DeMarrais, E., Gosden, C. & Renfrew, C. (2004) *Rethinking Materiality: The Engagement of Mind with the Material World*. Cambridge, McDonald Institute for Archaeological Research.

Gosden, C. (2004) Aesthetics, intelligence and emotions: Implications for archaeology. In DeMarrais *et al.* (2004), 33–42.

Gosden, C. (2004) *Archaeology and Colonialism: Cultural Contact from 5000 BC to the Present*. Cambridge, Cambridge University Press.

Gosden, C. (2004) Grid and group: An interview with Mary Douglas. *Journal of Social Archaeology* 4(3), 275–87.

Gosden, C. (2004) Shaping life in the late prehistoric and Romano-British periods. In R. M. Rosen (ed.), *Time and Temporality in the Ancient World*, 29–44. Philadelphia, PA, University of Pennsylvania Museum of Archaeology and Anthropology.

Knowles, C. & Gosden, C. (2004) A century of collecting: Colonial collectors in southwest New Britain. *Records of the Australian Museum*, Supplement 29, 65–74.

2003

Gosden, C. & Lock, G. (2003) Becoming Roman on the Berkshire Downs: The evidence from Alfred's Castle. *Britannia* 34, 65–80.

Gosden, C. & Lock, G. (2003). Frilford: A Romano-British ritual pool in Oxfordshire? *Current Archaeology* 184(4), 156–9.

Miles, D., Palmer, S., Lock, G., Gosden, C. & Cromarty, A. M. (2003). *Uffington White Horse and its Landscape: Investigations at White Horse Hill, Uffington, 1989-95, and Tower Hill, Ashbury, 1993–4*. Oxford, Oxford Archaeology.

2001

Gosden, C. (2001) Making sense: Archaeology and aesthetics. *World Archaeology*, 163–7.

Gosden, C. (2001) Postcolonial archaeology: Issues of culture, identity, and knowledge. In I. Hodder (ed.), *Archaeological Theory Today*, 251–66. Cambridge, Polity Press.

2000

Brown, A., Coote, J. & Gosden, C. (2000) Tylor's tongue: Material culture, evidence, and social networks. *Journal of the Anthropological Society of Oxford* 31(3), 257–76.

Gosden, C. (2000) On his Todd: Material culture and colonialism. In M. O'Hanlon & R. L. Welsch (eds), *Hunting the Gatherers: Ethnographic Collectors, Agents and Agency in Melanesia, 1870s–1930s*, 227–50. Oxford, Berghahn Books.

Gosden, C. (2000) Varieties of colonial experience: Material culture and colonialism in West New Britain Province, Papua New Guinea. In A. Anderson & T. Murray (eds), *Australian Archaeologist: Collected Papers in Honour of Jim Allen*, 161–70. Canberra, The Australian National University.

Knowles, C., Gosden, C. & Lienert, H. (2000) German collectors in south-west New Britain, 1884–1914. *Pacific Arts* 21/22, 39–52.

1999

Gosden, C. (1999) *Anthropology and Archaeology: A Changing Relationship*. London, Routledge.

Gosden, C. (1999) *Prehistory: A Very Short Introduction*. 1st ed. Oxford, Oxford University Press.

Gosden, C. (1999) The organization of society. In G. Barker (ed.), *Companion Encyclopedia of Archaeology*. London, Routledge.

Gosden, C. & Hather, J. G. (eds) (1999) *The Prehistory of Food: Appetites for Change*. London, Routledge.

Gosden, C. & Head, L. (1999) Different histories: A common inheritance for Papua New Guinea and Australia? In Gosden & Hather 1999, 227–45.

Gosden, C. & Marshall, Y. (1999) The cultural biography of objects. *World Archaeology* 31(2), 169–78.

1998

Gosden, C. & Lock, G. (1998) Prehistoric histories. *World Archaeology* 30(1), 2–12.

Summerhayes, G. R., Bird, J. R., Fullagar, R., Gosden, C., Specht, J. & Torrence, R. (1998) Application of PIXE-PIGME to archaeological analysis of changing patterns of obsidian use in West New Britain, Papua New Guinea. In M. S. Shackley (ed.), *Archaeological Obsidian Studies*, 129–58. Springer.

1997

Gosden, C. (1997) Iron Age landscapes and cultural biographies. In A. Gwilt & C. Haselgrove (eds), *Reconstructing Iron Age Societies*, 303–7. Oxford, Oxbow Books.

Gosden, C. (ed.) (1997) Special issue: Culture contact and colonialism. *World Archaeology* 28(3).

Matthews, P. J. & Gosden, C. (1997) Plant remains from waterlogged sites in the Arawe Islands, West New Britain Province, Papua New Guinea: Implications for the history of plant use and domestication. *Economic Botany* 51, 121–33.

Specht, J. & Gosden, C. (1997) Dating Lapita pottery in the Bismarck Archipelago, Papua New Guinea. *Asian Perspectives* 36(2), 175–99.

Terrell, J. E., Hunt, T. L. & Gosden, C. (1997) Human diversity and the myth of the primitive isolate. *Current Anthropology* 38(2), 155–95.

1996

Allen, J. & Gosden, C. (1996) Spheres of interaction and integration: Modelling the culture history of the Bismarck Archipelago. In J. Davidson, G. Irwin, F. Leach, A. Pawley & D. Brown (eds), *Oceanic Culture History: Essays in Honour of Roger Green*, 183–97. Dunedin, New Zealand Journal of Archaeology Special Publication.

Gosden, C. (1996) Can we take the Aryan out of Heideggerian? *Archaeological Dialogues* 3(1), 22–5.

Gosden, C. (1996) Transformations: History and prehistory in Hawaii. *Archaeology in Oceania* 31(3), 165–72.

Harris, D. & Gosden, C. (1996) The beginnings of agriculture in western Central Asia. In D. Harris (ed.), *The Origins and Spread of Agriculture and Pastoralism in Eurasia*, 370–89. Washington D.C., Smithsonian Institution Press.

1995

Gosden, C. (1995) Long term trends in the colonisation of the Pacific: Putting Lapita in its place. In P. Bellwood (ed.), *Indo-Pacific Prehistory 1990, Vol. 2*, 333–38. Bundoora, La Trobe University.

Gosden, C. (1995) Arboriculture and agriculture in coastal Papua New Guinea. *Antiquity* 69(265), 807–17.

1994

Gosden, C. (1994) *Social Being and Time*. Oxford, Blackwell.

Gosden, C. & Head, L. (1994) Landscape: A usefully ambiguous concept. *Archaeology in Oceania* 29(3), 113–16.

Gosden, C. & Webb, J. (1994) The creation of a Papua New Guinean landscape: Archaeological and geomorphological evidence. *Journal of Field Archaeology* 21(1), 29–51.

Gosden, C., Webb, J., Marshall, B. & Summerhayes, G. R. (1994) Lolmo Cave: A mid- to late Holocene site, the Arawe Islands, West New Britain Province, Papua New Guinea. *Asian Perspectives* 33(1), 97–119.

Pavlides, C. & Gosden, C. (1994) 35,000-year-old sites in the rainforests of West New Britain, Papua New Guinea. *Antiquity* 68(260), 604–10.

Pavlides, C. & Gosden, C. (1994) Are islands insular? Landscape vs. seascape in the case of the Arawe Islands, Papua New Guinea. *Archaeology in Oceania* 29(3), 162–71.

1993

Harris, D. R., Masson, V. M., Berezkin, Y. E., Charles, M. P., Gosden, C., Hillman, G. C., Kasparov, A. K., Korobkova, G. F., Kurbansakhatov, K., Legge, A. J. & Limbrey, S. (1993) Investigating early agriculture in Central Asia: New research at Jeitun, Turkmenistan. *Antiquity* 67(255), 324–38.

Summerhayes, G., Gosden, C., Fullagar, R., Specht, J., Torrence, R., Bird, J. R., Shagholi, N. & Katsaros, A. (1993) West New Britain obsidian: Production and consumption patterns. In B. L. Fankhauser & J. R. Bird (eds), *Archaeometry: Current Australasian Research*, 57–68. Canberra, Department of Prehistory, Research School of Pacific Studies, The Australian National University.

1992

Enright, N. J. & Gosden, C. (1992) Unstable archipelagos – southwest Pacific environment and prehistory since 30,000 BP. In J. Dodson (ed.), *The Naïve Lands: Prehistory and Environmental Change in Australia and the Southwest Pacific*, 160–98. Melbourne, Longman Cheshire.

Gosden, C. (1992) Dynamic traditionalism: Lapita as a long term social structure. In J. C. Galipaud (ed.), *Poterie Lapita et peiplement: Actes du Colloque Lapita, Nouméa, Nouvelle Calédonie, Janvier 1992*, 21–7. Nouméa, Orstom Nouméa.

Gosden, C. (1992) Endemic doubt: Is what we write right? *Antiquity* 66(252), 803–8.

Gosden, C. (1992) Production systems and the colonization of the Western Pacific. *World Archaeology* 24(1), 55–69.

1991

Allen, J. & Gosden, C. (eds) (1991) *Report of The Lapita Homeland Project*. Occasional Papers in Prehistory 20. Canberra, The Australian National University.

Gosden, C. (1991a) Towards an understanding of the regional archaeological record from the Arawe Islands, West New Britain, Papua New Guinea. In Allen & Gosden 1991, 205–16.

Gosden, C. (1991b) Learning about Lapita in the Bismarck Archipelago. In Allen & Gosden 1991, 260–68.

Gosden, C. & Robertson, N. (1991) Models for Matenkupkum: Interpreting a late Pleistocene site from Southern New Ireland, Papua New Guinea. In Allen & Gosden 1991, 20–45.

Gosden, C & Specht, J. (1991) Diversity, continuity and change in the Bismarck Archipelago, Papua New Guinea. *Bulletin of the Indo-Pacific Prehistory Association* 11, 276–80.

1990

Gosden, C. (1990) Archaeological work in the Arawe Islands, West New Britain Province, Papua New Guinea, December 1989–February 1990. *Australian Archaeology*, 37–44.

1989

Allen, J., Gosden, C. & White, J. P. (1989) Human Pleistocene adaptations in the tropical island Pacific: Recent evidence from New Ireland, a Greater Australian Outlier. *Antiquity* 63(240), 548–61.

Gosden, C. (1989) Debt, production, and prehistory. *Journal of Anthropological Archaeology* 8(4), 355–87.

Gosden, C. (1989) Prehistoric social landscapes of the Arawe Islands, West New Britain Province, Papua New Guinea. *Archaeology in Oceania* 24(2), 24–58.

Gosden, C., Allen, J., Ambrose, W., Anson, D., Golson, J., Green, R., Kirch, P., Lilley, I., Specht, J. & Spriggs, M. (1989) Lapita sites of the Bismarck Archipelago. *Antiquity* 63(240), 561–86.

1988

Allen, J., Gosden, C., Jones, R. & White, J. P. (1988) Pleistocene dates for the human occupation of New Ireland, northern Melanesia. *Nature* 331, 707–9.

1985

Gosden, C. (1985) Gifts and kin in early Iron Age Europe. *MAN* 20(3), 475–93.

1984

Gosden, C. & Allen, J. (1984) The Lapita Homeland Project. *Bulletin of the Indo-Pacific Prehistory Association* 5, 104–9.

1983

Gosden, C. (1983) *Iron Age pottery trade in central Europe*. Unpublished PhD thesis, University of Sheffield.

1982

Gosden, C. (1982) The recognition and interpretation of the exchange of pottery in the Baringo District of Kenya: Some preliminary results. *Archaeological Review from Cambridge* 1(2), 13–29.

2

Reflections on Populating the Western Pacific

Glenn Summerhayes

Archaeological excavations led by Chris Gosden during the mid–late 80s and early 90s in southwest New Britain, Papua New Guinea, uncovered evidence of human occupation from the mid-Holocene (Lolmo cave; Gosden et al. 1994). Of importance are major settlement sites spanning 800 years from 3300–2500 BP. These sites, known as Lapita settlements, provided glimpses into the Austronesian-speaking colonisation of the western Pacific and its movement eastwards over 4000 km from Papua New Guinea to Samoa. Occupation east of the Solomon Island Chain (Reef Islands and Santa Cruz, Vanuatu, Fiji, Tonga, and Samoa) was the first evidence of the peopling within this region. This paper will place Gosden's excavations into a regional perspective, outlining the important role that the discoveries had for interpreting the major social and technological changes happening in the western Pacific during the Late Holocene.

Keywords: Lapita, Arawe Islands, Holocene, Pottery

Introduction

Lapita is the term coined for colonising populations that moved out of Near Oceania by 3100 cal BP (the area of New Guinea, the Bismarck Archipelago, and the main Solomon Island chain) into what is termed Remote Oceania (the areas to the east of the main island chain), to colonise for the first time the small islands known today as Vanuatu, New Caledonia, Fiji, Tonga, and eventually Samoa by 2850 cal BP (see Fig. 2.1). The archaeological signature of this early colonisation is pottery known as Lapita, named after the site excavated in New Caledonia by Gifford and Shutler in the 1950s (Gifford & Shutler 1956). The term 'Lapita' has now also extended to the material cultural complex. The conundrum is that whilst Remote Oceania marked the initial footsteps of humanity into that region, Near Oceania had already been occupied for 50,000 years (Summerhayes et al. 2010). Modelling Lapita colonists' movements into a new territory, previously unoccupied, is one thing, but understanding what Lapita was within a region that already had a deep human history is another thing entirely. Models to account for Lapita's appearance ranged from an indigenous growth arising out of economic and social institutions with a deep past within the Bismarck Archipelago, to one of external colonising peoples from southeast Asia bringing a Neolithic way of life (Bellwood 2013).

In the mid–late 20th century, when archaeological programmes in the Pacific were first being developed, the problem in modelling Lapita in Near Oceania and elucidating its origins was the lack of archaeological research in the Bismarck Archipelago (see Fig. 2.2). The Lapita Homeland Project was set up in the 1980s to redress the imbalance by focusing on the Bismarck Archipelago region, the area where Lapita supposedly developed. Shortly after taking up fellowship at the Australian National University (ANU), Gosden was thrust into a managing role of the Lapita Homeland Project, led by Jim Allen then of the ANU, involving leading archaeologists of the day. Both Gosden and Allen moved to La Trobe University shortly afterwards and built an impressive research-led department known for its strength and leadership in New Guinea archaeology. Unfortunately, within ten years both Allen and Gosden had left La Trobe and Pacific archaeology ceased there. As part of the Lapita Homeland Project, Gosden conducted fieldwork in the Arawe island group, which had seen little previous archaeological work. Excavations and survey began in the region in 1985, with Gosden leading five

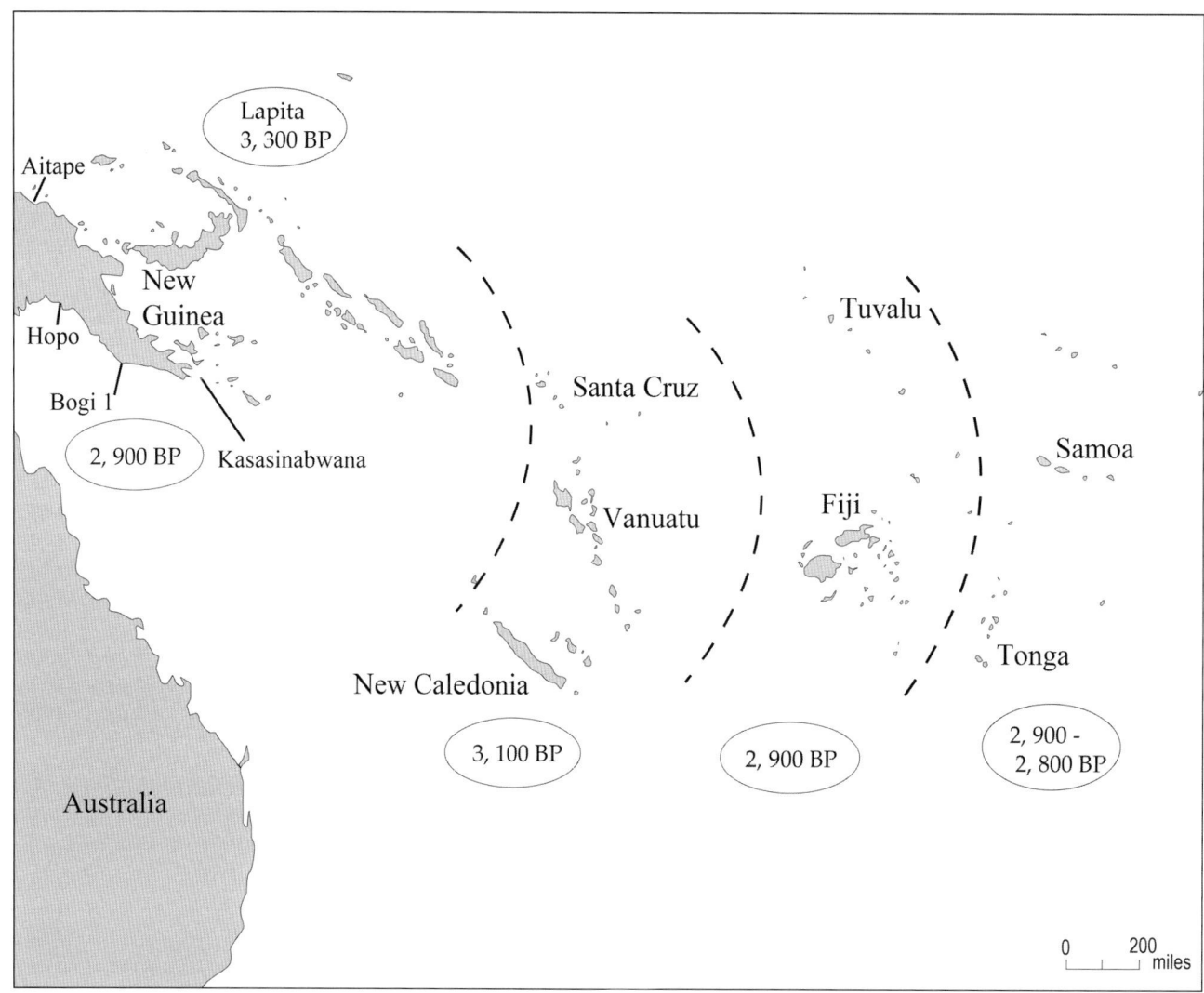

Figure 2.1 Spread of Lapita over the western Pacific.

further field seasons (1986–7; 1987–8; 1989–90, 1990–1, and 1991–2). This paper briefly reviews Gosden's important role in Lapita studies by focusing on his field research and the development of models used to account for his material. This review is followed by more recent interpretations of the archaeology of this region based on subsequent research of Gosden's excavated materials.

Gosden and Lapita

Gosden's fieldwork in the Arawe Islands was a landmark for Lapita studies. He excavated one of the most important Lapita areas known to date. Sites of Paligmete on Pililo Island, Apalo on Kumbun Island, Makekur on Adwe Island, and Amalut on the south coast opposite Pililo Island, yielded some of the richest archaeological deposits known. This remains the case. The pottery was excavated from waterlogged deposits in a freshwater ghyben-herzberg lens, resulting in remarkable preservation of material culture, including organics such as nut remains and wooded structures.

Faced with some of the richest sites known, it was a challenge to fit this into what was known at the time and to develop models to understand the material culture. Of Gosden's many insights into Lapita studies, two stand out as significant.

1. Understanding Geomorphology

The first insight concerns understanding how the archaeological record was created in context of its preservation and relationship to past landscapes. Gosden (1991a, 205) noted that we 'cannot understand an archaeological record unless we know the processes responsible for their formation'. He took the geomorphologist John Webb to the Arawes where they detailed the geomorphological history of the sites including major environmental changes that followed human occupation as outlined in their paper 'The creation

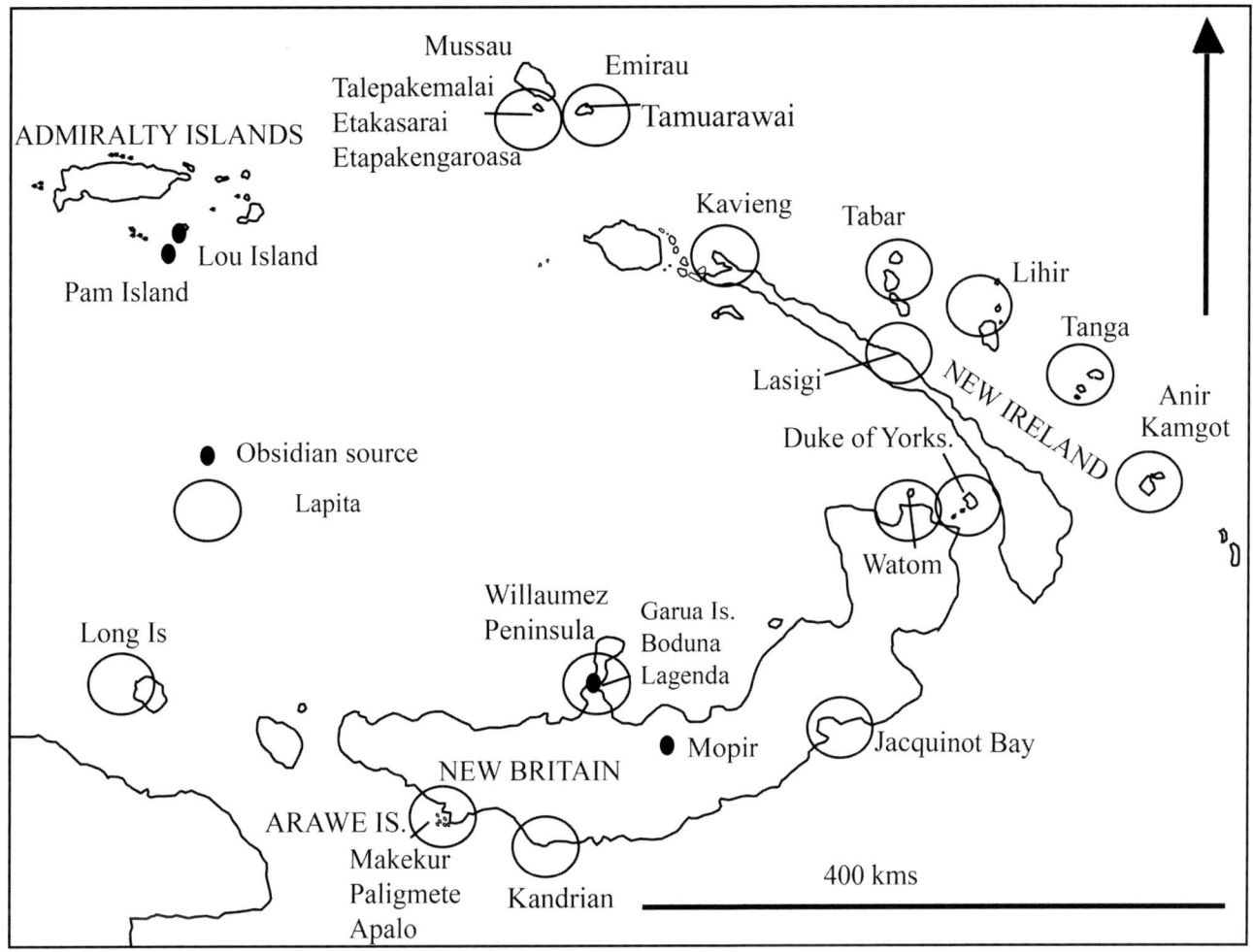

Figure 2.2 Lapita sites in the Bismarck Archipelago.

of a Papua New Guinean landscape: Archaeological and geomorphological evidence' (Gosden & Webb 1994). As argued in his book *Social Being and Time*, the landscapes of the past were different to those witnessed today. Gosden (1994, 29) posited that human action during and after Lapita times created the landscape of the present day. Importantly, Gosden and Webb identified that the initial Lapita settlements were stilt houses located on reef flats in low energy environments. The sea was, at the time of early Lapita occupation, between 1.5 and 2 m higher than today. Over time the reef flats filled up with erosional material from the coast of the associated islands, and as the sea level dropped to modern levels, the earlier materials which were deposited directly into the underlying waters were subsequently covered, thus sealing with remarkable preservation some of the richest Lapita materials known.

Gosden laid out a methodology and basic description on the distribution of sites, with basic summaries of material classes published. He questioned previous interpretations that these sites were large, long-term, stable villages, arguing instead that these interpretations were based on an over emphasis of the rich cultural deposits. He noted that the rich Lapita material was more visible due to deposition from stilt houses (Gosden 1991b, 265; see also Gosden 1994, 29), and argued that discard rates were low and 'there may not have been that many pots or stone tools in use at any one time'. Lastly, he posited that settlements were not settlements as we see today, but instead 'spots on the landscape to which people return to on a regular basis' (Gosden 1994, 29; Gosden & Pavlides 1994, 169).

2. Social Modelling – Making Sense of Lapita

The second crucial impact that Gosden made to Lapita studies was his attempt to understand what Lapita represented in terms of 'social forms'. His 1991 article 'Learning about Lapita' (Gosden 1991b, 264) was an attempt to look at the 'social forms' of Lapita, noting that we may be witnessing something very different to the social dynamics of the recent ethnographic past. He even postulated that Lapita had been over emphasised and was not 'necessarily directly ancestral to either Melanesian or Polynesian societies in the present' (Gosden 1991b, 264). His early articles were an attempt to come to grips with, and make sense of, what he had found. What did he come up with?

As a starting point, Gosden assumed, firstly, that social groups of the Lapita period were different in their essentials to any groups in the present and, secondly, that the archaeological record is hard to understand and therefore 'we need to develop methods which encompass a series of scales of analysis' (Gosden 1991b, 267). To do this, he suggested a three-scale approach. First, begin at local level: 'individual regions need to be understood as a totality' looking at geomorphology, 'and how far a distribution of artefacts is influenced by factors of preservation and destruction' (Gosden 1991b, 267). Secondly, from this local landscape understanding, he suggested combining them into a broader social whole looking at differences throughout larger areas such as the Bismarck Archipelago (Gosden 1991b, 268). Thirdly, once this was done, one could then look at the western Pacific as a whole.

An attempt at understanding a local landscape was made in his 1989 paper, 'Prehistoric social landscapes of the Arawe Islands' (Gosden 1989). Here he used the term 'social construct', where social groups operated in a landscape to sustain and provide a social environment (Gosden 1989, 45). He was adventurous and used the concept of 'debt' as an explanatory device to understand the archaeological record. Such a term was based on social anthropological models developed in select modern Papua New Guinea societies, but although relevant in assessing modern configurations of trade in southwest New Britain and mainland New Guinea (Gosden 1989, 48), it was limited when applied to Lapita studies.

Gosden did attempt to place the Arawe Islands into both a regional and Pacific-wide framework based on the distribution of dentate-stamped pottery and the broader processes of colonisation. In his ground-breaking article with Christina Pavlides on whether islands were insular, they argued for the development of social strategies that connected large regions as part of the process of colonisation (Gosden & Pavlides 1994, 162). Gosden and Pavlides argued against viewing the sea as an isolating barrier between societies. It begs the question: how much physical isolation was accountable for cultural divergence? They drew on the important work of Terrell (1986) who made the argument that it is not only physical distance that leads to isolations, as 'communities can become relatively isolated within a dense set of connections, making the general point that we need an idea of the processes by which connections are maintained or broken' (Gosden & Pavlides 1994, 162). From this point of the argument, they use observations derived from understanding site depositional history and low discard rates at Arawe Lapita sites and argue that these places are not settlements identical to coastal villages today, but 'spots on the landscape to which people returned on a regular basis' (Gosden 1994, 29; Gosden & Pavlides 1994, 169). That is – Lapita people were mobile.

Gosden and Pavlides (1994, 168) reiterated that the similarities in dentate-stamped pottery across the Lapita distribution were representative of a social universe where similarities were maintained over time by contemporaneous changes in a widely distributed cultural region. Here they refer to Kirch's (1990, 121–23) observation of gross changes in Lapita from complex dentate decorations to simpler decoration. What these technological changes over time meant for these societies was never discussed or outlined in any detail.

One of the major limitations on the research of this time was the focus on 'Lapita groups' and dentate-stamped pottery, which are viewed as a homologous whole – a single social universe (see Gosden 1989, 52). He, and others at the time, focused on modelling the meaning of pottery defined by dentate-stamping. There was a lack of understanding of what made up a Lapita ceramic assemblage. Another problem was looking at changes within the Arawe sites in terms of only gross changes, such as between mid-Holocene occupation to late Holocene Lapita occupation and then to post-Lapita societies, rather than changes within Lapita itself. Yet, the term social landscape is an important one, and the true value of Gosden's theoretical concepts were only realised after detailed archaeological analysis of the assemblages from the Arawes and other regional sites, such as those from the St Matthias Group excavated by Pat Kirch and the Anir Islands by Glenn Summerhayes. As outlined in Gosden's three-scale approach to Lapita studies, only once a regional understanding of spatial and temporal changes was developed could larger inferences be made on the colonisation process and subsequent changes to Lapita in the western Pacific.

New Perspectives on Lapita and its Place in the Western Pacific

Over the last 30 years, the assemblages excavated by Gosden have been re-analysed from many angles, adding much to his earlier observations. These analyses have allowed the identification of, firstly, a temporal change in pottery vessel form, decoration, and production within the Arawe assemblages. It gave a time depth to archaeological sequences allowing comparisons between the assemblages in the Bismarck Archipelago (Mussau and Anir) and those in Remote Oceania.

These new data allow a new look at the nature of the society that colonised the western Pacific in the 2nd millennium BC. Pottery, in particular dentate-stamped pottery, has held primacy in developing models to account for this expansion. The riddle that confronts archaeologists is the so-called rapid disappearance of complex dentate-stamped designs after the initial introduction of pottery to areas east of the Solomon Islands chain. To model why dentate-stamping disappeared needs a fuller understanding of what makes a Lapita pot assemblage.

Pottery Diversity – What Was a Lapita Assemblage?

One of the problems in understanding Lapita has been the over emphasis on dentate-stamped pottery. Indeed, the Lapita assemblages contained rich material culture including obsidian, shell technology, fauna, and archaeobotanical remains. Yet, the focus of earlier research and indeed peoples' perceptions of the Lapita material culture can be termed 'dentate-centric' (Summerhayes 2001, 54). Although much variation exists within a Lapita pottery assemblage, it is the complex patterns formed by dentate stamping that is often emphasised when describing the assemblages. Yet, after a detailed analysis of the Paligmete, Adwe, and Apalo pottery assemblages, a more complex picture emerged.

First, the dentate stamped sherds are only one component of the assemblages. Dentate-stamped vessels mostly consist of bowls and stands. The forms are also made primarily from a slab constructed technique, with thick vessel walls. Plain, undecorated vessels make up half the assemblage, while other types of decoration such as incision, punctation, and so on, make up the rest. These latter non-dentate-stamped vessels are mostly jars and globular pots, many having lip modification, and constructed using the coil-and-ring technique. Both dentate-stamped and non-dentate had been finished off with a paddle and anvil (for more detail see Summerhayes 2000a & b; 2001). The point made is that the assemblage was not just dentate-stamped forms, but many others that used different construction techniques. It is also clear that the functions of these pottery forms varied, with dentate-stamped vessel forms having a non-utilitarian social use, and the non-dentate vessels a more utilitarian use. The dentate designs have been interpreted as tattooing on the pot (Green 1979, 16; Kirch 1997, 142; Summerhayes 1998) – perhaps signifying clan/social markers (Summerhayes 2000b), or, as Chiu (2005) described them, as reflecting membership of house societies. Dentate marks were thus social/ideological signifiers that were socially active, conveying information and maintaining social boundaries. They fostered group identity.

Secondly, temporal changes can be seen in the assemblages, not only in the decoration conventions but also in how the proportion of vessel forms changed over time. Temporal trends were identified within the Arawe assemblage and the categories 'Early Lapita', 'Middle Lapita', and 'Late Lapita' were defined as heuristic devices to help model the changes (Summerhayes 2000a). Summerhayes (2001) noted that Early Lapita had a higher proportion of dentate-stamped bowls and stands. Middle and Later Lapita, on the other hand, had fewer bowls and stands and have a slightly higher proportion of dentate-stamped jars. The rest of the assemblage in terms of vessel forms remains the same (Summerhayes 2000a, chapter 10).

The dentate-stamped decoration, as well, becomes less complex and a coarse open design over time (see Figs 2.3 and 2.4). Different rates of change are seen in the two sets of vessels (dentate versus non-dentate). The non-dentate stamped ware changed little over Lapita's duration (Summerhayes 2000b; 2001). This slow rate of change was argued to be related to an ongoing domestic/utilitarian role. The change in dentate-stamped vessels is argued to be related to its lessening importance within the wider society that used it, as evidenced from Mussau, the Arawe Islands, and Anir (Summerhayes 2001, 62).

A petrographic and physico-chemical analysis also demonstrated distinct production changes over time within the Arawe assemblages regardless of vessel form or decoration used (Summerhayes 2000a). The study showed that any similarities between sites from the Arawe Islands and elsewhere in the Bismarck Archipelago, including the north coast of New Britain, were not due to pottery exchange, because most pottery was made locally. A reduction in the number of production centres was identified over time, which correlates with the production of more uniform and standardised jars at Apalo (Summerhayes 2000a). Summerhayes (2000a), on the basis of these results, argued that there had been a change in settlement patterns from a highly mobile pattern to a more sedentary one.

Regional Patterns

Similar changes are now seen in other parts of the Bismarck Archipelago. Major Lapita sites from Mussau, excavated by Pat Kirch (1990, 123; 2021), and those from Anir, excavated by Summerhayes (2001; Summerhayes *et al.* 2019a & b; see also Hogg *et al.* 2021), showed similar trends with a reduction in the number of production centres at the same time that the dentate-stamped decoration becomes less complex. Kirch also tied this change in decoration, from highly labour intensive to less labour intense techniques, to a change in the socio-economic role of pots over time (Kirch 1988, 335; 1990, 123). A model of mobility in Early Lapita to one of more settled communities during Middle and Later Lapita was put forward to explain these changes over time within the Bismarck Archipelago (Summerhayes 2000a). Similarities in material culture were not seen as part of trade and exchange but were explained by these geographically located sites as part of a social universe where change in material culture is best explained by mobility between communities. These results fit in well with Gosden's 'dots on the landscape' model, but only for the Early Lapita settlements. Later settlements are more fixed, no matter what the nature of that settlement may be.

How do these data fit in with the wider western Pacific and Lapita colonisation and subsequent developments? Research over the last 30 years provided a more precise and tighter chronology than that available to Gosden in the

Figure 2.3 Early Lapita pottery (from the site of Kamgot, Babase Island, Anir Island Group).

80s and 90s. The spread of Lapita across the western Pacific during this colonisation phase is now seen as taking a few hundred years, from the first appearance in the Bismarck Archipelago, to the Reef Islands and Santa Cruz sites by 3100 BP, to Vanuatu and New Caledonia by 3000/2900 BP, to Fiji by 2900 BP, and eventually to Tonga and then Samoa by 2800–2700 BP (Summerhayes 2010; Bedford *et al.* 2019). The distribution is also now extended along the south

Figure 2.4 Later Lapita pottery (from the Arawe Islands).

Papuan Coast (Negishi & Ono 2009; McNiven *et al.* 2011; Chynoweth *et al.* 2020) and the Massim of southeast Papua (Shaw *et al.* 2022). The distribution of Lapita pottery and the socially fuelled dentate-stamped motifs that accompanied the earliest colonisers suggest socially related groups that kept strong communication ties and fostered group identity during the colonisation phase. This is evident in the movement of resources, such as obsidian from sources in New Britain and the Admiralty Islands out into Remote Oceania reaching the Lau Island group of Fiji and island chains in between (Summerhayes 2009). Pottery on the other hand, although identical in design across 4000 km, were mostly made where they were found (see Summerhayes 2000a; Dickinson 2006). Similarities in design were, as Gosden predicted, the product of interaction between originally socially similar groups.

Yet major changes in the ceramic archaeological record are soon seen between Near and Remote Oceania, with the function of dentate-stamped pottery as a social signifier drastically changing. The problem is that the social role of Lapita dentate-stamped decoration was not the same across the Lapita universe. There were major differences. Within Near Oceania, Lapita dentate-stamped decoration lasts over 600 years, although changes in complexity reflect its lessening role in society. The situation in Remote Oceania, on the other hand, is completely different with dentate-stamping dropping out of use within a few generations of occupation (see Summerhayes 2019 for a review).

The longevity of dentate-stamped pottery as a signifier of social boundaries was no doubt related to the different social conditions faced by these Austronesian speaking communities in the Bismarck Archipelago where they had to share space with inhabitants whose ancestry in the region went back 50,000 years. This possible conflict between peoples led to the reinforcement of the demarcation of social boundaries as reflected in the continued use of dentate-stamped pottery.

Within Remote Oceania, on the other hand, people were entering unoccupied lands. There were *no* previous peoples in this seascape. Lapita dentate-stamped pottery dropped out of use after a few generations, although other forms persisted with new decoration appearing on jars and bowls. These new pottery wares are localised and normally named after a particular area or feature peculiar to where they were found. The dropping out of dentate-stamped pottery, and the development of new regional styles, indicates the social transformation occurring within these regions. The disappearance of dentate-stamped motifs equates with the changing social boundaries as these social markers were no longer needed, or perhaps were replaced by other social markers.

The changing nature of the 'social' importance of dentate-stamped, socially charged decoration is important. In an article by Summerhayes and Allen (2007) it was argued that early elaborate material is a reflection of the homeland culture, and that pottery decoration is elaborated internally as part of the colonising process (see Summerhayes & Allen 2007 for references to the theory of behavioural ecology that it derived from and more recent uses in anthropology). They used the concept of 'costly signalling' to explain the elaboration of pottery decoration from the earliest colonising phases in areas of New Guinea involving Neolithic societies. The use of fine, non-utilitarian pottery with or without decoration, and used in social and ritual use, would have reinforced group identity of any colonising group entering the domain of incumbent groups. Summerhayes and Allen (2007, 116–7) argued that although the colonists 'may have superior technology it is in the best long term interests of colonists to avoid conflict with incumbent groups when, by the very nature of the colonising act, the newcomers will inevitably compete for land and resources with existing groups'; thus, by 'elaborating their material culture the colonists signal their own strength or fitness and provide objects that by exchange will confer prestige or other more utilitarian values on the recipients'. The use of 'costly signalling' to already populated areas explains the continued use of elaborate decorated vessel forms with the Austronesian diaspora into Near Oceania. But what happens when a colonising group enters an area where no previous people exist? To whom are they signalling? As seen in Remote Oceania, the colonising populations lost their complex dentate designs and vessel forms (bowls and stands) soon after entering areas where no prior people lived. Yet, the utilitarian pottery forms continued (plain cooking and water storage vessels).

Conclusions

Gosden's fieldwork in southwest New Britain over 30 years ago has made significant contributions to our knowledge of Lapita. The advances in understanding what constituted a Lapita ceramic assemblage coupled with fine-tuning changes over time within that assemblage has built on the structures of interpretation he set up. Gosden's legacy continues to play a fundamental role in modelling the development of Lapita societies in a cross-cultural world and can also be seen in the advances made by the continued analyses of this important material.

Acknowledgements

I would like to thank Chris Gosden for not only inviting me into the field, but for his generosity and friendship over the last 35 years. I would also like to thank Ben Shaw and Dylan Gaffney for their insightful comments on this paper.

References

Bedford, S. & Spriggs, M. (eds) (2019) *Debating Lapita: Chronology, Society and Subsistence.* Canberra, Terra Australis.

Bedford, S., Spriggs, M., Burley, D., Sand, C., Sheppard, P. & Summerhayes, G. (2019) Debating Lapita: Distribution, chronology, society and subsistence. In Bedford & Spriggs 2019, 5–33.

Bellwood, P. (2013) *First Migrants: Ancient Migration in Global Perspective.* Chichester, Wiley Blackwell.

Chiu, S. (2005) Meanings of a Lapita face: Materialized social memory in ancient house societies. *Taiwan Journal of Anthropology* 3, 1–47.

Chynoweth, M., Summerhayes, G. R., Ford, A. & Negishi, Y. (2020) Lapita on Wari Island: What's the problem? *Asian Perspectives* 59, 100–16.

Dickinson, W. R. (2006) *Temper Sands in Prehistoric Oceanian Pottery: Geotectonics, Sedimentology, Petrography, Provenance*. Special Paper 406. Boulder, CO, The Geological Society of America.

Gifford, E. W. & Shutler, D. (1956) *Archaeological Excavations in New Caledonia*. Anthropological Records 18(1). Berkeley, CA, University of California Press.

Gosden, C. (1989) Prehistoric social landscapes of the Arawe Islands, West New Britain Province, Papua New Guinea. *Archaeology in Oceania* 24, 45–58.

Gosden, C. (1991a) Towards an understanding of the regional record from the Arawe Islands, West New Britain, Papua New Guinea. In J. Allen & C. Gosden (eds), *Report of the Lapita Homeland Project*, 205–16. Occasional Papers in Prehistory 20. Canberra, Department of Prehistory, Research School of Pacific Studies, The Australian National University.

Gosden, C. (1991b) Learning about Lapita in the Bismarck Archipelago. In J. Allen & C. Gosden (eds), *Report of the Lapita Homeland Project*, 260–68. Occasional Papers in Prehistory 20. Canberra, Department of Prehistory, Research School of Pacific Studies, The Australian National University.

Gosden, C. (1994) *Social Being and Time*. Oxford, Blackwell.

Gosden, C. & Pavlides, C. (1994) Are islands insular? Landscape vs. seascape in the case of the Arawe Islands, Papua New Guinea. *Archaeology in Oceania* 29, 162–71.

Gosden, C. & Webb, J. (1994) The creation of a Papua New Guinean landscape: Archaeological and geomorphological evidence. *Journal of Field Archaeology* 21, 29–51.

Gosden, C., Webb, J., Marshall, B. & Summerhayes, G. R. (1994) Lolmo Cave: A mid to late Holocene site, the Arawe Islands, West New Britain Province, Papua New Guinea. *Asian Perspectives* 33, 97–119.

Green, R. C. (1979) Early Lapita art from Polynesia and Island Melanesia. In S. M. Mead (ed.), *Exploring the Visual Art of Oceania*, 13–31. Hawaii, HI, University of Hawaii Press.

Hogg, N. W. S., Summerhayes, G. R. & Chen, Y. (2021) Moving on or settling down? Studying the nature of mobility through Lapita pottery from the Anir Islands, Papua New Guinea. *Technical Reports of the Australian Museum Online* 34, 71–86.

Kirch, P. V. (1988) The Talapakemalai Lapita site and Oceanic prehistory. *National Geographic Research* 4(3), 28–42.

Kirch, P. V. (1990) Specialization and exchange in the Lapita complex of Oceania. *Asian Perspectives* 29, 117–33.

Kirch, P. V. (1997) *The Lapita Peoples: Ancestors of the Oceanic World*. Oxford, Blackwell.

Kirch, P. V. (ed.) (2021) *Talepakemalai Lapita and Its Transformations in the Mussau Islands of Near Oceania*. Los Angeles, CA, Cotsen Institute of Archaeology Press.

McNiven, I., David, B., Richards, T., Aplin, K., Asmussen, B., Mialanes, J., Leavesley, M., Faulkner, P. & Ulm, S. (2011) New direction in human colonisation of the Pacific: Lapita settlement of South Coast New Guinea. *Australian Archaeology* 72, 1–6.

Negishi, Y. & Ono, R. (2009) Kasasinabwana shell midden: The prehistoric ceramic sequence of Wari Island in the Massim, eastern Papua New Guinea. *People and Culture in Oceania* 25, 23–52.

Shaw, B., Hawkins, S., Becerra-Valdivia, L., Turney, C. S. M., Coxe, S., Kewibu, V., Haro, J., Miamba, K., Leclerc, M., Spriggs, M., Privat, K., Haberle, S., Hopf, F., Hull, E., Pengilley, A., Brown, S., Marjo, C.E., Jacobsen, G. & Brooker and Panaeati Island communities, Papua New Guinea (2022) Frontier Lapita interaction with resident Papuan populations set the stage for initial peopling of the Pacific. *Nature Ecology & Evolution* 6, 802–12.

Summerhayes, G. R. (1998) The face of Lapita. *Archaeology in Oceania* 33, 100.

Summerhayes, G. R. (2000a) *Lapita Interaction*. Terra Australis 15. Canberra, Australian National University.

Summerhayes, G. R. (2000b) What's in a pot? In A. J. Anderson & T. Murray (eds), *Australian Archaeologist: Collected Papers in Honour of Jim Allen*, 291–307. Canberra, Coombs Academic Publishing, Australian National University.

Summerhayes, G. R. (2001) Lapita in the far West: Recent developments. *Archaeology in Oceania* 36, 53–64.

Summerhayes, G. R. (2009) Obsidian network patterns in Melanesia – sources, characterisation and distribution. *Bulletin of the Indo-Pacific Prehistory Association* 29, 110–24.

Summerhayes, G. R. (2010) Lapita interaction – an update. In M. Gadu & Hsiu-man Lin (eds), *2009 International Symposium on Austronesian Studies*, 11–40. Taitong, National Museum of Prehistory.

Summerhayes, G. R. (2019) The Archaeology of Melanesia. In E. Hirsch & W. Rollason (eds), *The Melanesian World*, 43–62. London, Taylor & Francis/Routledge.

Summerhayes, G. R. & Allen, J. (2007) Lapita write small. In S. Bedford, C. Sand & S. Connaughton (eds), *Oceanic Explorations: Lapita and Western Pacific Settlement*, 97–122. Canberra, Australian National University E-Press.

Summerhayes, G. R., Leavesley, M., Fairbairn, A., Mandui, H., Field, J., Ford, A. & Fullagar, R. (2010) Human adaptation and use of plants in highland New Guinea 49,000–44,000 years ago. *Science* 330, 78–81.

Summerhayes, G. R., Szabo, K., Leavesley, M. & Gaffney, D. (2019a). Kamgot at the lagoon's edge: Site position and resource use. In Bedford & Spriggs 2019, 89–103.

Summerhayes, G. R., Szabo, K., Fairbairn, A., Horrocks, M., McPherson, S. & Crowther, A. (2019b) Early Lapita subsistence: Evidence from Kamgot, Anir. In Bedford & Spriggs 2019, 379–402.

Terrell, J. (1986) *Prehistory in the Pacific Islands*. Cambridge, Cambridge University Press.

3

Diversity and Difference in New Britain, Papua New Guinea: Seeking Indigenous Communities in the Archaeological Record

Jim Specht and Robin Torrence

Western Pacific communities from New Guinea to New Caledonia display a remarkable diversity of languages and cultures, whose origins are generally attributed to the period of the Lapita cultural complex initiated by migrants from Island Southeast Asia. Archaeological evidence from New Britain in Papua New Guinea points to earlier variability among the indigenous communities. These groups then selectively participated in the Lapita complex while maintaining somewhat independent identities that lasted into the post-Lapita world.

Keywords: New Britain, Bismarck Archipelago, Lapita Cultural Complex, Cultural Diversity, Exchange Networks

Contemporary communities of the Melanesian islands in the western Pacific Ocean are renowned for their cultural diversity as expressed through language, material culture, social practices, and biology (*e.g.* Chowning 1977; Kirch 2017, 5–7) (Fig. 3.1). According to a widely accepted narrative, this pattern began *ca.* 3300–3250 BP when immigrants from Island Southeast Asia (ISEA) to the Bismarck Archipelago in Papua New Guinea introduced pottery and other new technologies, domestic animals, novel genetic lineages, and languages ancestral to the Oceanic branch of the Austronesian (AN) family (Kirch 1997; Pawley 2003). In his 'Triple–I' model, Green (1992; 2000; 2003) argued that interaction between the migrants and indigenous Bismarck communities led to the 'integration' of the two groups to form the 'Lapita cultural complex' (hereafter, Lapita), the implication being that the indigenous communities were absorbed into a broader, homogeneous cultural entity. Around 3100 BP, Lapita spread southwards and eastwards into the uninhabited islands of the Solomon Islands, Vanuatu, New Caledonia, and western Polynesia, though people maintained connections across this vast region through long-distance exchange networks that made communities feel they were 'part of "a people" in the cultural sense' (Sand 2017, 364). According to the dominant paradigm, these networks were replaced after *ca.* 2750 BP by smaller, local interaction spheres, leading to cultural divergence through 'contingent history and cultural drift' (Spriggs 1997, 152–86).

Recent developments in biological anthropology have seriously challenged the accepted narrative. Ancient DNA studies of Lapita and later burials in Vanuatu, several thousand kilometres south of the Bismarck Archipelago, concluded that the earliest colonists belonged to East Asian genetic lineages and that contemporary 'Papuan' lineages were introduced several hundred years later from the New Guinea–Bismarck Archipelago region (Spriggs & Reich 2019). These results strengthened the findings of a craniofacial study of the earliest Lapita burials of Vanuatu that concluded they were not the direct ancestors of the present-day people. Valentin *et al.* (2016, 296) were cautious in their conclusions, stating with respect to Near Oceania:

> Conflicting with what is implied in the Triple I model for Lapita, incorporation of Near Oceanian biological features and the spread of Lapita culture could therefore have been time-dissociated processes. Nevertheless, absence of evidence of Near Oceanian admixture in our sample does not rule out previous expectations of heterogeneity among the early populations of the region and among Lapita-associated groups themselves [references omitted].

While it is inappropriate to connect genomes and skeletal morphology with specific languages and cultures and far too early to assume that all early Lapita sites in Near

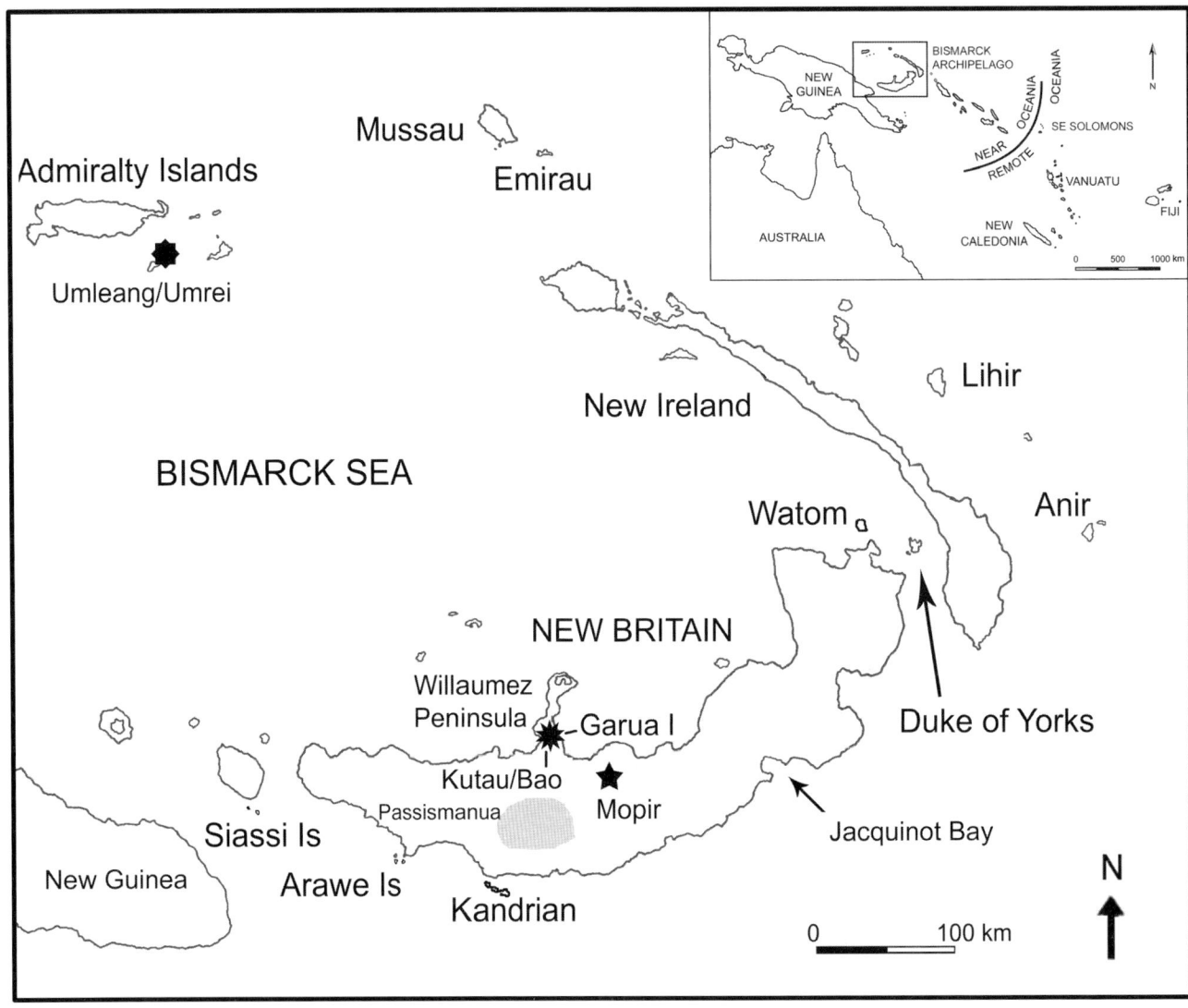

Figure 3.1 The Bismarck Archipelago in relation to other western Pacific Islands, the main obsidian source regions, and other locations cited in the text.

Oceania were initially occupied by people of un-admixed East Asian lineages, the notion of 'time-dissociated processes' in Green's integration stage is important. As Gosden (1991, 261) succinctly noted, '[e]xactly how the boundaries between Lapita and the others is manifest and was maintained has never been adequately argued'; in our view, this is largely because the Bismarck Archipelago has been primarily conceived of only as a linking zone between ISEA and the remote Pacific islands of Polynesia. Here, we review research from New Britain, the main island of the Bismarck Archipelago, to argue that the cultural diversity of the island was not solely due to the end of the Lapita cultural complex. While this is not an entirely new idea (*e.g.* Pawley 1981; 2003, 26–7), we bring together a range of archaeological data showing that post-Lapita cultural diversity is characterised by the persistence of indigenous cultural behaviour initiated before the ISEA migrants moved south into previously uninhabited islands.

New Britain Before Lapita

Scholars (*e.g.* Pawley 2003) have suggested that the introduction by pottery-bearing ISEA migrants of AN languages, which later developed as the Oceanic branch, gradually displaced most of the indigenous non-Austronesian (NAN) languages. Of the 51 languages of New Britain today, only 12 are NAN (Summer Institute of Linguistics 2015). Five of these constitute the Baining family and seven are isolates, four indigenous to New Britain and three recently arrived from New Ireland (Stebbins *et al.* 2018, map 7.1). Apart from these late arrivals, the NAN languages have no confirmed relatives beyond New Britain. It is impossible to calculate how many NAN languages were spoken on the island before the introduction of AN languages, but the absence of links between the indigenous isolates and the Baining family, and the influence of NAN languages on aberrant Oceanic languages in areas of New Britain where no NAN languages survive (Blust 2013, 697–700), imply there were many more in the past. Following on from language changes, scholars have assumed that the ISEA arrivals also caused a reduction of variation in other cultural practices.

Prior to the beginning of Lapita pottery production, cultural practices in New Britain were highly diverse. The most striking pre-Lapita archaeological finds in New Britain are stone mortars and pestles of various styles and stemmed and waisted tools made of obsidian or chert. The mortars and pestles were introduced from New Guinea, where some dated to the middle Holocene were used for processing nuts and tubers (Field *et al*. 2020). The Bismarck Archipelago mortars and pestles are undated, but their stylistic similarities with those of New Guinea suggest a similar age. Swadling (2021) divides them into three groups: a) bowl forms stylistically linked to New Guinea; b) tall conical forms with a pedestal-foot; and c) squat conical forms with a nipple-like base. Form a occurs in the western half of New Britain, whereas forms b and c are in east New Britain and New Ireland. The absence of stone mortars and pestles from Lapita contexts suggests they went out of fashion before or on the arrival of the ISEA migrants, though production of the pedestal-foot type in wood has persisted into the present (Swadling 2021).

Stemmed obsidian tools dated to at least the middle Holocene were produced at the Mopir and Willaumez Peninsula source areas of New Britain by indigenous artisans using technologies that required advanced skill and training (Fig. 3.2; Araho *et al*. 2002; Mulrooney *et al*. 2016; Torrence *et al*. 2022). Most were consumed on New Britain, where specimens with distinctive forms have been found across the island, but some carefully executed examples from the Willaumez Peninsula were exported to New Guinea and the Bismarck Archipelago (Torrence *et al*. 2013a). Their remarkable forms, complexity of production, attractiveness of the raw material, and state of preservation favour their role as meaningful items, perhaps denoting prestige, that were curated across generations, though not necessarily signifying shared values throughout their distribution (Torrence *et al*. 2013b).

Contemporary with the obsidian tools, the indigenous communities of the Passismanua region in the southern foothills of the Whiteman Range, central New Britain, made waisted and stemmed chert tools (Pavlides 2006). Although these have a superficial resemblance to the obsidian stemmed tools, technological and morphological differences define them as a separate, but not necessarily unrelated, tradition. They include many adze forms, whereas large obsidian choppers were made on the Willaumez Peninsula (Kononenko *et al*. 2015). There is also a marked contrast in the distribution and frequency of the two series of artefacts. Obsidian stemmed tools and expedient artefacts are found across New Britain and its island neighbours (Specht 2005; Torrence & Swadling 2008) and testify to extensive exchange networks. In contrast, chert artefacts are only common near the inland quarries with their distribution restricted to the Passismanua–Kandrian–Arawe Islands region. One broken stemmed tool is reported from east of Kandrian (Guillot 2014, 15–17) and only a few chert flakes of unknown origin have been found in northern New Britain (Torrence *et al*. 2004, 118).

Production of the obsidian stemmed tools was severely affected by the W-K2 eruption of Mount Witori (3480–3150 cal BP), which caused abandonment of the Willaumez Peninsula for *ca*. 150 years (Petrie & Torrence 2008, tables 5 & 7). In the more distant Passismanua area, the impact was less, but the abandonment was possibly longer (Specht 2011); there is no evidence for abandonment of the Kandrian coast or the Arawe Islands.

The Arrival of Lapita

On New Britain and neighbouring islands, there are 71 locations with Lapita pottery from all phases (Author's data; see also Bedford *et al*. 2019, table 1). These form eight clusters (Arawe Islands, Kandrian, Gasmata, Duke of York Islands, Watom Island, Willaumez Peninsula isthmus, Talasea/Garua Harbour, and Kove Islands) and eight isolates (Tuam Island, Kreslo, Liton River, Bialla, Walai Island, Walindi/Puro, Valahia Island, and Murikape) (Fig. 3.3). Only three sites – FYS (Garua Island), Lagenda (Talasea/Garua Harbour), and Paligmete (Arawe Islands) – clearly belong to the early Lapita phase, with probably several others in the Duke of York Islands (Summerhayes 2000; Specht & Torrence 2007; White 2007).

Each cluster appears to have had a limited number of production centres using local materials, with the wares being exchanged locally and only occasionally over long distances (*e.g.* Summerhayes 2000, fig. 11.36 for within New Britain). As many authors have pointed out, such exchanges were not necessarily between members of the same ethnolinguistic groups. In recent times in the New Guinea–Bismarck Archipelago region, exchange relationships frequently crossed language and cultural boundaries without impediment, as shown by exchanges between the AN-speaking Motu people of southern New Guinea and their inland NAN Koiari neighbours, and between the Motu and NAN communities of the Papuan Gulf (Allen 1977; Skelly & David 2017).

When people returned to the Willaumez Peninsula obsidian source areas after the W-K2 eruption, they settled at previously occupied locations and introduced Lapita pottery to some of them, though whether these were ISEA migrants, 'integrated' indigenous people, or a mix is open for debate. The presence of descendants of the indigenous people who made stemmed tools before W-K2 is very likely, as small stemmed tools were used on Garua and Boduna islands in Garua Harbour in the same way as the pre-W-K2 specimens (Kononenko *et al*. 2010, 26). The dominance

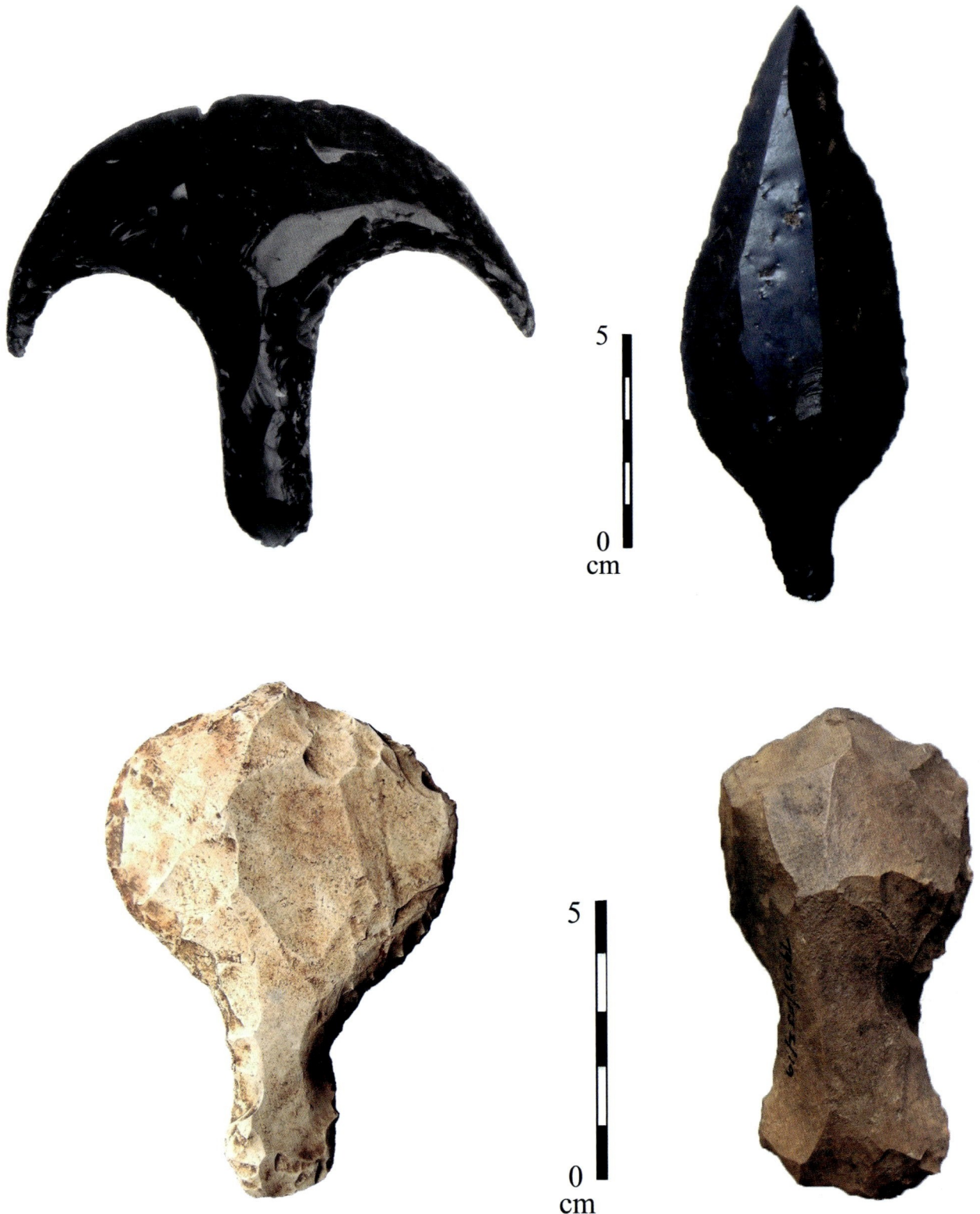

Figure 3.2 Stemmed tools from surface collections in New Britain: obsidian stemmed tools from Boku Hill, Willaumez Peninsula isthmus (top left) and Kandrian coast (top right); chert stemmed tool from Yombon (bottom left) and waisted tool from Asaihi (bottom right), Passismanua region, interior Kandrian.

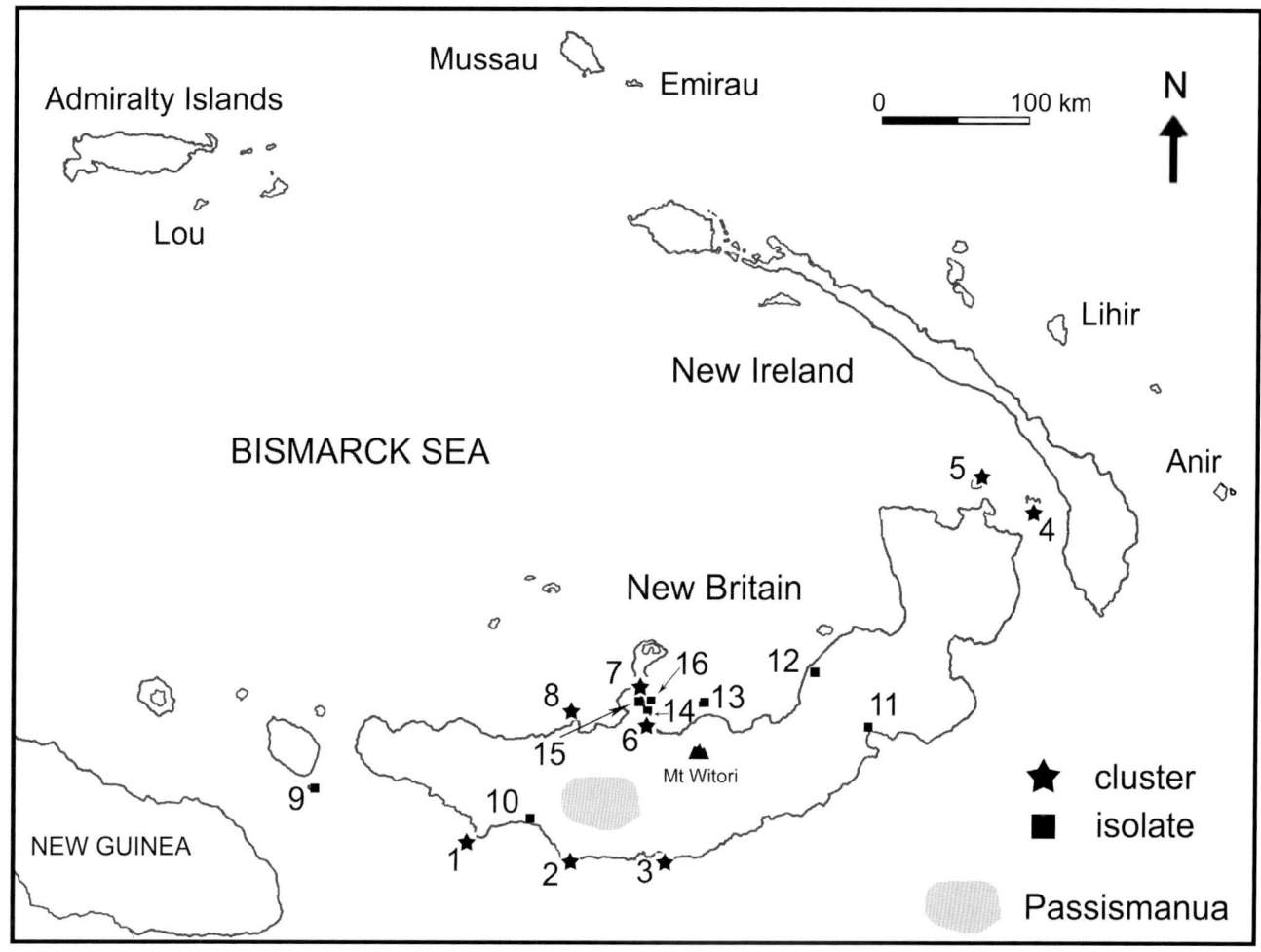

Figure 3.3 Distribution of Lapita sites and site clusters on New Britain and neighbouring islands. Clusters: 1 – Arawe Islands; 2 – Kandrian; 3 – Gasmata; 4 – Duke of York Islands; 5 – Watom Island; 6 – Willaumez Peninsula isthmus; 7 – Talasea/Garua Harbour; 8 – Kove Islands. Isolates: 9 – Tuam Island; 10 – Kreslo; 11 – Liton River/Bain Village; 12 – Bialla area (no location data); 13 – Walai Island; 14 – Walindi/Puro; 15 – Valahia Island; 16 – Murikape.

of obsidian from the Kutau–Bao source at sites across the region (Torrence 2004) indicates a different kind of exchange network that might have helped to integrate or possibly differentiate newcomers with indigenous groups.

Since the impact of the W-K2 eruption was less severe or non-existent around Kandrian and the Arawe Islands, the ISEA migrants were confronted by established indigenous communities. We speculate that their arrival was resolved by granting them access to islands that had marine resources but less productive land than the mainland, providing opportunities for interaction through exchange of marine products for wild and managed forest products of mainland New Britain. The indigenous side of this interaction is indicated by a small, stemmed obsidian tool found in a middle Lapita context at the unpublished site of FFT on Apugi Island, Kandrian, which resembles those of the north coast (Kononenko *et al*. 2010). This tool arguably arrived through exchange networks involving indigenous rather than migrant hands following cross-island routes established in pre-Lapita times. Its most likely origin is from the Willaumez Peninsula or Mopir sources; Admiralty Islands' sources are unlikely as their only recorded occurrence on the south coast is a few pieces in questionable late and post-Lapita contexts in the Arawe Islands (Sutton 2014, table 5.1; Henderson 2017, 62). Elsewhere on New Britain, Admiralty Island obsidian has been recorded only in Lapita sites on Watom Island and the Duke of York Islands (Summerhayes 2009).

Previous work on diversity between Lapita pottery sites identified marked differences between sites on the south coast of New Britain, with shell fishhooks and production waste present in the Arawe Islands but not in the Kandrian sites (Specht *et al*. 2016). Similarly, the Watom Island Lapita sites yielded only one fishhook and no production waste, whereas the early Lapita sites on Emirau Island and in the

Anir and Mussau Islands produced numerous fishhooks at all stages of production (Szabó 2007; Summerhayes *et al.* 2010; Kirch 2021, tables 7.1 & 13.4). The contrasts between sites could reflect limited reef development around Kandrian and Watom compared to the other sites or different cultural influences within local Lapita communities. Shell fishhooks do not feature in pre- or post-Lapita assemblages in New Britain. Their absence or scarcity at some Lapita sites might indicate the presence of indigenous people rather than ISEA migrants, as late Lapita burials on Watom resembled those of modern day and recent inhabitants of the region (Pietrusewsky *et al.* 2014, 15).

Conclusions

Green's (2000, 373) most extended discussion of the integration issue sought to demonstrate that Lapita's origins were in ISEA and not solely or primarily in the Bismarck Archipelago, while acknowledging that the indigenous participation should not be neglected. The evidence for this participation is now sufficient to explore the issue further and to make the indigenous communities visible in our interpretations. In this paper we have summarised some of this evidence for New Britain, revealing clues suggesting that indigenous communities persisted alongside the ISEA migrants and continued their cultural diversity into the post-Lapita period. The appropriateness of continuing to subsume these communities within the Lapita cultural complex is debatable, as this privileges one over the other rather than granting them equal standing. After the migrants with East Asian ancestry moved southwards to colonise uninhabited islands, communities in the Bismarck Archipelago went their own way, and production of pottery and some shell artefacts eventually ceased, except in the Admiralty Islands (Wahome 1997; Cath-Garling 2017; Kirch 2021).

Acknowledgements

We have been privileged to be colleagues and friends of Chris Gosden over four decades in both his English (Sheffield & Oxford) and Australian (Canberra & Melbourne) incarnations. The Pacific chapter of his career took off in the 1985 Lapita Homeland Project, when he began an extended fieldwork program in the Arawe Islands of New Britain. In addition to remarkable material results this produced a flow of theoretical contributions to understanding the significance of Lapita sites on New Britain and in the broader Pacific that continue to inform debates, largely because Chris always placed people and not potsherds or stone flakes as his central concern. We wish him a safe journey wherever retirement takes him.

This paper is based on many fieldwork seasons across New Britain since 1972. We thank sincerely Papua New Guinea's National Research Institute, National Museum and Art Gallery and University of Papua New Guinea, and successive East and West New Britain provincial governments, and their agencies. We give special thanks to the many people on whose lands we worked for their permission to undertake research. Funding was received from the Australian Research Council and its predecessors, the Australian Museum, La Trobe University, the Pacific Biological Foundation, and Earthwatch Foundation, with substantial logistic and other support from Kimbe Bay Shipping Co., New Britain Palm Oil Ltd., and Walindi Plantation Resort.

References

Allen, J. (1977) Fishing for wallabies: Trade as a mechanism for interaction, integration and elaboration on the central Papua coast. In M. Rowlands & J. Friedman (eds), *The Evolution of Social Systems*, 419–55. London, Duckworth.

Araho, N., Torrence, R. & White, J. P. (2002) Valuable and useful: Mid-Holocene stemmed obsidian artefacts from West New Britain, Papua New Guinea. *Proceedings of the Prehistoric Society* 68, 61–81. doi:10.1017/S0079497X00001444.

Bedford, S., Spriggs, M., Burley, D. V., Sand, C., Sheppard, P. & Summerhayes, G. R. (2019) Debating Lapita: Distribution, chronology, society and subsistence. In S. Bedford & M. Spriggs (eds), *Debating Lapita: Distribution, Chronology, Society and Subsistence*. Terra Australis 52, 5–33. Canberra, ANU Press.

Blust, R. (2013) *The Austronesian Languages.* Revised edition. Pacific Linguistics A–PL 008. Canberra, Asia-Pacific Linguistics, Australian National University. Available at: <https://openresearch-repository.anu.edu.au/bitstream/1885/10191/6/Blust-2013-AustronesianLanguages.pdf> [accessed 27 March 2022].

Cath-Garling, S. (2017) *Evolutions or Revolutions? Interaction and Transformation at the 'Transition' in Island Melanesia.* University of Otago Studies in Archaeology 27. Dunedin, Department of Anthropology and Archaeology.

Chowning, A. (1977) *An Introduction to the Peoples and Cultures of Melanesia.* 2nd edition. Menlo Park CA, Cummings Publishing Company.

Field, J. H., Summerhayes, G. R., Luu, S., Coster, A. C. F., Ford, A., Mandui, H., Fullagar, R., Hayes, E., Leavesley, M., Lovave, M. & Kealhofer, L. (2020) Functional studies of flaked and ground stone artefacts reveal starchy tree nut and root exploitation in mid-Holocene highland New Guinea. *The Holocene* 30, 1360–7. doi:10.1177/0959683620919983.

Gosden, C. (1991) Learning about Lapita in the Bismarck Archipelago. In J. Allen & C. Gosden (eds), *Report of the Lapita Homeland Project*. Occasional Papers in Prehistory 20, 260–8. Canberra, Department of Prehistory, Research School of Pacific Studies, Australian National University.

Green, R. C. (1992) Definitions of the Lapita cultural complex and its non-ceramic component. In J. C. Galipaud (ed.), *Poterie Lapita et Peuplement,* 7–20. Noumea, ORSTOM.

Green, R. C. (2000) Lapita and the cultural models for intrusion, integration and innovation. In A. Anderson & T. Murray (eds), *Australian Archaeologist: Collected Papers in Honour of Jim Allen*, 372–92. Canberra, Coombs Academic Publishing, Australian National University.

Green, R. C. (2003) The Lapita horizon and traditions – Signature for one set of Oceanic migrations. In C. Sand (ed.), *Pacific Archaeology: Assessments and Prospects. Les Cahiers de l'Archéologie en Nouvelle-Calédonie* 15, 95–120. Noumea, Département d'Archéologie, Service des Musées et du Patrimoine de Nouvelle-Calédonie.

Guillot, F. (2014) Archéologie. In *IOWA: Expédition National FFS 2014: Massif de Iowa, Nakanaï Range, Nouvelle-Bretagne, Papouasie Nouvelle-Guinée*, 15–17. Lyon, Fédération Française de Spéléologie.

Henderson, R. (2017) *The changing nature of Lapita mobility and interaction*. Unpublished thesis, University of Otago.

Kirch, P. V. (1997) *The Lapita Peoples: Ancestors of the Oceanic World*. Oxford, Blackwell.

Kirch, P. V. (2017) *On the Road of the Winds: An Archaeological History of the Pacific Islands Before European Contact*. 2nd edition. Oakland, CA, University of California Press.

Kirch, P. V. (ed.) 2021. *Talepakemalai. Lapita and Its Transformations in the Mussau Islands of Near Oceania*. Monumenta Archaeologica 47. Los Angeles, Cotsen Institute of Archaeology Press, UCLA.

Kononenko, N., Specht, J. & Torrence, R. (2010) Persistent traditions in the face of natural disasters: Stemmed and waisted stone tools in late Holocene New Britain, Papua New Guinea. *Australian Archaeology* 70, 17–28. doi:10.1080/03122417.2010.11681908.

Kononenko, N., Torrence, R. & White, P. (2015) Unexpected uses of obsidian: Experimental replication and use-wear analyses of chopping tools. *Journal of Archaeological Science* 54, 254–69. doi:10.1016/j.jas.2014.11.010.

Mulrooney, M., Torrence, R. & McAlister, A. (2016) The demise of a monopoly: Implications of geochemical characterisation of a stemmed obsidian tool from the Bishop Museum collections. *Archaeology in Oceania* 51, 62–9. doi:10.1002/arco.5069.

Pavlides, C. (2006) Life before Lapita: New developments in Melanesia's long-term history. In I. Lilley (ed.), *Archaeology of Oceania: Australia and the Pacific Islands*, 205–27. Oxford, Blackwell.

Pawley, A. (1981) Melanesian diversity and Polynesian homogeneity: A unified explanation for language. In J. Hollyman & A. Pawley (eds), *Studies in Pacific Languages and Culture in Honour of Bruce Biggs*, 269–309. Auckland, Linguistic Society of New Zealand.

Pawley, A. (2003) Locating Proto Oceanic. In M. Ross, A. Pawley & M. Osmond (eds), *The Lexicon of Proto Oceanic. The Culture and Environment of Ancestral Oceanic Society. 2 The Physical Environment*. Pacific Linguistics 545, 17–34. Canberra, Pacific Linguistics, Australian National University.

Petrie, C. A. & Torrence, R. (2008) Assessing the effects of volcanic disasters on human settlement in the Willaumez Peninsula, Papua New Guinea: A Bayesian approach to radiocarbon calibration. *The Holocene* 18, 729–44. doi:10.1177/0959683608091793.

Pietrusewsky, M., Buckley, H., Anson, D. & Toomay Douglas, M. (2014) Polynesian origins: A biodistance study of mandibles from the Late Lapita site of Reber-Rakival (SAC), Watom Island, Bismarck Archipelago. *Journal of Pacific Archaeology* 5, 1–20.

Sand, C. (2017) Disentangling the Lapita interaction spheres. In T. Hodos, A. Geurds, P. Lane, I. Lilley, M. Pitts, G. Shelach, M. Stark & M. J. Verluys (eds), *The Routledge Handbook of Archaeology and Globalization*, 354–68. London, Routledge.

Skelly, R. J. & David, B. (2017) *Hiri. Archaeology of Long-distance Maritime Trade Along the South Coast of Papua New Guinea*. Honolulu, University of Hawai'i Press.

Specht, J. (2005) Obsidian stemmed tools in New Britain: Aspects of their role and value in mid-Holocene Papua New Guinea. In I. Macfarlane, M. J. Mountain & R. Paton (eds), *Many Exchanges: Archaeology, History, Community and the Work of Isabel McBryde*. Monograph 11, 373–92. Canberra, Aboriginal History.

Specht, J. (2011) Diversity in lithic raw material sources on New Britain, Papua New Guinea. *Archaeology in Oceania* 46, 54–66. doi:10.1002/j.1834-4453.2011.tb00099.x.

Specht, J., Gosden, C., Pavlides, C., Richards, Z. & Summerhayes, G. R. (2016) Exploring Lapita diversity on New Britain's south coast, Papua New Guinea. *Journal of Pacific Archaeology* 7, 20–29. Available at: <https://pacificarchaeology.org/index.php/journal/article/view/175> [accessed 04 October 2022].

Specht, J. & Torrence, R. (2007) Archaeological studies of the middle and late Holocene, Papua New Guinea. Part IV. Pottery sites of the Talasea area, Papua New Guinea. *Technical Reports of the Australian Museum Online* 20, 131–96. doi:10.3853/j.1835-4211.20.2007.1476.

Spriggs, M. (1997) *The Island Melanesians*. Oxford, Blackwell.

Spriggs, M. & Reich, D. (2019) An ancient DNA Pacific journey: A case study of collaboration between archaeologists and geneticists. *World Archaeology* 51, 620–39. doi:10.1080/00438243.2019.1733069.

Stebbins, T., Evans, B. & Terrill, A. (2018) The Papuan languages of Island Melanesia. In B. Palmer (ed.), *The Languages and Linguistics of the New Guinea Area. A Comprehensive Guide*, 775–894. Berlin, De Gruyter Mouton.

Summerhayes, G. (2000) *Lapita Interaction*. Terra Australis 15. Canberra, Department of Archaeology and Natural History, and Centre for Archaeological Research, Australian National University.

Summerhayes, G. R. (2009) Obsidian network patterns in Melanesia – Sources, characterisation and distribution. *Bulletin of the Indo–Pacific Prehistory Association* 29, 110–24. doi:10.7152/bippa.v29i0.9484.

Summerhayes, G. R., Matisoo-Smith, E., Mandui, H., Allen, J., Specht, J., Hogg, N. & McPherson, S. (2010) Tamuarawai (EQS): An early Lapita site on Emirau, New Ireland, PNG. *Journal of Pacific Archaeology* 1, 62–75. Available at: <https://pacificarchaeology.org/index.php/journal/article/view/10> [accessed 04 October 2022].

Summer Institute of Linguistics (2015) Language maps for East and West New Britain Provinces, PNG. Available at: <https://pnglanguages.sil.org/resources/language_maps> [accessed 04 October 2022].

Sutton, P. (2014) *Lapita exchange across the Solomon and Bismarck Seas: pXRF sourcing and morphological analysis of obsidian artefacts from Apalo, West New Britain*. Unpublished thesis, University of Sydney.

Swadling, P. (2021) Mortars and pestles make the mid-Holocene occupation of New Guinea and the Bismarck Archipelago visible. In I. J. McNiven & B. David (eds), *The Oxford Handbook of the Archaeology of Indigenous Australia and New Guinea.* doi:10.1093/oxfordhb/9780190095611.013.26.

Szabó, K. (2007) An assessment of shell fishhooks of the Lapita cultural complex. In A. Anderson, K. Green & F. Leach (eds), *Vastly Ingenious: The Archaeology of Pacific Material Culture in Honour of Janet Davidson*, 227–41. Dunedin, Otago University Press.

Torrence, R. (2004) Now you see it, now you don't: Changing obsidian source use in the Willaumez Peninsula, Papua New Guinea. In J. Cherry, C. Scarre & S. Shennan (eds), *Explaining Social Change: Studies in Honour of Colin Renfrew*, 115–25. Cambridge, McDonald Institute for Archaeological Research.

Torrence, R. & Swadling, P. (2008) Social networks and the spread of Lapita. *Antiquity* 82, 600–16. doi:10.1017/S0003598X00097258.

Torrence, R., Neall, V., Doelman, T., Rhodes, E., McKee, C., Davies, H., Bonetti, R., Guglielmetti, A., Manzoni, A., Oddone, M., Parr, J. & Wallace, C. (2004) Pleistocene colonisation of the Bismarck Archipelago: New evidence from West New Britain. *Archaeology in Oceania* 39, 101–30. doi:10.1002/j.1834-4453.2004.tb00568.x.

Torrence, R., Kelloway, S. & White, P. (2013a) Stemmed tools, social interaction, and voyaging in early-mid Holocene Papua New Guinea. *Journal of Island and Coastal Archaeology* 8, 287–310. doi:10.1080/15564894.2012.761300.

Torrence, R., White, P. & Kononenko, N. (2013b) Meaningful stones: Obsidian stemmed tools from Barema, New Britain, Papua New Guinea. *Australian Archaeology* 77, 1–8. doi:10.1080/03122417.2013.11681974.

Torrence, R., Kononenko, N. & Dickinson, P. (2022) Crafting social networks: The production of obsidian stemmed tools in the Willaumez Peninsula, Papua New Guinea. *Journal of Archaeological Method and Theory* 29. doi:10.1007/s10816-021-09545-3.

Valentin, F., Détroit, F., Spriggs, M. J. T. & Bedford, S. (2016) Early Lapita skeletons from Vanuatu show Polynesian craniofacial shape: Implications for Remote Oceania settlement and Lapita origins. *Proceedings of the National Academy of Sciences* 113, 282–97. doi:10.1073/pnas.1516186113.

Wahome, E. W. (1997) Continuity and change in Lapita and post-Lapita ceramics: A review of evidence from the Admiralty Islands and New Ireland, Papua New Guinea. *Archaeology in Oceania* 32, 118–23.

White, J. P. (2007) Archaeological Studies of the middle and late Holocene, Papua New Guinea. Part I. Ceramic sites on the Duke of York Islands. *Technical Reports of the Australian Museum Online* 20, 3–50. doi:10.3853/j.1835-4211.20.2007.1473.

4

Why the Concept of Near and Remote Oceania Fails Island Melanesian Prehistory

Christophe Sand and Jim Allen

Introduced into Pacific archaeology in the 1970s by Roger Green and Andrew Pawley, the concept of Near and Remote Oceania has mostly been accepted and adopted without question. In this paper, we argue that this division satisfies Polynesian archaeology but has served to divide Melanesian archaeology, obscuring continuities between east and west in Melanesia.

Keywords: Melanesia, Near Oceania, Remote Oceania, Nomenclature, Unity and Diversity, Genetics, Linguistics, Social Organisation

Introduction

…while a Near Oceania-Remote Oceania divide accurately charts a boundary separating landmasses occupied in the Pleistocene from islands first settled in the late Holocene, it does not adequately explain when and how human diversity developed within the Western archipelagos of Remote Oceania (Clark 2003a).

Humans organise the physical world by naming and renaming geographical features, localities, and regions. The delineation of Melanesia (Fig. 4.1), the region we are concerned with here, has a long and varied history. Parts of northern Melanesia were named by Spanish explorers as early as the 16th century AD. For example, Manus Island was named Urays La Grande by Saavedra in 1528 and New Guinea (Nueva Guinea) by de Retes in 1545 (Spate 1979; Douglas 2010). Further east, Mendaña called the archipelago he visited in 1568 and in 1595 the Solomon Islands. In 1804, French geographers Mentelle and Malte-Brun suggested the term *Océanique* (soon changed to *Océanie*) to represent Island Southeast Asia, Australia, New Zealand, and all the Pacific Islands. Within this region, they maintained an earlier term, *Polynésie*, originally used by de Brosse to encompass what is now Polynesia, Micronesia, and some of Island Melanesia. Melanesia, together with Australia, was henceforth separated from Polynesia (Douglas 2010).

In 1832, the French navigator Jules Dumont d'Urville (1832; 2003) published a paper 'On the Islands of the Great Ocean', where he divided Oceania into four parts, maintaining Polynesia and separating Micronesia from it ('as it contains only very small islands'), identifying Melanesia, and adding Malaysia, a name already in use for the Dutch East Indies.

In Dumont d'Urville's classification, Melanesia comprised Australia, Tasmania, and New Guinea, and all nearby islands, reaching to the fringes of Micronesia and Polynesia. This area he called Melanesia because 'this is the home of the black race of Oceania', amending the name *Mélanians* previously offered by Bory de Saint-Vincent, who had sailed as naturalist with the Baudin expedition to Australia in 1798. Despite the borrowing and adaptation of previous terms (detailed by Tcherkézoff 2003), Dumont d'Urville is popularly attributed with creating the tripartite division of Polynesia, Micronesia, and Melanesia because, with some changes – most notably the subsequent exclusion of Australia and Tasmania – it is his division that continues in use today. Meanwhile the term Oceania has shrunk from Dumont d'Urville's originally expanded term to be a popular collective noun for Polynesia, Micronesia, and Melanesia.

Douglas (2010) argues that a racial taxonomy for the region developed with the introduction of the term *Oceania* in 1804, when Mentelle and Malte-Brun differentiated the

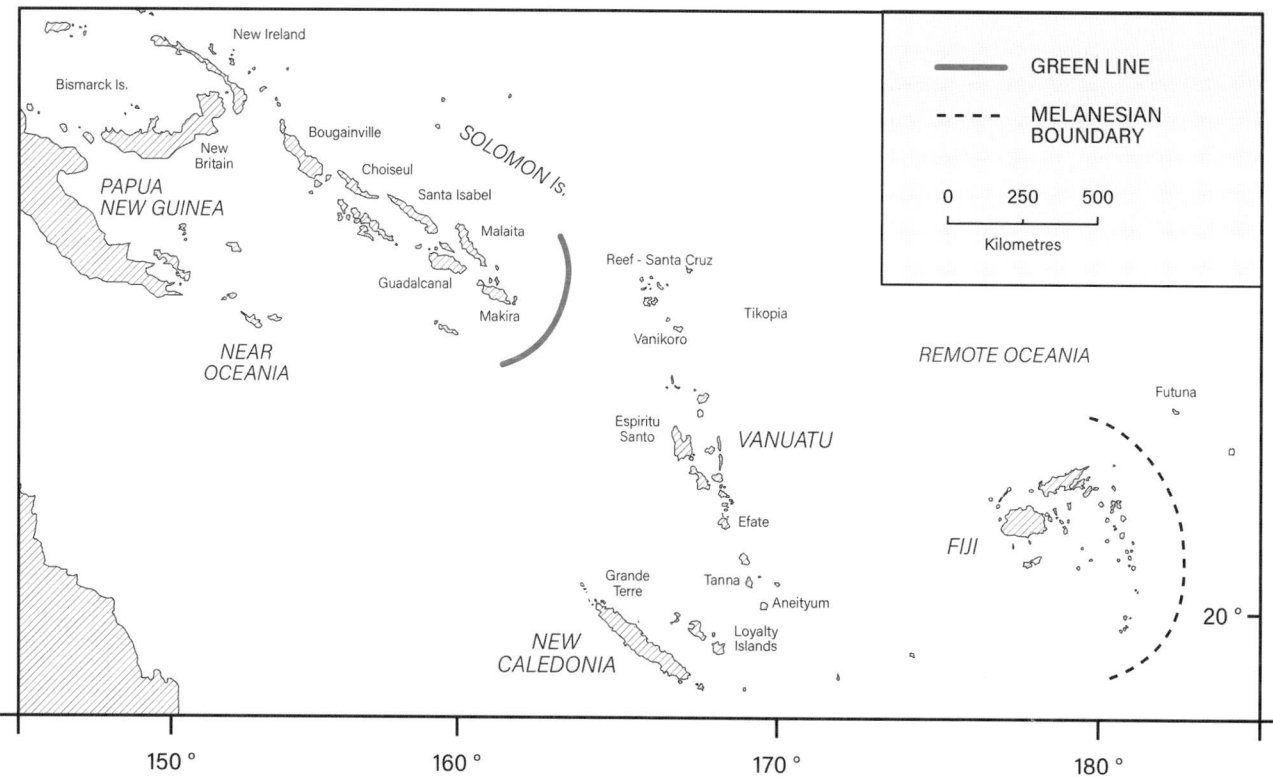

Figure 4.1 Map of Melanesia showing boundaries and localities mentioned in the text.

'beautiful' and 'copper-coloured' Polynesians, Micronesians, and Malays from the 'Oceanic Negroes' of New Guinea and Island Melanesia. Dumont d'Urville (1832; 2003) elaborated the division. For him Melanesians were in every way inferior to his three other groups, not only in physical appearance and in languages, but also in their aptitudes and intelligence, and their treacherous dealings with outsiders. He claimed that they possessed no government, law, nor formal religion. On the scale of barbarism to civility, Dumont d'Urville ranked Melanesians according to their proximity to and interactions with Polynesians. Melanesians in Fiji were most advanced, while those from the Bismarck Archipelago and New Guinea were inferior to all their Island Melanesian counterparts.

Early in the mid-19th century, anthropological perceptions led to New Guinea being separated from the Melanesian archipelago. Ballard (2008) argues that the racial distinction between Melanesians and Polynesians was provided by comparisons using Melanesians from New Caledonia, Vanuatu, and the Solomon Islands. In contrast, the common comparison for 'black-skinned, frizzy-haired Papuans' was with the lighter skinned Malays to the west. Authors like Crawfurd (1820), Earl (1853), and Wallace (1869) promoted a perception of 'Papuans' that linked them to sub-Saharan Africans. As noted, they were also referred to as 'Oceanic Negroes' a term that transferred emotively negative associations (Ballard 2008, 165).

This history of negative race-based assessments of Melanesians and invidious comparisons with Polynesians has led in the last 50 years to widespread reservations among academics about using the term Melanesia/Melanesian (*e.g.* Hau'ofa 1975; Thomas 1989 and comments; Clark 2003b; Lawson 2013). Paradoxically, at the same time, non-academic Melanesians have embraced the term in political and social contexts, such as The Melanesian Way (Otto 1997) as a focus of identity, where people continue to see their future as Melanesian (see also Flexner *et al.* 2019, 405 on this point).

Near and Remote Oceania

While in general archaeologists have been less concerned than historians and social anthropologists with the term Melanesia and its connotations, Roger Green (*e.g.* 1991) saw it as having little relevance to archaeology in the Western Pacific, a position that reflected the history of his own Melanesian research. In 1970, Green began preliminary fieldwork for his Southeast Solomon Islands Cultural History project, centred on the isolated, small island archipelago known as the Reef-Santa Cruz Islands. The Solomons were a blank spot linking then-recent archaeological work in Fiji, New Caledonia, and Vanuatu with early excavations in the Bismarck Archipelago. Green's choice of the Reef-Santa Cruz group over the main Solomons chain (Green 1976a;

Green & Creswell 1976) emphasised the former's linguistic, cultural, and ecological diversity, but Green's research interests and questions (Green 1976a, 10–13) were predominantly Polynesian in intent, and the biogeographic ocean divide separating the Reef-Santa Cruz Islands from San Cristobal (see below) linked this research much more directly with the Eastern Melanesian Islands (EMI) and Western Polynesia.

Remarkably, Green (1976a) did not refer directly to Lapita pottery – the archaeological harbinger of Polynesians-to-be – in his introduction, although this pottery, its widespread Melanesian distribution, its antiquity, and its links to Austronesian speakers and hence proto-Polynesians had at that time been widely discussed for more than a decade (*e.g.* Golson 1961). But Lapita pottery was not far from Green's mind, as demonstrated by the first of his research questions: 'what is the nature of the earliest cultural assemblages from Island Melanesia which may be ancestral to those already known for West Polynesia?' (Green 1976a, 10). Equally, Green noted that the Reef-Santa Cruz group marked the eastern limit of non-Austronesian languages, that the archipelago was separated from the main Solomon Islands chain by 350 km of ocean that formed a significant biogeographical boundary, and that this boundary also marked a change from tool kits predominantly made on chert to ones predominantly made on shell.

By 1973, with these data supplemented by the discoveries in the Reef-Santa Cruz excavations, the concept of Near and Remote Oceania was unveiled in a paper primarily concerned with linguistics (Pawley & Green 1973, 4) and subsequently developed in mainly environmental terms to 'disestablish' the concept of Melanesia (Green 1991).

Problems With This Concept

Fifty years later, should this matter? The geographical division and renaming of Island Melanesia on either side of what we refer to here as the 'Green Line' (Fig. 4.1) has been met with general acceptance and little critical assessment by Pacific archaeologists. However, we argue that this framework does damage to Melanesian archaeology in several different ways. Essentially it is a Polynesia-driven creation predicated on Lapita archaeology. Green concluded in the Southeast Solomon Islands Cultural History report 'the eastern end of Western Melanesia [we take this to mean the Reef-Santa Cruz Islands] … is firmly joined to Eastern Melanesia and Polynesian (sic) archaeologically, just as it is linguistically and culturally' (Green 1976b, 270). We argue that this does not hold for the post-Lapita period in the EMI. As exemplified by the quotation from Geoff Clark (2003a) at the head of this paper, the Near and Remote idea disenfranchises the EMI in the post-Lapita prehistory of the region. The Green Line separates it from the prehistories to the west and obscures later interactions that crossed it.

As the Clark quotation also exemplifies, the Green Line has, for many archaeologists working in Melanesia, been seen principally as a division that 'encapsulates two major epochs in the history of the Pacific islanders' (Kirch 2017, 5). Defining this boundary is nonetheless important because it puts the Western Melanesian Islands into an understandable historical context different from the EMI. One difficulty east of the Green Line is the absence of any further frontier in Remote Oceania. There remains no historical post-Lapita context for the EMI, except as part of a developing Polynesia. However argumentative the division between Fiji and Western Polynesia might be (for a detailed account see Burley 2013 and associated comments), we see it as providing a useful boundary for defining a pan-Melanesian coherence reflecting a dynamism and complexity different to Polynesia. As such, our objective in the remainder of the paper is less to disestablish Near and Remote Oceania and more to re-establish Melanesia as a distinctive cultural region.

Case Studies

Two obvious strategies to re-establish Melanesia are to show that once breached, the Green Line proved no barrier to the continuing transfers of people, goods, and ideas across it, and secondly to show subsequent differences between Melanesia and Polynesia, mostly changing across the traditional boundary between Fiji and Tonga/Samoa. Many aspects of transfer and difference might be considered, for example in material culture, such as portable art and artefacts, canoes, and domestic and ceremonial architecture. Limits of space have instead encouraged us to look both briefly and generally at several other markers: genetics, language, social unity and diversity, and social organisation.

The Complex Genetic History of the Melanesians

Mendaña's 1568 colonising fleet left Spain in search of the Great Southern Continent but instead fetched up on the Solomon Islands. On Santa Isabel Island, they encountered people 'of different complexions … some are of the same colour as those of Peru, others are black, and a few are quite fair' (Amherst & Thomson 1901, 133–4). Less than 40 years later, de Quiros failed to settle Santo Island in northern Vanuatu, noting that:

> the population there is as numerous as it can be. Its inhabitants are of diverse colours: we meet there blacks, whites, tanned & mulattos; some have black, long & smooth hair, the others have them red & curly. This diversity of colour testifies the blend of races (Dalrymple 1774, 283–4, our translation).

These two testimonies from opposite sides of the Green Line indicate that around AD 1600, there was no observable change in the skin pigment characteristics of the populations between the Solomon Islands and Northern Vanuatu. Recent genetic studies have given scientific flesh to these phenotypic observations made four centuries ago by identifying a complex diversity of genetic maternal DNA and paternal Y-chromosome inputs. These are evidently related to different episodes of human settlement and their subsequent combinations, as has been illustrated in detail for Vanuatu (*e.g.* Lipson *et al.* 2018; 2020; Posth *et al.* 2018; Choin *et al.* 2021). The oldest set of genetic markers relate to the Pleistocene settlement of Papua New Guinea and the western Melanesian islands (Pedro *et al.* 2020; Choin *et al.* 2021), with a first diversification over at least 30,000 years, permitting the build-up of regionally specific characteristics. The arrival of populations from Island Southeast Asia after 3400 cal BP in the Bismarck Archipelago and the subsequent spread of the Lapita Cultural Complex over a distance of 4500 km, reaching Tonga and Samoa in the Central Pacific, is identifiable by the mitochondrial lineage B4a1a1 (Skoglund *et al.* 2016).

The Lapita peoples represent the earliest settlers of the EMI, but their gene pool was subsequently significantly diversified by the arrival of populations carrying Papuan-derived genes, whose origins appear to be mostly restricted to the Bismarck Archipelago (Lipson *et al.* 2020). Debates remain unsettled about the exact timing of this 'Melanisation' of the EMI (*e.g.* Arauna *et al.* 2022), which may have been a slow process starting during Lapita times and continuing for at least a millennium.

A final major genetic contribution reflects the settlement of different parts of the Melanesian Crescent ('the Polynesian Outliers'), commonly accepted to be Polynesian groups out of the Central Pacific during the 2nd millennium AD (Spriggs 1997, chapter 7; Kirch & Swift 2017; Zinger *et al.* 2020). Early European explorers noticed the clear-cut difference of these groups from other populations in the archipelagos they visited (Spriggs 1997, 228–31). Accordingly, and different to Eastern Polynesia (*e.g.* Hudjashov *et al.* 2018), any discussion of an early genetic 'founder effect' has no relevance in Melanesia, as the regular input of new genes from distant regions has long since overwritten the original phenotypic characteristics of any first discoverers. However, multiple localised episodes of genetic 'bottleneck' processes might have repeatedly occurred after cataclysmic natural events, such as earthquakes, volcanic eruptions, or tsunamis. Population reduction linked to such events in New Britain (Torrence 2002), in the Solomon Islands, and in Vanuatu thus contributed to phenotypic diversity in the region. Accordingly, genetic patterns were related both to long-distance population movements and also contacts between neighbouring islands.

The major genetic inputs that characterise the history of the Southwest Pacific were associated with successive movements of specific populations through the Melanesian archipelagos. These generated innumerable localised combinations, depending on a wide variety of events and circumstances at any specific location (Terrell 2010). On small islands, phases of 'population replacement' or 'elimination of the former male component' may have happened (Friedlaender 2007, 233). On larger islands, the terrain often led to marked genetic differences, for example between groups settled on each side of a mountain range. Large scale diversity occurred even within coherent cultural groups, such as the Baining, living inland on the Gazelle Peninsula in northeast New Britain. Genetic studies indicate that 'the composite picture that emerges ... is that of a group divided into extremely differentiated sub-populations. This differentiation rivals that observed between populations sampled from different continents' (Wilder & Hammer 2007, 205; see Bergström *et al.* 2017 for a similar case study). While some islands appear to have a long tradition of regularly accepting new arrivals, others like Bougainville 'remained surprisingly isolated until relatively recently' (Friedlaender 2007, 232). However, beyond the eastern limits of Fiji, genetic diversity drops significantly. Polynesians have overall a more homogeneous signature between archipelagos (Friedlaender *et al.* 2008).

Linguistic Diversity

Melanesia has been identified as one of the most diverse linguistic regions in the world, with a total of over one thousand different languages. The oldest language group is formed by the Papuan languages, related to the Pleistocene settlement of Near Oceania (*e.g.* Ross 2005). Unsurprisingly, the vast majority of the Papuan languages are spoken on the main island of New Guinea, with a restricted number of groups spread along the Melanesian Crescent, mainly in New Britain and Bougainville. But some linguistic studies identify three of the languages spoken in the Santa Cruz group and those spoken on nearby Vanikoro, in the EMI, as Papuan languages (Lincoln 1978; Wurm 1978; Blench 2014). In addition, a reanalysis of Kanak languages of Grande Terre in New Caledonia classifies them as 'Austronesian, but [that] may have experienced bilingualism with Papuan', while those of the nearby Loyalty Islands are 'classified as Austronesian but [are] perhaps actually Papuan' (Blench 2014, 11 & fig. 5).

We do not discuss here the debated geographical homeland of proto-Oceanic (see Pawley 2008; 2010; Donohue & Denham 2012; *cf.* Terrell *et al.* 2002). It is sufficient to note that the Oceanic subgroup of Austronesian languages spread into the Pacific from Southeast Asia from Lapita times onward, obviously crossing the Green Line. Significantly, the diversification of the Oceanic languages was exponential in the Melanesian Crescent. In comparison, it was exceedingly reduced further east (*e.g.* Pawley 2007). In the modern country of Papua New Guinea (including Bougainville), there are *ca.* 250 Austronesian languages; in the Solomon Islands, *ca.* 80; and in Vanuatu 107, making this archipelago the most linguistically diverse country in the world per head of population (Blust 2013, 94–109). There were at least 34 languages in New Caledonia before colonial times, while only two clearly separated language groups are recognised in Fiji (Blust 2013, 110 & 120). Three thousand years later, there are at least two languages, Eastern and Western Fijian, and on the grounds of cognate percentages and degree of mutual intelligibility, a case can be made for distinguishing five or six Eastern Fijian languages and at least two Western languages (Pawley 2007, 25). In terms of numbers of languages, the Green Line has little pertinence, since Vanuatu on the eastern side of the divide developed proportionally more languages than the Solomon Islands, its counterpart on the western side.

To highlight this point, several specific characteristics of possible Papuan derivation are identified across the Melanesian Crescent. The most unexplained is the presence of a 'phonic tone' in languages from the northern half of the Grande Terre of New Caledonia. This phonological feature is present in only 1% of the Oceanic languages and is mostly restricted to mainland New Guinea (Schapper 2020, 4 & map 3). Other examples include the 'dominant order of noun and numeral' for most languages of Vanuatu (Schapper 2020, map 8), the 'colexification of "fire" and "firewood"' (Schapper 2020, map 9), and specific sets of numeral bases and counting systems (Schapper 2020, map 11).

Unity and Diversity: Melanesia in the Post-Lapita Era

Archaeological research in Melanesia has been significantly influenced in the second part of the 20th century by debates about population origins and the link with Polynesia. Pioneer archaeologist Edward Gifford (1951, 189) justified his first expedition to Fiji in 1947 by noting that, since:

> tropical Polynesia has yielded archaeologically only the early phase of the local cultures which were flourishing at the time of discovery, I decided to look farther west for a succession of cultures. Fiji seemed a likely place, and moreover I reasoned that it might show traces of early Polynesians, if they had come via Fiji.

Many Melanesian research programmes have targeted this problem, usually relying in large part on typological studies of ceramics. Pottery was introduced into Oceania from Island Southeast Asia by Lapita settlers at the end of the 2nd millennium BC. It vanished from the cultural productions of a large number of islands of the Western Pacific over time, disappearing in Western Polynesia around two thousand years ago (Sand 2007; Burley *et al.* 2018). It was thus not part of later Polynesian traditions. In contrast, it continued to flourish until European arrival in many parts of Melanesia and in the archipelagos of Palau and Yap in Micronesia (Kirch 2017, 172–3).

In all the islands where pottery continued to be produced, studies have shown typological evolutions over time, highlighting cultural processes of change (*e.g.* Irwin 1985; Wickler 2001; Bedford 2006; Sand *et al.* 2011; Cochrane 2018). Furthermore, provenance studies have demonstrated that non-ceramic communities imported pots from nearby ceramic production centres across Melanesia, some groups becoming specialised long-distance traders (*e.g.* Allen 2017).

Aside from ceramics, the study of petroglyphs offers another example of post-Lapita Melanesian unity. Rock engravings are numerous in the Melanesian Crescent, from Eastern New Guinea to the Grande Terre of New Caledonia (Specht 1979; Wilson 2003; Sand 2012; Wilson & Ballard 2019) but are very rare in Fiji and West Polynesia (Egan & Burley 2009; Cruz Berrocal & Millerstrom 2013). The Melanesian petroglyphs have specific typological characteristics, with numerous curvilinear geometric motifs (crosses, concentric forms, spirals), few figurative designs, and a propensity for containing the central motif in an envelope (Monnin & Sand 2004, 217–22) (Fig. 4.2). The chronology for the production of this regional art form is still debated but is thought to span several millennia, to judge from the carvings themselves. Typological comparisons with engravings produced in other cultural regions, like Aboriginal Australia and Eastern Polynesia, set Melanesian petroglyphs apart. Meanwhile, the Melanesian examples exhibit a common coherence along a 3000 km island axis.

At the same time, the cultural diversity of Melanesia and its long chronology often make it difficult to

Figure 4.2 Enveloped crosses on petroglyphs from Poro, New Caledonia. These are a recurrent motif in engraved sites across the Melanesian Crescent (Photo: C. Sand).

Figure 4.3 Megalithic ceremonial platform at Nasara Peterhil on Vao Island, Vanuatu (Photo: S. Bedford).

distinguish between independent, localised cultural traits and long-distance historical connections. An example is a set of megalithic constructions (Riesenfeld 1950) comprising raised monoliths and capstones present in some islands of the Melanesian Crescent that are widely separated geographically. Structures of this type have been recorded on the Island of Uneapa (aka Unea) in the Bismarck Archipelago, where andesite 'seats' and 'tables' (Byrne 2005, 98) are present (see also Torrence *et al.* 2002, 4; Bickler 2006; Byrne 2013). Typologically related constructions of 'capstones' are present in some regions of southern Bougainville (Terrell 1976; 1986, 225–38), but are also a central component of the meeting-places (*Nasara*) built to celebrate the passage of chiefly grades on Malekula and neighbouring small islands in northern Vanuatu (Bedford 2019) (Fig. 4.3). Cultural influences from Western Melanesia appear to best explain the emergence of the *Nasara* tradition in Malekula and the significant changes in the political systems during the centuries preceding European settlement of this archipelago (Bedford & Spriggs 2008). These are only several examples of many that might be drawn from archaeological contexts that crossed the Green Line between Near and Remote Oceania.

Melanesian 'Big Men' in Historical Perspective

Among the most important anthropological examples drawn from Pacific ethnography and used worldwide in theoretical historical models is the dichotomy between simple traditional Melanesian societies, characterised by low hierarchies and political systems centred on grades producing 'big men', and complex Polynesian polities organised in hierarchical, hereditary chiefdoms, that sometimes reach a level of 'archaic states' (Kirch 2017, chapter 8). As noted in our introduction, the Melanesia/Polynesia division originated on a racial basis: dark-skinned people being considered by early European explorers as necessarily more 'primitive' than lighter-skinned oceanic populations (Douglas 2010). The identification of two different basic systems of political organisation in the Pacific was one extension of this mindset. In particular, it was used to provide ethnographic examples to build early social models for the prehistory of Europe, a science then in its infancy (see Gosden 1999). Sahlins (1963), amongst others, popularised this anthropological model, globalising in its wake the idea that Melanesian societies were organised on one supposedly homogeneous low-hierarchical political system. One argument to explain the putative existence of a political dichotomy in Oceania depends in part on an overall low population density in the Southwestern Pacific, said to result from the presence of malaria in Western Melanesia as far east as central Vanuatu (Groube 1993; Bayliss-Smith & Hyiding 2014). In this model, the absence of malaria further east allowed higher population densities and more complex political systems.

Proposing relatively low Melanesian populations in pre-European times as an explanation for their egalitarian political systems is questionable on several counts. One prime ethnographic example concerns the first Europeans entering the New Guinea Highlands in 1933, where they found dense human populations with little or no hierarchical social organisation, supported by the basic staple sweet potato – transported from the Americas – taro, and a wide range of lesser crops. Archaeological research (Golson *et al.* 2017) documents the development of this system over the last 10,000 years, demonstrating that grade-systems have unquestionably a deep history in parts of Melanesia and that these cannot be explained only by demographic constraints.

Various archaeological studies have also recently challenged the preconception of low population densities in Island Melanesia in the centuries predating the first European intrusions by describing the presence of intensified cultural landscapes in many of the Melanesian archipelagos, especially in the interior regions of large islands (*e.g.* Field 2005; Sand *et al.* 2016; Bedford *et al.* 2017; Bayliss-Smith *et al.* 2019). Accordingly, the probable existence of large populations in much of Melanesia up to the 18th century, echoing the observations of the first Western explorers who sailed across the region (see Sand 2023), challenges the orthodox historical discourse about the deep past of this region. It also reinforces the proposal that a diversity of political systems, often of complex hierarchical nature, were created over the last millennia in Island Melanesia, repeatedly altered by multiple influences between non-Austronesian and Austronesian polities.

Nineteenth-century accounts attest to the extent of contemporary population collapse everywhere in the Southwest Pacific that eliminated between 70 and >95% of the indigenous inhabitants of Island Melanesia. This was the result of epidemics, endemic wars using European muskets, and low fertility resulting from venereal diseases or the coercion and sometimes forced transfer of Island Melanesians ('blackbirding') to work the sugarcane and cotton fields of Queensland and other places (Rivers 1922; Harrisson 1937; McArthur 1974; Bayliss-Smith 2019).

While population collapse obliterated part of the region's genetic diversity that so impressed the first Spanish sailors, it also reduced a range of complex political systems that had developed in the different Melanesian islands over time. It led, for example, to a change from chiefdom to big-man at the beginning of the 20th century on Uneapa Island in the Bismarck Archipelago (Blythe 2018). On nearby southern Bougainville, Richard Thurnwald recorded in 1907 that the Buin society was organised on a hierarchical chiefdom originating from 'black navigating conquerors' (Thurnwald 1951, 138). When he returned to visit the Buin in 1933–34, 'class differences had diminished' and 'differences in status among the three main groups had been mitigated considerably' (see Terrell 1986, 222–40). On Santo Island in Northern Vanuatu, the mountain-tribes asserted in the 1970s 'that the *mele* (pig festivals that … were an integral part of the "graded society") was a fairly recent institution, imported to Santo from other areas after the coming of the Europeans' (Ludvigson 1981, paragraph 1.9) and during a period of massive population reduction. These examples illustrate the devolution of former complex traditional political systems, with a reduction in hierarchy and the build-up of individual power through the grade-system, as an adaptive response in the context of a catastrophic demographic decline associated with European contact and colonisation. Even so, elsewhere in Melanesia various forms of chiefly social hierarchy remain, although modified by the colonial presence. Mosko (1992, 714) cites multiple cases from the New Guinea mainland and nearby islands, including groups that speak Austronesian languages and others that are Papuan speakers. Further afield, hierarchical chiefdoms are reported in the Solomon Islands (*e.g.* Terrell 1978), as well as in Southern Vanuatu, New Caledonia, and Fiji (Guiart 1992). While these Melanesian chiefdoms may differ from their Polynesian counterparts in terms of their prerogatives and obligations between clans, they are definitely not 'big man' polities. Analysis by Scaglion (1996) suggests a strong correlation between Melanesian chiefdoms and Austronesian speakers but much less so with Papuan speakers. Interestingly, he notes an associated loss of 'strong elements of ascriptive leadership' among some Austronesian-speaking groups (Scaglion 1996, 11).

Spriggs (2008) argues that changes in social organisation driven by the massive dislocation of traditional ways caused by European arrival were often studied in a synchronic and ahistorical way, which represented them as appropriate testimonies of the normal functioning of Melanesian societies before European times. In this region of large islands where westerners first ventured inland, sometimes centuries after 'first contact', they failed to recognise this demographic loss (*e.g.* Spriggs 1997, 232–4). In many of these places, the historicity of the different Melanesian political systems remains to be properly understood. For example, as noted, the Buin of southern Bougainville retained a hierarchical chiefdom system at the beginning of the 20th century, while their immediate neighbours, the Siwai (aka Siuai) were described by anthropologist Douglas Oliver in 1938–39 as organised on a grade model (Oliver 1955; Terrell 1986, 197–212). The Siwai were used by Sahlins (1963) as the archetype of the Melanesian 'big man society' for his 'political types' in Oceania, but archaeological evidence of megaliths and burial patterns (Terrell 1978) indicates the probable prehistoric devolution of political structures on the south Bougainville plain over the last millennium.

Discussion

When the ships of Quiros stopped for a few days at Taumako Island in the southeast Solomons in 1606, on the immediate eastern side of the Green Line, Chief Tumai identified for the Spanish over 60 islands in different directions around them, spanning a distance of about 700 km:

> from Ticopia [Tikopia] you can reach Manicolo in five days of sailing, a land of immense expanse and very populated. … He spoke only with admiration of the size of the population and of the fertility of this land (Dalrymple 1774, 278, our translation).

The examples used here, and many others, could be used to show that the Green Line has undisputed historical value in separating that part of Melanesia settled in Pleistocene times from the Lapita-period expansion into uninhabited islands east of this limit. However, no clear-cut post-Lapita division between Western and Eastern Melanesia can be observed. Despite the dubious history that originally separated Melanesia and Polynesia (Thomas 1989; Tcherkézoff 2003), and allowing that this division disguises some of the overarching pan-Pacific similarities that exist between the cultural groups of Oceania, it is at the same time clear that 'Melanesia' remains a coherent entity of geographically related cultural diversity, especially when compared to Polynesia and Micronesia.

Distinctive phenotypic differences are identified between and inside archipelagos, linked to the diversity of local historical, linguistic, and artistic trajectories between Melanesian islanders. Fijians as a related group have a different morphology to Papuans, inhabitants of Malaita can be distinguished from those of nearby Guadalcanal, as can be the inhabitants from Lifou Island from those of Maré in New Caledonia. But despite these differences, modern studies have revived the evident reality of Melanesian populations having distinctive genetic markers that can be distinguished from those present in other populations of Oceania. And finally, contrary to common belief, science shows that Polynesians and Micronesians were not the only long-distance voyagers of Oceania. In the past, the dark-skinned islanders of the Southwest Pacific have also enacted episodes of expansion, towards the north into Micronesia (Blust 2010; Athens 2018, 276), towards the east and towards the west, leaving their genetic signature among Polynesians (Hudjashov et al. 2018, 7) and some Southeast Asian populations (Purnomo et al. 2021, 981).

Conclusion

By 'disestablishing' Melanesia for the sake of highlighting the importance of the Lapita expansion in Polynesian prehistory, the Green Line artificially divides a coherent cultural region. It is time to reattach Near Oceania to the southwestern part of Remote Oceania, to recognise the specific historical path of Melanesia as a coherent entity with a unique diversity of cultures, languages, and political systems that developed in these large archipelagos. In the early 19th century, the mapmakers drawing the division between Polynesia and Melanesia did not know where to locate Fiji, the archipelago 'in between' (Kirch 2017, 155). The evidence would now support that it be kept in Melanesia, re-establishing the logical cultural entity.

References

Allen, J. (2017) *Excavations on Motupore Island, Central District, Papua New Guinea*. University of Otago Working Papers in Anthropology 4. Dunedin, Department of Anthropology and Archaeology, University of Otago.

Amherst, W. & Thomson, B. (1901) *The Discovery of the Solomon Islands by Alvaro de Mendana in 1568*. London, Hakluyt Society.

Arauna, L.R., Bergstedt, J., Choin, J., Mendoza-Revilla, J., Harmant, C., Roux, M., Mas-Sandoval, A., Lémée, L., Colleran, H., François, A., Valentin, F., Cassar, O., Gessain, A., Quintana-Murci, L. & Patin, E. (2022) The genomic landscape of contemporary western Remote Oceanians. *Current Biology* 32(21), 4565–75.

Athens, S. (2018) Archaeology of the eastern Caroline Islands, Micronesia. In E. Cochrane & T. L. Hunt (eds), *The Oxford Handbook of Prehistoric Oceania*, 271–301. Oxford, Oxford University Press.

Ballard, C. (2008) 'Oceanic Negroes': British anthropology of Papuans, 1820–1869. In B. Douglas & C. Ballard (eds), *Foreign Bodies: Oceania and the Science of Race 1750-1940*, 157–201. Canberra, ANU E Press.

Bayliss-Smith, T. (2019) Population decline in Island Melanesia. Aphrodisian cultural practices, sexually transmitted infections, and low fertility. In S. Szreter (ed.), *The Hidden Affliction: Sexually Transmitted Infections and Infertility in History*, 188–218. Rochester, University of Rochester Press.

Bayliss-Smith, T. & Hviding, E. (2014) Taro terraces, chiefdoms and malaria: Explaining Landesque Capital formation in the Solomon Islands. In N. T. Hakansson and M. Widgren (eds), *Landesque Capital: The Historical Ecology of Enduring Landscape Modifications*, 75–97. Walnut Creek, CA, Left Coast Press.

Bayliss-Smith, T., Prebble, M. & Manebosa, S. (2019) Saltwater and bush in New Georgia, Solomon Islands: Exchange relations, agricultural intensification and limits to social complexity. In M. Leclerc & J. Flexner (eds), *Archaeologies of Island Melanesia: Current Approaches to Landscapes, Exchange and Practice*, 35–52. Terra Australis 51. Canberra, ANU Press.

Bedford, S. (2006) *Pieces of the Vanuatu Puzzle: Archaeology of the North, South, and Centre*. Terra Australis 23. Canberra, Australian National University.

Bedford, S. (2019) The complexity of monumentality in Melanesia: Mixed messages from Vanuatu. In M. Leclerc & J. Flexner (eds), *Archaeologies of Island Melanesia. Current Approaches to Landscape, Exchange and Practice*, 67–79. Terra Australis 51. Canberra, ANU Press

Bedford, S., Siméoni, P. & Lebot, V. (2017) The anthropogenic transformation of an island landscape: Evidence for agricultural development revealed by LIDAR on the island of Efate, Central Vanuatu, South-West Pacific. *Archaeology in Oceania* 53(1), 1–14.

Bedford, S. & Spriggs, M. (2008) Northern Vanuatu as a Pacific crossroads: The archaeology of discovery, interaction and emergence of the 'ethnographic present'. *Asian Perspectives* 47(1), 95–120.

Bergström, A., Oppenheimer, S. J., Mentzer, A. J., Auckland, K., Robson, K., Attenborough, R., Alpers, M. P., Koki, G., Pomat, W., Siba, P., Xue, Y., Sandhu, L. S. & Tyler-Smith, C. (2017). A Neolithic expansion, but strong genetic structure, in the independent history of New Guinea. *Science* 357, 1160–63.

Bickler, S. H. (2006) Prehistoric stone monuments in the northern region of the Kula Ring. *Antiquity* 80, 38–51.

Blench, R. (2014) Lapita canoes and their multi-ethnic crews: Marginal Austronesian languages are non-Austronesian. *Workshop on the Languages of Papua 3*. Manokwari, Indonesia, 20–24 January 2014. Unpublished manuscript.

Blust, R. (2010) Malaita – Micronesian once again. *Oceanic Linguistics* 49(2), 559–67.

Blust, R. (2013) *The Austronesian Languages*. Revised ed. Asia-Pacific Linguistics. Canberra, Australian National University.

Blythe, J. (2018) Uneapa Island society in the 19th century: A reconstruction. *Journal of the Polynesian Society* 127(4), 425–49.

Burley, D. V. (2013) Fijian polygenesis and the Melanesian/Polynesian divide. *Current Anthropology* 54(4), 436–62.

Burley, D., Connaughton, S. P. & Clark, G. (2018) Early cessation of ceramic production for ancestral Polynesian society in Tonga. *PLoS ONE* 13(2), e0193166.

Byrne, S. (2005) Recent survey and excavation of the monumental complexes on Uneapa Island, West New Britain, Papua New Guinea. *Papers for the Institute of Archaeology* 16, 95–101.

Byrne, S. (2013) Rock art as material culture: A case study on Uneapa Island, West New Britain, Papua New Guinea. *Archaeology in Oceania* 48, 63–77.

Choin, J., Mendoza-Revilla, J., Arauna, L. R., Cuadros-Espinoza, S., Cassar, O., Larena, M., Min-Shan Ko, A., Harmant, C., Laurent, R., Verdu, P., Laval, G., Boland, A., Olaso, R., Deleuze, J-F., Valentin, F., Ko, Y.-C., Jakobsson, M., Gessain, A., Excoffier, L., Stoneking, M., Patin, E. & Quintana-Murci, L. (2021) Genomic insights into population history and biological adaptation in Oceania. *Nature* 592(7855), 583–9.

Clark, G. (2003a) Shards of meaning: Archaeology and the Melanesia-Polynesia divide. *The Journal of Pacific History* 38(2), 197–215.

Clark, G. (ed.) (2003b) Dumont d'Urville's Divisions of Oceania: Fundamental Precincts or Arbitrary Constructs. *Journal of Pacific History* 38(2) Special Issue, 155–287.

Cochrane, E. (2018) Ancient Fiji. Melting pot of the Southwest Pacific. In E. Cochrane & T. L. Hunt (eds), *The Oxford Handbook of Prehistoric Oceania*, 206–30. Oxford, Oxford University Press.

Crawfurd, J. (1820) *History of the Indian Archipelago: Containing an Account of the Manners, Arts, Languages, Religions, Institutions, and Commerce of its Inhabitants*. 3 vols. Edinburgh, Archibald Constable.

Cruz Berrocal, M. & Millerstrom, S. (2013) The archaeology of rock art in Fiji: Evidence, methods and hypotheses. *Archaeology in Oceania* 48(3), 63–77.

Dalrymple, A. (1774) *Voyage dans les Mers du Sud, par les Espagnols et les Hollandois*. Translated to English by M. De Fréville. Paris, Saillant & Nyon et Pissot.

Donohue, M. & Denham, T. (2012) Lapita and Proto-Oceanic. *The Journal of Pacific History* 47(4), 443–57.

Douglas, B. (2010) Terra Australis to Oceania. *The Journal of Pacific History* 45, 179–210.

Dumont d'Urville, J-S-C. (1832) Sur les îles du Grande Océan. *Bulletin de la Société de Géographie* 17, 1–21.

Dumont d'Urville, J. S. C. (2003) On the islands of the Great Ocean. Translated by I. Olivier, A. de Biran & G. Clark. *The Journal of Pacific History* 38(2), 163–74.

Earl, G. W. (1853) *The Native Races of the Indian Archipelago: Papuans*. London, Hippolyte Baillière.

Egan, S. & Burley, D. V. (2009) Triangular men on one very long voyage: The context and implication of a Hawaiian-style petroglyph site in the Polynesian kingdom of Tonga. *Journal of Polynesian Society* 118(3), 209–32.

Field, J. S. (2005) Land tenure, competition and ecology in Fijian prehistory. *Antiquity* 79(305), 586–600.

Flexner. J. L., Bedford, S. & Valentin, F. (2019) Who was Polynesian? Who was Melanesian? Hybridity and ethnogenesis in the South Vanuatu Outliers. *Journal of Social Archaeology* 19(3), 403–26.

Friedlaender, J. S. (2007) Conclusion. In J. S. Friedlaender (ed.), *Genes, Language, and Culture History in the Southwest Pacific*, 231–7. Oxford, Oxford University Press.

Friedlaender, J. S., Friedlaender, F. R., Reed, F. A., Kidd, K. K., Kidd, J., Chambers, G., Lea, R. A., Loo, J.-H., Koki, G., Hodgson, J. A., Merriwether, D. A. & Weber, J. L. (2008) The genetic structure of Pacific Islanders. *PLoS Genetics* 4(1), e19.

Gifford, E. W. (1951) *Archaeological Excavations in Fiji*. Anthropological Records 13, 189–288. Berkeley, University of California Press.

Golson, J. (1961) Report on New Zealand, Western Polynesia. New Caledonia and Fiji. *Asian Perspectives* 5, 166–80.

Golson, J., Denham, T., Hughes, P., Swadling, P. & Muke, J. (2017) *Ten Thousand Years of Cultivation at Kuk Swamp in the Highlands of Papua New Guinea*. Terra Australis 46. Canberra, ANU Press.

Gosden, C. (1999) *Anthropology & Archaeology. A Changing Relationship*. London & New York, Routledge.

Green, R. C. (1976a) An introduction to the Southeast Solomons Culture History Programme. In Green & Creswell 1976, 9–17.

Green, R. C. (1976b) Conclusion. In Green & Creswell 1976, 267–70.

Green, R. (1991) Near and Remote Oceania – disestablishing 'Melanesia' in culture history. In A. Pawley (ed.), *Man and a Half: Essays in Pacific Anthropology and Ethnobiology in Honour of Ralph Bulmer*, 491–502. Auckland, The Polynesian Society.

Green, R. C. & Creswell, M. M. (eds) (1976) *Southeast Solomon Islands Cultural History: A Preliminary Survey*. Bulletin of the Royal Society of New Zealand 11. Wellington, The Royal Society of New Zealand.

Groube, L. (1993) Contradictions and malaria in Melanesian and Australian prehistory. In M. A. Smith, M. Spriggs & B. Frankhauser (eds), *Sahul in Review: Pleistocene Archaeology in Australia, New Guinea and Island Melanesia*. Occasion Papers in Prehistory 24, 164–86. Canberra, Department of Prehistory, Australian National University.

Guiart, J. (1992) *La Chefferie Mélanésienne*. Paris, Institut d'ethnologie.

Harrisson, T. (1937) *Savage Civilisation*. New York, Alfred A. Knopf.

Hau'ofa, E. (1975) Anthropology and Pacific Islanders. *Oceania* 45(4), 283–9.

Hudjashov, G., Endicott, P., Post, H., Nagle, N., Ho, S. Y. W., Lawson, D. J., Reidla, M., Karmin, M., Rootsi, S., Metspalu, E., Saag, L., Villems, R., Cox, M. P., Mitchell, R. J., Garcia-Bertrand, R. L., Metspalu, M. & Herrera, R. J. (2018) Investigating the origins of eastern Polynesians using genome-wide data from the Leeward Society Isles. *Nature: Scientific Reports* 8, 1823, 1–12.

Irwin, G. (1985) *The Emergence of Mailu as a Central Place in Coastal Papuan Prehistory*. Terra Australis 10. Canberra, Research School of Pacific Studies, Australian National University.

Kirch, P. V. (2017) *On the Road of the Winds*. Berkeley, University of California Press.

Kirch, P. V. & Swift, J. A. (2017) New AMS radiocarbon dates and re-evaluation of the cultural sequence of Tikopia Island, Southeast Solomon Islands. *Journal of the Polynesian Society* 126(3), 313–36.

Lawson, S. (2013) Melanesia. *The Journal of Pacific History* 48(1), 1–22.

Lincoln, P. C. (1978) Reef-Santa Cruz as Austronesian. In S. A. Wurm & L. Carrington (eds), *Second International Conference on Austronesian Linguistics: Proceedings 2*, 929–67. Canberra, Pacific Linguistics.

Lipson, M., Skoglund, P., Spriggs, M., Valentin, F., Bedford, S., Shing, R., Buckley, H., Phillip, I. Ward, G. W., Mallick, S., Rohland, N., Broomandkhoshbacht, N., Cheronet, O., Ferry, M. Harper, T. K., Michel, M. Oppenheimer, J. Sirak, K., Stewardson, K., Auckland, K., Hill, A. V. S., Maitland, K., Oppenheimer, S. J., Parks, T., Robson, K., Williams, T. N., Kennett, D. J., Mentzer, A. J., Pinhasi, R. & Reich, D. (2018) Population turnover in Remote Oceania shortly after initial settlement. *Current Biology* 28(7), 1157–65.

Lipson, M., Spriggs, M., Valentin, F., Bedford, S., Shing, R., Zinger, W., Buckley, H., Petchey, F., Matanik, R., Cheronet, O., Rohland, N., Pinhasi, R. & Reich, D. (2020) Three phases of ancient migration shaped the ancestry of human populations in Vanuatu. *Current Biology* 30, 1–11.

Ludvigson, T. (1981). *Kleva: Some healers in central Espiritu Santo, Vanuatu*. Unpublished PhD thesis, Auckland University.

McArthur, N. (1974) *Population and prehistory: The Late Phase on Aneityum*. Unpublished PhD thesis, Australian National University.

Monnin, J. & Sand, C. (2004) *Kibo, le Serment Gravé. Essai de Synthèse sur les Pétroglyphes Calédoniens*. Les Cahiers de l'Archéologie en Nouvelle-Calédonie 16. Nouméa, Département Archéologie, Service des Musées et du Patrimoine de Nouvelle-Calédonie.

Mosko, M. (1992) Motherless sons: 'Divine Kings' and 'Partible Persons' in Melanesia and Polynesia. *Man* 27, 697–717.

Oliver, D. L. (1955) *A Solomon Island Society: Kinship and Leadership Among the Siuai of Bougainville*. Cambridge, Mass.: Harvard University Press.

Otto, T. (1997) The Melanesian Way. In T. Otto & N. Thomas (eds), *Narratives of Nation in the South Pacific*, 33–64. Amsterdam, Harwood Academic Publishers.

Pawley, A. (2007) Why do Polynesian island groups have one language and Melanesian island groups have many? Patterns of interaction and diversification in the Austronesian colonisation of Remote Oceania. *Workshop 'Migration'*, 5–7 September 2007, Île de Porquerolles, France. Unpublished manuscript.

Pawley, A. (2008) Where and when was Proto-Oceanic spoken? Archaeological and linguistic evidence. In Y. A. Lander & A. K. Oglobin (eds), *Language and Text in the Austronesian World: Studies in Honour of Ülo Sirk*, 47–71. Münich, Lincom Europa.

Pawley, A. (2010) The origins of the early Lapita culture: The testimony of historical linguistics. In S. Bedford, C. Sand & S. P. Connaughton (eds), *Oceanic Explorations: Lapita and Western Pacific Settlement*. Terra Australis 26, 17–49. Canberra, ANU E Press.

Pawley, A. & Green, R. C. (1973) Dating the dispersal of the Oceanic languages. *Oceanic Linguistics* 12(1), 1–67.

Pedro, N., Brucato, N., Fernandes, V., André, M., Saag, L., Pomat, W., Besse, C., Boland, A., Deleuze, J.-F., Clarkson, C., Sudoyo, H., Metspalu, M., Stoneking, M., Cox, M. P., Leavesley, M., Pereira, L. & Ricaut, F.-X. (2020) Papuan mitochondrial genomes and the settlement of Sahul. *Journal of Human Genetics* 65, 875–87.

Posth, C., Nägele, K., Colleran, H., Valentin, F., Bedford, S., Kami, K. W., Shing, R., Buckley, H., Kinaston, R., Walworth, M., Clark, G. R., Reepmeyer, C., Flexner, J., Maric, T., Moser, J., Gresky, J., Kiko, L., Robson, K. J., Auckland, K., Oppenheimer, S. J., Hill, A. V. S., Mentzer, A. J., Zech, J., Petchey, F., Roberts, P., Jeong, C., Gray, R. D., Krause, J. & Powell, A. (2018) Language continuity despite population replacement in Remote Oceania. *Nature Ecology Evolution* 2(4), 731–40.

Purnomo, G. A., Mitchell, K. J., O'Connor, S., Kealy, S., Taufik, L., Schiller, S., Rohrlach, A., Cooper, A., Llamas, B., Sudoyo, H., Teixeira, J. C. & Tobler, R. (2021) Mitogenomes reveal two major influxes of Papuan ancestry across Wallacea following the Last Glacial Maximum and Austronesian contact. *Genes* 12, 965–86.

Riesenfeld, A. (1950) *The Megalithic Culture of Melanesia*. Leiden, E.J. Brill.

Rivers, W. H. R. (ed.) (1922) *Essays on the Depopulation of Melanesia*. Cambridge, Cambridge University Press.

Ross, M. (2005) Pronouns as a preliminary diagnostic for grouping Papuan languages. In A. Pawley, R. Attenbrough, J. Golson & R. Hide (eds), *Papuan Pasts: Cultural, Linguistic and Biological Histories of Papuan-Speaking Peoples*. Pacific Linguistics, 15–66. Canberra, Australian National University.

Sahlins, M. D. (1963) Poor man, rich man, big-man, chief: Political types in Melanesia and Polynesia. *Comparative Studies in Society and History* 5(3), 285–303.

Sand, C. (2007) The eastern frontier: Lapita ceramics in the Fiji-West Polynesia region. In S. Chiu & C. Sand (eds), *From Southeast Asia to the Pacific. Archaeological Perspectives on the Austronesian Expansion and the Lapita Cultural Complex*, 214–42. Taipei, Academia Sinica.

Sand, C. (2012) Southern Melanesian rock art: The New Caledonian case. In J. McDonald & P. Veth (eds), *A Companion to Rock Art*, 160–78. Oxford, Blackwell.

Sand, C. (2023) *Hécatombe Océanienne (XVIIe-XXe siècle). La Dépopulation des Peuples du Pacifique et ses Conséquences*. Papeete, Au Vent des Iles.

Sand, C., Bolé, J. & Ouetcho A. J. (2011) A revision of New Caledonia's ceramic sequence. *Journal of Pacific Archaeology* 2(1), 56–68.

Sand, C., Ouetcho, A. J., Bolé, J., Baret, D. & Gony, Y.-B. (2016) Traditional Kanak landscapes: An assessment of settlement pattern studies in New Caledonia (Southern Melanesia). *Bulletin de la Société Préhistorique Française* 112, 31–48.

Scaglion, R. (1996) Chiefly models in Papua New Guinea. *The Contemporary Pacific* 8(1), 1–31.

Schapper, A. (2020) Linguistic Melanesia. In E. Adamou & Y. Matras (eds), *Routledge Handbook of Language Contact*, 480–502. London, Routledge.

Skoglund, P., Posth, C., K. Sirak, K., Spriggs, M., Valentin, F., Bedford, S., Clark, G. R., Reepmeyer, C., Petchey, F., Fernandes, D., Fu, Q., Harney, E., Lipson, M., Mallick, S., Novak, M., Rohland, N., Stewardson, K., Abdullah,S., Cox, M. P., Friedlaender, F. R., Friedlaender, J. S., Kivisild, T., Koki, G., Kusuma, P., Merriwether, D. A., Ricaut, F.-X., Wee, J. T. S., Patterson, N., Krause, J., Pinhasi, R. & David Reich, D. (2016) Genomic insights into the peopling of the Southwest Pacific. *Nature* 538(7626), 510–13.

Spate, O. H. K. (1979) *The Spanish Lake*. Canberra, Australian National University Press.

Specht, J. R. (1979) Rock art in the Western Pacific. In S. M. Mead (ed.), *Exploring the Visual Art of Oceania: Australia, Melanesia, Micronesia, and Polynesia*, 58–82. Honolulu, University Press of Hawaii.

Spriggs, M. (1997) *Island Melanesians*. Oxford, Blackwell.

Spriggs, M. (2008) Ethnographic parallels and the denial of history. *World Archaeology* 40(4), 538–52.

Tcherkézoff, S. (2003) A long and unfortunate voyage towards the 'invention' of the Melanesia/Polynesia distinction 1595–1832. *The Journal of Pacific History* 38(2), 175–96.

Terrell, J. (1976) *Perspectives in the prehistory of Bougainville Island, Papua New Guinea: A study in the human biogeography of the southwestern Pacific*. Unpublished PhD thesis, Harvard University.

Terrell, J. (1978) Archaeology and the origins of social stratification in Southern Bougainville. In *Rank and Status in Polynesia and Melanesia, Essays in Honour of Professor Douglas Oliver*. Publication de la Société des Océanistes 39, 23–43. Paris, Musée de l'Homme.

Terrell, J. (1986) *Prehistory in the Pacific Islands*. Cambridge, Cambridge University Press.

Terrell, J. (2010) Social networks analysis of the genetic structure of Pacific Islanders. *Annals of Human Genetics* 74(3), 211–32.

Terrell, J., Hunt, T. L. & Bradshaw, J. (2002) On the location of the Proto-Oceanic homeland. *Pacific Studies* 25(3), 57–93.

Thomas, N. (1989) The force of ethnology: Origins and significance of the Melanesia/Polynesia division. *Current Anthropology* 30(1), 27–41.

Thurnwald, R. (1951) Historical sequences on Bougainville. *American Anthropologist* 53(1), 137–9.

Torrence, R. (2002) What makes a disaster? A long-term view of volcanic eruptions in Papua New Guinea. In R. Torrence & J. Grattan (eds), *Natural Disasters and Cultural Change*, 292–312. London, Routledge.

Torrence, R., Specht, J. & Vatete, B. (2002) *Report of an archaeological survey of the Bali-Witu islands, West New Britain Province, Papua New Guinea*. Unpublished report, West New Britain Provincial Government.

Wallace, A. R. (1869) *The Malay Archipelago: The Land of the Orang-Utan, and the Bird of Paradise. A Narrative of Travel, with Studies of Man and Nature*. Volume 1. London, Macmillan and Co.

Wickler, S. (2001) *The Prehistory of Buka: A Stepping Stone Island in the Northern Solomons*. Terra Australis 16. Canberra, Australian National University.

Wilder, J. A. & Hammer, M. F. (2007) Extraordinary population structure among the Baining. In J. S. Friedlaender (ed.), *Genes, Language, and Culture History in the Southwest Pacific*, 199–207. Oxford, Oxford University Press.

Wilson, M. (2003) Rock-art transformations in the western Pacific. In C. Sand (ed.), *Pacific Archaeology: Assessments and Prospects*. Les Cahiers de l'Archéologie en Nouvelle-Calédonie 15, 265–84. Nouméa, Département Archéologie, Service des Musées et du Patrimoine de Nouvelle-Calédonie.

Wilson, M. & Ballard, C. (2019) Rock art of the Pacific: Context and intertextuality. In B. David & I. McNiven (eds), *Oxford Handbook of the Archaeology and Anthropology of Rock Art*, 221–52. Oxford, Oxford University Press.

Wurm, S. A. (1978) Reefs-Santa Cruz: Austronesian, but! In S. A. Wurm & L. Carrington (eds), *Second International Conference on Austronesian Linguistics: Proceedings 2*, 969–1010. Canberra, Pacific Linguistics.

Zinger, W., Valentin, F., Flexner, J., Bedford, S., Detroit, F. & Grimaud-Hervé, D. (2020) How to explain Polynesian Outliers' heterogeneity? In A. Hermann, F. Valentin, C. Sand & E. Nolet (eds), *Networks and Monumentality in the Pacific*, 62–77. Oxford, Archaeopress Publishing Ltd.

5

Storied Landscapes in the Palaeolithic? The View from the Cave

Graeme Barker and Chris O. Hunt

It has long been accepted that European Palaeolithic societies of the last interglacial/glacial cycle were likely linked in social networks that connected individuals and groups in information flows to spread risk and provide access to resources and mates. Building on this, Michelle Langley (2013) argued that European Neanderthals inhabited 'social landscapes' of this kind, but Modern Humans imbued their physical environments with symbolic meaning to create 'storied landscapes'. In this paper, we consider these arguments in terms of the archaeological records of three caves we have investigated, all outside Europe: the Niah Cave in Borneo used by anatomically modern humans ('Modern Humans' or simply 'Moderns' in archaeological parlance) since ca. *50,000 years ago; the Haua Fteah in Libya used by Moderns from* ca. *140,000 years ago; and Shanidar Cave in Iraqi Kurdistan used by Neanderthals until* ca. *45,000 years ago and then by Moderns. Reviewing the evidence in terms of Langley's principal criteria of landscape marking, personal identities, raw material transport, and norms and customs tied to the landscape, we conclude that the evidence, whilst often ambiguous, serves to widen the debate about Palaeolithic social networks and 'storied landscapes'. At least for the Palaeolithic people using these three caves, there were different ways of being human and different ways of envisaging the landscape beyond, which do not map onto the Archaic/Modern dichotomy that is such a cornerstone of evolutionary studies based on the European archaeological record.*

Keywords: Haua Fteah Cave, Modern Humans, Neanderthals, Niah Cave, Shanidar Cave, Social Landscapes, Storied Landscapes

Introduction

It is an enormous pleasure to contribute to a volume dedicated to Chris Gosden's contributions to archaeology. Earlier in our careers, we (the authors) worked together on three landscape archaeology projects: the Biferno Valley Survey in central-southern Italy (Barker 1995), the UNESCO Libyan Valleys Survey in Tripolitania, northwest Libya (Barker 1996), and the Wadi Faynan Landscape Survey in southern Jordan (Barker *et al.* 2007). In the past two decades we have collaborated in the re-excavations of three caves with deep habitation sequences (Fig. 5.1): the Niah Caves in Sarawak, the Haua Fteah in Libya, and currently Shanidar Cave in Iraqi Kurdistan famously excavated several decades ago by, respectively, Tom and Barbara Harrisson (B. Harrisson 1967; T. Harrisson 1970), Charles McBurney (1955) and Ralph Solecki (1971). But whether we have been walking around a landscape, or looking out at one from a cave, our shared abiding interest has been in how people have shaped landscapes and landscapes have shaped people. Cultural landscapes of this kind have been central to Chris Gosden's research. As a Sheffield undergraduate he was a member of the team of students who mapped the ploughzone archaeology of the Biferno Valley, so perhaps Barker can claim some small credit for launching him on his landscape trajectory. And 30 years later all three of us worked together in the Kelabit Highlands of Sarawak in the Cultured Rainforest Project (significantly, Chris Gosden came up with the title) that was the successor to the Niah Cave excavations (Barker *et al.* 2017). Whether reading his books and papers or talking archaeology with him as we walked down a rainforest

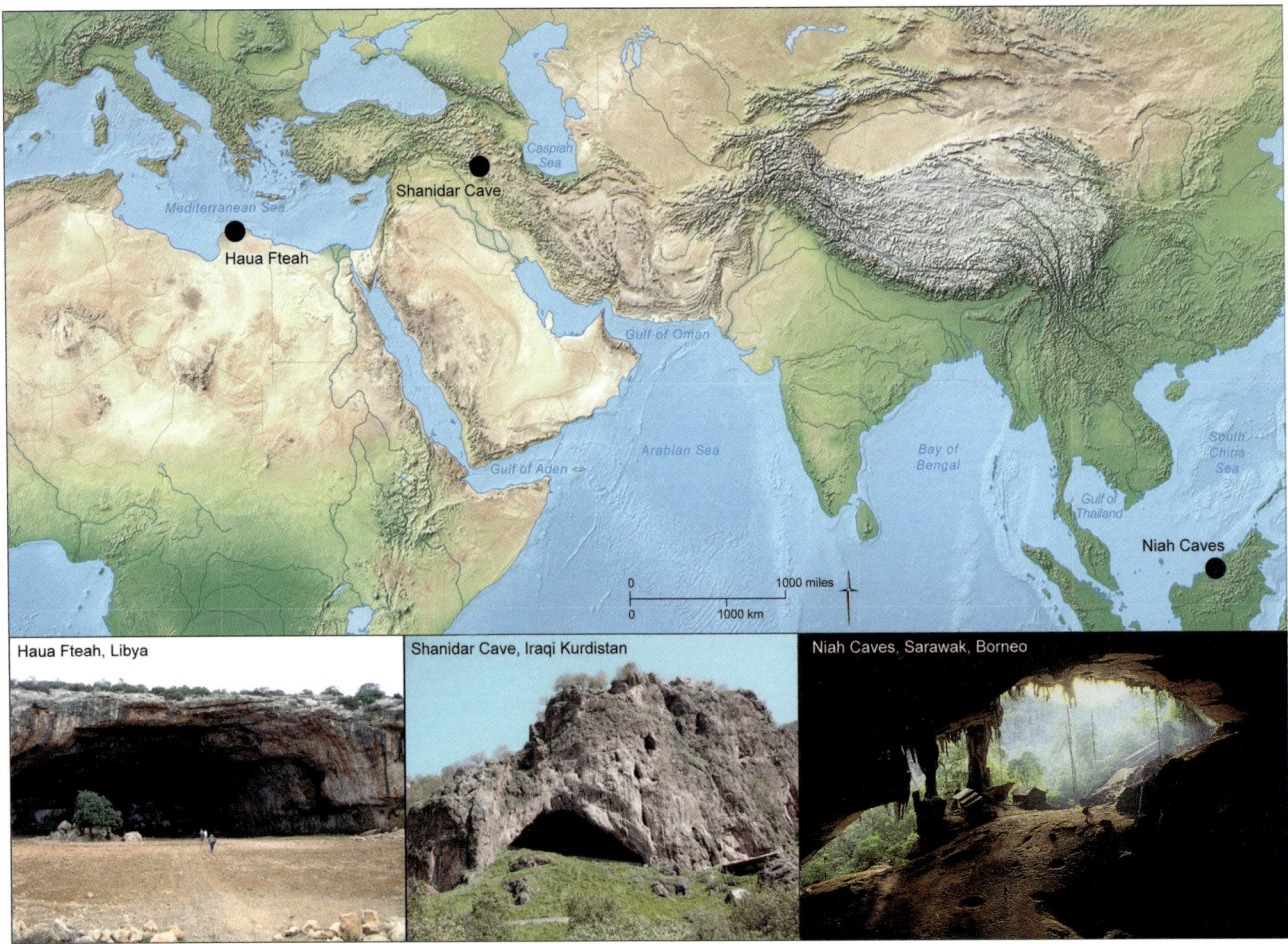

Figure 5.1 The locations of the Haua Fteah cave (Libya), Shanidar Cave (Iraqi Kurdistan), and the Niah Caves (Sarawak) with images of the caves (Map: Vicki Herring; Photos: Graeme Barker).

path (Fig. 5.2), like all the contributors to this collection of essays, we have been hugely influenced by his ideas and by the ways in which he has put those ideas into practice and – jargon-free – communicated them to the rest of us. Stimulated by these, and in the spirit of this book, in this paper we offer some reflections on the challenges of trying to access the 'sentient landscapes' of far distant Palaeolithic peoples, in particular 'Modern Humans' or 'Moderns' – the Pleistocene members of our own species *Homo sapiens* – and Neanderthals, our closest evolutionary cousins, through the fragmentary archaeological materials they have left behind in the three caves we have explored. We use the term 'Moderns' hereafter as a convenient and common descriptor for a physical type without any *a priori* connotations of 'modern' versus 'ancient' or 'archaic' behaviour.

Social Landscapes and Storied Landscapes

It has long been argued that both Neanderthals and Moderns were likely linked in a variety of social networks that connected individuals and groups in information flows that spread risk and provided access to resources and mates. Gamble (1998), for example, postulated that by the last interglacial 100,000 years ago Palaeolithic societies would have shared in 'intimate' networks of 3–7 persons, 'effective' networks of 10–25 people, and 'extended' networks of 100–400 persons. In an influential review, Langley (2013) argued that both European Neanderthals and Modern Humans inhabited 'social landscapes' or 'socialised landscapes' of this kind but that Modern Humans actively *socialised* their landscapes to create what she terms 'storied landscapes'. She defined 'social landscapes' as 'people to people interactions…mapped onto and over the physical landscape and which join various locales together through paths and trackways' and 'landscape socialisation' as 'the direct social interaction between people and topography where meaning is imbued into the physical features of the terrain by its human viewers and inhabitants' (Langley 2013, 615). She cited examples from the ethnographic record of *Homo sapiens*' propensity to 'turn the wilderness into our friend or enemy through imparting thoughts, feelings and meanings into it', attaching stories specially

Figure 5.2 The sentient archaeologist in the field: Chris negotiating a plank bridge in the Borneo rainforest during the Cultured Rainforest Project fieldwork in 2008 (Photo: Graeme Barker).

Table 5.1 Langley's proposed potential archaeological differences between Palaeolithic populations that participate in landscape socialisation and those that use symbolic material culture solely for the mediation of interpersonal social interaction (adapted from Langley 2013, table 1).

	Evidence for 'storied landscapes'	*Evidence for 'social landscapes'*
Marking the landscape	Rock art, cairns, monuments, scar trees	Absent
Movement of landscape/ raw materials	Frequent transport of distinct (often distant) raw materials tied to specific locations; evidence of increased value with distance	Infrequent long-distance transport of raw materials tied to specific locations
Group and individual identity, inhabiting and using a specific landscape	Personal ornamentation (beads, clothing, body painting, etc.), mobiliary art, distinctive and/or decorated weapons and/or other artefacts for individual and/or group use likely to be viewed by both ingroup and outgroup members, which may transmit information regarding the social interaction of individuals/groups with landscape features	Distinctive and/or decorative artefacts for personal use (beads, clothing, body painting, etc.) to be viewed by ingroup and outgroup members
Norms and customs tied to landscape	Repetitive pattern and high density of symbolic artefacts with emblematic 'style'	Low density of symbolic items; no or minimal repetitive use of symbolic elements

to outstanding topographical features such as mountain peaks, rivers, and prominent caves. We encountered such 'storied landscapes' in the Cultured Rainforest Project's fieldwork (Barker *et al.* 2017): the Kelabit, who combine foraging with rice farming, distinguish between the 'Big Forest' (primary forest) that is imbued with a great spirit that only males can enter with safety, and the 'Small Forest' (secondary forest) where women and children as well as men were safe to hunt and gather (Janowski 2003); for the Penan foragers, in contrast, the entire forest world in which they live and of which they are a part is a spirit-animated universe to nurture through their stewardship or *molong* (Janowski & Langub 2011).

How might we distinguish between 'social landscapes' and 'storied landscapes' in the Palaeolithic archaeological record? Langley (2013) focuses on the major categories of material culture such as site types and distributions, landscape modification, the transport of material from source to point of use, and items of personal ornamentation (Table 5.1). She acknowledges that several of the 'archaeological signatures' of the two landscape types overlap, in particular the use of personal ornamentation and the long-distance transport of raw materials, but argues that cumulatively they divide in terms of scale and frequency. She concludes that, whilst Neanderthals had great technical skill, were capable of surviving in extremely difficult climatic conditions, and showed some but limited evidence of ritual behaviour and the use of ornament, their archaeology in Europe suggests a different approach to interaction with the physical landscape than that of Moderns. In this conclusion she builds on Burke's (2006; 2012) argument that European Neanderthals likely relied on detailed local knowledge for moving around their habitual territories and on local patterns of social interaction, whereas Moderns dispersing from Africa must have been able to maintain spatially extensive well-integrated social networks.

There are obvious dangers in generalising about the behaviour of Neanderthals, given their chronological range (getting on for half a million years) and geographical range (extending from the Atlantic to the Urals). Yet, there is general agreement that they were physically, genetically, and cognitively capable of language, and that some Neanderthals engaged in a degree of symbolic thinking. This is based on the evidence for the occasional use of pigments, the making of abstract marks on various raw materials including cave walls, examples of the use of bone and shell beads, eagle talons, and bird feathers for personal ornamentation, and ritual behaviours associated with the dead, such as the much disputed 'Flower Burial' in Shanidar Cave (Solecki 1971; Leroi-Gourhan 1975; 1998) and cases of cannibalism. But even cumulatively, Langley argues, this record of Neanderthal 'behavioural complexity' is strikingly less abundant than the evidence of similar behaviours in the archaeological record associated with contemporary or near contemporary populations of Moderns in Africa, Eurasia, and Australasia. These relative differences in the frequency, regularity, and abundance of indicators of symbolic behaviour in the archaeological record, she argues, cannot be explained by factors such as differential survival or methods of excavation. In short, she concludes, Neanderthals may have had social landscapes, but their Modern contemporaries had – just as we have – 'storied' (*i.e.* meaning-imbued) landscapes.

Caves in Palaeolithic Landscapes

The assumption is that most Neanderthal and Modern lives were highly mobile, because except in very unusual circumstances a foraging group of around 25–50 individuals staying in one place would have exhausted gatherable resources including fuelwood as well as food sources and depleted or scared off game. In these mobile lives, caves would have been fixed and dependable points in seasonal rhythms of movement. In practice, in most environments, both Neanderthals and Moderns would have spent virtually all their lives under the sky, because except in some limestone landscapes, habitable caves are very unusual landforms. Open air Palaeolithic lives are extremely difficult to recognise, of course, because of the geomorphological disruption of soils and their contained archaeologies by the repeated climatic events of the Late Pleistocene stadials (cold phases). As a result, a very large proportion of what we know about Palaeolithic people comes from excavations in caves. Most of these were carried out a generation or more ago, when resources permitted large-scale excavations (as in the case of the three caves we have re-investigated) but when few of what we would recognise as modern excavation or scientific techniques were available.

Dateable, stratified sequences with great time depth, rich in material culture and biotic remains, are preserved in caves. Through most of the Palaeolithic, most caves were prosaic shelters for living, sleeping, and repairing equipment. In the European Upper Palaeolithic, however, a few caves seem to have been significant places for many people, for instance as meeting points (Conkey *et al.* 1980; Bourdier 2013), burial sites (*e.g.* Aldhouse-Green & Pettit 1998; Geiling & Marin-Arroyo 2015), and shamanistic locations (*e.g.* Clottes & Lewis-Williams 1998; Lewis-Williams 2002), activities that have resulted in material culture sets that fit Langley's criteria for the 'storied lives' of the people who used them. She argues that the reasons for their paucity amongst European Neanderthals could be threefold (singly and in combination): taphonomic, with evidence for such activities having been removed by the geomorphological consequences of the climatic events of the Last Glacial Maximum; analytical, if storied lives were expressed through types of material culture that we do not recognise or which were made of impermanent materials that have not survived; or behavioural, the factor that she prefers, *i.e.* that the capacity to live storied lives had not fully manifested itself amongst Neanderthals. Here we reflect on these arguments from the perspective of the three deeply stratified Palaeolithic caves that we have investigated, all outside Europe (Fig. 5.1): the Niah Cave in Island Southeast Asia, occupied by Moderns equipped with stone tools that poorly fit the expectations of the European Middle and Upper Palaeolithic; the Haua Fteah cave in North Africa, containing European-type Middle and Upper Palaeolithic material culture but all manufactured by Moderns; and Shanidar Cave in the Zagros mountains of Southwest Asia, containing European-type Middle Palaeolithic material culture manufactured by Neanderthals and European-type Upper Palaeolithic material culture manufactured by Moderns.

The Cave Occupation Sequences

The Niah Cave

The Niah Cave is in fact a complex of caves dominated by a series of interlinked cathedral-like caverns, some 20 km from the present coast of the South China sea in Sarawak, Malaysian Borneo. Between 1954 and 1965 Tom Harrisson, the Director of Sarawak Museum, assisted after the initial season by his second wife Barbara Harrisson, explored many of these caves, but their main focus was the West Mouth of the magnificent Niah Great Cave, where they exposed several metres of guano-rich sediment that contained a rich habitation and burial archaeology. The lithic implements were mostly crude flakes that compared poorly with the fine flint-dominated technologies of Europe, and famously separated from the latter by the 'Movius Line' (Movius 1948). They were associated with the skeletal remains of Modern Humans, notably the 'Deep Skull', so called because of its location in the basal levels they investigated about 5 m below the present ground surface. Radiocarbon dating of

Table 5.2 The major cultural phases of the Haua Fteah, Shanidar Cave, and the West Mouth of Niah Great Cave and their approximate chronologies in dates ka (thousands of years before the present).

	Haua Fteah	Shanidar Cave	Niah Great Cave
Neolithic	7.5–3.5		4.0–2.0
Final Palaeolithic/ Epipalaeolithic	Capsian 14–7.5	Zarzian 12.0–10.5	11.5–4.5
Upper Palaeolithic	Oranian 17–14 Dabban 45–19	Baradostian 45–33	Palaeolithic* 50–11.5
Middle Palaeolithic	Levalloiso-Mousterian 80–45* Pre-Aurignacian 140–80	Mousterian 120–45*	

*Associated fossils: Haua Fteah: two Modern Human (*Homo sapiens*) mandibles *ca.* 80 ka; Shanidar Cave: Neanderthal skeletal remains *ca.* 75–55 ka; Niah Cave: Modern Human skull fragments containing resin *ca.* 42 ka, Modern Human skull ('Deep Skull') and limb bones *ca.* 35 ka.

adjacent charcoal suggested it was some 40,000 years old, the oldest *Homo sapiens* skull in the fossil record of the time (Harrisson 1958; Brothwell 1960).

Our own excavations in the West Mouth and other entrances were undertaken in four campaigns between 2000 and 2003 (Barker 2013; Barker & Farr 2016). New radiocarbon dates on charcoal using ABOX pre-treatment indicated that occupation in the West Mouth began around 50,000 years ago, or 50 ka (Higham *et al.* 2008), and we were able to obtain a direct U-series date on the Deep Skull of *ca.* 35.2 ka (Table 5.2). The first phase of Palaeolithic occupation in the West Mouth (*ca.* 50–35 ka) falls within the climatic phase termed Marine Isotope Stage (MIS) 3 that is dated globally to 57–29 ka and was generally a period of significant cooling and drying. This was succeeded by evidence for denser occupation within the markedly cooler Marine Isotope Stage 2 (dated globally to 29–16 ka), even during the extreme phase of glaciation termed the Last Glacial Maximum *ca.* 20 ka, when plant and animal species now restricted to Mount Kinabalu *ca.* 4000 m above sea level were around Niah. Occupation continued at similar density into the Early Holocene (11.4–8.2 ka) but amidst the occupation layers now was a series of extended burials. This burial form continued into the Mid- and Late Holocene when these burial types, together with new crouched forms, were associated with Neolithic pottery.

The Haua Fteah Cave

The Haua Fteah cave was excavated between 1951 and 1955 by Charles McBurney (1955) of the University of Cambridge. The cave is a handsome hangar-like karstic cavern looking northwards to the Mediterranean Sea about a kilometre distant, with an entrance about 20 m high and 60 m wide and with an interior roofed area about 80 m across. It lies on the maritime edge of the Gebel Akhdar (the 'Green Mountain'), an isolated massif in the middle of the North African coast that rises to almost 1000 m above sea level and measures some 350 km west/east and 50–100 km north/south, forming an island of green surrounded by desert on its landward sides. McBurney excavated a stepped trench that eventually reached some 14 m below the present ground surface, exposing a deep sequence of occupation that he divided into seven major phases: A. earlier Middle Palaeolithic ('Pre-Aurignacian'); B. later Middle Palaeolithic ('Levalloiso-Mousterian'); C. Upper Palaeolithic ('Dabban'); D. Late Upper Palaeolithic or Epipalaeolithic ('Oranian'); E. Mesolithic ('Capsian'); F. Neolithic ('Neolithic of Capsian Tradition'); and G. Historic. In the terminology of African prehistory, Phases A and B would now be classified as Middle Stone Age (MSA) and Phases C–E as Late Stone Age (LSA). Two mandibles in Phase B were originally classified as 'like Neanderthal' but were later shown to belong to archaic *Homo sapiens* (Hublin 1992). Radiocarbon dates indicated that the Dabban began around 40 ka and with dating earlier than 50 ka beyond the reach of radiocarbon dating, McBurney estimated an age for the start of the Middle Palaeolithic at the site of perhaps 80 ka.

In the new excavations (2007–2014), we emptied the McBurney trench of the backfill placed there at the end of the 1955 season and collected sediment samples down the trench walls for re-dating and for palaeoecological data such as pollen and land snails to inform on climate and environment. We then excavated a *ca.* 2 m × 1 m trench from top to bottom on the southern side of the McBurney trench to collect larger sets of chronological and palaeoecological data and collect archaeological materials such as stone tools and food refuse (*e.g.* butchered animal bone, marine molluscs, plant remains), to compare with the very large datasets from the original excavations curated in Cambridge's Museum of Archaeology and Anthropology. We were able to extend the 14 m deep sequence downwards by about a metre, and our basal OSL (optically stimulated luminescence) dates on feldspars indicate that initial occupation began around 140 ka (Douka *et al.* 2014; Jacobs *et al.* 2017). Although hominin fossils were not found in the LSA layers by McBurney or in the new excavations, the assumption is that both the MSA and LSA occupations can be ascribed to Moderns, whose origins in Africa on current dating can be placed at around 350 ka (Hublin *et al.* 2017).

The occupation of the cave was in fact highly episodic. The main early phase of MSA occupation ('Pre-Aurignacian') dates to MIS 5 and especially to MIS 5e (*ca.* 130–123 ka), the period when the earth's climate was significantly wetter and warmer than today and when the present-day Saharan desert was transformed into grassland interspersed with lakes and rivers (Drake *et al.* 2011). The main MSA 'Levalloiso-Mousterian' occupation was *ca.* 80–67 ka, across the MIS 5a/4 boundary. The two human mandibles date to *ca.* 80 ka at the start of MIS 4, a period when the world's climate began to trend towards drier and cooler conditions. After a significant hiatus there was further, less intensive, MSA occupation *ca.* 45–38 ka that transitioned into the first phase of the Dabban *ca.* 38–29 ka, both of them falling within the cooler and drier MIS 3. After a hiatus of several thousand years there was another Dabban phase *ca.* 24.2–23 ka. The Oranian Epipalaeolithic falls within MIS 2, the phase of maximum glacial conditions, and consists of a series of short but very intense occupations separated by significant gaps within the overall period 19–14.1 ka.

Shanidar Cave

Shanidar Cave, similar in size to the Haua Fteah, is located at around 800 m above sea level in the western foothills of the Zagros Mountains and faces south to the valley of the Great Zab River, a tributary of the Euphrates. Between 1951 and 1960, Ralph Solecki of Columbia University excavated a trench of similar depth to that of the Haua Fteah, exposing a Middle and Upper Palaeolithic/Epipalaeolithic sequence that he termed Layers D, C, and B respectively. Spectacular discoveries of the skeletal remains of several Neanderthals indicated that Neanderthals were the makers of the Layer D Middle Palaeolithic material (Solecki 1971). No human fossils were found within the Upper Palaeolithic Layer C (the material from which was called Baradostian from the name of a local mountain), but the similarities between these Baradostian lithics and Aurignacian lithics in Europe and elsewhere in the Middle East indicated that they were made by Moderns. Solecki's radiocarbon dates suggested that the latest Neanderthal skeletal remains dated to around 50 ka and that there was a 10,000-year hiatus between the Middle and Upper Palaeolithic occupations, the latter beginning around 35 ka. The lower Neanderthal layers could not be dated.

Our own excavations began in 2015 and still continue, using the same methods as for the Haua Fteah. So far, we have exposed parts of Solecki's trench wall down to about 10 m below the present ground surface. An OSL date places this level at around 83 ka, at the end of the MIS 5 interglacial. Photographs in the Solecki archive indicate that the 4 m of sediments he exposed below where we have currently reached probably formed in the interglacial conditions of MIS 5, implying that the length of occupation of the cave may be not so different from that of the Haua Fteah. Unexpectedly we have found further Neanderthal remains, including articulated bones, about 6 m below the present ground surface that we have shown belong to Solecki's 'Shanidar Neanderthal no. 5' (Pomeroy *et al.* 2017) and, about 9 m below the present ground surface, the crushed but articulated upper body (skull, upper limbs, thorax) of a new individual we have termed Shanidar Z (Pomeroy *et al.* 2020a). The latter was positioned immediately adjacent to where Solecki found a group of Neanderthal skeletal remains including Shanidar 4, the skeleton famously identified from pollen in its surrounding sediment as having been buried with flowers (Leroi-Gourhan 1975). Shanidar 5 dates to around 55–60 ka and Shanidar Z to *ca.* 73 ka. The latter date places the Shanidar 4 cluster of skeletal remains at the end of MIS 5, and our various palaeoenvironmental proxies (sediments, land snails, microfauna) indicate a climatic regime somewhat similar to that of today. The sediments at this depth also contain evidence for quite intensive occupation compared with much more ephemeral occupation evidence associated with the period of the upper Neanderthal skeletal remains within MIS 3. Our radiocarbon dates and stratigraphic evidence indicate a 'blurred' transition from the Mousterian to Baradostian occupations around 45–40 ka within MIS 3, with no evidence of a significant hiatus. The Baradostian occupation evidence consists of short-period camps (single-use hearths) dating to *ca.* 45–30 ka, especially to 42–38 ka, also within MIS 3. The lack of evidence for the use of the cave in MIS 2 suggests that, with the high Zagros mountains glaciated, the cave and its surrounding landscape were too marginal to access.

The View from the Cave
Marking the Landscape

Langley's suggested evidence for 'storied landscapes' under this heading includes rock art, cairns, monuments, and scar trees. Interestingly, the Kelabit forager-farmers of interior Sarawak have traditionally marked their presence in the forest through such activities: carving prominent stones, building a variety of burial monuments and cairns, and cutting ditches and forest breaks across ridges, as well as cutting clearings in the forest for growing hill rice and making wet rice fields. This is in contrast with the Penan foragers in the same part of Borneo, who aim just to 'leave footprints' (Janowski & Langub 2011), though they do in fact change the landscape by their protection (*molong*) of the sago plants that are their primary source of carbohydrate – the removal of competitor vegetation, for example – activities that have created distinctive sago groves that can be identified in 1940s and 1960s air photographs. Likewise the Moderns who used the Niah Caves, from the earliest evidence of their presence around 50 ka, made marks on the landscape by using fire to enhance clearings in the rainforest to encourage the growth of the tuberous plants that they consumed, and to attract to those clearings the main game they hunted, bearded pig (*Sus barbatus*), with traps and nets at the clearing edges, as well as pursuing them with spears (Barton *et al.* 2013; Reynolds *et al.* 2013; Piper & Rabett 2016). There is no evidence at Niah for painting cave walls at this time, though hand stencils and animal motifs elsewhere in Borneo and in Sulawesi have been dated to around 40–35 ka (Aubert et al. 2014; 2018).

We cannot discern such 'landscape marking' activities by the MSA and LSA foragers using the Haua Fteah cave, though the unusual dominance of pine in the cave pollen throughout the Pleistocene sequence hints at vegetation-burning regimes. In both periods everyday activities extended across a broad segment of terrain: people hunted a variety of game and foraged for plants on the northern slopes of the Gebel Akhdar and coastal plain (the extent of which was little affected by sea level lowering), collected land snails around the cave and shellfish and crustaceans from the coast, fished for species that included deep water ones, collected fuelwood and, around the time of the two human mandibles, brought large quantities of grass into the cave probably for bedding.

The Shanidar Cave data likewise provide no clues as to the physical impact on the surrounding landscape of the Neanderthals and Moderns using the cave, whose subsistence activities included a similar range of hunting, gathering, and fishing to those at the Haua Fteah, the fishing in this case involving the capture of large species from the Greater Zab river. On the other hand, there are indications that the location of the Shanidar Z body was marked by special stones (Pomeroy et al. 2020a), as Solecki observed for some of the Neanderthal skeletal remains he found, and the Shanidar Z/4 cluster of bodies was all placed within touching distance of a prominent rock pillar (fallen from the cave roof before the burial activities) that would have been a prominent landmark within the cave and likely visible from its entrance.

It should also be noted that like many caves preferred for repeated occupation by Palaeolithic people, these three caves all have spectacular entrance arches that are very prominent landmarks (Fig. 5.1). Another characteristic of long-inhabited caves can be soot-stained or soot-encrusted ceilings from campfires, as in all three of our caves, but especially Shanidar Cave, and sometimes this firing extends to the cliffs above the entrance and is visible from a distance. Shanidar Cave in spring is also a good example of how a prominent feature of such caves can be brighter and/or thicker vegetation growing on the talus below the entrance, enriched and fertilised by organic-rich midden that has cascaded or been throw down from the entrance rampart (Fig. 5.3).

Personal Identities

The intensive flotation and residue searching regime practised in the new excavations in the Haua Fteah and Shanidar Cave has yielded a series of tiny beads from both sites. In the Haua Fteah shell beads occur as early as in the MSA Pre-Aurignacian levels dating to MIS 5, ascribed to archaic *Homo sapiens*, significantly earlier than the larger perforated shell beads from sites such as Blombos Cave

Figure 5.3 The entrance to Shanidar Cave in spring 2022, showing dark green vegetation growing on the talus below the entrance arch. The image also shows the wealth of wildflowers at this season, including several of those identified by Leroi-Gourhan (1975) in the Shanidar 4 'Flower Burial' (Photo: Chris Hunt).

in South Africa (Henshilwood et al. 2004) and Grotte des Pigeons in Morocco (Bouzouggar et al. 2007) and about the same age as the beads found more recently in Bizmoune Cave in Morocco (Sehasseh et al. 2021). In the case of Shanidar Cave most are in Baradostian layers, but the earliest ones found so far are from around the level of the Shanidar 5 Neanderthal skeletal remains, dating to before 50 ka. They occur in secure stratigraphic contexts in which it is very difficult to dismiss them as Baradostian artefacts that have slipped downwards into Neanderthal layers as a result of bioturbation, water flows, burrowing animals, etc. No such beads or similar artefacts of personal ornamentation have been found in the Pleistocene occupation levels in the Niah Caves, but cutmarks on bones suggest the taking of birds of paradise, presumably for their feathers for adorning headdresses or personal equipment (Piper & Rabett 2016). Feather fragments were also found within organic residues attached to stone flakes (Barton 2016). The continued preference of both MSA and LSA hunters using the Haua Fteah to focus on the pursuit of the highly agile *Ammotragus lervia* (Barbary sheep) rather than on antelopes and bovids could conceivably be an indication of similar person-centred conspicuous display, and the same might apply to the focus of both Neanderthals and Moderns using Shanidar Cave on hunting ibex.

Raw Material Transport

The stone tools used in the Niah Caves were from locally available cherts, and there is little indication that tools were extensively curated and carried around. The main technology was probably of organic materials: alongside the crude stone flakes are pieces of bone and pig tusk fashioned into points (Rabett 2016), stingray barbs were fashioned into harpoons in the Late Pleistocene, and usewear studies of the stone tools and attached organic residues indicate the likelihood of an elaborate hunting and gathering technology of organic materials collected from the forest (Barton 2016; Barton *et al.* 2016; Rabett 2016). Evidence for long distance transport and/or exchange networks consists of clear shiny quartz crystals found inside the Deep Skull that were collected from a source in inland Borneo hundreds of kilometres away from Niah (Hunt & Barker 2014). Even more remarkably, the palaeoecological record indicates the very long-distance translocation of starchy plants across the Wallace Line from Australasia to northern Borneo, including to the interior highlands, by 25 ka and possibly well before (Hunt 2020).

The stone tools in the Haua Fteah, both MSA and LSA, were mainly from chert outcrops on the northern slopes of the Gebel Akhdar, and there are few indications of cultural linkages beyond the Gebel Akhdar until the Oranian, after the peak of glacial aridity (see below). In the case of Shanidar Cave both Neanderthals and Moderns mainly used what we assume were local river cobbles from the Greater Zab, but artefacts made of bright-coloured stones that may be exotic occur from the Shanidar Z layers upwards. The sources of the latter are unknown as yet, but the indications from the plant and animal remains are that both Neanderthals and Moderns largely used the cave in the spring and autumn months, in climate conditions much like those of today, probably moving to lower valleys in the winter and into the high Zagros in the summer (Reynolds *et al.* 2018; 2022). These raw materials may derive from the latter. The best indication of long-distance transport of raw material to Shanidar Cave is tiny pieces of obsidian in Baradostian layers that have been sourced to eastern Turkey and Armenia (Reynolds *et al.* 2018).

Norms and Customs Tied to the Landscape

Langley suggests that highly socialised landscapes in the Palaeolithic might be identified by material culture, including artefacts identified as redolent of symbolism, of an emblematic and geographically circumscribed style. There were fragments of human skull and turtle shell dated to *ca.* 42 ka in the West Mouth of Niah Great Cave that appear to have been used as some kind of palettes as they are stained red from tree resin (Pyatt *et al.* 2005; 2010). There are no other Palaeolithic sites in Borneo of comparable richness, but the broad similarities of the material culture and subsistence data from Tabon Cave and other caves in Palawan, the island of the southern Philippines that was connected to Borneo by sea level lowering at the time of Niah Cave's Palaeolithic occupation, might indicate some kind of broadly coastal inter-linked cultural entity (Dizon *et al.* 2002). Certainly, Niah was linked to its local landscape, from coast to inland hills, in terms of the sources of the animals, molluscs, and plants brought to it for consumption by Palaeolithic foragers (Barker 2013). This linkage extended into the symbolic realm as well: in addition to the human skull and turtle shell 'resin palettes', the Deep Skull and associated limb bones are likely to be the remains of a secondary burial placed near the lip of the West Mouth around 35 ka, perhaps from primary burial activity involving exposure of the corpse in the forest as the Penan do today (Hunt & Barker 2014).

The rounded condition of the human jaws from the Haua Fteah dating to *ca.* 80 ka suggests some kind of recycling, but their mortuary significance is unclear. What is striking about most of the Palaeolithic record of the Haua Fteah and other sites in the Gebel Akhdar, though, is their cultural distinctiveness with respect to the rest of North Africa: whilst the Gebel Akhdar was accessible to other parts of North Africa during the 'Green Sahara' phases of MIS 5 (Drake *et al.* 2011), the dominant characteristic of the Pre-Aurignacian lithic technology of the period is its lack of similarities with the contemporary Aterian technologies that were widespread across North Africa. The same applies to the Levalloiso-Mousterian technologies that were used as aridity developed in MIS 5a and MIS 4 (Scerri 2013; 2017; Scerri *et al.* 2014), and even more so to the ensuing blade-based 'Dabban' industry: especially in its early manifestation *ca.* 38–29 ka, it is in many respects a mix of MSA and LSA technologies quite unlike the contemporary industries of the rest of North Africa. Significant linkages with the Maghreb (northwest Africa) only became apparent with the development of the Oranian after the Last Glacial Maximum.

The Zagros Mousterian and the Baradostian are also recognised as distinct cultural entities straddling the mountain range (Reynolds *et al.* 2018; 2022), with the Shanidar Cave skeletal remains providing unique evidence of Neanderthals' ties to landscape. Our new work at the site suggests that the distinction that Solecki drew between individuals accidentally killed by, and buried underneath, rockfalls and individuals buried with funerary rites does not hold, the likelihood being that bodies, or parts of bodies, were in most cases carefully placed in restricted areas (Pomeroy *et al.* 2020a; 2020b). The accumulating evidence, including newly found fragmentary remains underneath Shanidar Z, accords with Pettitt's argument that for the Neanderthals who brought their dead to them, sites with multiple skeletal remains like Shanidar Cave, Krapina in Croatia, and L'Hortus and La Ferrassie in France suggested 'the transmission of mortuary tradition…centred around a fixed point in the landscape that could be used, if not exclusively, to hide, process, and bury the dead' (Pettitt 2011, 122). 'To the groups of La Ferrassie and Shanidar', he commented in another paper, 'the dead had not quite departed [implying that] religious thought *sensu lato* emerged prior to, or at least not exclusive to, *Homo sapiens*' (Pettitt 2015, 273–74). Whether the placing of the Shanidar Z/4 cluster of bodies spanned days, months, years, decades, or centuries (even many centuries), the rock pillar and grave marker stones, along with the evocative characteristics of the cave entrance itself (Fig. 5.3), look to be strong candidates for components of a storied, memory-imbued, Neanderthal landscape.

Conclusions

There is ample evidence, well summarised in Langley's 2013 paper, in support of the argument that, whilst European Neanderthals may have engaged in symbolic behaviour especially in the later millennia of their long history, the archaeological record associated with the Upper Palaeolithic Moderns that entered Europe around 45–40 ka represents a step change in the range and diversity of symbolic indicators compared with the evidence ascribed to Neanderthals. These changes underpin the arguments of Burke (2006; 2012), that Neanderthals operated within spatially more constrained social networks than Moderns, and of Langley (2013), that Moderns not only lived within far more extensive social networks than Neanderthals ('social landscapes') but also imbued their landscapes with symbols and stories ('landscape socialisation'). Outside Europe, though, at least in terms of the archaeology of the three caves we have explored, the evidence for how Upper Pleistocene humans (that is, humans living through the MIS 5–2 interglacial/glacial cycle) related to the landscapes in terms of the main categories identified by Langley (Table 5.1) is more ambiguous.

Palaeolithic Moderns using the Niah Cave in MIS 3–2 certainly marked their landscape by the impact of their foraging activities on the surrounding rainforest. Whether the Moderns using the Haua Fteah in MIS 5–2, the Neanderthals using Shanidar Cave in MIS 4–3, and the Moderns using Shanidar Cave in MIS 3, did so is less clear, though all three caves are prominent places in the landscape, and likely all the more visible at the time of Palaeolithic occupations from soot-staining and richly vegetated taluses. Langley draws the distinction between Neanderthal personal ornamentation primarily for ingroup and outgroup members within social networks and landscapes, and Moderns' personal ornamentation serving to transmit information about their interactions with landscape features within storied landscapes, but in the case of the three caves it is difficult to see significant temporal or between-species differences in the complex ways that personal identities seem to have been marked by the Moderns of Niah Cave and the Haua Fteah and the Neanderthals and Moderns of Shanidar Cave. Distant raw materials were acquired by the Niah Cave Moderns and the Shanidar Cave Moderns and perhaps also by the Shanidar Cave Neanderthals. There is clear evidence for symbolic behaviour linked to the placement of bodies, or parts of bodies, practised by the Niah Cave Moderns around 40 ka and the Shanidar Cave Neanderthals around 75–55 ka, and the two human mandibles dated to *ca.* 80 ka in the Haua Fteah might hint at something similar. The Niah Cave Moderns were likely part of a distinct sociocultural entity that linked northern Borneo to modern Palawan at the times of lowered sea levels and the Haua Fteah Moderns appear to have developed successive technologies in MIS 5–3 that were markedly distinct from those used beyond the Gebel Akhdar. The Neanderthal and Baradostian assemblages of Shanidar Cave have both been regarded as specifically Zagros manifestations despite their broad linkages with, respectively, Mousterian and Aurignacian technologies further afield (Reynolds *et al.* 2018; 2022).

Much of the commentary in this paper is avowedly speculative, but we hope it serves to widen the debate about Palaeolithic social networks and 'storied landscapes' beyond the Neanderthal/Modern dichotomy that is such a cornerstone of evolutionary studies based on the European archaeological record. That dichotomy is further questioned by archaeobotanical evidence that Neanderthals and Moderns in Shanidar Cave both processed – cooked – plant foods in the same way (Kabucku *et al.* 2022). As well as widening the focus geographically and chronologically, we have also tried to emphasise that all facets of the archaeological record can hold potential information about symbolic behaviour and not just the 'usual suspects' of art, beads, and burials. In the case of Niah, marking the landscape has been inferred from the cave's palynology, the long-distance transport of valued plant resources (and presumably the knowledge of how to use them) has been inferred from estuarine sediments, and personhood display from cutmarks on bird bones. Perhaps most intriguing – and challenging – of all regarding our attempts to capture how these distant Palaeolithic societies thought about the landscapes they inhabited is the evidence observed in the Niah Cave butchery practices that these foragers divided up the animal kingdom in ways quite alien to our own Linnean taxonomies (Piper & Rabett 2016). The evidence of our three caves emphasises that, at least for the Palaeolithic people using them, there were different ways of being human and different ways of envisaging the landscape beyond.

References

Aldhouse-Green, S. & Pettitt, P. (1998) Paviland Cave: Contextualizing the 'Red Lady'. *Antiquity* 72, 756–72.

Aubert, M., Brumm, A., Ramli, M., Sutikna, T., Saptomo, E., Hakim, B., Morwood, M., van der Bergh, G., Kinsley, L. & Dosseto, A. (2014) Pleistocene cave art from Sulawesi, Indonesia. *Nature* 514, 223–7.

Aubert, A., Setiawan, P., Oktaviana, A., Brumm, A., Sulistyarto, P., Saptomo, E., Istiawan, B., Ma'rifat, T. A., Wahyuono, V., Zhao, J.-X., Huntley, J. & Taçon, P. S. (2018) Palaeolithic cave art in Borneo. *Nature* 564, 254–7.

Barker, G. (1995) *A Mediterranean Valley: Landscape Archaeology and Annales History in the Biferno Valley*. London: Leicester University Press.

Barker, G. (ed.) (1996) *Farming the Desert: The UNESCO Libyan Valleys Archaeological Survey*. Paris: UNESCO (with the Society for Libyan Studies, London and the Department of Antiquities, Tripoli).

Barker, G. (ed.) (2013) *Rainforest Foraging and Farming in Island Southeast Asia: The Archaeology of the Niah Caves, Sarawak*. Cambridge, McDonald Institute for Archaeological Research, McDonald Institute Monographs.

Barker, G. & Farr, L. (eds) (2016) *Archaeological Investigations in the Niah Caves, Sarawak*. Cambridge, McDonald Institute for Archaeological Research, McDonald Institute Monographs.

Barker, G., Gilbertson, D. & Mattingly, D. (eds) (2007) *Archaeology and Desertification: The Wadi Faynan Landscape Survey, Southern Jordan*. Oxford/London: Oxbow/Council for British Research in the Levant.

Barker, G., Hunt, C., Barton, H., Jones, S., Lloyd-Smith, L., Farr, L., Nyirí, B. & O'Donnell, S. (2017) The 'cultured rainforests' of Borneo. *Quaternary International* 448, 44–61.

Barton, H. (2016) Functional analysis of stone tools from the West Mouth. In Barker & Farr 2016, 279–300.

Barton, H., Barker, G., Gilbertson, D., Hunt, C., Kealhofer, L., Lewis, H., Paz, V., Piper, P., Rabett, R., Reynolds, T. & Szabó, K. (2013) Late Pleistocene foragers, *c.* 35,000–11,500 years ago. In Barker 2013, 173–215.

Barton, H., Paz, V. & Carlos, J. (2016) Plant food remains from the Niah Caves: macroscopic and microscopic approaches. In Barker & Farr 2016, 455–68.

Bourdier, C. (2013) Rock art and social geography in the Upper Paleolithic: Contribution to the socio-cultural function of the Roc-aux-Sorciers rock-shelter (Angles-sur-l'Anglin, France) from the viewpoint of its sculpted frieze. *Journal of Anthropological Archaeology* 32(4), 368–82.

Bouzzouggar, A., Barton, N., Vanhaeren, M., d'Errico, F., Collcutt, S., Higham, T., Hodge, E., Parfitt, S., Rhodes, E., Schwenniger, J.-L., Stinger, C., Turner, E., Ward, S., Moutmir, A. & Stambouli, A. (2007) 82,000-year-old shell beads from North Africa and implications for the origins of modern human behaviour. *Proceedings of the National Academy of Sciences USA* 104, 9964–9.

Brothwell, D. R. (1960) Upper Pleistocene human skull from Niah Caves. *Sarawak Museum Journal* 9(n.s. 15–16), 323–49.

Burke, A. (2006) Neanderthal settlement patterns in Crimea: A landscape approach. *Journal of Anthropological Archaeology* 25, 510–23.

Burke, A. (2012) Spatial abilities, cognition and the pattern of Neanderthal and modern human dispersals. *Quaternary International* 247, 230–5.

Clottes, J. & Lewis-Williams, D. (1998) *The Shamans of Prehistory: Trance and Magic in the Painted Caves*. New York, NY, Harry N. Abrams.

Conkey, M. W., Beltrán, A., Clark, G. A., González Echegaray, J., Guenther, M. G., Hahn, J., Hayden, B., Paddayya, K., Strauss, L. G. & Valoch, K. (1980) The identification of prehistoric hunter-gatherer aggregation sites: The case of Altamira [and Comments and Reply]. *Current Anthropology* 21(5), 609–30.

Dizon, E., Détroit, F., Sémah, F., Falguères, C., Hameau, S., Ronquillo, W. & Cabanis, E. (2002) Notes on the morphology and age of the Tabon Cave fossil *Homo sapiens*. *Current Anthropology* 43, 660–6.

Douka, K., Jacobs, Z., Lane, C., Grün, R., Farr, L., Hunt, C., Inglis, R. H., Reynolds, T., Albert, P., Aubert, M., Cullen, V., Hill, E., Kinsley, L., Roberts, R. G., Tomlinson, E. L., Wulf, S. & Barker, G. (2014) The chronostratigraphy of the Haua Fteah cave (Cyrenaica, Northeast Libya). *Journal of Human Evolution* 66, 39–63.

Drake, N. A., Blench, R. M., Armitage, S. J., Bristow, C. S. & White, K. H. (2011) Ancient watercourses and biogeography of the Sahara explain the peopling of the desert. *Proceedings of the National Academy of Sciences of the United States of America* 108(2), 458–62.

Gamble, C. (1998) Palaeolithic society and the release from proximity: A network approach to intimate relations. *World Archaeology* 29(3), 426–49.

Geiling, J. M. & Marín-Arroyo, A. B. (2015) Spatial distribution analysis of the Lower Magdalenian human burial in El Mirón Cave (Cantabria, Spain). *Journal of Archaeological Science* 60, 47–56.

Harrisson, B. (1967) A classification of Stone Age burials from Niah great cave, Sarawak. *Sarawak Museum Journal* 15(n.s. 30–31), 126–200.

Harrisson, T. (1958) The caves of Niah: A history of prehistory. *Sarawak Museum Journal* 8(n.s. 12), 549–95.

Harrisson, T. (1970) The prehistory of Borneo. *Asian Perspectives* 13, 17–45.

Henshilwood, C., D'Errico, F., Vanhaeren, M., Van Niekerk, K. & Jacobs, Z. (2004) Middle Stone Age shell beads from South Africa. *Science* 304, 404.

Higham, T. F. G., Barton, H., Turney, C. S. M., Barker, G., Bronk Ramsey, C. & Brock, F. (2008) Radiocarbon dating of charcoal from tropical sequences: Results from the Niah Great Cave, Sarawak, and their broader implications. *Journal of Quaternary Science* 24, 189–97.

Hublin, J.-J. (1992) Recent human evolution in Northwestern Africa. *Philosophical Transactions of the Royal Society B* 337(1280), 185–91.

Hublin, J.-J., Ben-Ncer, A., Bailey, S. E., Freidline, S. E., Neubauer, S., Skinner, M. M., Bergmann, I., Le Cabec, A., Benazzi, S., Harvati, K. & Gunz, P. (2017) New fossils from Jebel Irhoud, Morocco and the pan-African origin of *Homo sapiens*. *Nature* 546, 289–92.

Hunt, C. (2020) Agroforestry and its impact in Southeast Asia. [online] *Oxford Encyclopaedia of Agriculture and the Environment*. Available at: <https://oxfordre.com/environmen-

talscience/view/10.1093/acrefore/9780199389414.001.0001/acrefore-9780199389414-e-170>.

Hunt, C. & Barker, G. (2014) Missing links, cultural modernity and the dead: Anatomically modern humans in the Great Cave of Niah (Sarawak, Borneo). In R. Dennell & M. Porr (eds), *Southern Asia, Australia, and the Search for Human Origins*, 90–107. Cambridge, Cambridge University Press.

Jacobs, Z., Li, B., Farr, L., Hill, E., Hunt, C., Jones, S., Rabett, R., Reynolds, T., Roberts, R. G., Simpson, D. & Barker, G. (2017) The chronostratigraphy of the Haua Fteah cave (Cyrenaica, northeast Libya) II – optical dating of early human occupation during Marine Isotope Stages 4, 5 and 6. *Journal of Human Evolution* 105, 69–88.

Janowski, M. (2003) *The Forest, Source of Life: The Kelabit of Sarawak*. London/Sarawak, British Museum/Sarawak Museum.

Janowski, M. & Langub, J. (2011) Footprints and marks in the forest: The Penan and Kelabit of Borneo. In G. Barker & M. Janowski (eds), *Why Cultivate? Anthropological and Archaeological Approaches to Foraging-Farming Transitions in Southeast Asia*, 121–32. Cambridge, McDonald Institute for Archaeological Research, McDonald Monograph.

Kabucku, C., Hunt, C., Hill, E., Pomeroy, E., Reynolds, T., Barker, G. & Asouti, E. (2022) Cooking in caves: Palaeolithic carbonised plant food remains from Franchthi and Shanidar. *Antiquity*, doi:10.15184/aqy.2022.143.

Langley, M. (2013) Storied landscapes make us (Modern) Human: Landscape socialisation in the Palaeolithic and consequences for the archaeological record. *Journal of Anthropological Archaeology* 32, 614–29.

Leroi-Gourhan, A. (1975) The flowers found with Shanidar IV, a Neanderthal burial in Iraq. *Science* 190, 562–64.

Leroi-Gourhan, A. (1998) Shanidar et ses fleurs. *Paléorient* 24(2), 79–88.

Lewis-Williams, D. (2002) *The Mind in the Cave*. London, Thames and Hudson.

McBurney, C. B. M. (1955) *The Haua Fteah, Cyrenaica, and the Stone Age of Northeast Africa*. Cambridge, Cambridge University Press.

Movius, H. (1948) The Lower Palaeolithic cultures of southern and eastern Asia. *Transactions of the American Philosophical Society* 38, 329–420.

Pettitt, P. (2011) *The Palaeolithic Origins of Human Burial*. London, Routledge.

Pettitt, P. (2015) Landscapes of the dead: The evolution of human mortuary activity from body to place in Palaeolithic Europe. In F. Coward, R. Hosfield, M. Pope & F. Wenbow-Smith (eds), *Settlement, Society and Cognition in Human Evolution: Landscapes in Mind*, 258–74. Cambridge, Cambridge University Press.

Piper, P. & Rabett, R. (2016) Vertebrate fauna from the Niah Caves. In Barker & Farr 2016, 401–38.

Pomeroy, E., Lahr, M. M., Crivellaro, F., Farr, L., Reynolds, T., Hunt, C. O. & Barker, G. (2017) Newly-discovered Neanderthal remains from Shanidar Cave, Iraqi Kurdistan, and their attribution to Shanidar 5. *Journal of Human Evolution* 111, 102–18.

Pomeroy, E., Bennett, P., Hunt, C., Reynolds, T., Farr, L., Frouin, M., Holman, J., Lane, R., French, C. & Barker, G. (2020a) New Neanderthal remains associated with the 'Flower Burial' at Shanidar Cave, Iraqi Kurdistan. *Antiquity* 94(373), 11–26.

Pomeroy, E., Hunt, C., Reynolds, T., Abdulmutalb, D., Asouti, E., Bennett, P., Bosch, M., Burke, A., Farr, L., Foley, R., French, C., Frumkin, A., Goldberg, P., Hill, E., Kabukcu, C., Mirazón Lahr, M., Lane, R., Marean, C., Maureille, B., Mutri, G., Miller, C. E., Mustafa, K., Nymark, A., Pettitt, P., Sala, N., Sandgathe, D., Stringer, C., Tilby, E. & Barker, G. (2020b) Issues of theory and method in the analysis of Palaeolithic mortuary behavior: A view from Shanidar Cave. *Evolutionary Anthropology* 29, 263–79.

Pyatt, B., Wilson, B. & Barker, G. (2005) The chemistry of tree resins and ancient rock paintings in the Niah Caves, Sarawak (Borneo): Some evidence of rainforest management by early human populations. *Journal of Archaeological Science* 32, 897–901.

Pyatt, B., Barker, G., Rabett, R., Szabó, K. & Wilson, B. (2010) Analytical examination of animal remains from Borneo: The painting of bone and shell. *Journal of Archaeological Science* 37, 2102–5.

Rabett, R. (2016) Bone and tusk tools from the West Mouth and Lobang Hangus. In Barker & Farr 2016, 301–24.

Reynolds, T., Barker, G., Barton, H., Cranbrook, G., Farr, L., Hunt, C., Kealhofer, L., Paz, V., Pike, A., Piper, P., Rabett, R., Rushworth, G., Stimpson, C. & Szabó, K. (2013) The first Modern Humans at Niah, c.50,000–35,000 years ago. In Barker 2013, 135–72.

Reynolds, T., Farr, L., Hunt, C., Gratuze, B., Hill, E., Abdulmutalb, D., Nymark, A. & Barker, G. (2018) Shanidar Cave and the Baradostian, a Zagros Aurignacian industry. *L'Anthropologie* 122, 737–48.

Reynolds, T., Hunt, C., Hill, E., Tilby, E., Pomeroy, E., Burke, A. & Barker, G. (2022) Le Moustérien du Zagros: le panorama depuis la grotte de Shanidar. *L'Anthropologie* 126, doi:10.1016/j.anthro.2022.103045.

Scerri, E. (2013) The Aterian and its place in the North African Middle Stone Age. *Quaternary International* 300, 111–30.

Scerri, E. (2017) The North African Middle Stone Age and its place in recent human evolution. *Evolutionary Anthropology* 26, 119–35.

Scerri, E., Drake, N. A., Jennings, R. & Groucutt, H. S. (2014) Earliest evidence for the structure of *Homo sapiens* populations in Africa. *Quaternary Science Reviews* 101, 207–16.

Sehasseh, E. M., Fernandez, P., Kuhn, S., Stiner, M., Mentzer, S., Colarossi, D., Clark, A., Lanoe, F., Pailes, M., Hoffmann, D., Benson, A., Rhodes, E., Benmansour, M., Laissaoui, A., Ziani, I., Vidal-Matutano, P., Morales, J., Djellal, Y., Longet, B., Hublin, J.-J., Mouhiddine, M., Rafi, F.-Z., Worthey, K. B., Sanchez-Morales, I., Ghayati, N. & Bouzouggar, A. (2021) Early Middle Stone Age personal ornaments from Bizmoune Cave, Essaouira, Morocco. *Science Advances* 7(39), doi:10.1126/sciadv.abi8620.

Solecki, R. S. (1971) *Shanidar: The First Flower People*. New York, NY, Alfred A. Knopf.

6

A Circular Tomb with 'Stones' of Clay: The Tomb of Lord Bai of Zhongli, Anhui Province, Central China, Early 6th Century BC

Jessica Rawson

A circular tomb dating ca. *early 6th century BC at Bengbu (Anhui Province) astounded those excavating in central China[1] when it was found intact in 2006–08, as no one had seen anything like it before. Inscribed bronze vessels identified the occupant as Lord Bai of Zhongli. The only other circular tombs known from early China belong to his spouse and son, whereas most 1st millennium BC tombs are rectangular. In addition, those of the principal Zhou lineage align north–south, but Lord Bai's tomb was aligned east–west. The burial of attendants, the stone wall recreated in earth near the top of the pit, and the large mound covering the tomb also echo burial practices from regions north of Anhui, where people from the Mongolian Plateau and southern Siberia, regions with circular tombs, made incursions over several centuries. This chapter explores the contribution of these people to the burial choices made by Lord Bai and his court.*

Keywords: China's Eastern Seaboard, Circular Tombs, Steppe, Stone, Burial Rites

Introduction

Lord Bai's circular tomb is unique. The only other examples known to date are those of his spouse and son. These extraordinary tombs chimed with those that I and members of the School of Archaeology at the University of Oxford, along with our collaborators from Peking University, visited in the Valley of the Kings in the Tuva Republic (Russian Federation), thanks to expert planning by Peter Hommel. The mounds in Siberia and their circular plans were one of the inspirations for Chris's important article, 'Making Mounds: Monuments in Eurasian Prehistory' (Gosden *et al.* 2019). His confident descriptions of how mounded tombs spread across northern Eurasia matched my interest in the similarities between Lord Bai's extraordinary tomb in central China and the great kurgans of the Eurasian Steppe. Our many conversations since the visit to Siberia, usually by phone in Covid times, have led us both to explore the seminal interactions between China and its northern neighbours.

Lord Bai's Tomb

Lord Bai's tomb and that of his spouse are under large mounds, still standing today at about 9 m in height, at a village named Shuangdun, or 'pair of tombs', on the edge of the city of Bengbu in Anhui Province (API & BMM 2010). The only other excavated circular tomb is that of his son (API & FCA 2018). Lord Bai's tomb was excavated in 2008–09 and revealed one of the most unusual major tombs of the Zhou period (*ca.* 1045–221 BC). Below the massive mound, which alone had a diameter of 60 m, the circular burial pit measured 20 m at the mouth and 14 m at the base (API & BMM 2013). At the centre was the coffin of Lord Bai (Fig. 6.1), orientated east–west, with slightly bowed outlines on all four sides. Little could be made of the skeletal remains, though from the teeth, the excavators estimated that he was 40 years old when he died. Arranged neatly around this large coffin on three sides were groups of three attendants, each with their own coffin. A further coffin lay between the main coffin and a large wooden frame or chest with ritual bronze vessels, bells (on which was found the names of Lord Bai and the small state of Zhongli) and chime stones, horse and chariot equipment, as well as a large group of painted ceramics that probably held grains and vegetables. The lord had a sword, the origins of which lie to the north in the tombs of the Upper Xiajiadian (1000–600 BC) (IMI & NLM 2009), and a pair

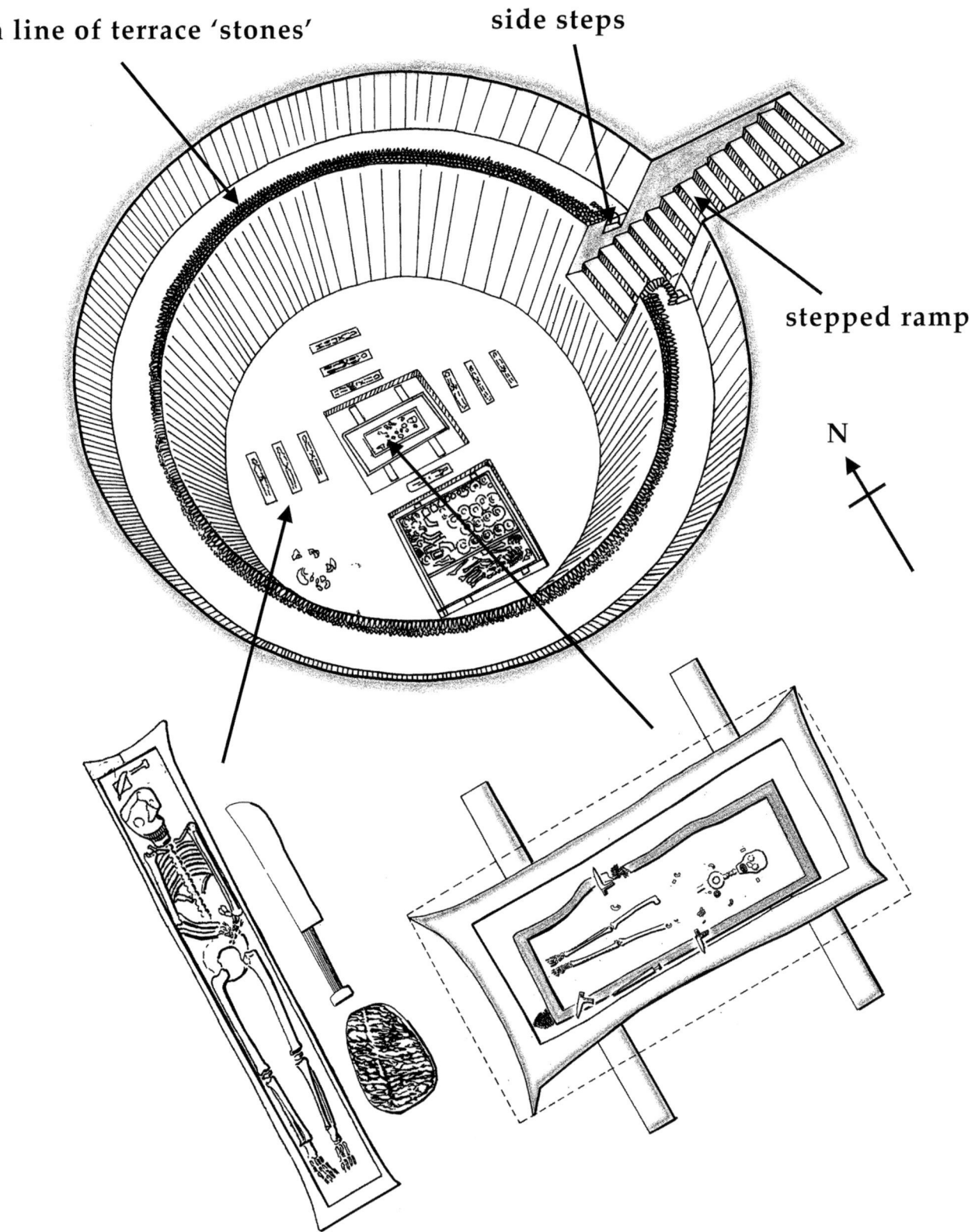

Figure 6.1 The tomb of Lord Bai, early 6th century BC. The central grave and one of the subordinate ones are enlarged to illustrate their bowed sides (after API & BMM 2013, vol. 1, 37–42, modified by John Rawson).

of steppe-type bowed knives with rings on the grips (API & BMM 2013, vol. 1, figs 71 & 95:1–2), which not only appear in the Jundu Mountains (BMI 2010) but were also widely used in the steppe. The lord had exceptional armour in bronze covered with gold foil (API & BMM 2013, vol. 1, 185–92). The gleaming metal rendered complex patterns, including a few in the shape of tigers, a motif particularly favoured in the steppe and by their neighbours in northern China that are frequently copied in jade. The coffins of Lord Bai and his attendants are also telling. Their bowed outlines can be compared with those of northern ones from the southern and eastern Mongolian steppe, whose people were significant contributors to Shang and Western Zhou engagement with the north (Rawson *et al.* 2020). Compared with other constructions in central China, everything about this tomb is very unusual (see *e.g.* Rawson *et al.* 2021, 499–514), from the overall tomb construction, to the placing of the coffins and the separate compartments for the bronzes, ceramics, and animal parts. Yet it is a major construction, built to a plan that must have been familiar to its makers; we can see in its formal arrangement that it is extremely unlikely to have been created fortuitously.

For parallels for some aspects of Lord Bai's tomb structure, we can look nearby just to the northeast, to the 7th- and 6th-century BC tombs of the Yi people in Shandong Province. The Yi were eastern outsiders to the Zhou realm (Shaughnessy 1999), who started to take names recorded in characters at this time, calling themselves the Ju (Wu & Zhang 1978) and Xue (Gong *et al.* 1991). Their tombs, though almost invariably rectangular, share three of the significant features evident in Lord Bai's tomb that are not part of the Zhou tradition: they often had large mounds above the grave pits, with the central occupant neatly surrounded by attendants in their own coffins and separate compartments for goods. We see these also at a site called Fenghuangling (Yanshi Railway Cultural Heritage Team of Shandong Province 1988). The strength of the shared northeastern burial tradition is attested by its reach south, where it appears in the form of the tomb of the spouse of Wu State's ruler, buried in the late 6th century BC during the campaign at Hougudui in Henan (Henan Provincial Institute of Cultural Relics and Archaeology 2004). A much earlier tomb of the small Huang State (conquered in the mid-7th century BC) also shows traces of a mound (Ou 1984) and exhibits signs of contact with the north through a bowed, steppe-type knife and jade imitations of tigers, which are images more typical of steppe tastes and are found in bronze in Shaanxi (Li 2006, fig. 29) and gold in the Jundu Mountains near Beijing (BMI 2010, vol. 4, plate 47). Although local people further south along the Eastern Seaboard took up the practice of covering burials with mounds, it looks highly likely that the origin of the mounds placed by the Lord of Bai over his tomb and that of his wife first emerged in the southern Henan region with stimulus from further north.

Contacts with northern outsiders were especially prevalent in the area that the Chinese archaeologist, Tong Enzheng, called the 'crescent-shaped region', which I have translated as the 'Arc' (Tong 1986; Rawson 2017). This key region is often overlooked, particularly as it is not singled out in transmitted texts, in spite of its size. Over many centuries from the late Shang (*ca.* 1250 BC) to the 3rd century BC, different groups of agro-pastoralists occupied these extensive lands – 2000 km from Lanzhou in the west to Shenyang in the east – between the steppe and the agricultural basins of the Yellow River and its tributaries. Particularly in the 9th century BC, people known to the Zhou as the Rong brought military and political unrest to the Wei River valley in particular (Rawson 2017; Rawson *et al.* 2021), especially around modern Xi'an (Li 2006, 141–92). Unless we take the diverse inhabitants of the Arc into account, we cannot arrive at a picture of the forms of communication between the north and the agricultural areas, including, in this case, Anhui Province.

Contacts with the steppe were driven by people known to the Zhou as the Rong, Di, and Yi, who provided the rulers of the agricultural basins with horses. Horses were essential to the livelihoods of the steppe people and were introduced to central China with the chariot in the late Shang (Rawson *et al.* 2020). This was a turning point for the rulers of central China and led to constant contact with the north to acquire horses. It has emerged that the main agricultural areas in the Yellow River Basin and the Wei River were generally unsuitable for breeding horses, owing to a shortage of selenium in the soil (Sun *et al.* 2016; Rawson *et al.* 2021). This deficiency was a very serious problem, leading to weaknesses in bone and muscle for horses and causing extreme disease for humans, known as Kashin Beck Disease (Stone 2009). Although the problems of selenium were recognised in eastern Russia (which like northern Korea is affected by the same geological and climate issues), only recently has the significance of this shortage in nutrition for horses in China been described (Whitfield 2020; Rawson *et al.* 2021). From the Shang period down to the 19th century AD, to contend with the strong horses of Mongolia, China's dynastic rules had to seek horses from the north. In consequence, the Zhou and the rulers of the polities engaged closely with northerners to benefit from their horses and their horse management (Rawson *et al.* 2021; Rawson 2023).

The movement of northern horse owners was also encouraged by climate and weather challenges (Huang *et al.* 2003; Geel *et al.* 2004; Struck 2022). They penetrated westwards to the mountains of present-day Xinjiang Province (Xinjiang Institute of Cultural Relics and Archaeology 2015) and east to the north of what is now North Korea (Zhu 2017,

301–21). We know from extensive and revealing settlements of the Upper Xiajiadian culture groups in Inner Mongolia and Liaoning (IMI & NLM 2009), on the promontory of Liaodong (Institute of Archaeology, Chinese Academy of Social Sciences 1996), and in the Judun Mountains (BMI 2010) that the steppe people arrived there also. In all these areas, variations of unusual tomb types with extensive use of stone have been excavated. The most northerly, those in Xinjiang, northern Liaoxi, and the Liaodong peninsula, had the most in common with the kurgans of the Altai-Sayan regions (Chugunov 1998). Like their steppe contemporaries, these diverse peoples used stone in or on their tombs, and buried animal heads, especially those of horses, in their graves. These people then brought these customs southwards across present-day Shaanxi, Shanxi, and Hebei (Zhang et al. 2018). This continuous movement south brought them into contact with the Yi in Shandong Province and the Huai Yi in the region where Lord Bai's tomb was constructed. We can thus see the primary sources of the features of his tomb as coming from this contact with north-eastern peoples who had moved out of the steppe, bringing with them the use of stone structures, deposits of grave goods in separate compartments, neatly arranged coffins for attendants, and the burial of horses heads, the latter being the prototype for the animal parts in the lord's tomb.

The circular shape of Lord Bai's tomb references above all the kurgans and even minor tombs of the steppe, including those in the Chinese Altai and along the Tianshan. This is strongly supported by the recreation of stone boulders in earth that line the terrace above the pit; it is very likely that Lord Bai or his assistants have some direct contact with steppe leaders. This becomes especially clear if we consider the purpose of the terrace. To access this terrace from ground level, people walked down a stepped ramp then turned left or right onto small flights of steps on either side. Stepped ramps like this have been discovered in Shang and early Zhou tombs of outsiders on the Loess Plateau (Wang et al. 2021; Zhao et al. 2021), but they were not combined with terraces. In Lord Bai's tomb, a wall of earthen 'stones' prevented the people on the terrace from slipping into the pit, or, if the pit had already been filled, from walking on to the filled surface. This is a clear imitation in earth of a wall built of stone. From inside the main pit, it looks like a wall, whereas seen from the terrace, the pointed ends of some of the 'boulders' are more conspicuous (Fig. 6.2). The creation of a structure in one material – clay or earth – to represent something quite different – a stone wall – is unprecedented in this region. There are few, if any, walls built of stone in central China, and without stone walls as part of daily life, imitation is unexpected. Such a terrace with clear access cannot have simply been made coincidentally during construction; it must have had a purpose. It feasibly

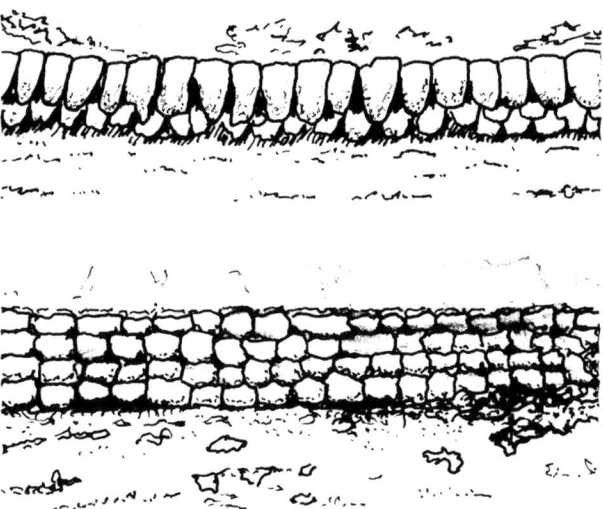

Figure 6.2 Drawings of the terrace wall around the tomb of Lord Bai: (top) the replica earthen boulders seen from inside the terrace; (bottom) the wall made from earthen boulders as seen from inside the tomb pit (after API & BMM 2013, vol. 3, plate 43, modified by John Rawson).

allowed onlookers, well organised, and with familiarity or understanding of the structure, to take part in one or more stages of the burial ritual.

This observation encourages us to look in more detail at the kurgans in the Altai-Sayan region, especially at Arzhan 2. The leader of the Russo-German excavation team, Konstantin Chugunov of the State Hermitage Museum, has recently argued that it is possible to see deliberate preparation for the funeral rituals within Arzhan 2 (Chugunov 2020). He suggests that clay deposits were carefully laid as an access path (A in Fig. 6.3) and two bands either side of the central section (B in Fig. 6.3), delineated by a fence. The access path may have been used by mourners to enter the central area of the kurgan, where they stood on the two curved bands. This allows us to consider the possibility that the terrace at Lord Bai's tomb offered similar locations for mourners to stand. This explains both the access afforded by the ramp and the two small side stairs allowing entry onto both sides to the terrace. Moreover, in making the wall of clay or earthen boulders in imitation of stone, it seems that the people planning and building the tomb had a clear vision of what they were creating, *i.e.* they held the northern tradition in mind.

This imitation, or even simulation, of a tomb type and funeral practice from elsewhere was continued further with the next phase, when the central pit and the terrace around it were both filled. Once the lower pit was fully filled, the 20-m wide pit projected a new image, and small mounds or half mounds were raised around the edge. These were made up of layers of different-coloured earth, as was the

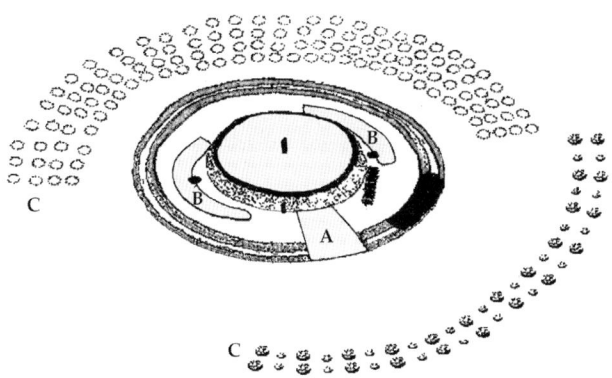

Figure 6.3 Plan of Arzhan 2. (a): the access pathway; (b): the side positions for people attending the ritual; and (c): the subsidiary mounds and deposits (after Chugunov 2020, fig. 21, modified by John Rawson).

main tomb mound. From this comparison, we recognise that these were deliberately manufactured and were not a few piles of meaningless soil that someone forgot to clean up. Scattered across the surface of the filled main pit were small replicas of angular fragmented stones made of earth (API & BMM 2013, vol. 1, fig. 16; vol. 3, plates 37–45). A stony landscape was thus recreated, in which the tomb was encircled by small mounds intended as representations of other small tombs or ritual deposits. This simulates the small mounds around the kurgans, as at Arzhan 2 (C in Fig. 6.3; see Chugunov *et al.* 2010). It seems possible that further stages in the funerary rituals were carried out on this level of fill before a final layer was put on top. This last phase involved an astonishing disc-shaped layer of quartzite, in some parts several centimetres thick, which covered not just the mouth of the tomb pit but the whole 60-m area under the mound (API & BMM 2013, vol. 3, plate 21). At the centre, directly over the pit itself, with its stepped ramp, was an area of yellowish earth, arranged in a radial pattern a bit like a dartboard. We have no idea what this huge investment in a thick quartzite layer could have signified. The sides of the pit were also covered in quartzite. Quartzite is found in southern Anhui (Gao *et al.* 2020) from where the rock must have come. The mound was then built over this layer, with several different colours of earth. These different features are evidence of plans and calculations for and execution of a massive burial.

At first sight, the mound brings to mind the large mounds found in eastern China that were often constructed over small house-shaped graves, built up from the ground. This might have been a possible source, but such graves are usually equipped with high-fired ceramics from the area. The structures and tomb contents do not correspond to Lord Bai's tomb. Instead, we should think about the mounds and much else as products or part of a band of communication along what I have elsewhere called the 'Eastern Seaboard' (Rawson 2023, chapter 1). There are two earlier phenomena that demonstrate that the eastern area of central China supported distinct cultural traits, which thrived and developed as a result of such lines of contact. The earliest phenomenon was the series and sequences of large middens, predominantly comprising shells, that were the product of long-lasting maritime cultures from at least 4000 BC (Zhao 2019). Further inland, completely different cultural groups worked jade and related hardstones into personal ornaments, above all for pieces of great symbolic significance, such as the coiled dragons of the Hongshan culture in Inner Mongolia and Liaoning and the *bi* and *cong* of the Liangzhu south of Shanghai (Rawson 2023, chapter 1). The spread of both material cultures by contact along the coastal area brought both middens and jade to different groups down the east, where people used them in their own separate styles. This Eastern Seaboard contact area did not disappear, and we can see it as one of the possible routes by which the notion of using mounds was carried south, in addition to swords, daggers, knives, mirrors, and horse tack also being brought along this route.

Conclusions

The peoples of these eastern regions were recognised as powerful groups of outsiders by the leaders of the central Chinese states and named the Eastern Yi, or sometimes even the Huai Yi. Their contacts spread north and south. They engaged in many battles with the Zhou and smaller states, exchanging weaponry, prisoners, and materials from the 7th to the 5th century BC. The tomb of Lord Bai was thus built at a time when different forces and influences were at work, possibly providing direct communication through the northeast with peoples from the Mongolian steppe and even South Siberia. Lord Bai's tomb illustrates one unusual example of the gradual combination of two important trends in this period of change and even upheaval. Local traditions, often embracing features of northern customs were entwined with the ancestral rituals and the use of Chinese characters, spreading from Zhou centres in the Yellow River basin, north and south into the lands infiltrated by the Rong, Di, and Yi, and south to the states of Chu and Wu. Moreover, the northern features of neatly placed attendants, some stone encasements and separate frames or chests for burial goods were maintained in the east and exploited in the Qi state (Shandong Provincial Museum 1977). Even more telling of this steppe connection is the burial of *ca.* 600 horses around another Qi tomb in Shandong (Zhang & Luo 1984), an attempt at a steppe-like display of wealth. However, even in this wider context of exchange and hybridity, Lord Bai's tomb was exceptional.

Note

1. Throughout this paper, the word China refers to the Loess Plateau and the central regions along the Yellow and Yangtze Rivers (in which Chinese characters were used) in the territory of today's People's Republic of China, though of course, no area called China existed in the period described.

References

Abbreviations

API   Anhui Provincial Institute of Cultural Relics and Archaeology
BMM   Bengbu Municipal Museum
BMI   Beijing Municipal Institute of Cultural Relics
FCA   Fengyang County Cultural Relics Administration
IMI   Inner Mongolia Autonomous Region Institute of Cultural Relics and Archaeology
NLM   Ningcheng County Liaozhongjing Museum

Secondary Sources

API & BMM (2010) The Spring and Autumn tomb no.1 at Shuangdun, Bengbu City, Anhui. *Chinese Archaeology* 10(1), 31–7.

API & BMM (2013) *The Tomb of Lord Bai of the Zhongli State*, 3 Volumes. Beijing, Cultural Relics Publishing House (in Chinese).

API & FCA (2018) *The Dadongguan and Bianzhuang Sites in Fengyang*, Anhui. Beijing, Sciences Press (in Chinese).

BMI (2010) *The Jundushan Cemeteries: Yuhuangmiao*, 4 Volumes. Beijing, Cultural Relics Publishing House (in Chinese).

Chugunov, K. V. (1998) Der Skythenzeitliche Kulturwandel in Tuva. *Eurasia Antiqua* 4, 273–307.

Chugunov, K. V. (2020) The Arzhan-2 funerary commemorative complex: Stages of function and internal chronology. In S. V. Pankova and St. J. Simpson (eds), *Masters of the Steppe: The Impact of the Scythians and Later Nomad Societies of Eurasia: Proceedings of a Conference Held at the British Museum, 27–29 October 2017*, 80–104. Oxford, Archaeopress.

Chugunov, K. V., Parzinger, H. & Nagler, A. (2010) *Der Skythenzeitliche Fürstenkurgan Aržan 2 in Tuva*. Mainz, Philipp von Zabern.

Gao, S., Yan, L., Chen, Z., Zhan, J., Jiao, L., Zhou, W. & Han, L. (2020) Preliminary division of the quartz vein belts in China. *China Non-Metallic Minerals Industry* 5, 5–9 (in Chinese).

Geel, B. van, Bokovenko, N. A., Burova, N. D., Chugunov, K. V., Dergachev, V. A., Dirksen, V. G., Kulkova, M., Nagler, A., Parzinger, H., Plicht, J. van der, Vasiliev, S. S. & Zaitseva, G. I. (2004) Climate change and the expansion of the Scythian culture after 850 BC: A hypothesis. *Journal of Archaeological Science* 31(12), 1735–42.

Gong, Y., Xie, H. & Hu, X. (1991) Explorations of the ruined Xue State City and excavations of its tombs. *Acta Archaeologica Sinica* 4, 449–95 (in Chinese).

Gosden, C., Hommel, P. N. & Nimura, C. (2019) Making mounds: Monuments in Eurasian prehistory. In T. Romankiewicz, M. Fernández-Götz, G. R. Lock & O. Buchsenschutz (eds), *Enclosing Space, Opening New Ground: Iron Age Studies from Scotland to Mainland Europe*, 141–52. Oxford, Oxbow Books.

Henan Provincial Institute of Cultural Relics and Archaeology (2004) *Tomb M1 at Hougudui, Gushi*. Zhengzhou, Elephant Press (in Chinese).

Huang, C., Zhao, S., Pang, J., Zhou, Q., Chen, S., Li, P., Mao, L. & Ding, M. (2003) Climatic aridity and the relocations of the Zhou culture in the southern Loess Plateau of China. *Climatic Change* 61, 361–78.

IMI & NLM (2009) *The Excavation of the Upper Xiajiadian Culture Site at Xiaoheishigou*. Beijing, Sciences Press (in Chinese).

Institute of Archaeology, Chinese Academy of Social Sciences (1996) *Shuangtuozi and Gangshang: Discovery and Study of Prehistoric Culture in the Liaodong Peninsula*. Beijing, Sciences Press (in Chinese).

Li, F. (2006) *Landscape and Power in Early China: The Crisis and Fall of the Western Zhou, 1045–771 BC*. Cambridge, Cambridge University Press.

Ou, T. (1984) The excavation of an early spring and autumn period tomb belonging to Huang-state Lord Meng and his spouse. *Archaeology* 4, 302–32.

Rawson, J. (2017) China and the steppe: Reception and resistance. *Antiquity* 91, 375–88.

Rawson, J. (2023) *Life and Afterlife in Ancient China*. London, Allen Lane.

Rawson, J., Chugunov, K. V., Grebnev, Y. & Huan, L. (2020) Chariotry and prone burials: Reassessing late Shang China's relationship with its northern neighbours. *Journal of World Prehistory* 33, 135–68.

Rawson, J., Huan, L., & Taylor, W. T. T. (2021) Seeking horses: Allies, clients and exchanges in the Zhou period (1045–221 BC). *Journal of World Prehistory* 34, 489–530.

Shandong Provincial Museum (1977) Excavation of Eastern Zhou tomb no. 1 with human sacrifices at Langjiazhuang, Linzi, Shandong. *Acta Archaeologica Sinica* 1, 73–104 (in Chinese).

Shaughnessy, E. L. (1999) Western Zhou history. In M. Loewe & E. L. Shaughnessy (eds), *The Cambridge History of Ancient China: From the Origins of Civilization to 221 BC*, 293–351. Cambridge, Cambridge University Press.

Stone, R. (2009) Diseases. A medical mystery in middle China. *Science* 324(5933), 1378–81.

Struck, J., Bliedtner, M., Strobel, P., Taylor, W., Biskop, S., Plessen, B., Klaes, B., Bittner, L., Jamsranjav, B., Salazar, G., Szidat, S., Brenning, A., Bazarradnaa, E., Glaser, B., Zech, M. & Zech, R. (2022) Central Mongolian lake sediments reveal new insights on climate change and equestrian empires in the Eastern Steppe. *Scientific Reports* 12, 2829.

Sun, G., Meharg, A., Li, G., Chen, Z., Yang, L., Chen, S. and Zhu, Y. (2016) Distribution of soil selenium in China is potentially controlled by deposition and volatilization? *Scientific Reports* 6, 20953.

Tong, E. (1986) On the crescent-shaped cultural-contact belt from Northeast to Southwest China. In Cultural Relics Publishing House Editorial Office (ed.), *Essays on Archaeology and Culture: The Thirtieth Anniversary of the Cultural Relics*

Publishing House, 17–43. Beijing, Cultural Relics Publishing House (in Chinese).

Wang, Y., Zhong, J., Lei, X. & Wang, Z. (2021) The excavation of the Western Zhou tomb M10 at Kongtougou, Qishan County, Shaanxi. *Archaeology* 9, 24–42.

Whitfield, S. (2020) Alfalfa, pasture and the horse in China: A review. *Quaderni di Studi Indo-Mediterranei* 12, 503–18.

Wu, W. & Zhang, Q. (1978) Spring and Autumn period human-sacrificial tombs of the Ju State at Dadian, Ju'nan, Shandong. *Acta Archaeologica Sinica* 3, 259–88 (in Chinese).

Xinjiang Institute of Cultural Relics and Archaeology (2015) *Collected Works on Archaeology and History in Altay Prefecture, Xinjiang*. Beijing, Cultural Relics Publishing House (in Chinese).

Yanshi Railway Cultural Heritage Team of Shandong Province (1988) *Eastern Zhou Tomb at Fenghuangling, Linyi*. Jinan, Shandong Qilu Press (in Chinese).

Zhang, C., Qi, R., Chang, H. & Yan W. (2018) The Gujun site of the Eastern Zhou period in Xingtang County, Hebei. *Archaeology* 7, 44–66 (in Chinese).

Zhang, X. & Luo, X. (1984) The excavation of the large horse pit no. 5 in the Linzi site of the Qi state. *Cultural Relics* 9, 14–9 (in Chinese).

Zhao, H., Liu, J., Yan, Y., Feng, G., Xue, Y., Liu, Y., Bai, C., Wang, X., Liang, X., Hou, K. & Wu, J. (2021) Excavation of Shang tombs at Houshi Village in Lishi, Shanxi, *Journal of National Museum of China* 12, 6–15 (in Chinese).

Zhao, L. (2019) Subsistence patterns associated with shell middens from the pre-Qin period in the coastal region of China. In C. Wu & B. V. Rolet (eds), *Prehistoric Maritime Cultures and Seafaring in East Asia*, 89–101. Singapore, Springer.

Zhu, Y. (2017) *The Archaeology of Northeast China Before the Han*. Beijing, Sciences Press (in Chinese).

7

Agricultural Places as Processes

Amy Bogaard

In this paper, I consider how process ontology can shed fresh light on long-term change and continuity in agriculture. Farming over the long term in western Eurasia is conventionally framed as a sequence of revolutions in technology and practice, from the Palaeolithic Broad Spectrum Revolution to the Neolithic, Secondary Products, Urban, medieval (early and late), 18th century, Green, and Super-Green revolutions. Recent archaeological work on the earlier 'revolutions', however, reveals their processual character, being not only extended in time but also reflecting complex, diffuse causality beyond any single 'technological' development. Perhaps the best way of apprehending these processes is to think of how particular agricultural 'places' persist and evolve over time, often featuring remarkable continuity, as well as distinctive episodes of local change. I develop this argument using two case studies: Çatalhöyük, south-central Turkey (late 8th–mid-6th millennium cal BC), and Knossos, Crete (early 7th–late 2nd millennium cal BC).

Keywords: Process, Time, Agriculture, Revolution, Place, Neolithic, Bronze Age

Introduction

Chris Gosden's *Social Being and Time* (1994) drew attention to different species of practice and time. If all practice produces time, there are as many kinds of time as there are of practice (Gosden 1994, 125). 'Habitual time' is the product of unconscious, routine action. 'Public time' refers to conscious interventions in the landscape – monuments, field systems, and so on – that seek to resolve contradictions and tensions arising from habitual time. A provocation of the book was to insist that habitual time must have priority if we are to grasp what is at stake in public time, though much archaeological research has focused on the latter (Gosden 1994, 126 & 184). Reading this book as a graduate student, exciting potential opened up for placing the habitual time of agriculture as a foundation for approaching the public time of monuments. That excitement drives this essay.

Time is the currency of process, and Gosden and Malafouris (2015) have drawn attention to the possibilities of process philosophy for archaeology, including its elastic timescales. This relevance is especially obvious when thinking about agricultural practice. Farming is keyed into the inherently processual nature of organisms – through their life-cycles, for example. But it also participates in longer term evolutionary processes arising from webs of interaction between crops, herds, humans, and the wider macro- and microscopic ecosystems of which they are a part (Bogaard *et al.* 2021a).

Agriculture is also relevant for process philosophy in another sense, outlined explicitly in the work of an early 20th-century process philosopher, Alfred North Whitehead. His philosophy of process – or 'philosophy of organism' – was a reaction to science that defined itself on the basis of fixed, preordained laws: recurrent, persistent observations that apparently enable accurate forecasting (Whitehead 1925; 1929). Whitehead pointed out that agriculture confronted past societies with a central paradox, and one crucial also to his philosophy (Whitehead 1933). Agriculture depends on habitual responses to recurrent seasonal cues: an autumn rain that soaks the soil ready for sowing, for example, or a spell of warm, dry weather that hastens crop ripening and the urgency of harvest. These recurrences of cause and effect could be thought of as the 'laws' of agriculture; ancient writers from Hesiod in his *Works and Days* (8th century BC) to Walter of Henley in *Husbandry* (written AD 1276–90) sought to capture this experience in their advice for sound farming practice. But the recurrence of seasonal changes from one year to the next and their apparent predictability belies what Whitehead (1933, 111) called 'the interweaving of law and capriciousness in the mystery of things'. Past experience, in the end, does not enable farmers to predict the future; no two years are ever the same, and the complexity of what transpires depends not only on the weather but also on a myriad of influences from the welter of substances and organisms involved, of which humans are only a part. The art of farming is one of hedging bets between what is likely to happen and what alternatives are anticipated based on experience (Halstead 2014). Such 'high grade reflection upon the course of events' (Whitehead 1933, 110) links back to Chris Gosden's public time. The establishment of farming in western Asia and Europe unleashed a wave of new forms of conspicuous, monumental construction, from the settlement mounds of western Asia and southeast Europe to the longhouse enclaves of central Europe and the funerary mounds and enclosures of the Atlantic façade. These new forms of public time insisted on recurrence and continuity in agricultural worlds that were fundamentally uncertain.

Finally, process ontology is useful for thinking through long-term sequences of agricultural practice. Farming over the long term in western Eurasia is conventionally framed as a series of revolutionary steps in technology, from the Upper Palaeolithic Broad Spectrum Revolution to the Neolithic, Secondary Products, Urban, medieval (early and late), 18th-century, later 20th-century Green, and now Super-Green revolutions. Two kinds of problem are raised by the rhetoric of agricultural 'revolutions'. One is that key developments are presumed to emerge at a specific time and place, in a fixed sequence and with a consistent causal background. As explored further below, archaeological evidence overturns these expectations for the first 'revolution' in the sequence – the Broad Spectrum Revolution (Flannery 1969). Subsequent 'revolutions' have similarly dissolved into complex processes in the light of archaeological evidence, as shown elsewhere for the Neolithic (*e.g.* Bogaard *et al.* 2021a; Weide *et al.* 2022), secondary products (*e.g.* Halstead 1995; Isaakidou 2006), urban (*e.g.* Bogaard *et al.* 2018; Styring *et al.* 2022), and early medieval revolutions (Hamerow *et al.* 2022; McKerracker & Hamerow 2022). Historical scrutiny of the 18th-century Agricultural Revolution has revealed both its deeper agricultural roots and *politically* 'revolutionary' inflection (Bloch 1931).

A second problem is that the rhetoric of 'revolutionary' steps in agriculture drives a progressivist narrative in which the past is irrelevant to the unfolding future and its new technology. This modernist perspective underlies how later 20th-century developments in agriculture, driven by mechanisation, fossil fuels, and chemical fertilisers, have been framed as a 'Green Revolution'. Lynn Meskell (2022) has shown how mid to later 20th-century developments in archaeological science – an 'atomic archaeology' of radiocarbon dating, geophysical prospection, surveillance, and other methodological innovations stemming from military technologies – emerged as spin-offs of WW2 and its Cold War aftermath (see also Meskell & LaPorte 2022). Agricultural practice was similarly in step with the development of 20th-century military technologies, perhaps most famously in the dual consequence of chemical fertilisers and chemical weapons as outcomes of artificial atmospheric nitrogen fixation through the Haber-Bosch process, initially scaled up during WW1 (Friedrich *et al.* 2017). It is now obvious that the spread of these technologies, together with high-yielding, 'elite' crop varieties engineered by seed companies to demand copious inputs of chemical fertiliser and water (Glaeser 2010), have accelerated the climate and food crisis (West *et al.* 2014). Current discussion of a Super-Green Revolution perpetuates the claim that technological improvements, such as the engineering of crop varieties with greater water and nitrogen use efficiencies, will be sufficient to address this crisis (Li *et al.* 2018).

In this paper, I first sketch out recent work that overturns simple notions of agricultural practice over the *longue durée* as a sequence of revolutionary step-changes. In the final section, I outline two site-based case studies in the investigation of agricultural places as processes. My aim is to advocate a shift from abstract revolutions to the evolution and affordances of places from the distant past and into the future.

The Broad Spectrum Revolution

The 'Broad Spectrum Revolution' (BSR) was a term coined by Flannery (1969) to characterise widening subsistence practices of growing Late Pleistocene forager populations as they spread into more 'marginal' settings. The food spectrum was characterised as broad relative to the narrower ranges of species targeted by agricultural communities. The BSR was primarily conceived, and subsequently discussed, in terms of 'high-turnover' faunal resources, such as shellfish, reptiles, and small mammals (*e.g.* Stiner 2001), though Flannery (1969, 79) supposed that plant foods were also involved. Weiss and colleagues attempted to bring plants into the discussion by associating small-grained grasses with low-ranked resources and wide resource spectra, though grasses represent only part of the potential plant food spectrum (Weiss *et al.* 2004).

Recent synthesis of western Asian archaeobotanical data before and after the agricultural transition demonstrates that, in fact, the plant food spectrum remained broad, rather than narrowing with the establishment of farming (Wallace *et al.* 2018). Flannery himself had anticipated this for the early Neolithic, pointing to what he regarded as evidence of a broad plant food spectrum at later Pre-Pottery Neolithic (*ca.* 7500–7000 BC) Ali Kosh in lowland southwest Iran (Flannery 1969, 86 & 88; Helbaek 1969). Subsequent interpretation of this assemblage, however, suggested that the Ali Kosh plant material likely derived at least partly from the burning of dung from herded caprines as fuel, and so reflected caprine rather than (purely) human diet (Miller 1996; Charles 2007). A caprine origin, for example, is likely for the small-seeded, clover-like legumes that dominate the early Ali Kosh assemblage and would survive small ruminant digestion (Wallace & Charles 2013). Flannery had specifically highlighted the abundance of these small-seeded legumes as evidence that the BSR continued into the earliest Neolithic (Flannery 1969, 86 & 88). Importantly, a broad plant food spectrum in the Neolithic can now be demonstrated *despite* the clear fuel/forage-derivation of some plants and deposits and continuing into the later Pottery Neolithic (7th–6th millennia BC) in western Asia (Wallace *et al.* 2018; Bogaard *et al.* 2021b).

Not only does current evidence of Neolithic plant use overturn the idea that farming necessarily diminished the breadth of the plant food spectrum; it is also likely that the breadth we can observe is, if anything, an underestimate. Virtually all plant food evidence from western Asia is preserved by charring (or carbonisation). This form of preservation biases the representation of plants towards those that were stored year-round (like crops, as opposed to other plants consumed in season), as well as towards plants used as/derived from fuel (Green 1982). For Neolithic Europe, bioarchaeological work on waterlogged lakeshore sites in the Alpine foreland (beginning in the late 5th millennium BC) has repeatedly demonstrated the broader spectrum of human plant use as evidenced by uncharred plants compared with exclusively charred evidence from dryland sites (*e.g.* Jacomet *et al.* 1989; Schibler *et al.* 1997). New, ongoing work on lakeshore sites of northern Greece and the southwest Balkans (6th–5th millennia BC) is similarly widening the true spectrum of Neolithic plant use in southeast Europe (Holguin *et al.* in press).

In sum, the BSR has general relevance beyond 'marginal' settings of the Late Pleistocene and as a precursor to farming. Equally, diverse plant use is a familiar practice of recent subsistence farmers (*e.g.* Halstead 2014) and is now advocated more widely for future food security (*e.g.* Food and Agriculture Organization of the United Nations 2016). As Flannery (1969, 74) himself conceded, the BSR 'continued long after cultivation had begun', and contemporary concerns over agrobiodiversity are prompting a renewed focus on widening food spectra, with particular interest in plants, insects, and other 'high-turnover' food types. All of this runs counter to Flannery's claim that the BSR was a development of marginal settings in the Late Pleistocene that became obsolete during the Neolithic.

Places as Processes

Perhaps the best way of apprehending agriculture-as-process is to think of how particular 'places' and landscapes persist and evolve over time, often featuring remarkable continuity, as well as episodes of (rapid or slow) change. I consider two site sequences – Neolithic to early Chalcolithic Çatalhöyük in central Anatolia and Neolithic to Bronze Age Knossos on Crete – that interweave continuity and change over millennia of agriculture at particular locales. Notably, local episodes of punctuated change do not easily align with classic 'revolutions'.

The archaeobotany of Çatalhöyük (*ca.* 7100–5500 BC), in the western Konya plain of south–central Turkey, problematises the classic 'revolutions' of agriculture in multiple ways. First, it provides a detailed case for continuation of a broad plant food spectrum through its entire 1500-year sequence (Bogaard *et al.* 2017; 2021b; Wallace *et al.* 2018). Second, Çatalhöyük upsets the orthodoxy that the Neolithic Revolution was based on a canonical set of 'founder crops', presumed to emerge in a single centre of origin (Zohary 1996). The Neolithic sequence presents abrupt local changes in favoured crops through time, some of them in conjunction with other shifts in food-related practice and material culture as in the mid-7th millennium BC, others in concatenation with other dimensions of daily life. Furthermore, local cultivators recruited some of their own crops. These comprised a tiny-seeded, oil-rich annual mustard (flixweed, *Descurainia sophia*) that was never formally 'domesticated', and an early form of Timopheev's wheat (*Triticum timopheevii* group) that lost its wild, shattering habit and wild grain shape/size through several centuries of the later Neolithic sequence (Bogaard *et al.* 2021b; Charles *et al.* 2021; Roushannafas *et al.* 2022).

The zooarchaeological spectrum is equally distinctive at Çatalhöyük. Key features include a strong emphasis on sheep- and (to a lesser extent) goat-herding throughout, veneration and hunting of wild cattle alongside gradual infiltration of domesticated cattle midway through the Neolithic sequence, and a lack of interest in boar hunting or pig keeping (Twiss *et al.* 2021; Wolfhagen *et al.* 2021). The core combination of sheep-herding and diverse cropping maintains agropastoral continuity throughout the sequence. Sheep dung would enrich cultivation plots, an effect picked up in stable nitrogen isotope values of crop remains and likely most marked near penning areas in and around the settlement (Stroud *et al.* 2021), while the arable landscape could supplement the sheep diet through grazing of stubble and early vegetative crops or foddering (*cf.* Halstead 2006). The botanical composition of preserved dung fuel deposits charts the scope of sheep grazing from the local riverine landscape to the wider steppe (Bogaard *et al.* 2021b). Increasingly 'steppic' signatures in sheep dung through time support zooarchaeological inferences that the scale of sheep herding increased through the Neolithic sequence (Twiss *et al.* 2021), while the arable catchment waxed and waned in response to the scale of the human community (Green *et al.* 2018; Stroud *et al.* 2021).

The so-called Neolithic Revolution in the western Konya plain was a creative, local process of shaping agricultural practices to the possibilities of particular landscapes (see also Baird *et al.* 2018). Çatalhöyük was positioned alongside a 1 km-wide, braided river channel system that presented a shifting gradient of wetter conditions within and near the channel belt to drier conditions as the groundwater table dropped off around 100 m from its edge (Ayala *et al.* 2022). In this low-rainfall, drought-prone region, a landscape of agricultural resilience developed around this 'sweet spot' of hydrological variability, with a diverse repertoire of crops, other food plants, herded and hunted fauna constructing different niches and affordances.

By the mid-7th millennium BC, some five centuries after Çatalhöyük was founded, the unique long-term agricultural sequence at Knossos, Crete, begins. Compared with the protracted process of agricultural beginnings on the Konya plain (Baird *et al.* 2018), agriculture at Knossos starts abruptly, with the establishment of a Neolithic crop-and-sheep 'package' through colonisation. But the local evolution of prehistoric agriculture in the Knossos valley took its own course, traceable over a remarkable *ca.* 5500-year-long period to the end of the Bronze Age. Recent work on the stable carbon and nitrogen isotope ecology of livestock and crops through these millennia reveals startling continuity, as well as distinctive local episodes of change (Isaakidou *et al.* 2022). The initial Neolithic 'hamlet' of perhaps 30 people is characterised by an isotopically homogeneous feeding ecology across sheep, goats, pigs, and cattle, kept locally at a small scale. Sheep track the nitrogen values of crops over the longer term, diverging from goats as the scale of herding expanded and reflecting particularly close integration with arable land, on which sheep periodically grazed and provided manure. A fall in manuring effects on stable nitrogen isotope ratios in crops and sheep in the later Neolithic is followed by a 'bounce' back upwards in the latest Neolithic phases. We hypothesise that this restoration of intensive manuring was enabled by increasing use of early cattle traction, as evidenced by a growing relative abundance of cattle and severity of traction-related pathologies (Isaakidou *et al.* 2022). This form of unspecialised cattle traction, featuring cows, 'stretched' the range of intensive land management as the Neolithic settlement expanded. Ultimately, however, the settlement expanded beyond the spatial range of intensive gardening, even with the help of cow-traction. A process of extensification – expansion of low-input cultivation and herding – continued from the latest phases of the Neolithic through the Bronze Age.

Expansion of the Neolithic settlement was the start of a protracted urbanisation process at Knossos, a multi-millennial development culminating in the *ca.* 100-hectare sprawl of the Neopalatial city in the mid-2nd millennium BC (Whitelaw *et al.* 2019). The stable isotope data chart a unique 'stop-motion' sequence of agropastoral change through this extended demographic process, overturning any simple agroecological expectations of an 'Urban Revolution' (Isaakidou *et al.* 2022). Moreover, a distinctive emphasis on pulse crops persists through the palatial period and beyond, presenting a thread of continuity from the initial Neolithic settlement to the end of the Bronze Age (Nitsch *et al.* 2019). While isotopic evidence suggests that pulses were grown in rotation with cereals through the Neolithic, however, by the Final Palatial (Mycenaean) period (15th–14th century BC) they were grown in separate 'garden' plots (Nitsch *et al.* 2019; Isaakidou *et al.* 2022). The Knossos sequence thus reveals a re-emergence of intensive gardening alongside expansive cereal production, folding together agricultural strategies conventionally assigned to different phases of agricultural development.

Concluding Thoughts

These sketches of long-term agricultural process at Çatalhöyük and Knossos reveal similarities and differences. A broad similarity is that the key Neolithic combination of sheep-herding and diverse cropping proved sufficiently flexible in both landscapes to be resilient on a millennial scale. But the divergences are equally clear. At Çatalhöyük a local combination of food-related practices fostered a distinctive social morphology of aggregated households, renewed with mostly incremental modifications over many human generations. Its location on a broad, braided river channel belt offered continual possibilities for adjusting and experimenting with different food strategies. Despite the presence of domesticated cattle through the later Neolithic–early Chalcolithic sequence, the radical potential for change through animal-powered traction is hardly perceptible at Çatalhöyük (see also Stroud *et al.* 2021). The much longer sequence at Knossos charts more dramatic changes through time: an abrupt Neolithic beginning in the Knossos valley is followed by an extended series of shifts in agropastoral strategy, some hinging on cattle traction, culminating in larger, urban scales of nucleation. The possibilities of mixed Neolithic agropastoralism shaped social life at Çatalhöyük and Knossos in unforeseen ways. The rhetoric of agricultural 'revolutions' only obscures these distinctive social sequences.

The present-day agricultural landscapes surrounding Çatalhöyük and Knossos are dramatically altered relative to the prehistoric sequences traced here. And yet, both retain a sense of the possibilities developed by early cultivators, despite large-scale landscape engineering towards industrial cereal farming (in the Konya Basin) and olive cultivation (in the Knossos valley). The backyard gardens of local villagers harbour diverse ranges of crops, now incorporating cultivars from the western hemisphere (such as runner beans and chilli peppers) alongside indigenous pulses (such as lentils, broad beans, peas) and herbs. This is where a broad plant food spectrum continues to thrive. As the climate and biodiversity crisis deepens, it remains to be seen how long it will take for the habitual practices of these gardeners to redirect the public discourse and 'monumental' landscapes of farming.

References

Ayala, G., Bogaard, A., Charles, M. & Wainwright, J. (2022) Resilience and adaptation of agricultural practice in Neolithic Çatalhöyük, Turkey. *World Archaeology.*

Baird, D., Fairbairn, A., Jenkins, E., Martin, L., Middleton, C., Pearson, J., Asouti, E., Edwards, Y., Kabukcu, C., Mustafaoğlu, G., Russell, N., Bar-Yosef, O., Jacobsen, G., Wu, X., Baker, A. & Elliott, S. (2018) Agricultural origins on the Anatolian plateau. *Proceedings of the National Academy of Sciences* 115(14), E3077.

Bloch, M. (1931) *Les Caractères originaux de l'Histoire rurale française.* Oslo, H. Aschehoug.

Bogaard, A., Filipović, D., Fairbairn, A., Green, L., Stroud, E., Fuller, D. & Charles, M. (2017) Agricultural innovation and resilience in a long-lived early farming community: The 1500-year sequence at Neolithic-early Chalcolithic Çatalhöyük, central Anatolia. *Anatolian Studies* 67, 1–28.

Bogaard, A., Styring, A., Whitlam, J., Fochesato, M. & Bowles, S. (2018) Farming, inequality and urbanization: A comparative analysis of late prehistoric northern Mesopotamia and south-west Germany. In T. A. Kohler & M. E. Smith (eds), *Ten Thousand Years of Inequality: The Archaeology of Wealth Differences*, 201–29. Tucson, University of Arizona Press.

Bogaard, A., Allaby, R., Arbuckle, B. S., Bendrey, R., Crowley, S., Cucchi, T., Denham, T., Frantz, L., Fuller, D., Gilbert, T., Karlsson, E., Manin, A., Marshall, F., Mueller, N., Peters, J., Stépanoff, C., Weide, A. & Larson, G. (2021a) Reconsidering domestication from a process archaeology perspective. *World Archaeology* 53, 56–77.

Bogaard, A., Charles, M., Filipović, D., Fuller, D.Q., Gonzalez Carretero, L., Green, L., Kabukcu, C., Stroud, E. & Vaiglova, P. (2021b) The archaeobotany of Çatalhöyük: Results from 2009-2017 excavations and final synthesis. In I. Hodder (ed.), *Peopling the Landscape of Çatalhöyük: Reports from the 2009-2017 Seasons*, 91–123. London, British Institute at Ankara.

Charles, M. (2007) East of Eden? A consideration of neolithic crop spectra in the eastern Fertile Crescent and beyond. In S. Colledge & J. Conolly (eds), *The Origins and Spread of Domestic Plant in Southwest Asia and Europe*, 37–51. Walnut Creek, CA, Left Coast Press.

Charles, M., Fuller, D. Q., Roushannafas, T. & Bogaard, A. (2021) An assessment of crop plant domestication traits at Çatalhöyük. In I. Hodder (ed.), *Humans and Environments of Çatalhöyük: Reports from the 2009-2017 Seasons*, 125–36. London, British Institute at Ankara.

Flannery, K. V. (1969) Origins and ecological effects of early domestication in Iran and the Near East. In P. J. Ucko & G. W. Dimbleby (eds), *The Domestication and Exploitation of Plants and Animals*, 73–100. London, Duckworth.

Food and Agriculture Organization of the United Nations (2016) *The State of Food and Agriculture.* Rome, FAO.

Friedrich, B., Hoffmann, D., Renn, J., Schmatlz, F. & Wolf, M. (eds) (2017) *One Hundred Years of Chemical Warfare: Research, Deployment, Consequences.* Cham, Springer.

Glaeser, B. (ed.) (2010) *The Green Revolution Revisited: Critique and Alternatives.* London, Routledge.

Gosden, C. (1994) *Social Being and Time.* Oxford, Blackwell.

Gosden, C. & Malafouris, L. (2015) Process archaeology (P-Arch). *World Archaeology* 47, 701–17.

Green, F. J. (1982) Problems of interpreting differentially preserved plant remains from excavations of medieval urban sites. In A. Hall & H. Kenwards (eds), *Environmental Archaeology in an Urban Context.* CBA Research Report 43, 40–46. London, Council for British Archaeology.

Green, L., Charles, M. & Bogaard, A. (2018) Exploring the agroecology of Neolithic Çatalhöyük, central Anatolia: An archaeobotanical approach to agricultural intensity based on functional ecological analysis of arable weed flora. *Paléorient* 44(2), 29–43.

Halstead, P. (1995) Plough and power: The economic and social significance of cultivation with the ox-drawn ard in the Mediterranean. *Bulletin on Sumerian Agriculture* 8, 11–22.

Halstead, P. (2006) Sheep in the garden: The integration of crop and livestock husbandry in early farming regimes of Greece and southern Europe. In D. Serjeantson & D. Field (eds) *Animals in the Neolithic of Britain and Europe*, 42–55. Oxford, Oxbow.

Halstead, P. (2014) *Two Oxen Ahead: Pre-Mechanised Farming in the Mediterranean.* Oxford, Wiley-Blackwell.

Hamerow, H., Zerl, T., Stroud, E. & Bogaard, A. (2022) The 'cerealisation' of the Rhineland: Extensification, crop rotation and the medieval 'agricultural revolution' in the longue durée. *Germania* 99(2021), 157–84.

Helbaek, H. (1969) Plant collecting, dry-farming and irrigation agriculture in prehistoric Deh Luran. In F. Hole, K. V. Flannery & J. V. Neely (eds), *Prehistory and Human Ecology of the Deh Luran Plain. An Early Village Sequence from Khuzistan.* Memoirs Museum 1, 383–426. Ann Arbor, University of Michigan.

Holguin, A., Antolín, F., Charles, M., Jesus, A., Martínez Grau, H., Soteras, R., Steiner, B., Stroud, E. & Bogaard, A. (in press) Archaeobotanical investigations at the mid-5th millennium BCE pile-dwelling site of Ploča, Mičov Grad, Lake Ohrid, North Macedonia. In A. Ballmer, A. Hafner & W. Tinner (eds), *Prehistoric Wetland Sites of Southern Europe. Archaeology, Chronology, Palaeoecology and Bioarchaeology.* Cham, Springer.

Isaakidou, V. (2006) Ploughing with cows: Knossos and the Secondary Products Revolution. In D. Serjeantson & D. Field (eds), *Animals in Neolithic Britain and Europe*, 95–112. Oxford, Oxbow.

Isaakidou, V., Halstead, P., Stroud, E., Sarpaki, A., Hatzaki, E., Nitsch, E. & Bogaard, A. (2022) Changing land use and political economy at Neolithic and Bronze Age Knossos, Crete: Stable carbon ($\delta^{13}C$) and nitrogen ($\delta^{15}N$) isotope analysis of charred crop grains and faunal bone collagen. *Proceedings of the Prehistoric Society.*

Jacomet, S., Brombacher, C. & Dick, M. (1989) *Archäobotanik am Zürichsee. Ackerbau, Sammelwirtschaft und Umwelt von neolithischen und bronzezeitlichen Seeufersiedlungen im Raum Zürich*. Zurich, Orell Füssli Verlag.

Li, S., Tian, Y., Wu, K., Ye, Y., Yu, J., Zhang, J., Liu, Q., Hu, M., Li, H., Tong, Y., Harberd N. P. & Fu, X. (2018) Modulating plant growth–metabolism coordination for sustainable agriculture. *Nature* 560, 595–600.

McKerracher, M. & Hamerow, H. (eds) (2022) *New Perspectives on the Medieval 'Agricultural Revolution': Crop, Stock and Furrow*. Liverpool, Liverpool University Press.

Meskell, L. (2022) Atomic archaeology: Italian innovation and American adventurism. *American Anthropologist*.

Meskell, L. & LaPorte, S. (2022) 'Your mysterious instruments': American devices and imperial designs in Cold War archaeology. *Journal of Field Archaeology* 47, 212–27.

Miller, N. F. (1996) Seed eaters of the ancient Near East: Human or herbivore? *Current Anthropology* 37, 521–8.

Nitsch, E. K., Jones, G., Sarpaki, A., Hald, M. M. & Bogaard, A. (2019) Farming practice and land management at Knossos, Crete: New insights from $\delta^{13}C$ and $\delta^{15}N$ analysis of Neolithic and Bronze Age crop remains. In D. Garcia, R. Orgeolet, M. Pomadère & J. Zurbach (eds), *Country in the City. Agricultural Functions of Protohistoric Urban Settlements (Aegean and Western Mediterranean)*, 152–68. Oxford, Archaeopress.

Roushannafas, T., Bogaard, A. & Charles, M. (2022) Geometric morphometrics sheds new light on the identification and domestication status of 'new glume wheat' at Neolithic Çatalhöyük. *Journal of Archaeological Science* 142, 105599.

Schibler, J., Hüster-Plogmann, H., Jacomet, S., Brombacher, C., Gross-Klee, E. & Rast-Eicher, A. (1997) *Ökonomie und Ökologie neolithischer und bronzezeitlicher Ufersiedlungen am Zürichsee*. Zurich, Kantonsarchäologie Zürich und Egg.

Stiner, M. C. (2001) Thirty years on the 'Broad Spectrum Revolution' and paleolithic demography. *Proceedings of the National Academy of Sciences* 98, 6993.

Stroud, E., Bogaard, A. & Charles, M. (2021) A stable isotope and functional weed ecology investigation into Chalcolithic cultivation practices in Central Anatolia: Çatalhöyük, Çamlıbel Tarlası and Kuruçay. *Journal of Archaeological Science: Reports* 38, 103010.

Styring, A. K., Carmona, C. U., Isaakidou, V., Karathanou, A., Nicholls, G. K., Sarpaki, A. & Bogaard, A. (2022) Urban form and scale shaped the agroecology of early 'cities' in northern Mesopotamia, the Aegean and central Europe. *Journal of Agrarian Change* 22(4), 831–54.

Twiss, K.C., Wolfhagen, J., Demirergi, G.A. & Mulville, J.A. (2021) Macromammals of Çatalhöyük: New practices and durable traditions. In I. Hodder (ed.), *Peopling the Landscape of Çatalhöyük: Reports from the 2009–2017 Seasons*, 145–80. London, British Institute at Ankara.

Wallace, M. & Charles, M. (2013) What goes in does not always come out: The impact of the ruminant digestive system of sheep on plant material, and its importance for the interpretation of dung-derived archaeobotanical assemblages. *Environmental Archaeology* 18, 18–30.

Wallace, M., Jones, G., Charles, M., Forster, E., Stillman, E., Bonhomme, V., Livarda, A., Osborne, C. P., Rees, M., Frenck, G. & Preece, C. (2018) Re-analysis of archaeobotanical remains from pre- and early agricultural sites provides no evidence for a narrowing of the wild plant food spectrum during the origins of agriculture in southwest Asia. *Vegetation History and Archaeobotany* 28, 449–63.

Weide, A., Green, L., Hodgson, J.G., Douché, C., Tengberg, M., Whitlam, J., Dovrat, G., Osem, Y. & Bogaard, A. 2022. A new functional ecological model reveals the nature of early plant management in southwest Asia. *Nature Plants* 8, 623–34.

Weiss, E., Wetterstrom, W., Nadel, D. & Bar-Yosef, O. (2004) The broad spectrum revisited: Evidence from plant remains. *Proceedings of the National Academy of Science* 101, 9551–5.

West, P. C., Gerber, J. S., Engstrom, P. M., Mueller, N. D., Brauman, K. A., Carlson, K. M., Cassidy, E. S., Johnston, M., MacDonald, G. K., Ray, D. K. & Siebert, S. (2014) Leverage points for improving global food security and the environment. *Science* 345, 325–8.

Whitehead, A. N. (1925) *Science and the Modern World*. London, Macmillan.

Whitehead, A. N. (1929) *Process and Reality: An Essay in Cosmology*. New York, Free Press.

Whitehead, A. N. (1933) *Adventures of Ideas*. New York, Free Press.

Whitelaw, T., Bredaki, M. & Vasilakis, A. 2019. The long-term dynamics of Knossos in context. In C. Mitsotaki, L. Tzedaki-Apostolaki & S. Giannadaki (eds), *Proceedings of the 12th International Congress of Cretan Studies, Heraklion, 21–25 September 2016*. Available at: <https://12iccs.proceedings.gr/el/proceedings/category/39/35/816> [accessed 06 October 2022].

Wolfhagen, J., Twiss, K. C., Mulville, J. A. & Demirergi, G. A. (2021) Examining caprine management and cattle domestication through biometric analyses at Çatalhöyük East (North and South Areas). In I. Hodder (ed.), *Peopling the Landscape of Çatalhöyük: Reports from the 2009–2017 Seasons*, 181–98. London, British Institute at Ankara.

Zohary, D. (1996) The mode of domestication of the founder crops in Southwest Asian agriculture. In D. R. Harris (ed.), *The Origins and Spread of Agriculture and Pastoralism in Eurasia*, 142–58. London, University College London Press.

8

A *Viereckschanze* in Oxfordshire, England? Enclosure and Memory at Marcham

Gary Lock and Sheila Raven

The extensive archaeological site at Marcham includes a Romano-British temple complex overlying a series of Iron Age ritual features. One of these is an intriguing and unusual barrel-shaped enclosure. Here we describe the enclosure based on geophysical and excavated evidence and place its interpretation within the context of European and British ritual enclosures.

Keywords: Marcham, Iron Age, Romano-British, *Viereckschanze*, Enclosure, Memory

Background

The extensive archaeological site at Marcham[1] has a long history of investigation. In summary – as a result of quarrying in the mid-19th century a late Romano-British and early Anglo-Saxon cemetery was discovered north of the A338 road to Wantage (Akerman 1865; Rolleston 1869; 1880), with further excavation and analysis of the numerous graves continuing just after the First World War (Dudley Buxton 1921). More recent archaeological evaluation has confirmed the extent of the cemetery together with Saxon occupation (Cass & Ford 2008). In the 1930s, acting on the advice of Arthur Evans, Bradford and Goodchild (1939) excavated within the garden of the Noah's Ark Inn just to the south of the road and cemetery. Here they uncovered a series of Iron Age features, including an unusual 'shrine' with votive offerings, directly overlain by a stone-built Romano-British temple (Bradford & Goodchild 1939). Parts of the temple were explored further by Harding in the 1960s (Harding 1987). The surrounds of the temple, including the southern half of the Inn's garden and the adjoining field known as Trendles, were the focus of fieldwork by Hingley (1982; 1985), who pulled together the aerial photographic evidence and conducted surface survey. The importance of Hingley's work was in showing the extent of the archaeological remains, consisting of a series of buildings including a circular stone-built arena to the east of the temple, which he confirmed by excavation.

This accumulating evidence together with conflicting aspects of the site's interpretation established Marcham as a site of national importance and, encouraged by the generosity of the landowners, initial large scale geophysical survey and then excavations were conducted for 11 seasons starting in 2001, co-directed by Gary Lock and Chris Gosden.[2] These acted as training excavations for full- and part-time students at the University of Oxford as well as for many other people. In total, nearly 1000 were trained over the 11 years. With over 50 trenches excavated and a large amount of material culture recovered, the post-excavation and publication tasks are considerable and ongoing. Much of it so far has been supported by the National Lottery Fund, thus offering opportunities for local volunteers to be involved.

The extent of the evidence at Marcham was indicated by the geophysical survey (Kamash *et al.* 2010, fig. 2), although the core of the site comprises the areas in the garden of the Noah's Ark and in Trendles to the east. The circular arena has a stone wall *ca.* 40 m in diameter surrounded to the east, south, and west by a low earthen bank. To the north, excavated features suggest a possible platform giving rise to its designation as a semi-amphitheatre, a cross between an amphitheatre and a theatre, a form known in central and northern Gaul (Golvin 1988; for a fuller discussion of the Marcham amphitheatre see Kamash *et al.* 2010, 106). The main arena entrance is to the west and this and a series of other buildings and shrines were probably connected by pathways to the temple eastern entrance through the *temenos* wall. To the east of the arena were a series of small circular and rectangular ditched Iron Age enclosures, with much material culture in the ditches and internal pits. These were clearly not domestic and appear to have a ritual function, as does a nearby defined area of intercutting pits which seems to represent a formal area of burial, containing both human and unusual animal remains, plus a wealth of Iron Age finds.

In summary, during the Roman period Marcham was a thriving rural sanctuary located in a busy landscape of villas across the Vale of the White Horse, close to a Roman road running to Wantage and close to the larger Roman settlement at Abingdon. While the Roman remains make Marcham an important site, it is their relationship with the extensive underlying prehistoric features which have placed it at the centre of discussion regarding continuity and change between the two periods. In this short paper, we will continue with this theme by concentrating on aspects of the temple and nearby prehistoric features, including a large Iron Age barrel-shaped enclosure.

Continuity and Change

Bradford and Goodchild (1939) argued for direct continuity between the Iron Age and Roman remains in both time and use as a persistent religious/ritual/sacred place. Their interpretation was based on three main lines of argument. Firstly, that many of the Iron Age features, specifically postholes and ditches, were intentionally back filled in the early Roman period when the temple was built, suggesting that they were surviving holes or hollows in the ground. Secondly, the spatial correlation of the Roman and Iron Age structures, specifically the 'rotunda' being constructed over the ditched 'shrine', and the temple itself being built over a post-built circular building. Thirdly, the early dating of the construction of the temple to *ca.* AD 80–90, based on coins found under the first phase path leading from the eastern entrance.

These arguments and counter arguments have been fully explored by Kamash and colleagues who have downplayed the need for direct continuity and shown that 'commemoration of the past' can be a different form of relating to past people, places, and events (Kamash *et al.* 2010). This happens by the significance of a place being passed down through social memory aided by the physical remnants of past activities surviving on the ground such as partially filled pits and ditches. One remarkable example of this is the small Roman shrine constructed over 60 m directly east of the temple *temenos* and linked to the eastern entrance by a long and aligned pathway. The shrine comprised a small square stone-built enclosure surrounding a 'ritual shaft' containing offerings including 4th-century coins and cockerel bones. This Roman shaft was created with great effort inside the fill of an earlier pit, which was flanked by two other large Iron Age pits, equidistant from the central pit and its later 'ritual' shaft. The fact that it would have been easier to dig the shaft straight into the ground rather than elaborately supporting a cylindrical shape inside the larger pit with layers of stones, may suggest the central pit was already significant and may also have been prehistoric. The later construction of the shaft inside it destroyed any dating evidence or indication of its original use. Close to this shrine was an unusual Iron Age crouched burial with animal skulls arranged round the grave also indicating the early ritual importance of this part of the site. The centre line of the temple runs through the *temenos* gateway and straight through the centre of this Roman shrine and its underlying pit, showing that the alignment of the temple was based on these pits, the outer two of which were still hollows in the ground when the temple was constructed based on Roman pottery in their upper fills and clear weathering cones (Kamash *et al.* 2010, fig. 4).

Excavations at Marcham subsequent to Kamash and colleagues' 2010 paper have yielded further evidence relevant to the discussions of continuity and the relationships in time, space, and function between Iron Age and Roman features. Central to this are three enclosures within and close to the Noah's Ark garden, and it is to these that we now turn. To the north is the rectangular *temenos* enclosure of the temple, shown by excavation to have been stone-built but now mostly robbed. South-west of the *temenos* wall, and very close to it, is a large barrel-shaped enclosure, while to the south-east are two sides of a smaller rectangular enclosure (Fig. 8.1). Based on the aerial photographic and geophysical evidence, the south-eastern enclosure appears to have been the earliest as it is cut and overlain by both of the other two thus destroying its northern and eastern sides and much of its interior. The early date was confirmed by excavation which revealed 273 sherds of an almost complete Middle Bronze Age bucket urn laid near the base of a deep V-shaped ditch. Intersecting with this was a circular, small Iron Age enclosure, reminiscent of Bradford and Goodchild's Iron Age 'shrine' in their Site C (1939, fig. 5), which started as a ring of pits later joined together by a ditch. The latter partially cut into the filled Bronze Age ditch, indicating recognition of its existence (Kamash *et al.* 2010, fig. 3). The line of the Bronze Age ditch was then recut in the Roman period.

The barrel-shaped enclosure is next in the sequence and details have been established through aerial photography, geophysics, and the excavation of three trenches, although it can be seen from Figure 8.1 that the enclosure is not complete as the River Ock and an associated reed bed cuts off the south-western half of it (Wintle 2007; Wintle *et al.* 2010). This is probably due to the Ock having changed its course and moved eastwards, as indicated by a series of dry oxbow lakes a few hundred metres to the south visible on aerial photographs. If the known shape of the enclosure is extrapolated, it could have originally been approximately

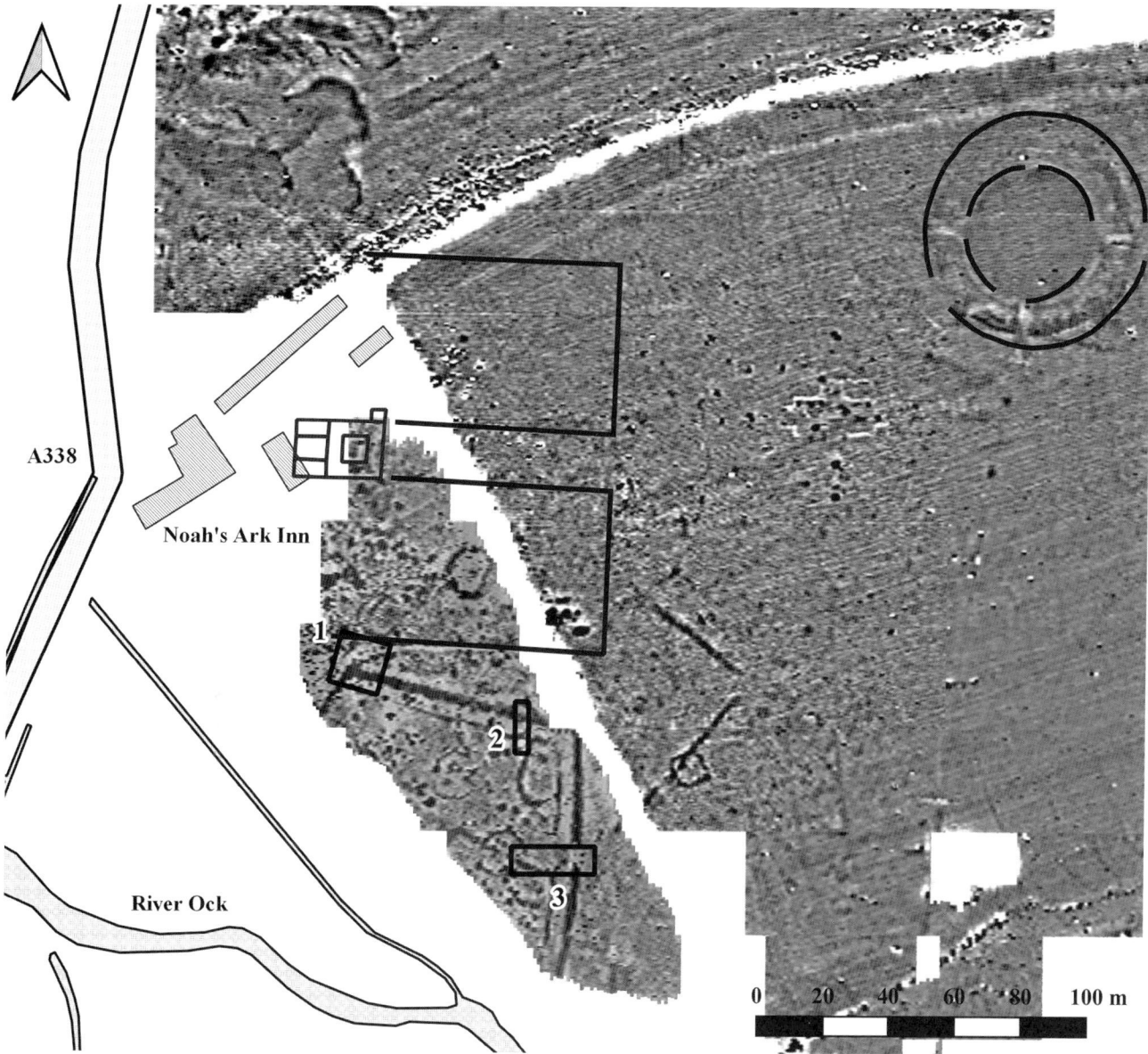

Figure 8.1 Geophysics of the western half of the Marcham site showing the three enclosures. The Roman temple and temenos is to the north, the barrel-shaped enclosure south of it with excavation trenches marked, and the two sides of the rectangular enclosure to the east (Courtesy of William Wintle; Contains OS data © Crown copyright and database right 2022. Geophysical survey by William Wintle & Tony Johnson).

140 by 85 m and enclosed 1.2 ha. The main enclosure ditch was a stepped V-shape cut through bedrock a maximum of *ca.* 5 m wide and 2 m deep with middle Iron Age pottery (some of it very early Middle Iron Age, possibly 400–350 BC), contained within its primary fills in Trench 2. The bedrock here is unstable and collapsed into the ditch, which was then re-cut to the same shape but smaller. The geophysics clearly shows a narrower linear feature running parallel with the ditch *ca.* 4 m on its inside, which excavation revealed to be 0.65 m wide and 0.95 m deep with steep sides and a flat base (Fig. 8.2). This is probably a slot to hold a wooden inner revetment of the bank and it contained similar pottery to the outer ditch.

The enclosure had an entrance through the eastern side, the southern half of which was excavated (Fig. 8.3). The terminus end of the ditch was surrounded by an arc of four shallow pits/postholes, all with steep sides and flat bottoms measuring *ca.* 1 m in diameter and a maximum of 0.3 m deep. This unusual configuration was clearly related to the entrance architecture and may have held posts that were fixed together rather than being embedded in the ground. The geophysics and excavation show a series of small, ditched enclosures of various shapes within the enclosure, together with a series of pits/large postholes. Although it is difficult to make sense of many of these features, within the area excavated a row of at least five postholes running from

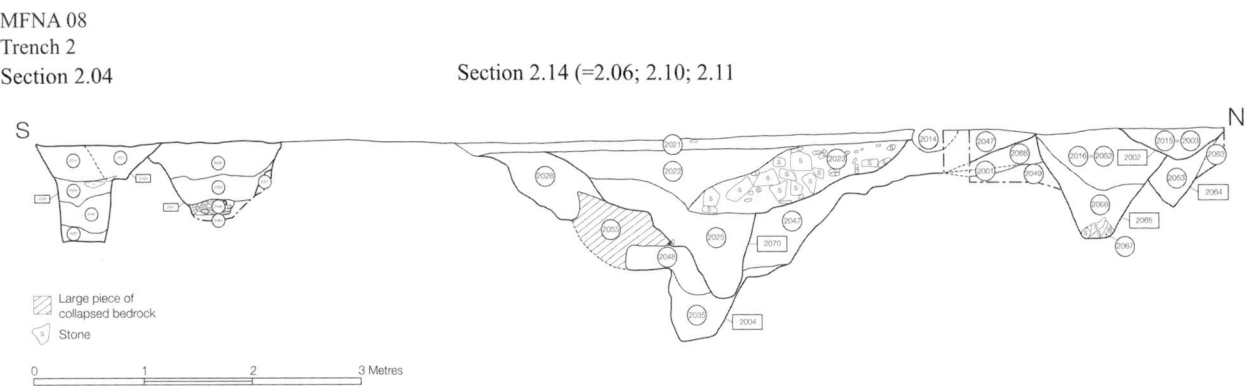

Figure 8.2 Section drawn across Trench 2 showing cut features (from left to right): the Iron Age bank revetment ditch, a ring ditch, the barrel-shaped enclosure ditch, and the ring ditch.

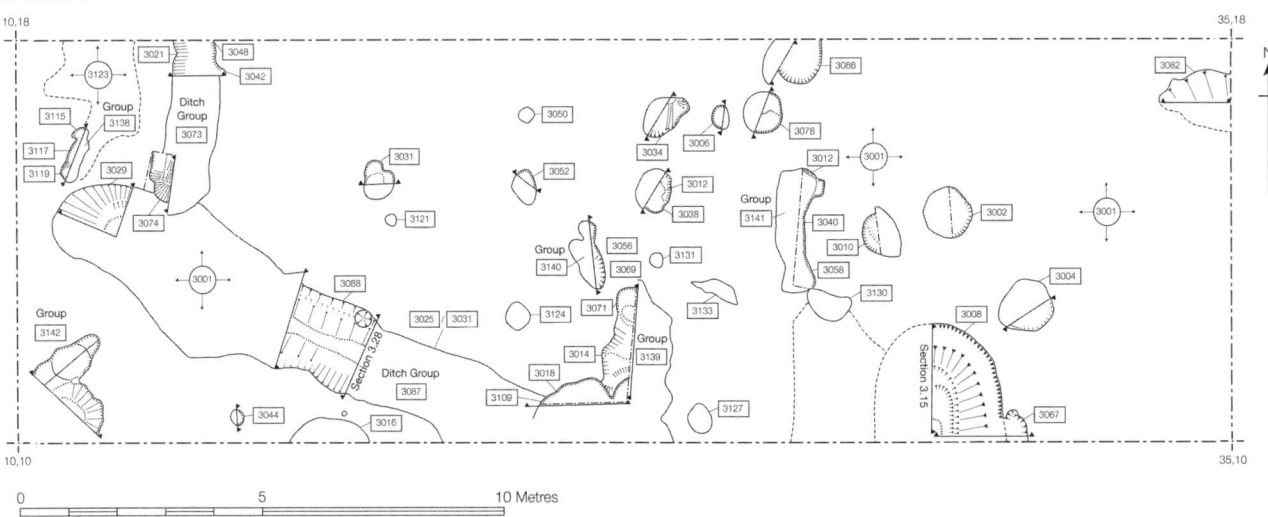

Figure 8.3 Trench 3 showing the barrel-shaped enclosure entrance with internal cut features. The main ditch terminal is to the bottom right with the four shallow postholes in an arc around it. Inside the enclosure are the central line of postholes and ditches representing smaller internal enclosures.

the entrance westwards seems to have divided the interior into two. These were spaced *ca.* 2.8 m apart and averaged 0.7 m diameter and 0.45 m deep. Other groups of postholes running north–south may have formed radial divisions from the central line. The whole area of the enclosure was a complex of inter-cutting ditches, some of which predated the barrel-shaped enclosure, based on stratigraphy and Early Iron Age/early Middle Iron Age pottery in upper fills.

The *temenos* wall of the temple enclosure at its southwestern end runs very close to the northern ditch of the barrel-shaped Iron Age enclosure. The upper fills of the Iron Age ditch in this area, and only in this area, produced a very large assemblage of 1st century AD Roman artefacts including late Iron Age and very early Roman pottery, two asses (one of Domitian and the other a Claudian copy), and four early brooches (a Nauheim derivative, a T-brooch, a penannular brooch, and a Hod Hill type). It may be that a break in the *temenos* wall at this point was another entrance giving access to the area of the earlier enclosure, the ditch of which was still evident as a hollow in the ground.

A *Viereckschanze*?

It seems clear from the evidence that the barrel-shaped enclosure is not domestic, as its characteristics are so different to those of many enclosed farmsteads/settlements known in this area (for example Lambrick & Robinson 2009). Caution is advisable when applying the term *viereckschanze* however. Bradley (2003; 2005) amongst others has shown how its use can lead to confusion and complexity based on morphological evidence alone (at its simplest level it means 'rectangular enclosure'). Also, the term, its use, and interpretation, has become more confused rather than clarified, through increasing amounts of excavated evidence. Interpretation (again at its simplest level as a cult/ritual enclosure) can raise increasingly complex issues of what is meant by prehistoric ritual and how it can be recognised.

It seems useful here to briefly review some of the discussion concerning the term *viereckschanze*. Their recognition and original interpretation are usually accredited to Schwartz (1959), whose excavations of one of the two enclosures at Holzhausen, Bavaria, in the 1950s established it as the 'type site' for the 'Celtic cult enclosure model' (Buchsenschutz 1989). The usually cited characteristics of *viereckschanzen* in their core area of Bavaria, Bohemia, and southern Germany are rectangular, sometimes trapezoidal, enclosures defined by one or more bank and ditch, although not large enough to be defensive, with typically 0.4–1.0 ha enclosed area and 100 × 80 m maximum dimensions, often located away from settlements with a single entrance typically on the eastern side. Details established through excavation, again for the 'type site', include enhancement of the entrance with various posts and timber structures, a lack of material culture generally and domestic features in particular, one or two simple wooden square or rectangular buildings represented by postholes and/or bedding trenches and interpreted as temples, and pits/shafts, some up to 20–30 m deep, for the deposition of votive offerings. Subsequent to work at Holzhausen and the establishment of the model, some other excavated sites have reproduced some of these elements, for example Dornstadt-Tomerdingen also in southern Germany (Zürn & Fischer 1991).

With increasing evidence from excavations the 'type site' model within the core area has become challenged due to the difference between sacred and profane activities within the enclosures becoming blurred. Venclová (1993) uses evidence from excavated sites to argue against simplifying and generalising models showing that the evidence is much more complex and difficult to interpret. This is explored by Wieland (1999) who provides a detailed discussion of *viereckschanzen* in southern Germany and the Czech Republic, with a catalogue of 24 excavated sites and plans of their internal structures. Their changing interpretation is reviewed by Rieckhoff and Biel (2001) who suggest that many are more likely to be high status settlements with elements of ritual activity rather than purely ritual enclosures, a view supported by Von Nicolai (2009) and Webster (1995).

Feeding into the doubts regarding the ritual function of *viereckschanzen* is the recognition of enclosures which appear more convincingly to have been used for ritual purposes, for example the cult centres of northern France such as Gournay-sur-Aronde and Ribemont-sur-Ancre. Here the form of the enclosure itself, the internal structures, and the deposition of large amounts of material including weaponry and human remains indicate a different sort of site to *viereckschanzen* (Arcelin & Brunaux 2003). In Brunaux's (1988) overview of Celtic religious beliefs and practices he recognises four types of sanctuaries of which *viereckschanzen* and Belgic, *i.e.* the Gournay and Ribemont type, are two. He questions why the former show so few offerings and why they never developed into Gallo-Roman sanctuaries as the Belgic ones often did and suggests that *viereckschanzen* are either a special site limited to southern Germany and nearby, or they are not cult sites but places for assemblies partly of cult character. A further complicating factor is the increasing number of Iron Age enclosures known as shown by Haselgrove (2007) for northern France where, since 1992, 263 Late Iron Age sites have been excavated of which 70% are settlements with the rest being sanctuaries, cemeteries, and cult enclosures. The complication as shown here is that some sites categorised as domestic settlements show evidence of ritual practices (Haselgrove 2007, 501).

There are well-established and long-recognised religious/ritual structures in Britain (Cunliffe 2005, 561) including the shrines within Danebury, South Cadbury, and Maiden Castle hillforts, at Lancing Down and Stanstead and the temples of Hayling Island and Heathrow. The identification of purely ritual enclosures of the Iron Age, however, is more difficult. There are often-cited possibilities, with one of the more remarkable being the Phase III enclosure at Fison Way, Thetford (Gregory 1991), and others being the large ritual enclosure with burials replaced by a Romano-British temple at St Albans, Folly Lane (Niblett 1999) and also possibly Gosbecks Farm, Colchester (Crummy 1980). As well as the often-cited examples it must be recognised that with the increasing numbers of rectilinear enclosures known from aerial photography there is a possibility that some of them are of a ritual character. It is interesting to note that nearly 50 years ago, Harding (1974, 111) drew attention to a series of enclosures 'possibly related' to *viereckschanzen* within Berkshire and not far from Marcham. These are a list of nearly 30 sites produced by Cotton (1961), together with discussion of continental parallels, and three ditched enclosures at Long Wittenham possibly containing ritual shafts (Harding 1974, pl. XIX).

What's in a Name?

It can be argued that square/rectangular Iron Age enclosures form a continuum in terms of the ratio of sacred/profane activities carried out within them based on excavated evidence; even so, a defining characteristic is enclosure. The *viereckschanze* debate has shown how difficult it is to characterise the function of an enclosure based on morphology alone, although it is accepted that the act of enclosing is a key element of Iron Age sacred sites alongside natural locations such as groves and watery places (Webster 1995). Delimiting a specific space is seen by Brunaux (1988, 7) as an 'essential feature' in creating a sacred place of 'assembly, passage and encounter', with entrances having a particular importance where 'the holiest forces meet the pressure of the profane' (Brunaux 1988, 27), hence their often-preferred orientation to the east, like the Marcham big enclosure and the later temenos enclosure, together with special treatment. Webster's (1995, 447) survey of Classical accounts of Celtic sacred places concludes that, despite any detail on construction and use of enclosures as cult sites, 'the formalized structuring of space, incorporating enclosing works, is suggested by most texts' reinforcing that 'enclosure was the primary and indispensable feature of Celtic cult sites'.

With the Marcham enclosure in mind, we can suggest that a second important factor in identifying ritual enclosures is 'sequence' as argued by Bradley (2003; 2005). He explores the confusion surrounding the evidence for ritual and the differences between ritual and profane practices and suggests that ritualisation is a process dependent on social context. So rather than isolated individual ritual acts, they fit together temporally and become an historical process through a 'prescribed liturgy' being passed on by social memory. The three enclosures at Marcham form such a sequence, the Bronze Age rectangular enclosure replaced by the Iron Age barrel-shaped one, which is then replaced by the Romano-British temple *temenos*. The earlier Middle Iron Age features, pits, and circular ditches, show a tighter sequence of activity during this period culminating in the construction of the large enclosure. The Iron Age/Roman continuity ending in a temple is not unusual with well-known examples being Hayling Island (King & Soffe 1998) and Thetford (Gregory 1991).

The question posed in the title of this paper is to a large extent 'tongue in cheek' and intended only to draw attention to the unusual characteristics of the Marcham barrel-shaped enclosure. Whether or not we call it a *viereckschanze* or just a ritual enclosure, there is no doubt that it is a significant feature within the Marcham Iron Age landscape. The shape of the enclosure is unusual, as are its internal bank revetment and its entrance architecture, together with the arrangement of posts leading from the eastern entrance and dividing the internal space. The lack of evidence for roundhouses or domestic activity suggests a ritual function and this can be seen as just one such element of the wider Marcham site, which has a series of small Iron Age enclosures and pit groups all appearing not to be domestic. As we have seen above, sequence is crucial here with Bronze Age features being within and around the barrel-shaped Iron Age enclosure, which is then replaced by the Roman temple complex. The deposition of very early Roman material in the upper fills of the enclosure's northern ditch is a powerful sign that the enclosure was still visible and maintained continuing significance. This adds to the arguments for continuity of this sacred place first put forward by Bradford

and Goodchild (1939) and supports their suggested early date for the Roman temple in the second half of the 1st century AD. Kamash and colleagues added to the evidence for a continuity based on the persistence of landscape features and the strength of social memory (Kamash *et al.* 2010), and the evidence from the barrel-shaped enclosure strengthens this further.

Acknowledgements

Thanks to William Wintle and Peter Davenport for producing the figures and to Lisa Brown for looking at the pottery and providing spot dates.

Notes

1. The site is sometimes known as Frilford which is the neighbouring parish, and sometimes as Marcham/Frilford.
2. Interim reports for the 11 seasons appeared in South Midlands Archaeology (32–42) and are available at: <https://www.arch.ox.ac.uk/vale-and-ridgeway-project> [accessed January 2022].

References

Akerman, J. (1865) Report of excavations in an ancient cemetery at Frilford, near Abingdon, Berks. *Proceedings of the Society of Antiquaries* 2(3), 136–9.

Arcelin, P. & Brunaux, J-L. (2003) Cultes et sanctuaries en France à l'âge du fer. *Gallia* 60, 1–268.

Bradford, J. & Goodchild, R. (1939) Excavation at Frilford, Berks, 1937-8. *Oxoniensia* 4, 1–70.

Bradley, R. (2003) A life less ordinary: The ritualization of the domestic sphere in later prehistoric Europe. *Cambridge Archaeological Journal* 13(1), 5–23.

Bradley, R. (2005) *Ritual and Domestic Life in Prehistoric Europe*. London, Routledge.

Brunaux, J-L. (1988) *The Celtic Gauls: Gods, Rites and Sanctuaries*. London, Seaby.

Buchsenschutz, O. (1989) Introduction. In O. Buchsenschutz & L. Olivier (eds), *Les Viereckschanzen et les enceintes quadrilaterales en Europe Celtique. Actes du IXe colloque de l'A.F.E.A.F., Chateaudun, 16–19 mai 1985. Association Française pour l'Etude de l'Age du Fer*, 4–9. Paris, Editions Errance.

Cass, S. & Ford, S. (2008) *Millett's Farm, Frilford, Oxfordshire. An archaeological evaluation for Hills Waste Solutions Ltd.* Unpublished report, Thames Valley Archaeological Services Ltd.

Cotton, M. (1961) Robin Hood's Arbour and rectilinear enclosures in Berkshire. *Berkshire Archaeological Journal* 59, 1–35.

Crummy, P. (1980) The temples of Roman Colchester. In W. J. Rodwell (ed.), *Temples, Churches and Religion in Roman Britain*, 6–10. Oxford, BAR British Series 77.

Cunliffe, B. (2005) *Iron Age Communities in Britain*. Fourth edition. London, Routledge.

Dudley Buxton, L. (1921) Excavations at Frilford. *Antiquaries Journal* 1, 87–97.

Golvin, A. L. (1988) *L'Amphithéâtre romain: essai sur la théorisation de sa forme et de ses fonctions*. Paris, Diffusion de Boccard.

Gregory, T. (1991) *Excavation in Thetford 1980–1982. Fison Way*. Norwich, East Anglian Archaeology 53.

Harding, D. (1974) *The Iron Age in Lowland Britain*. London, Routledge and Kegan Paul.

Harding, D. (1987) *Excavations in Oxfordshire 1964-66*. Department of Archaeology Occasional Paper No. 15. Edinburgh, University of Edinburgh.

Haselgrove, C. (2007) The age of enclosure: Later Iron Age settlement and society in Northern France. In C. Haselgrove & T. Moore (eds), *The Later Iron Age in Britain and Beyond*, 492–522. Oxford, Oxbow Books.

Hingley, R. (1982) Recent discoveries of the Roman Period at the Noah's Ark Inn, Frilford, South Oxfordshire. *Britannia* 13, 305–9.

Hingley, R. (1985) Location, function and status: A Romano-British 'Religious Complex' at the Noah's Ark Inn, Frilford. *Oxford Journal of Archaeology* 4, 201–14.

Kamash, Z., Gosden, C. & Lock, G. (2010) Continuity and religious practices in Roman Britain: The case of the Rural Religious Complex at Marcham/Frilford, Oxfordshire. *Britannia* 41, 95–125.

King, A. & Soffe, G. (1998) Internal organisation and deposition at the Iron Age temple on Hayling Island. *Proceedings of the Hampshire Field Club and Archaeological Society* 53, 35–47.

Lambrick, G. & Robinson, M. (2009) *The Thames Through Time. The Archaeology of the Gravel Terraces of the Upper and Middle Thames. The Thames Valley in Later Prehistory: 1500 BC–AD 50*. Thames Valley Landscapes Monograph 29. Oxford, Oxford Archaeology.

Niblett, R. (1999) *The Excavation of a Ceremonial Site at Folly Lane, Verulamium*. London, Britannia Monograph 14.

Rieckhoff, S. & Biel, J. (2001) *Die Kelten in Deutschland*. Stuttgart: Konrad Theiss.

Rolleston, G. (1869) Researches and excavations carried on in an ancient cemetery at Frilford, near Abingdon, Berks in the years 1867, 1868. *Archaeologia* 42, 417–85.

Rolleston, G. (1880) Further researches in an Anglo-Saxon cemetery at Frilford, with remarks on the northern limit of Anglo-Saxon cremation in England. *Archaeologia* 45, 405–10.

Schwartz, K. (1959) *Atlas der Spätkeltischen Viereckschanzen Bayerns*. Munich.

Venclová, N. (1993) Celtic shrines in Europe: A sceptical approach. *Oxford Journal of Archaeology* 12(1), 55–66.

Von Nicolai, C. (2009) La question des viereckschanzen d'Allemagne du sud revisitée. In I. Bertrand, A. Duval, J. Gomez de Soto & P. Maguer (eds), *Habitats et paysages ruraux en gaule et regards sur d'autres regions du monde celtique. Actes du XXXIe colloque international de AFEAF*, 245–80. Chauvigney, Mémoire XXXV.

Webster, J. (1995) Sanctuaries and sacred places. In M. J. Green (ed.), *The Celtic World*, 445–64. London, Routledge.

Wieland, G. (1999) *Keltische Viereckschanzen: einem Rätsel auf der Spur*. Stuttgart, Theiss.

Wintle, W. (2007) *Geophysical survey of the Noah's Ark Inn, Marcham, Oxfordshire, July and August 2007*. Unpublished report.

Wintle, W., Boyer, K., Hawes, J. & Levick, P. (2010) Geophysical survey at the Noah's Ark Inn, Marcham. *South Midlands Archaeology* 40, 75–80.

Zürn, H. & Fischer, F. (1991) *Die Keltische Viereckschanzen von Tomerdingen*. Stuttgart, Konrad Theiss.

9

A Landscape's Memory: The Long-term Impact of Proto-industrial Salt Extraction in the Seille Valley in France

Laurent Olivier

In the last decades, archaeologists have extensively studied what people have done to landscape and how landscapes have changed under such anthropic pressure. But how about what landscapes have done to people, in their own deep temporality? In this chapter, a case study is presented on the eco-human history of the upper Seille valley in Eastern France. It shows how archaeological events, which occurred nearly three millennia ago, have constantly shaped human activity and behaviour – including archaeological investigation itself.

Keywords: Iron Age, Briquetage, Archaeology of Salt, Geoarchaeology, Memory

A New Approach to Landscapes

In his recent research, our good friend Chris Gosden (Gosden *et al.* 2021) examined the history of English landscapes and the influence the changes they have undergone have had on the ways in which humans inhabited the land over long periods of time. He showed how a landscape's material constraints created structures that favoured the appearance and perpetuation of specific collective identities: that very unique way of being and feeling English, we should say. Chris demonstrated not only how the dynamics of a landscape interact with the changes affected by human intervention, but how, in a broader sense, the landscape itself actually creates history, how it is a full-fledged actor in our 'eco-human' history, far more than a simple backdrop to it.

One of Chris's greatest talents lies in his ability to sense the import of emerging fields of research, which is to say to see how new approaches lead to a new way of viewing the past and alter our relation to the vanished civilisations that are the subject of archaeological research. In effect, in the course of the last 30 years, the massive upheavals of the Great Acceleration of the Anthropocene have radically undermined our perception of 'landscapes as objects' (Steffen *et al.* 2011). The dramatic changes in Holocene climate, which had remained more or less stable for 11,700 years, have led us to understand that humans – their social structures and their activities – do not constitute an external force impacting some environmental system from which they are distinct (Oldfield & Steffen 2004), but rather that we are an integral part of the 'Earth system', bound to it and subject to its influence. As Bruno Latour (2017, 56) pointed out, 'the enormous change in thinking that we have to accept is that… the Earth plays an active, powerful role, that it no longer serves as mere background to human pursuits'.

Chris knows that I have long viewed archaeology as a means by which to uncover the *material memory* of places and things, whether it comes down to us in the form of human creations or land formations (Olivier 2011). In this modest tribute to Chris's work, I would like to present the principal findings of field research that was carried out between 2001 and 2017 on a particular landscape, that of the upper Seille valley in the Lorraine region of eastern France (Olivier 2015) (Fig. 9.1). The multidisciplinary work in which some 20 European research institutions and laboratories engaged totally changed our perception of the object of our study. We were compelled to recognise that the material remains uncovered in this valley did not allow us to reconstitute past human activity, as we had initially thought they would. We learned rather that they bore witness to the *effects* that thousands of years of human occupation had had on this particular natural setting, and, in return, the impact this environment's *response* to human's activities had on them.

Figure 9.1 Excavation at Marsal (Moselle) 'la Digue' in 2012 (Photo: Projet Briquetage de la Seille).

The *Briquetage* of the Seille Valley

In the upper Seille valley, between the villages of Salonnes and Marsal in the Moselle department in France, lie abundant vestiges of extensive salt extraction operations that date back to the last millennium before the Common Era (Fig. 9.2). These piles of terra cotta production waste – which the locals refer to as the *briquetage de la Seille* – whose total volume is estimated to be 1.5 million m³ (Gouhier 2021, table 5), were formed with the debris from salt moulds and ovens that were cast out to the periphery of Iron Age salt extraction sites. They can form veritable *tells* 30 ft high, such as those found today in the villages of Marsal and Moyenvic.

Scattered over a distance of more than six miles, these mounds of *briquetage de la Seille* first became an object of study in the early 18th century. The construction of fortifications ordered by Louis XIV around the city of Marsal uncovered huge masses of *briquetis* or *briquetage* upon which the village rested (La Sauvagère 1740). In the 19th century, numerous hypotheses were proposed to explain the nature of such a large-scale enterprise: some attributed it to the Romans, others to barbarian settlements of Antiquity, and others still to hunters of the Reindeer Age (Ancelon 1879).

Following the defeat of France in 1870, Germany annexed this part of Lorraine. In 1901, the German director of the Museum of Metz, Johan Baptist Keune (1858–1937), carried out extensive excavations that allowed him to establish that the *briquetage de la Seille* represented the remains of the process of heating brine for the large-scale extraction of salt. He was further able to date this activity back to the Iron Age (Keune 1901). On the basis of core samples, Jean-Paul Bertaux (1976), an archaeologist working with the Archaeological Services of the Lorraine Region in the 1970's, was able to elaborate a typology of the vats in which the brine was heated and the moulds in which blocks of salt were formed.

Salt Extraction Operations

Research carried out in the first two decades of the 21st century has allowed us to establish the chronology of these proto-historical operations. During an initial period, corresponding to the 6th century BC, a series of ateliers sprung up over an area of some 2 to 12 acres of the alluvial plain of the upper Seille valley. The ateliers were set up near saltwater springs whose brine emerged from dissolving blocks of rock salt located between 200 and 250 ft underground.

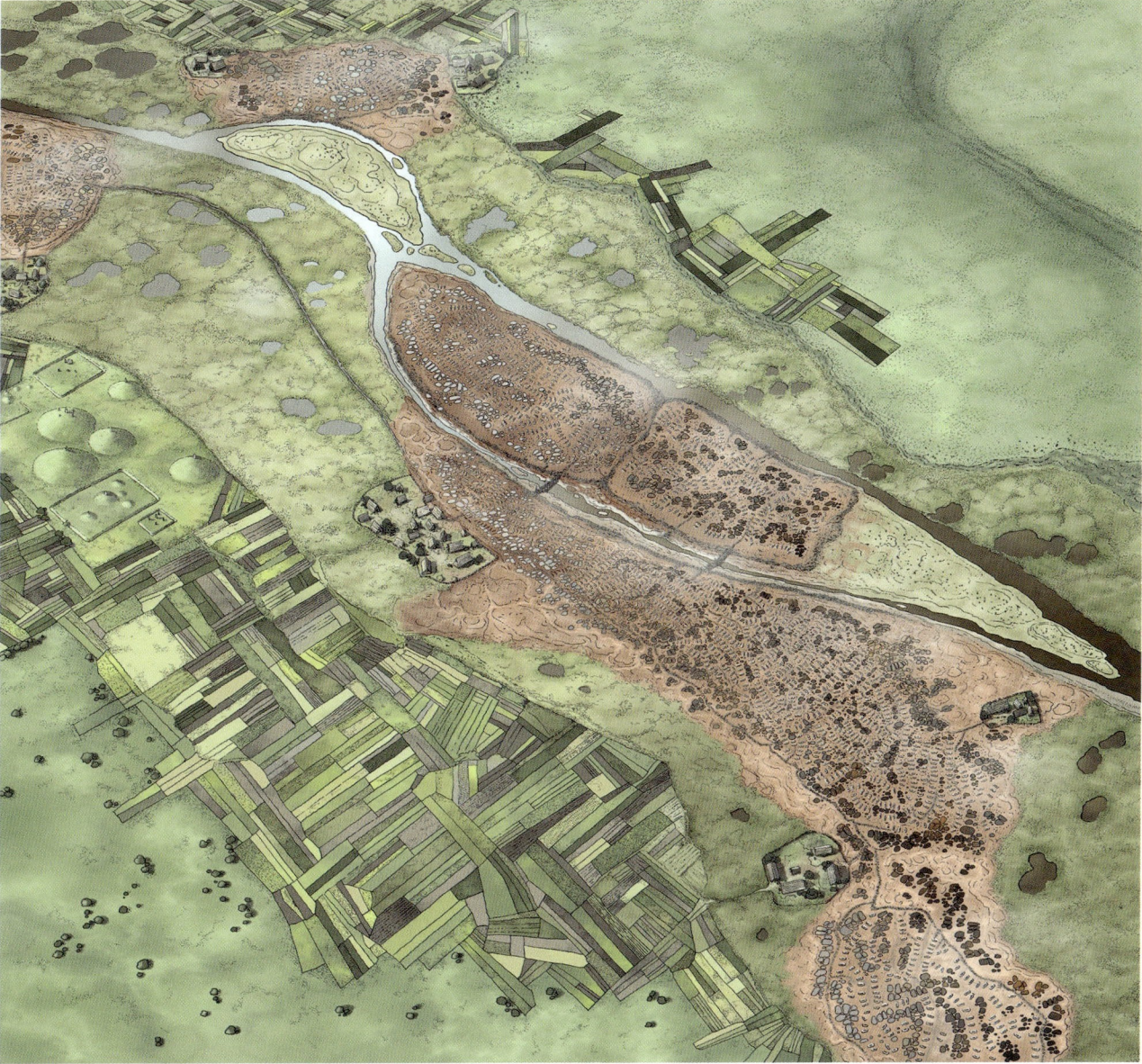

Figure 9.2 A reconstruction of the Iron Age salt industry in the upper Seille valley during the 6th century BC, around Marsal (Moselle) (Drawing: Loïc Derrien).

Surface water passing through geological rifts would come in contact with the 'roof' of a layer of rock salt while pressure from layers of sediment would cause the now salted water to rise to the surface, where it would pool periodically.

The Iron Age salt workers harvested the rising saltwater in all likelihood through 'saltwater wells'. Salt would be extracted from the brine through a series of processes, the first of which involved concentrating and decanting. Somewhere in the course of these operations, the brine was very likely enriched by leaching saline soil, after which it was brought to saturation by passing it through a series of flat-bottomed basins lined with impermeable clay.

During the second phase of the operation, the now-saturated brine would be taken to the atelier and poured into smaller-sized basins, undoubtedly to ensure that there would be a constant supply of it when, in this new stage, it was heated. It would be transferred to large, terracotta, flat-bottomed vats that would be loaded into ovens and heated at near-boiling temperatures. The salt crystals that formed and accumulated in the solution would be regularly removed and set aside to drain and dry.

In the third and final phase of the operation, the salt crystals were pressed together and dehydrated in porous terracotta salt moulds made from templates. Then the moulds would be broken and salt blocks of standard weight and volume would be removed, ready to be shipped. Salt production has been estimated at several thousand tonnes per year, far more than the local population required, and thus clearly destined for large populations elsewhere.

Impact on the Environment

The work was seasonal, most likely carried out in the summer, and it would leave behind large quantities of waste (Fig. 9.2). The leaching of saline soil generated mounds of sediment; removing the salt moulds from the ovens involved breaking the terracotta grills to which they had been attached to ensure their stability during the heating process; breaking moulds to remove the blocks of salt produced masses of unusable fragments; and firing both the brine and the blocks of salt left behind ashes and other combustion residue that had to be cleaned out regularly. All this waste was thrown outside the ateliers, often into the numerous channels of the Seille, gradually impeding its flow.

Undesired consequences first appeared before the end of the 6th century BC, and they proved to be recurrent. The ateliers in the alluvial plain were more and more frequently inundated, and the receding waters would leave behind layers of clayey soil. Iron Age salt workers responded both by using some of the accumulated waste to elevate the ground on which their equipment rested and by covering the area around the basins in which the brine was decanted with branches so that they could walk about dry-footed. The erosion that resulted from these periods of inundation was obviously favoured by the intensive deforestation of hills whose trees provided the wood charcoal that fired the ovens. By the time the Iron Age salt workers began responding to the situation that they had helped produce, the noble species of trees, such as oak and beech, had long disappeared from the limestone plateaus where they grew. The workers were left to burn local wetland species that were not nearly as good for generating heat, after having first reduced the size of their ovens to decrease the need for combustible matter.

By the first decades of the 5th century BC, the situation had become untenable. Little by little, ateliers were deserted, following which, salt production was relocated and concentrated on three sites that had the great advantage of remaining permanently dry. In the course of this second period of the *briquetage de la Seille*, which reached its peak at the end of the Late Iron Age, during the last two centuries before the Common Era, salt extraction, which had earlier been intensive, became 'proto-industrial' – its overall production was 10 times that of the preceding Early Iron Age period. Ever-increasing quantities of waste continued to be dumped into the channels of the Seille, irrepressibly turning what was once an industrial landscape into swampland.

The Roman conquest would put an end to these terribly polluting industrial-scale salt works, probably during the first decades of the first century AD, but small-scale operations in the valley persisted throughout the Middle Ages and into the Modern Era. Operations ceased entirely during the Industrial Revolution, in the mid-19th century, when salt extraction from the Seille valley was supplanted by direct extraction from rock salt in mines in Varangéville, in the Meurthe-et-Moselle department of eastern France.

A Never-Ending Tale

Through geo-archaeological core sampling and bio-stratigraphic samples, we have been able to situate these archaeological events in an environmental history of the Seille valley that dates back to the Last Ice Age (Riddiford 2009). Beginning around 13,000 BC, the evolution of the Seille valley can be divided into three major periods of erosion that alternate with periods of sedimentation. While these periods correspond closely to various climate regimes of the Holocene, the last phase of sediment accumulation that begins around 2050 BC clearly shows the impact of human activity. Alluvial deposits from this time are filled with both *briquetage* residue and combustion waste, and the rate of sedimentation appears to have been abnormally high, four times that of waterways in neighbouring areas. For the first time, the alluvial plain of the Seille was covered with argilous clay, burying what were once river terraces and reducing the incline of the plain by a factor of four, virtually eliminating the possibility for waters to run off.

More detailed study shows that a phase of river erosion followed the time during which the Early Iron Age ateliers were deserted and that it, in turn, was quickly followed by a long period of sedimentation, this latter phase beginning at the end of the Middle Iron Age and continuing mainly during the Early Middle Ages. Sediment continued accumulating from the 11th to the 14th centuries AD, and marshes reached their highpoint between the 17th and 18th centuries AD. The upper valley of the Seille was by then an immense swamp, six miles long. Occasional operations to drain the area gradually reduced this marshy expanse, but such efforts were long opposed by the army, which saw in the inundated plains of the Seille an added element of defence for the fortifications in Marsal. Standing alone in the midst of swampland, the town was impregnable (Fig. 9.3).

A Contaminated Legacy

Near the end of the 1780s, the winds of revolution were blowing across the kingdom of France, fanned by two years of poor harvests. In the provinces, people drew up lists of grievances that they sent to the king to inform him of their plight and the measures they felt needed to be undertaken to remedy the situation. Whereas elsewhere throughout the kingdom people were outraged by the heavy taxes they had to pay, the injustices of the bureaucracy, and the extravagant privileges enjoyed by a select few, the principal preoccupation of the inhabitants of the valley lay elsewhere: it was the Seille.

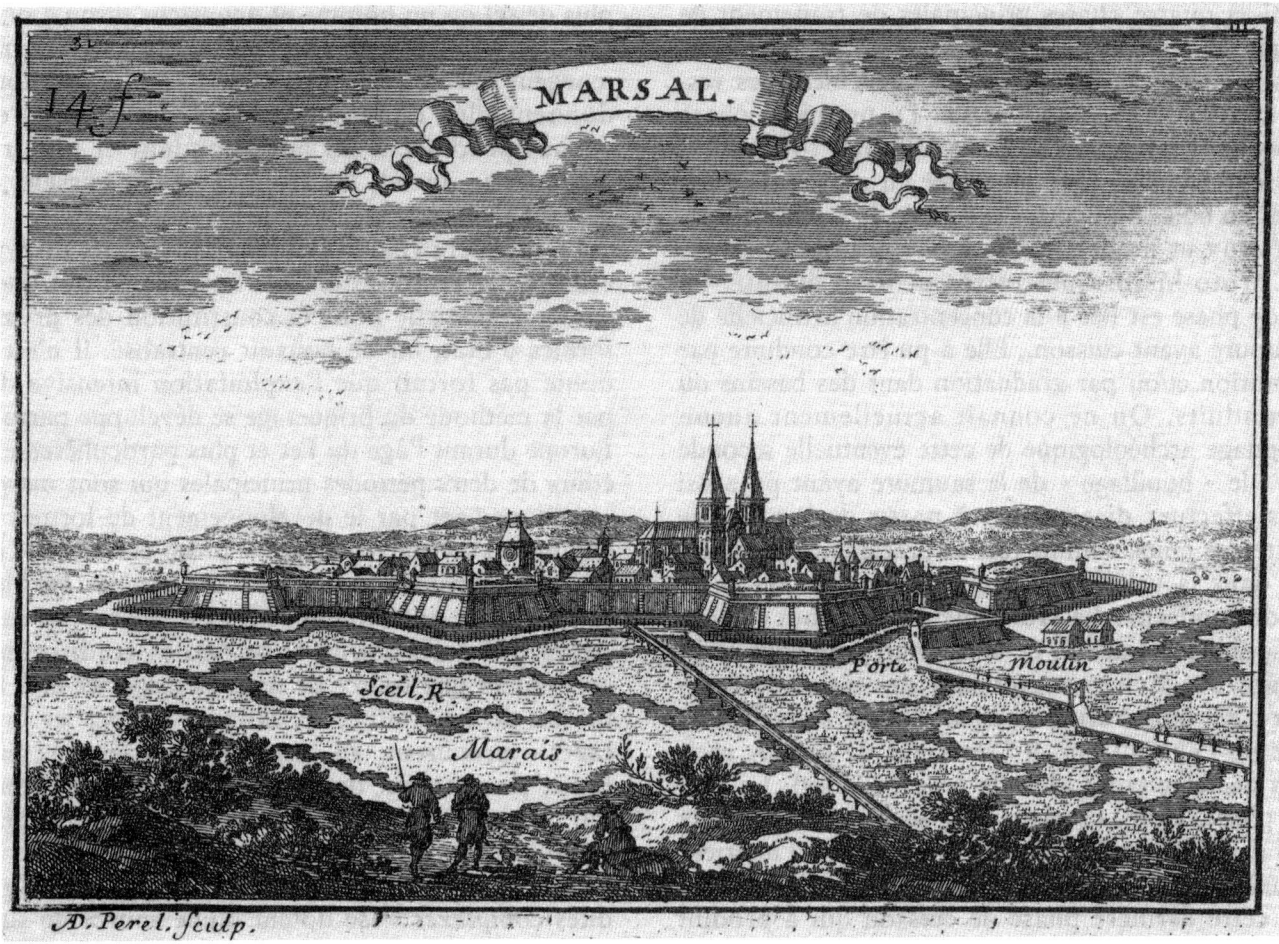

Figure 9.3 The city of Marsal in the late 17th century AD.

The residents of Marsal wanted the river to at long last be channelled, and the swamps into which it emptied to be drained. People living upstream, in Blanche-Église and Mulcey, wanted the mills on the Seille to be removed because they 'impeded the flow of the river, whose incline had virtually disappeared and whose bed was in many places extremely narrow and filled in, so that the overflow travelled large distances and sat there for a considerable time, ruining our lovely, vast prairies'; above all, 'so contaminated [was] the air' that for more than a century no one in Blanche-Église had lived to see sixty. Everyone in the valley called for a decrease in salt extraction activities because of 'the huge amount of wood it required', which had 'more than quadrupled the price of firewood' and forced people 'to devastate our forests' at the risk of incurring huge fines (Archives of the Moselle department, list of grievances of Blanche-Église, Marsal, Mulcey, and Saint-Médard).

The swamps of the Seille valley were a persistent source of public health problems. Life expectancy there was 10 years shorter than in neighbouring villages. Physicians noted that the population was especially prone to goitres and cretinism. In the 1830s, Dr Étienne-Auguste Ancelon (1806–86) frequently found himself a powerless witness to devastating episodes of typhoid fever that claimed the lives of his patients in a matter of weeks, and often just days. These epidemics would regularly recur in the summer, with the return of 'fever season'. As Dr Ancelon explained, the swamp waters would dry out in the sun and the wind, exposing large expanses of decomposing organisms and plant residue whose 'miasma' would be released into the air (Ancelon 1847). It would not be until the German annexation of 1871 that this stretch of the upper Seille valley was completely channelled, with the swampland eventually, finally, drained at the beginning of the 20th century.

What should we take away from this curious story? It shows how interference with a hydrological system that began in the 6th century BC with Iron Age salt workers of the Hallstatt period had massive repercussions that would reach their highpoint more than 2000 years later, contributing to a social and political crisis of unprecedented proportion that led to the French Revolution. Draining the swamps that were the result of intensive Iron Age salt production was not just a matter of public health, but also, and above all, of social justice. The 18th-century inhabitants of the

Seille valley had it right: the continual extraction of salt in an environment that had been rendered uninhabitable was the root cause of their misfortune.

The Iron Age Lives On

This research experience profoundly altered our grasp of human occupation of the upper Seille valley. Above all, it led us to see what archaeology can bring to the understanding of the past. It was a diachronic analysis, covering a period of thousands of years, that revealed the dynamics of what had been inherited from the past, which would have otherwise gone unnoticed. Without this chronological depth, we would not have become aware of the influence that constraints from distant times had had upon the various periods of the history of the Seille valley. These periods would rather have appeared to us as various human pursuits carried out in an inert and virtually immutable environment. Diachronic analysis gives this history its meaning, one which is not afforded when archaeology is 'cut into slices', with scholars limiting their research to specific time frames.

But that is not all. Once we began viewing archaeology as the study of the material impact of human enterprise on places and things, there appeared before our eyes, in this upper Seille valley, a particularly striking phenomenon: material creations from the Iron Age continue to leave their mark on the present-times that follow, including our own. We, the archaeologists and scholars of the 21st century, realised that we had been subject as well to the effects of Iron Age material heritage. We had been able to carry out vast geophysical studies and perform excavations on Early Iron Age ateliers precisely because these sites had been deserted *en masse* during the 5th century BC, never to be reoccupied because the valley had been turned into swampland. Conversely, we had not been able to study the major production areas of the Late Iron Age that had, over various periods of history, become home to urban environments. It was impossible to carry out large-scale, open-air excavations to the levels we needed because access lay buried under yards of urban sediment.

The salt workers of the Iron Age were not the only ones to have an impact on our work. It was possible to carry out excavations in the Seille valley, following the line of research opened by Keune and Bertaux, only because 19th-century engineers had partially drained it. We moreover concentrated our work in Marsal rather than in Moyenvic, even though the latter presents the largest accumulations of *briquetage*, for Marsal did not hold a strategic position commanding the valley and had not been repeatedly destroyed, as Moyenvic was in 1944. Whereas the atmosphere in Marsal was friendly and cheery, the people in Moyenvic were gloomy and withdrawn; they saw us as intruders disturbing their retirement. But our veritable partner in this undertaking, the one which gave us, or failed to give us, access to the information we sought was the Seille, the ancient *Salia*, this hybrid of environmental dynamics and human impact. We, too, were caught in this entanglement of multiple heritages that continue to affect us, often in hidden ways.

As long as archaeology continues to attempt to recreate the past 'as it was', it will be condemning itself to the pursuit of an unattainable object. It is an impossible undertaking, for the remains of the past can only come down to us as they have been transformed by their post-history, a truth of which the Seille valley offers glowing proof. Archaeologists will find themselves forced to seek theoretical models in disciplines foreign to their own, such as anthropology, philosophy, and sociology, if they fail to see that archaeological matter speaks not so much of the past alone, but rather of its transmission through transformation, its *transformission*, one might say. Archaeology deals far more with the *temporality* of the past than with its *historicity*. Archaeological matter is living material memory from which we have not been separated.

La vie est belle, Chris!

References

Ancelon, E. A. (1847) *Mémoire sur les fièvres typhoïdes périodiquement développées par les émanations de l'étang de Lindre-Basse*. Nancy, veuve Raibois et Cie.

Ancelon, E. A. (1879) Sur les habitations lacustres connues sous le nom de briquetages de la Seille. *Bulletin de la Société d'Anthropologie de Paris* 2, 620–38.

Bertaux, J.-P. (1976) L'archéologie du sel en Lorraine: 'Le Briquetage de la Seille' (état actuel des recherches). In J.-P. Millotte, A. Thévenin & B. Chertier (eds), *Livret guide de l'excursion A7 Champagne, Lorraine, Alsace, Franche-Comté. 9ème Congrès de l'Union Internationale des Sciences Préhistoriques et Protohistoriques*, 64–79. Nice, Éditions du CNRS.

Gosden, C., Green, C., Cooper, A., Creswell, M., Donnelly, V., Franconi, T., Glyde, R., Kamash, Z., Mallet, S., Morley, L., Stansbie, D. & ten Harkel, L. (2021) *English Landscape and Identities: Investigating Landscape Change from 1500 BC to AD 1086*. Oxford, Oxford University Press.

Gouhier, B. (2021) *Un anthroposystème du sel dans la vallée supérieure de la Seille depuis la Protohistoire*. Unpublished PhD thesis, Université de Tours.

Keune, J. B. (1901) Das Briquetage im oberen Seillethal. *Jahrbuch der Gesellschaft für lothringische Geschichte und Altertumskunde* XIII, 366–94.

Latour, B. (2017) *Où atterrir? Comment s'orienter en politique*. Paris, La Découverte.

Oldfield, F. & Steffen, W. (2004) The Earth System. In W. Steffen (ed.), *Global Change and the Earth System: A Planet under Pressure*. Berlin, Springer.

Olivier, L. (2011) *The Dark Abyss of Time: Archaeology and Memory*. Landham, MD, AltaMira Press.

Olivier, L. (2015) Iron Age proto-industrial salt mining in the Seille River Valley (France): Production methods and social organization of labour. In A. Danielisova & M. Fernandez-Götz (eds), *Persistent Economic Ways of Living: Production, Distribution and Consumption in Late Prehistory and Early History*, 69–89. Budapest, Archaeolingua.

Riddiford, N. (2009) *Palaeoenvironmental history and the public understanding of scientific research: Prehistoric salt production in the Seille Valley, northeast France*. Unpublished PhD thesis, University of London.

le Royer de la Sauvagère, F. (1740) *Recherches sur la nature et l'étendue de ce qui s'appelle communément Briquetage de Marsal avec un abrégé de l'histoire de cette ville, et une description de quelques antiquités qui se trouve à Tarquimpole*. Paris, Charles-Antoine Jombert.

Steffen, W., Grinevald, J., Crutzen, P. J. & McNeill, J. R. (2011) The Anthropocene: Conceptual and historical perspectives. *Philosophical Transactions of the Royal Society A* 369, 842–67.

10

Taking, Using, and Giving Back Again: The Deposition of Living Matter in Ancient Europe

Richard Bradley

The deposition of hoards began among hunter-gatherers in Europe, and similar practices can be traced from the Mesolithic period to the Viking Age. Because so much attention is paid to finds of metalwork in dry land and water, it is seldom appreciated that the same places can contain animal bones and other food remains. The latter may be older than the collections of metal objects, but they may also be more recent. Did the significance of depositional sites change over time, or is there one explanation that applies to all the evidence? Sacrifices are usually thought to consist of living matter, but in an animistic system of belief, portable artefacts would also have been imbued with life. Both derived their power from supernatural sources, and eventually they had to be returned. In this respect, metal objects were equivalent to the deposits of animal bones.

Keywords: Animal Bones, Metalwork, Hoards, Deposits, Sacrifice, Fertility, Animism

Llyn Cerrig Bach to Skedemosse

Chris Gosden has always favoured thought experiments. This paper follows his example and develops from a problem he encountered in studying Celtic Art. The starting point is the remarkable collection of metal artefacts from Llyn Cerrig Bach in Wales which he discussed in a book with Duncan Garrow (Garrow & Gosden 2012, 157). The assemblage was found under difficult conditions during the Second World War, and the excavators could not decide whether it should be interpreted as a hoard since individual items might have been made and deposited at different times. Even more problematical was the collection of animal bones recovered with the metalwork. The faunal remains are poorly dated, but Macdonald's (2007) study of copper alloy artefacts from the hoard raised the possibility that some of them were deposited before the Iron Age metalwork. It was hard to refine the dating of the animal bones, but the more recent project showed that metalwork of this kind had a longer history than previously supposed (Garrow *et al.* 2010).

Fortunately, the problem is not confined to this deposit. The artefacts from Llyn Cerrig Bach were discovered by accident in a peat bog, but those from the hillfort at South Cadbury in southern England were found by excavation.

In one part of the site there was evidence of a distinctive sequence (Barrett *et al.* 2000, 155–78, 281–2, 291–301). The earliest artefacts were Late Bronze Age. Although they were interpreted as scrap metal assembled by a smith, some of them could have had a more specialised character as they included parts of a bronze bucket and a cauldron, a barbed spearhead, and a gold bracelet. Activity continued at the same spot during the Iron Age when it was evidenced by bronze casting waste and iron slag. Horse fittings were made there, and other finds included weapons as well as buckets and cauldrons that recalled those in the Bronze Age phase. The definitive report could not decide whether they were 'votive deposits or the debris of metalworking' (Barrett *et al.* 2000, 301).

During the Late Iron Age animals were buried in the same area. They included the bodies of calves that had not been butchered. They were interpreted by the excavators as sacrificial deposits. During the 1st century AD a timber building of a kind usually interpreted as a shrine was erected close to these burials, and in the 2nd century AD it may have been replaced by a larger Roman temple. South Cadbury shared elements with Llyn Cerrig Bach – deposits of metalwork and faunal remains which could have accumulated episodically – but here they can be placed in sequence and related to the emergence of a temple complex. It seems unlikely that the earlier collection of metalworking debris was straightforwardly mundane in character, while the animal burials were religious deposits. Ritual practice may have been a common strand that connected them.

Similar problems are posed by another excavated deposit, at Broadward on the border between England and Wales (Rock & Barnwell 1872). Here, a large collection of Late Bronze Age weapons was found while draining a bog. It included a number of barbed spearheads not unlike the example from South Cadbury and was apparently associated with faunal remains, including the skull of a horse. Recent work (Bradley et al. 2015) established that the metalwork had been buried in a pit dug into the side of a spring, but when the animal bones were submitted for radiocarbon dating, they were assigned to quite different periods extending from the Early Bronze Age to medieval times. Other finds from the same location include an Early Bronze Age macehead, a shale bracelet of uncertain age, and a complete Roman pot. On this basis, it seems likely that varied deposits were made at this location over an extended time period and that animal bones were placed there on several occasions during this sequence.

The first account of Broadward assumed that the metalwork was lost by a smith (Rock & Barnwell 1872), but Mörtz (2018) has recently analysed the surviving weapons and concluded that they had been used in combat immediately before they were buried. The ceremony seems to have followed a battle. On this basis, Mörtz compared this collection with deposits of 'war booty' in Iron Age Scandinavia (Nørgård Jørgensen 2009). Again, animal remains and metalwork were deposited in the same places, but not necessarily at the same times.

At the Danish site of Vimose, Pauli Jensen (2009) identified a lengthy sequence extending from the end of the 1st millennium BC to the Viking Age. Six successive groups comprised damaged weapons, but two were of different character. One of these 'non-military' deposits dated from the beginning of the sequence, and the second from its end. Other bogs showed the same pattern, although some of them featured a transitional phase with smaller quantities of metalwork. Pauli Jensen related the groups of animal remains at such sites to Iron Age deposits of pots containing food that are typically interpreted as 'fertility sacrifices' (Kaul 2015). The same interpretation has been applied to Neolithic food deposits from bogs and lakes (Koch 1998). Such practices may even have begun in the Mesolithic period in this region (Bjørnevad-Ahlqvist 2020).

Problems and Interpretations

These observations raise two important issues.

Firstly, why were the same locations – often, but not exclusively, wetlands – used during different periods when their physical features were not especially distinctive? They were often unmarked, and David Fontijn (2007) has described them as 'invisible places'. Did it happen because these locations originally had names that might have endured long after any traces of human activity were lost (I owe this suggestion to David Mullin)? For instance, the names of many European rivers and lakes are of considerable antiquity, and some of them refer to ancient deities (Bradley 2017, 150–1).

Secondly, accounts of dryland hoards almost invariably pay more attention to the metal artefacts than the animal bones occasionally found with them, which are seldom dated. It is even more the case with discoveries in water. That is surprising, as by ignoring the bone deposits important chronological patterns can be overlooked. The deposits of faunal remains that researchers have associated with food production and fertility were most important during two distinct phases: the Neolithic period and the Early Iron Age. Metalwork on the other hand, played a prominent part in deposits of the Bronze Age and the later part of the Iron Age. Collections of both kinds can be found on the same sites, whether or not their histories overlapped. For example, animal bones were discovered to the north of the great timber causeway at Flag Fen in eastern England, but Bronze Age metalwork entered the water on the other side of this structure (Pryor 2001, 295–350). Animal bone deposits occurred in separate phases at the Swedish site of Skedemosse, but on some occasions, they were combined with the metalwork: 'Animals and humans were deposited not only at the time when weapon sacrifices were prevalent, but for an extended period … before and after this' (Monikander 2010, 96).

Separate explanations have been offered for each kind of deposit. The faunal remains are generally interpreted as evidence of sacrifice in which animals were killed but their bodies remained intact – that was what happened at South Cadbury. Alternatively, they might be the residue of feasts in which their meat was eaten on special occasions. The bones of butchered horses at Skedemosse provide a good example. In both cases accumulations of bones have been taken as evidence of conspicuous consumption. Another explanation for some of these collections is that they result from 'fertility sacrifices' (Lund 2002; Storå et al. 2020). The use of this term can be criticised because such deposits can overlap with finds of weapons (Monikander 2010), but in the Iron Age it would account for the presence not only of large amounts of meat, but also of wooden ploughs, agricultural tools, and ceramics containing food (Kaul 2015).

Metalwork deposits are sometimes explained in similar ways to those of animal remains. For instance, at one time the concentrations of broken metalwork found in rivers and lakes were also treated as evidence of conspicuous consumption. An obvious objection to this interpretation is that some of the prehistoric objects were discarded singly – in this respect accumulations of 'war booty' were exceptional

(Løvschal & Holst 2018). It is more likely that individual items were taken out of circulation because their roles had to be protected. One possible explanation for deposits of single metal objects is that they were funerary items which had been removed from a corpse before its burial. In that sense they can be considered as grave goods without graves (Cooper *et al.* 2020; Fontijn 2020). 'Burials' of this kind may have been a widespread practice in Late Bronze Age Europe (Sperber 2006). Another possibility is that they were sacred objects – *sacra* – whose distinctiveness needed to be maintained when their roles were over. This would prevent the raw material from entering the wider economy (Meillassoux 1968).

The problem with all these interpretations is that they do not explain why faunal remains should be found together with valuables, or why deposits of animal bodies and meat joints sometimes alternated with those of metalwork. These ideas cannot account for the recurrent use of *the same places* for deposits of such different kinds. Perhaps it would be better if all the deposits discussed here were viewed at a broad level as being part of a single phenomenon.

Nourishing the Earth

It is worth returning to the idea of 'fertility sacrifice'.

What exactly is sacrifice? Since the work of Hubert and Mauss (1964 [1898]), two features have played an important role in the thinking of anthropologists. Sacrifices should consist of living matter, and they must be violent. People and animals have to be killed in order to fulfil this criterion, but it is clear that the same approach might apply to the artefacts taken out of circulation. It could explain their destruction.

At one time it seemed important to distinguish between sacrifices of living beings and 'votive offerings' composed of inorganic items (Bradley 1990, 37). The latter were mainly metal artefacts, but we now appreciate that such interpretations are misleading since many societies consider objects to be alive (Insoll 2012). It is a familiar argument that, like the people who used them, weapons had histories and names. Those artefacts could exercise agency in their own right. Thus, it was the sword Excalibur that made Arthur into a king, just as his exploits added to its reputation. Their biographies were intertwined and their lives ended together. When Arthur died, his famous sword was deposited in a lake. During the Bronze and Iron Ages weapons were embellished with anthropomorphic designs (Pearce 2013), and some of the pieces of decorated metalwork deposited on sites like Llyn Cerrig Bach were embellished with motifs that suggested faces, animals, and imaginary creatures (Garrow & Gosden 2012). The same is true of the Viking swords from wetland deposits studied by Androschuk (2010). The difficulty of distinguishing between sacrifices of people or livestock and deposits of portable objects is illustrated by the pre-Roman and Roman Iron Age site of Oberdorla in northeast Germany where wooden sculptures interpreted as anthropomorphic figures were deposited in the same bog as faunal remains and the bodies of people who had been killed (Behm-Blanke 2003).

If diverse kinds of deposit can be considered as *sacrifices*, how were they connected with *fertility*? An insight is provided by Bloch and Parry in the introduction to their edited book *Death and the Regeneration of Life* (1982). There, they argue that some societies believe that fertility is a finite resource and can only be replenished when someone dies. Life itself is a 'limited good' but, they say, 'we do not use the term "fertility" in any restricted or technical way, but in the dictionary sense of "fecundity" or "productiveness". If death is often associated with a renewal of fertility, that which is renewed may be either the fecundity of people, of animals and crops, or of all three' (Bloch & Parry 1982, 7). One life replaces another, and balance is restored.

The anthropologist Mary Helms (2012) has taken a comparable approach to deposits of Bronze Age metalwork in Europe. The metal artefacts that play a pivotal role in prehistoric archaeology were the product of human workmanship, but they were also alive. Their production and deposition formed part of a more basic cycle. In her words:

> People in ethnographically and ethnohistorically known societies commonly believe that the basic principles of existence are organic in nature and that *all* things that exist, be they animate or inanimate, are basically alive ... It is widely understood that people ... have a fundamental responsibility to tend that basic flow of absolute life energy so that the order, stability, and dynamic potency of the entire cosmos may be protected and constantly renewed ... Typically this goal is achieved by 'nourishing' the universe with offerings of substances infused with life energy (Helms 2012, 106).

Her analysis is inspired by accounts of how metals were perceived and used in the Classical world, medieval Europe, and South America. Although they can be extracted by mining and worked by specialists, they grow like plants beneath the ground. The smiths not only make new objects out of the raw material, they accelerate 'the final maturation of the ore ... by removing accompanying elements or impurities ... *Future offerings of such power-filled metal would literally return to the earth a portion of the nutritive life force originally taken from it*' (Helms 2012, 108; my emphasis).

These ideas are inspired by everyday experience as well as abstract notions. When artefacts made of copper or bronze are buried they acquire a distinctive green patina similar to the colour of plants. The same happens after prolonged exposure to moisture, and this process has a special significance for societies in different parts of the world who compare the green corrosion with the appearance of vegetation after rain (Aldersley-Williams 2011, 239; Fors 2015). A similar notion applies to some of the most distinctive lithic artefacts, for they are often the same colour. A recent book edited by Rodríguez, Nelson, and Fàbregas (2020) draws attention to this phenomenon in the Old and New World archaeology.

Among the best examples are Neolithic axeheads, and particularly those of Alpine jadetite, which could be set upright in the ground as if they were growing there (Pétrequin *et al.* 2012). Again this suggests that these objects were animate, and illustrates a concern with fertility.

Whether they were made of metal or stone, special objects had to be returned to the beings who provided them. They might be placed in land or water, either in their original forms or as token deposits. That was the case from the Mesolithic period onwards, but such practices were very long-lived and some of the same ideas retained their significance until the early medieval period when Androschuk (2010) suggests that elaborately decorated swords were really the property of the gods. They had to be given back to them in order to maintain peace and order in the world. This was achieved by burying them or submerging them in rivers and lakes. Although Helms's study was limited to the European Bronze Age, her ideas have a wider application.

Marcel Mauss (2002[1925]) developed the idea that sacrifice is really a kind of exchange. The giving and returning of gifts are among the fundamental transactions on which social life depends. Because they carry an obligation to reciprocate, they provide a mechanism for building and maintaining relationships that extend across long periods of time. Sacrifices can be considered as gifts between the living and supernatural powers. But here there is a problem for if, as he believed, the gods and the dead were the real owners of all the possessions in the world, it is difficult to see how this system would work: 'Humans are at a disadvantage from the outset, and their gift almost by definition must be small when compared to the counter-gift from the gods and spirits. *It also implies people are giving something to gods and spirits that was theirs already*' (Fontijn 2020, 59; my emphasis). Perhaps there is an answer to this problem. Gift exchange may not be the right analogy to use, for it is open to manipulation and its scale can escalate over time. The relationship considered by Bloch and Parry (1982), and again by Helms, has more to do with *maintaining balance* and sustaining the continuity of life. That is the significance of fertility and how the circulation and deposition of metalwork ensure 'order, stability, and dynamic potency' (Helms 2012, 106).

These ideas apply most clearly to the earliest material, but also extend to pre-Roman societies who used imported metals. It is equally relevant to those that followed them, but there are complications. Iron might have been treated differently from copper and its alloys, and this framework does not account for the contractual nature of sacrifice in the Roman period (Bradley 1990, 188–9), or the treatment of defeated war bands and their possessions during the 1st millennium AD. On the other hand, the basic principles remained the same for millennia. Whether it involved the sacrifice of people, animals, or artefacts imbued with life, it was meant to maintain a balance between different worlds.

In a thought experiment conducted with Mark Pollard, Chris Gosden asked: 'Is the universe sentient?' (Gosden & Pollard 2021). The case sketched in this article might suggest part of the answer.

References

Aldersley-Williams, H. (2011) *Periodic Tales*. London, Penguin.

Androschuk, F. (2010) The gifts to men and the gifts to the gods: Weapon sacrifices and the circulation of swords. In C. Theune, F. Biermann, R. Struwe & G. Jeute (eds), *Zwischen Fjorden und Steppe*, 263–75. Rahden, Leidorf.

Barrett, J., Freeman, P. & Woodward, A. (2000) *Cadbury Castle, Somerset: The Later Prehistoric and Early Historic Period*. London, English Heritage.

Behm-Blanke, G. (2003) *Heiligtümer der Germanen und ihren Vorgängern in Thüringen: die Kultstätte Oberdorla*. Stuttgart, Theiss.

Bjørnevad-Ahlqvist, M. (2020) Ritualised hoarding in Mesolithic Scandinavia: An under-recognised phenomenon. *Current Swedish Archaeology* 28, 203–45.

Bloch, M. & Parry, J. (1982) Introduction: Death and the regeneration of life. In M. Bloch & J. Parry (eds), *Death and the Regeneration of Life*, 1–44. Cambridge, Cambridge University Press.

Bradley, R. (1990) *The Passage of Arms: An Archaeological Analysis of Prehistoric Hoards and Votive Deposits*. Cambridge, Cambridge University Press.

Bradley, R. (2017) *A Geography of Offerings: Deposits of Valuables in the Landscapes of Ancient Europe*. Oxford, Oxbow Books.

Bradley, R., Lewis, J., Mullin, D. & Branch, N. (2015) Where water wells up from the earth: Excavations at the findspot of the Late Bronze Age hoard from Broadward, Shropshire. *Antiquaries Journal* 95, 21–64.

Cooper, A., Garrow, D. & Gibson, C. (2020) Spectrums of depositional practice in later prehistoric Britain and beyond: Grave goods, hoards and deposits 'in between'. *Archaeological Dialogues* 27(2), 135–57.

Fontijn, D. (2007) The significance of 'invisible' places. *World Archaeology* 39, 70–83.

Fontijn, D. (2020) *Economies of Destruction: How the Systematic Destruction of Valuables Created Value in Bronze Age Europe, c.2300–500 BC*. Abingdon, Routledge.

Fors, H. (2015) *The Limits of Matter: Chemistry, Mining and Enlightenment*. Chicago, IL, University of Chicago Press.

Garrow, D., Gosden, C., Hill, J.D. & Bronk Ramsay, C. (2010) Dating Celtic Art: A major radiocarbon dating programme of Iron Age and Early Roman metalwork in Britain. *Archaeological Journal* 166, 79–123.

Garrow, D. & Gosden, C. (2012) *Technologies of Enchantment? Exploring Celtic Art 400 BC to AD 100*. Oxford, Oxford University Press.

Gosden, C. & Pollard, M. (2021) Is the universe sentient? What implications might this have for archaeology? In M. Boyd & R. Doonan (eds), *Far from Equilibrium: An Archaeology of Energy, Life and Humanity*, 313–23. Oxford, Oxbow Books.

Helms, M. (2012) Nourishing a structured world with living metal in Bronze Age Europe. *World Art* 2, 105–18.

Hubert, H. & Mauss, M. (1964[1898]) *Sacrifice: Its Nature and Functions*. Chicago, IL, University of Chicago Press.

Insoll, T. (2012) Sacrifice. In T. Insoll (ed.), *The Oxford Handbook of Ritual and Religion*, 152–65. Oxford, Oxford University Press.

Kaul, F. (2015) Changes in iconography: Votive practices and burial rites at 500 BC in southern Scandinavia. In F. Hunter & I. Ralston (eds), *Scotland in Later Prehistoric Europe*, 85–102. Edinburgh, Society of Antiquaries of Scotland.

Koch, E. (1998) *Neolithic Bog Pots from Zealand, Møn and Falster*. Copenhagen, De Kongelige Nordiske Oldskriftselskab.

Løvschal, M. & Holst, M. (2018) Governing martial traditions: Post-conflict ritual sites in Iron Age Northern Europe (200 BC–AD 200). *Journal of Anthropological Archaeology* 50, 27–39.

Lund, J. (2002) Forlev Nymølle – En offerplads fra yngre førromersk jernalder. *Kuml* 2002, 143–95.

Mauss, M. (2002[1925]) *The Gift: The Form and Reason for Exchange in Archaic Societies*. London, Routledge.

MacDonald, P. (2007) *Llyn Cerrig Bach: A Study of the Copper-alloy Artefacts from the Insular La Tène Assemblage*. Cardiff, University of Wales Press.

Meillassoux, C. (1968) Ostentation, destruction, reproduction. *Economie et Société* 1, 93–105.

Monikander, A. (2010) *Våld och vatten. Våtmarkskult vid Skedemosse under järnåldern*. Stockholm, University of Stockholm.

Mörtz, T. (2018) Violence and ritual in Late Bronze Age Britain: Weapon deposits and their interpretation. In C. Horn & K. Kristiansen (eds), *Warfare in Bronze Age Society*, 168–88. Cambridge, Cambridge University Press.

Nørgård Jørgensen, A. (2009) Weapon-offering types in Denmark 350 BC to 1200 AD. In U. Von Freeden, H. Friesinge & E. Wamers (eds), *Glaube, Kult und Herrschaft*, 37–51. Bonn, Habelt.

Pauli Jensen, X. (2009) From fertility sacrifices to weapon sacrifices: The case of the South Scandinavian bog finds. In U. Von Freeden, H. Friesinge & E. Warmers (eds), *Glaube, Kult und Herrenschaft*, 53–64. Bonn, Habelt.

Pearce, M. (2013) The spirit of the sword and spear. *Cambridge Archaeological Journal* 23, 55–67.

Pétrequin, P., Cassen, S., Errera, E., Klassen, L., Sheridan, A. & Pétrequin, A-M. (2012) *Jade. Grands haches alpins du Néolithique européen*. Besançon, Presses Universitaires de France-Comté.

Pryor, F. (2001) *The Flag Fen Basin: Archaeology and Environment of a Fenland Landscape*. London, English Heritage.

Rock, T. & Barnwell, E. (1872) The Bronze Age relics from Broadward, Shropshire. *Archaeologia Cambrensis* 4, 338–55.

Rodríguez Rellán, C., Nelson, B. & Fàbregas Valcarce, R. (eds) (2020) *A Taste for Green: A Global Perspective on Ancient Jade, Turquoise and Variscite Exchange*. Oxford, Oxbow Books.

Sperber, L. 2006. Bronzezeitliche Flussdeponierungen aus dem Altrhein bei Roxheim. *Archäologische Korrespondenzblatt* 36, 195–214.

Storå, J., Ullen, I. & Drenzel, L. 2020. Splitting bodies – a close-up study of a Swedish bog deposition site from the pre-Roman Iron Age. *Journal of Archaeological Science: Reports* 54, 102621.

11

Rock Art: A Marker of Concepts and Practices

Courtney Nimura, Rebecca O'Sullivan, and Peter Hommel

Rock art is found across the globe and has been used, with varying levels of success, as evidence of cultural connections between different communities. In this paper, we discuss some of the methods used to analyse rock art and cultural connectivity, drawing on examples from three regions: Northern Europe, the Eurasian Steppe, and East Asia. We present ideas for how existing methods for this type of analysis could be reworked, specifically discussing multi-media comparison within archaeological contexts.

Keywords: Rock Art, Theory of Connections, Material Agency, Scandinavia, Korea, Japan, Central Asia

Power and Potential

Chris Gosden has often emphasised the power of art to impact people's behaviour by drawing them into relationships with objects and the role of art in highlighting patterns of contact, communication, influence, and exchange (*e.g.* Garrow & Gosden 2012; Gosden 2020). He has been interested in understanding these kinds of cultural connections across both time and space and investigating how they can be discerned in the archaeological record, whether through the collective construction and use of monuments (Gosden *et al.* 2018) or the study of portable material culture (Gosden *et al.* 2020). In this short contribution, we extend Chris's line of questioning into the study of (mostly) immovable art emplaced on cliff-faces, outcrops, and boulders across northern Eurasia. Through a series of brief vignettes, we assess the potential of two key questions in wider discussions in prehistoric Eurasian studies: can rock art be used as a proxy for understanding connections in later prehistory (*ca.* 3000 BC–AD 100), and if so, how?

Why Rock Art?

There are several notable hotspots for later prehistoric rock art across continental Eurasia, from Fennoscandia in the west through southern Siberia to the Primorye, Sakhalin, and Japanese Archipelago in the east. Slightly further south, there are also large concentrations of rock art in Mongolia, Kazakhstan, northern China, and the Korean Peninsula. Rock art in these regions can be painted, but the majority is pecked or incised into the surface of the rock. These images are occasionally found within caves and on standing stones, but the majority were created on horizontal and vertical, natural, open-air rock surfaces. These were incorporated into funerary sites and other ceremonial places for which they may have been intentionally created or reused from other contexts.

Rock art from these regions has been studied for centuries, which has resulted in different interpretive histories. The geographical location of the sites also leaves them particularly affected by linguistic barriers, which often prevent rock art from being studied holistically across wide geographical areas. Ambiguous dates and fears of repeating diffusionist narratives or simplistic 'looks-like' analogies have deterred researchers from making direct comparisons. While we too hesitate to return to arguments of large scale 'genetic' connections on the basis of a few decontextualised images (*e.g.* Okladnikov 1971, 112–23), there are perhaps shared patterns of behaviour that can still be explored, and more local connections in contemporary material culture can be salvaged and developed as new dates become available and improved theory is applied.

Rock Art Across Eurasia

To draw out some key themes, we will use examples from three separate areas across a vast geographical span in which we have, singly or collectively, focused our research over the last 15 years: Northern Europe and Central and East Asia (Fig. 11.1). The time period covered in this paper is equally broad, from the 3rd millennium BC to the early 1st millennium AD. We will use this material to examine some of the ways in which rock art could be used to draw meaningful connections between different communities and reflect upon some of the challenges and possibilities these examples present.

Scandinavia has the largest concentration of rock art in prehistoric Europe, mainly dated to the Bronze Age around 1700–500 BC. Some rock art sites could have been created as early as the Stone Age around 8200 BC (Sognnes 2003), while others were made well into the Early Iron Age at least until 200 BC; there is regional chronological difference, with a greater concentration of earlier sites in the north and later sites in the south. Although rock art continued to be created into the Early Iron Age, its frequency diminished as societies turned to other modes of art production. Over 22,000 rock art sites have been registered in Denmark, Norway, and Sweden, totalling over 300,000 individual motifs (Nimura 2016). These include a wide variety of figurative, identifiable motifs and unidentifiable, abstract motifs from watercraft and animals to humans wielding weapons, and spirals or circles with dots in the centre. Across the Eurasian Steppe, there are many concentrations of prehistoric rock art, but the largest of these are found in eastern Kazakhstan, southern Siberia, and Mongolia.

Again, the majority of this art dates to a period between the 3rd millennium BC to the 1st millennium AD, although there are significant traditions of rock art both earlier and later in some areas, as well as a long tail of remarkable sites stretching back into the Holocene (*e.g.* Shulgan-Tash in Bashkhortostan, see Abramova 1997). The images at these sites are remarkably varied, with vehicles (including watercraft), anthropomorphic figures, stylised faces, fantastic deer, and other more alarming composite creatures being the most widely discussed. However, the majority of images are more routine and overwhelmingly dominated by wild and domesticated animals (elk, argali, ibex, sheep/goats, deer, horse, boar, and cattle). This wealth of images in continental Eurasia makes them useful sources of data against which rock art in neighbouring parts of East Asia can be compared (Gantulga 2020). There, rock art is mainly found across northern China, the Korean Peninsula, and the Russian Far East, with only two sites in Japan on the northern island of Hokkaido. Similar to the major Eurasian clusters, most rock art in East Asia spans the 3rd millennium BC to early 1st millennium AD, with some images dating back to the Holocene (*e.g.* Yinshan; see Gai 1986) or even earlier (Sikachi Alyan, see Okladnikov 1971; Laskin *et al.* 2019). In addition to unidentifiable motifs and geometric shapes, animals constitute the most abundant imagery, particularly goat, deer, and horse, with whale and fish appearing at coastal sites on the Korean Peninsula. Various stylised anthropomorphic faces and figures are also found across this region, notably in the Amur Basin and Primorye (Russian Far East), Inner Mongolia (China), and Hokkaido (Japan, see Ogawa 2014; 2019).

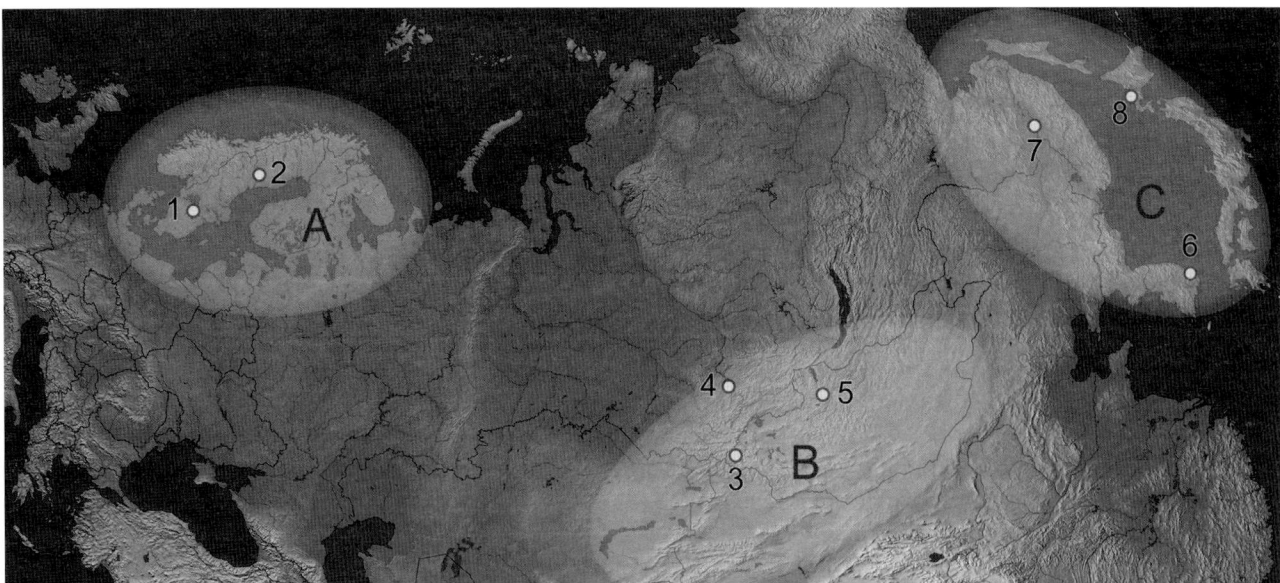

Figure 11.1 Map of Eurasia showing the three regions examined in the text: (a) Northern Europe; (b) Central Eurasia; (c) Eastern Asia. Sites: 1: Hästholmen; 2: Nämforsen; 3: Khatuugiin gol; 4: Minusinsk Basin; 5: Ushkiin Uver; 6: Cheonjeon-ri and Bangudae; 7: Sikachi Alyan; 8: Temiya and Fugoppe (Map: P. Hommel).

Methods for Analysing Connectivity with Rock Art

Different methods have been used to analyse rock art as evidence of cultural connectivity (or the lack thereof) across varying distances. In some cases, the very act of making rock art has been used to distinguish a society from its neighbours. In the European Bronze Age, for example, a cultural boundary line is drawn in northern Germany separating the Nordic Bronze Age, where rock art is abundant, from the Central European Bronze Age, where there is a rock art lacuna. This presence or absence of rock art is also used to identify anomalous cultural practices within a locale. There are only two rock art sites in the Japanese Archipelago, and both are located in coastal caves on Hokkaido. Dated to around the start of the 1st century AD, Temiya and Fugoppe caves are distinctive, in that the society that made the images clearly did not continue the practice for long. As if to highlight this fact, ceramics found at Fugoppe have been linked to types more common on the Eurasian mainland (Ogawa 2019), leading to the theory that the rock art was made by a seafaring society that did not remain on Hokkaido (Ogawa 2014).

Other studies have focused on analysing the occurrence and ubiquity of motifs. There is, perhaps unsurprisingly, a lexicon of motifs that seem to have been created in rock art over thousands of years and across the globe (*e.g.* Bertilsson 2013). From handprints to spirals, these pervasive motifs have been used as evidence of cultural connectivity across long distances. Such arguments are persuasive when applied to particularly distinctive motifs that appear suddenly within a set timeframe, such as depictions of vehicles that appear across Central Asia from the mid-3rd millennium BC, then spread across eastern Eurasia through the Altai Mountains into Siberia and Mongolia (Fig. 11.2) (Novozhenov 2012; Esin *et al.* 2021). Here, this specific motif has been interpreted as a marker of possible routeways, but also as evidence of long-distance technological transmission (Esin *et al.* 2021). The same phenomenon is exemplified by another form of transportation in Northern Europe: the

Figure 11.2 An eastern Eurasian vehicle motif. Two horses are connected by a shaft to a vehicle body with two four-spoked wheels and a driver. Khatuugiin gol, Bayan Ulgii aimag, Mongolia (Photo: R. O'Sullivan).

boat or ship. These vessels are the most abundant figurative motif in Scandinavia – over 19,000 are known from the three countries (Nimura 2016). They are found mainly in coastal regions or along inland waterways and waterbodies, potentially marking navigable routes, safe landing places/harbours, or meeting places (Nimura *et al.* 2019; Bradley *et al.* 2020). The pervasiveness of watercraft imagery is thought to stem from their vital importance for long-distance travel and economic success in the Bronze Age, which is inherently tied to the bronze trade, as copper had to be imported to Scandinavia from abroad (Ling *et al.* 2013; 2014; Melheim *et al.* 2018). Therefore, they too are thought to represent long-distance cultural connections.

These methods present some challenges. Comparing rock art separated by huge distances can lead to connections being identified when an alternative explanation grounded in regional archaeology may be more appropriate (Jang 2012). For example, grand theories that similarities between face or mask motifs on the Pacific coast from the Russian Far East to Vietnam (Okladnikov 1971, 114) and Eastern China to North America (Wang & Zhang 2014, 73–4) demonstrate long-distance connections between these societies rely heavily on one type of motif, ignoring the social, economic, and cultural variability indicated by other images at the same sites. The fallibility of long-distance links is also evident when pronounced differences between rock art within a locality are evident. Motifs along a stream called the Daegokcheon in southeast South Korea demonstrate this kind of distinction, where two major sites, related in terms of image style and execution, deviate distinctively in content. The majority of identifiable animal motifs at Cheonjeon-ri are land-based, hoofed animals or predators (Jeon *et al.* 2014), whereas barely 2 km downstream at Bangudae, marine creatures like whales, turtles, and fish constitute the greater number (Jeon *et al.* 2019). Bangudae also has depictions of sea-going vessels, indicating a closer relationship to the sea. Despite these clear differences in their artistic subjects, both groups engaged consistently in similar cultural practices by choosing to mark the landscape with rock art.

Many early studies of long-distance connections represented in rock art were also historically rooted in the idea that distinct cultural groups existed in Eurasian prehistory. In the past, archaeologists relied heavily on creating typologies of specific object types to draw cultural boundary lines on maps, which implied the existence of different 'cultures'/groups of people. We now recognise that these boundaries were blurred. Mobility and connectivity were key aspects of life in the past as they are today (Hahn & Weis 2013), and cross-cultural interaction and cultural permeability seem to have been inevitable. Therefore, we are presented with a methodological problem. If we have moved past the belief that material culture is a direct representation of a person/group; and we are aware that 'the exclusive reliance upon a single category of data in defining an archaeological culture has often been heralded as one of the main failures of this approach' (Roberts & Vander Linden 2011, 12), how can we proceed?

Chris has attempted to move beyond analyses of typology and style and instead focus on the function of objects in society (*e.g.* Gosden 2005). He has argued that the ability of art to affect humans (and their societies and cultures) is rooted in the myriad ways in which materials are deployed within local environments (Gosden & Marshall 1999). A piece of portable art, once removed from the place in which it was made, could migrate long distances to a new social-cultural context in which it may be perceived very differently or even hold a completely new meaning. In order to embrace this fluidity, Chris has asked: what was this object intended to do within this context? This question leads us to new ways of identifying cultural connectivity, where different groups with varied material culture traditions can be connected by shared concepts and practices.

Rock Art as a Marker of Concepts and Practices

Analysing different media together (*e.g.* rock art and portable art) within their archaeological contexts can shed light on the function of material culture in prehistoric societies (*e.g.* Bradley 2015). There is much potential for this type of analysis in our three case study areas.

In the Nordic Bronze Age, rock art and metalwork were closely linked and are thought to have played active roles in society, with metalwork, such as swords and axes, depicted in miniature form as part of rock art 'scenes' or at 1:1 scale (Fig. 11.3) (Skoglund 2017). This has led some rock art sites to be equated with hoards, in that they 'collect' multiple artefacts in one place (Bradley 2016). Watercraft were so significant to Bronze Age communities that they were not only depicted in rock art but were also etched onto bronze razors (Kaul 1998) and represented by stones arranged in the ground (Wehlin 2013). These stone 'ship settings' often contained cremation burials. The links between metalwork, rock art, and long-distance connections has been argued for in relation to specific motifs (Ling & Uhnér 2014), and as demonstrated by the relationship of the watercraft motif to burial rites, it is possible that these intimately linked yet distinct forms of material culture had synonymous social and cultural functions (Bradley 2009) connected to a Nordic Bronze Age cosmology.

The occurrence and ubiquity of specific motifs in other regions has also been interpreted as evidence of broader worldviews that could be shared across great distances. Early images of wheeled vehicles in eastern Eurasia not only extend our limited archaeological evidence for their dispersal across the continent, but they also provide critical information about their place within socio-symbolic systems. The conjunction of wheels and censers in Okunev art (Esin *et al.* 2021), for example, shows how playful

Figure 11.3 Examples of motifs found in southern Scandinavian rock art, from the site of Hästholmen, Sweden. The red paint was applied by the heritage agency (Swedish National Heritage Board [Riksantikvarieämbetet] number: Västra Tollstad 21:1): (a) axes (palstaves); (b) quadrupeds; (c) watercraft with humans and cupmarks (Photos: C. Nimura).

visual metaphors can create connections between physical and conceptual movement in the cosmology of this society. The subsequent explosion of chariot imagery, often showing consistent modes of representation, shows again how important technologies moved through societies, while also highlighting critical variability (Novozhenov 2012).

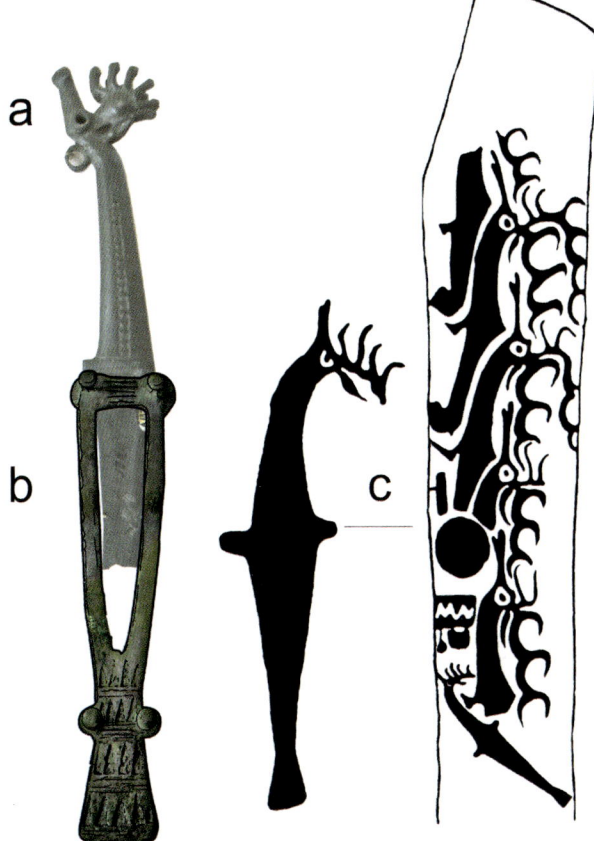

Figure 11.4 Animal-headed dagger/knife depicted on a Late Bronze Age deer stone in northern Mongolia, with comparanda in the metalwork of the Karasuk Culture. (a) animal-headed knife (chance find, Minusinsk Basin); (b) sheath fittings (Krasnopol'e, Minusinsk Basin) (Photos: P. Hommel 2017; Minusinsk Museum); (c) deer stone No. 15 from Ushkiin Uver, Murun, Khovsgol aimag, and detail (re-drawn after Volkov 1981).

Countless studies have drawn attention to parallels in knives and animal motifs depicted on deer stones typically found in Mongolia (Fig. 11.4) and actual examples from northern China (Chlenova 2000; Kovalev 2007; Fitzhugh 2009). Archaeologists have pointed to the unique nature of these stelae to support theories for long-distance trade networks between the nomadic people of these regions (Rawson *et al.* 2020), facilitated through, and in its turn perhaps strengthening, a shared concept of the deer as a spiritual being (Jacobson 1993).

Explorations of cultural connections that combine the study of different types of archaeological evidence can help us move past comparisons of traits within a single category of material culture, but they must still be interpreted within their archaeological context. This is especially true of the majority of rock art, which is typically fixed in the landscape.

The placing of concentrations of rock art in critical spaces within the landscape has been widely discussed (*e.g.*

Bradley 2000, 81; Tilley 2008; Nimura 2016; Brandišauskas 2020; O'Sullivan 2021). Analysing rock art within its environmental context has led to many theories for its purpose, whether as markers of valuable resources, places of veneration, or routeways of communication and movement. Rock art here may act as an index of particular concepts as much as particular practices. Such concepts may be very widely shared without any meaningful relationship between specific sites and images or their makers. For example, the reading of Scandinavian rock art within the context of an ethnographic understanding – a local historical analogy within structurally similar northern circumpolar cosmologies – draws attention to associations with water that were clearly important to the makers of both rock art and a variety of other forms of material culture and the decisions underlying their placement within the landscape. The making of rock art at points of turbulent water (*e.g.* Nämforsen) seems to emphasise other practices that give primacy to stillness or reflection, including various forms of hoarding or ritualised deposition of bodies in watery places. These acts of placemaking seem to encapsulate a shared symbolic logic that sees the surface of water as a boundary: transgressable and potentially dangerous. In much the same way that reflections speak of a world beneath into which objects could be placed (and perhaps returned), turbulence in the water may be understood as a weakening of the boundaries between worlds. In these spaces, rock art served as a marker, potentially used as a tool by the people who made it to negotiate connections to other worlds or beings (*e.g.* Lahelma 2005).

Drawing on well-established symbolic logics that have been very widely held may help us to better understand why rock art was placed in other locations, without claiming any direct connection between their makers. The siting of rock art boulders adjacent to rivers in places that are regularly encroached upon by water – as in the case of Sikachi Alyan in the Amur Basin – brings different dynamics and experiences into play, perhaps helping to explain some of the intriguing patterns of inversion and symmetry and even the placement of particular images, which would have been transferred through their reflections into the water. Clearly there is insufficient space here to discuss the myriad factors influencing the placement and content of rock art in detail, but these examples highlight the range of social, cultural, and aesthetic processes at play that must be considered, making any intra- and inter-group comparisons an incredibly complex but potentially rewarding task.

Summary

The examples from Northern Europe, the Eurasian Steppe, and East Asia highlighted in this chapter, though brief, are intended to demonstrate the insights that rock art affords into the variability and complexity of past societies. Moving beyond generic concepts of 'similarities' and 'connections' in examining northern Eurasian rock art greatly refines our picture of prehistoric societies. Not only does comparing it to artefacts excavated in the same region situate rock art within relative dating frameworks, it also embeds it within the context of local social, economic, and cultural processes. As Chris's work has argued with other forms of material culture, this more fluid perspective can broaden our perceptions of rock art as much more than static images: they were active agents that served to construct, while also being constructed by, the people around them. In this way, the commonalities between images and artefacts that are observed across huge distances become much more meaningful than simply reflecting the movement of people themselves; different types of material culture may have contributed to a communal sense of identity or shared worldview across prehistoric Eurasia between societies that were otherwise highly variable in terms of subsistence, economy, and culture. Recognising these connections in one field of material culture may have the power to extend or deepen our understanding of other patterns in the contemporary material world. This, we feel, is a direction worth pursuing in future research.

Acknowledgements

At different points in time, all three of the authors have worked, travelled, or collaborated with Chris Gosden; Courtney even lived in his basement, and he introduced her to her now-husband. Chris was Courtney's external examiner for her PhD thesis and appointed her to a postdoctoral position at Oxford, where she became fast friends with Peter and Rebecca. At the time, Peter was a postdoctoral researcher, and Rebecca was completing her doctoral work on rock art in the Altai mountains. Both travelled extensively with Chris, visiting some of the sites and regions mentioned here together. We are forever indebted to Chris for his ongoing support, generosity of spirit, intellectually challenging chats, and continued interest in our work on rock art, in honour of which we have collaborated to write this modest contribution.

References

Abramova, Z. A. (1997) Caves of the Palaeolithic Urals. In *Materials from the International Conference, Ufa, 9–15 September 1995*, 52–5. Ufa, IIYAL UFITS RAN (in Russian).

Bertilsson, U. (2013) Footprints on the rock faces: Following the tracks of cosmological archetypes and pictograms for millennia of prehistory. In S. Sabatini & S. Bergerbrant (eds), *Counterpoint: Essays in Archaeology and Heritage Studies in Honour of Professor Kristian Kristiansen*. British Archaeological Reports International Series 2508, 243–52. Oxford, Archaeopress.

Bradley, R. (2000) *An Archaeology of Natural Places*. London, Routledge.

Bradley, R. (2009) *Image and Audience: Rethinking Prehistoric Art*. Oxford, Oxford University Press.

Bradley, R. (2015) Mixed media, mixed messages: Religious transmission in Bronze Age Scandinavia. In P. Skoglund, J. Ling & U. Bertilsson (eds), *Picturing the Bronze Age*, 37–46. Oxford, Oxbow Books.

Bradley, R. (2016) *A Geography of Offerings: Deposits of Valuables in the Landscapes of Ancient Europe*. Oxford, Oxbow Books.

Bradley, R., Nimura, C. & Skoglund, P. (2020) Meetings between strangers in the Nordic Bronze Age: The evidence of Swedish rock art. *Proceedings of the Prehistoric Society* 86, 261–83.

Brandišauskas, D. (2020) Sensory perception of rock art in east Siberia and the Far East. *Sibirica* 19, 50–76.

Chlenova, N. L. (2000) Deer, horses and hooves (on links between Mongolia, Kazakhstan and Central Asia in the Scythian epoch). *Archaeology of Russia* 1, 90–106 (in Russian).

Esin, Yu. N., Magail, J., Gantulga, J. & Yeruul-Erdene, C. (2021) Chariots in the Bronze Age of Central Mongolia based on the materials from the Khoid Tamir River valley. *Archaeological Research in Asia* 27, 100304.

Fitzhugh, W. W. (2009) Stone shamans and flying deer of northern Mongolia: Deer goddess of Siberia or chimera of the steppe. *Arctic Anthropology* 46(1–2), 72–88.

Gai S.-L. (1986) *Petroglyphs of the Yin Mountains*. Beijing, Cultural Relics Publishing House (in Chinese).

Gantulga, J.-O. (2020) Ties between steppe and peninsula: Comparative perspective of the Bronze and Early Iron Ages of Mongolia and Korea. *Proceedings of the Mongolian Academy of Sciences* 60(4), 65–88.

Garrow, D. & Gosden, C. (2012) *Technologies of Enchantment? Exploring Celtic Art: 400 BC to AD 100*. Oxford, Oxford University Press.

Gosden, C. (2005) What do objects want? *Journal of Archaeological Method and Theory* 12(3), 193–211.

Gosden, C. (2020) Art, ambiguity and transformation. In C. Nimura, H. Chittock, P. Hommel & C. Gosden (eds), *Art in the Eurasian Iron Age: Context, Connections and Scale*, 9–22. Oxford, Oxbow Books.

Gosden, C. & Marshall, Y. (1999) The cultural biography of objects. *World Archaeology* 31(2), 169–78.

Gosden, C., Hommel, P. & Nimura, C. (2018) Making mounds: Monuments in Eurasian prehistory. In T. Romankiewicz, M. Fernández Götz, G. Lock & O. Büchsenschütz (eds), *Enclosing Space, Opening New Ground: Iron Age Studies from Scotland to Mainland Europe*, 141–52. Oxford, Oxbow Books.

Gosden, C., Chittock, H., Nimura, C. & Hommel, P. (2020) Context, connections and scale: An introduction. In C. Nimura, H. Chittock, P. Hommel & C. Gosden (eds), *Art in the Eurasian Iron Age: Context, Connections and Scale*, 1 8. Oxford, Oxbow Books.

Hahn, H.-P. & Weis, H. (eds) (2013) *Mobility, Meaning and Transformations of Things*. Oxford, Oxbow Books.

Jacobson, E. (1993) *The Deer Goddess of Ancient Siberia: A Study in the Ecology of Belief*. Leiden, Brill.

Jang, S.-H. (2012) Comparisons and researches between facial shapes of the petroglyphs in Inner Mongolia and main icons of the 'Yangjeun-dong'-style petroglyph in South Korea. *The Dongguk Historical Society* 53, 353–91 (in Korean).

Jeon, H.-T., Jang, M.-S., Gang, J.-H., Nam, Y.-U., & Yoon, H.-J. (2014) *The Cheonjeon-ri Petroglyphs in Ulsan*. Seoul, Hollym.

Jeon, H.-T., Rhee, H.-W., & Park, Y.-H. 2019. *The Bangudae Petroglyphs in Ulsan*. Carlsbad, CA, Hollym.

Kaul, F. (1998) *Ships on Bronzes: A Study in Bronze Age Religion and Iconography*. Copenhagen, National Museum.

Kovalev, A. A. & Erdenebaatar, D. (2007) The Mongolian Altai in the Bronze and Early Iron ages (results of the International Central Asian Archaeology Expedition of the Institute of History at Saint-Petersburg State University and National University of Mongolia). In V. V. Nevinskii & A. A. Tishkin (eds), *The Mountainous County of the Altai-Sayan and Its Historical Development by Nomads*, 80–85. Barnaul, Altai State University Press (in Russian).

Lahelma, A. (2005) Between the worlds: Rock art, landscape and shamanism in subneolithic Finland. *Norwegian Archaeological Review* 38(1), 29–47.

Laskin, A. R., Devlet, E. G., Svoyskiy, Yu. M., Romanenko, E. V. & Levanova. E. S. (2019) New petroglyphs at Sikachi Alyan. *Archaeology of Russia* 2019(3), 122–30.

Ling, J. & Uhnér, C. (2014) Rock art and metal trade. *Adoranten* 2014, 23–43.

Ling, J., Hjärthner-Holdar, E., Granding, L., Billström, K. & Persson, P.-O. (2013) Moving metals or indigenous mining? Provenancing Scandinavian Bronze Age artefacts by lead isotopes and trace elements. *Journal of Archaeological Science* 40, 291–304.

Ling, J., Stos-Gale, Z. E., Granding, L., Billström, K., Hjärthner-Holdar, E. & Persson, P.-O. (2014) Moving metals II: Provenancing Scandinavian Bronze Age artefacts by lead isotope and elemental analyses. *Journal of Archaeological Science* 41, 106–32.

Melheim, L., Grandin, L., Persson, P.-O., Billström, K., Stos-Gale, Z., Ling, J., Williams, A., Angelini, I., Canovaro, C., Hjärthner-Holdar, E. & Kristiansen, K. (2018) Moving metals III: Possible origins for copper in Bronze Age Denmark based on lead isotopes and geochemistry. *Journal of Archaeological Science* 96, 85–105.

Nimura, C. (2016) *Prehistoric Rock Art in Scandinavia: Agency and Environmental Change*. Swedish Rock Art Research Series 4. Oxford, Oxbow Books.

Nimura, C., Skoglund, P. & Bradley, R. (2019) Navigating inland: Bronze Age watercraft and the lakes of southern Sweden. *European Journal of Archaeology* 23(2), 186–206.

Novozhenov, V. A. (2012) *Communications and the Earliest Wheeled Transport of Eurasia*. Moscow, TAUS.

Ogawa, M. (2014) Dating petroglyphs from Fugoppe cave, Japan. *Arts* 3, 46–53.

Ogawa, M. (2019) Rock art in Japan. In J. Clottes & B. Smith (eds), *Rock art in East Asia: A Thematic Study*, 32–9. Charenton-le-Pont, ICOMOS.

Okladnikov, A. P. (1971) *Petroglyphs of the Lower Amur*. Leningrad, NAUKA (in Russian).

O'Sullivan, R. (2021) Replication in rock art past and present: A case study of Bronze and Iron age rock art in the Altai, eastern Eurasia. *Journal of Archaeological Method and Theory* 28, 387–412.

Rawson, J., Chugunov, K., Grebnev, Ye., & Huan, L.-M. (2020) Chariotry and prone burials: Reassessing Late Shang China's relationship with its northern neighbours. *Journal of World Prehistory* 33(2), 135–68.

Roberts, B. & Vander Linden, M. (2011) Investigating archaeological cultures: Material culture, variability, and transmission. In B. Roberts & M. Vander Linden (eds), *Investigating Archaeological Cultures*, 1–21. New York, NY, Springer.

Skoglund, P. (2017) Axes and long-distance trade. In P. Skoglund, J. Ling & U. Bertilsson (eds), *North Meets South*, 199–213. Oxford, Oxbow Books.

Sognnes, K. (2003) On shoreline dating of rock art. *Acta Archaeologica* 74, 189–209.

Tilley, C. (2008) *Body and Image: Explorations in Landscape Phenomenology 2*. Walnut Creek, CA, Left Coast Press.

Volkov, V. V. (1981) *Deer Stones of Mongolia*. Ulaanbaatar, AN MIR (in Russian).

Wang, X.-K. & Zhang, W.-J. (2014) The dissemination of human faces rock painting in China: The General Cliff and Chifeng as case studies. *Southeast Culture* 4, 70–75 (in Chinese).

Wehlin, J. (2013) *Osterjöns skeppssättingar. Monument och mötesplatser under yngre bronsålder*. Unpublished PhD thesis, University of Gothenburg Institute for Historical Studies.

12

Celtic Art Beyond Metal: Material Matters in Iron Age and Early Roman Southern England

Sarah Downum and Duncan Garrow

This chapter discusses decoration and visual culture across a range of materials, including antler/bone, ceramics, metal, stone, and wood. Building on recent arguments made specifically in relation to Celtic Art and much more generally across the social sciences, we argue that it is important to investigate materials relationally. Celtic Art on metal objects can only really be understood properly in relation to other decorated materials; the full 'stylistic universe' needs to be considered in our interpretations. Through consideration of a substantial database of decorative motifs found on over 2,000 artefacts in southern England ca. 600 BC–AD 100, we consider the 'regionality of style' alongside the different 'styles of material'. Our analysis demonstrates that, whilst a full range of motifs was available everywhere, different materials, especially metals and non-metals, were decorated in significantly different, if related, ways.

Keywords: Celtic Art, Iron Age, Britain, Materiality, Material Culture

Introduction

'You will need your sunglasses to look at that lot' began Garrow & Gosden's book about Celtic Art, referring to comments made by 'a seasoned curator' at the British Museum upon our visit to see the Snettisham torcs during the course of the Technologies of Enchantment project (Garrow & Gosden 2012, 1) – a wonderful sight indeed, and one which represents just a tiny proportion of the many fantastic, enchanting objects that make up the canon of insular Celtic Art. The chapter presented here, and the doctoral research from which it stems (undertaken by Downum), is concerned primarily with those Celtic Art objects for which sunglasses are very much *not* required.

Not Everything is Metal, but Metal is Everywhere

The title of this section is a quote from Deleuze & Guattari (1988, 411), one sentence forming part of their wide-ranging philosophical discussion of materials and matter. It could equally be applied to past discussions of Celtic Art and Iron Age visual culture. Deleuze & Guattari (1988, 411) themselves viewed metallurgy as the technology *par excellence*:

'what metal and metallurgy bring to life is a life proper to matter … a material vitalism that doubtless exists everywhere but is ordinarily hidden or covered' (see Conneller 2011, 13 for a more detailed discussion). As many authors writing relatively recently have pointed out, archaeologists have shown a comparable bias towards metallurgy and metals in past considerations of Celtic Art (Megaw & Megaw 2001, 16; Sharples 2008, 204; Bradley 2009, 11; Joy 2011, 205, Chittock 2014, 314; Nimura *et al.* 2020, 25). As several have emphasised, there has generally been a deep-seated bias in terms of the way in which different material expressions of Celtic Art have been discussed, with metal objects being viewed as 'high art' or artisanal, and non-metal objects being viewed as folk/peasant art or craft (Bradley 2009, 14–19; Joy 2011, 206; Chittock 2014, 314–15). Bradley's (2009, 18) survey of the types of artefact selected as examples of Celtic Art within book-length studies throughout the 20th century indicated that this 'problem' has actually worsened through time.

The 'Technologies of Enchantment' project, as some have (pointedly) pointed out, did little to rectify this situation. In fact, it undoubtedly exacerbated it in many ways.

Constrained by the need to lead with a 'total collection' database, we looked only at metalwork, and not even all of that, eschewing both coins and brooches (Garrow 2008; Garrow & Gosden 2012, 6), a situation partially resolved at least through the inclusion of the latter in Gosden's subsequent 'Early Celtic Art in Context' (ECAIC) project (Nimura *et al.* 2020). In setting off on that course very consciously at the time, we were of course fully aware of the wealth of information and intrigue that non-metal Celtic Art objects held. The decision was taken pragmatically; and also (and we are sure that the recipient of this volume would agree) a little disappointedly at the same time.

Prior to publication of the 'Technologies of Enchantment' project, others had certainly investigated Celtic Art-style decoration across a range of materials. Relatively recent examples include Evans's (1989) investigation of the kinds of decoration seen on different artefacts at Glastonbury in his study of perishables and worldly goods; Sharples's (2008, 209–10) consideration of the possible relationships between decoration on metalwork and ceramics; and Giles's (2008) study of the aesthetics of martial objects. Subsequent research has also attempted to put non-metals firmly back into the picture – Joy (2011), for example, considered the depositional contexts of a variety of materials, including ceramics and bone/antler as well as metal, in asking the question 'why decorate?', while Chittock (2014) also focused on the bone and antler combs from Glastonbury in an attempt to move discussions of Celtic Art beyond metal, and on a wide range of materials in her study of 'pattern and purpose' in Iron Age East Yorkshire (2017). All of these studies have necessarily been quite narrowly focused. The research summarised in this chapter (from Downum 2022) represents a wider ranging study of non-metal decoration and art across southern England (Fig. 12.1).

De-throning the Metaphysical Concepts of Substantial Essence and Form

The title of this section is borrowed from Conneller (2011, 19) and her articulation, drawing on Deleuze & Guattari (1988), of a 'rhizomatic' approach to materials and matter.

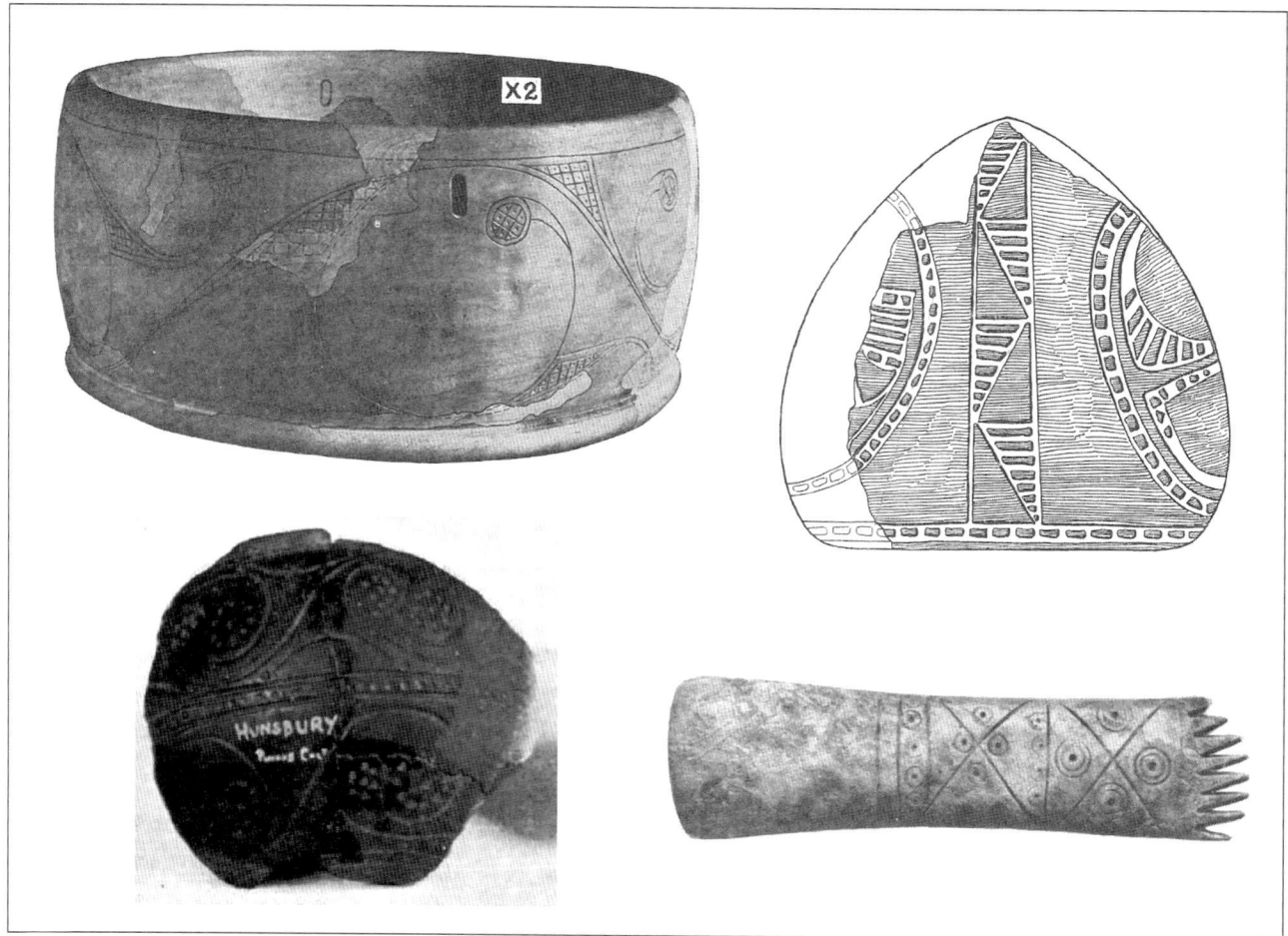

Figure 12.1 Celtic Art on 'other' materials. Clockwise from top left (not to scale): wooden bowl from Glastonbury (Bulleid & Gray 1917, plate LI); stone 'ritual' object from Barnwood (Anon 1934, 60, fig. 1); bone comb from Meare Lake Village West (Bulleid & Gray 1948, plate XXI, HH61); pottery from Hunsbury (Fell 1936, plate XI, D1).

Many past critiques of metal-centric approaches to Celtic Art have focused, whether explicitly or implicitly, on the fact that to ignore non-metals is to conduct a *partial* study. To elaborate this point just a little, in his seminal paper asking, 'what do objects want?', Gosden (2005, 193) explored the notion of a 'stylistic universe' created by groups of objects such as pots or metal ornaments acting together to exert agency over people and process. By this logic, to look at just metal is to investigate only part of that universe, and thus to come up with an only partial understanding of the past.

Before we get into the details of the material, it is important to delve a little deeper, drawing on Conneller's de-throning point above, into the reasons why it is important to consider a wider range of materials – beyond those fundamental arguments about partialness – particularly in relation to discussions concerned with the relationality of materials. In the *Technologies of Enchantment* book, we considered the fact that the rise of Celtic Art occurred around the same time as people's broader conceptualisations of materials appear to have changed (Garrow & Gosden 2012, 14). As Gosden (2020, 16) has subsequently put it, 'when iron started to become common, bronze also made a comeback (as did gold and silver). Celtic Art in Britain emerges at a time when metalwork becomes more important overall, so that bronze, iron, gold, and silver were all newly valued, a value which made them ideal media for complex decoration'. It is vital to note that, in all likelihood, other non-metal materials would have been implicated in this re-evaluation of matter as well: 'novel substances forge and fracture relationships between existing materials, and in so doing disrupt the taxonomies and the symbolic and practical repertoires which define and constitute the wider 'culture' of materials' (Shove *et al.* 2007, 104). As Shove and colleagues (Shove *et al.* 2007, 106) discuss in relation to a rather more recent innovation,

> the arrival of the plastic washing-up bowl arguably transformed both the practices of doing the dishes and the material identities of enamel and ceramic. Suddenly, washing up without one became associated with undesirable clatter… In the same move, enamel acquired undesirable characteristics: causing chips, scrapes, cracks, scratches and problems that no one had noticed before.

Any re-evaluation of metals during the Middle Iron Age is likely to have involved a knock-on re-evaluation and re-constitution of other materials as well. Alongside these realisations to do with the relationality of materials, it is also important to note two other key, related points: that the substances we view as 'metals' may not have been seen as a category or coherent group of materials in the prehistoric past (Sharples 2008, 205) and that the properties of 'the same' materials can vary in different situations when different qualities come to the fore (Conneller 2011, 22).

Motifs and Materials During the Iron Age and Early Roman Periods in Southern England

In anticipation of our summary of key elements of Downum's research presented in the remainder of this chapter, it is helpful first of all to raise two key points. As Sharples (2008, 205) simply put it, 'there is much to be learned by taking a more inclusive approach to the term art, and in the British Iron Age it is arguable that most of the art objects that conform to Megaw & Megaw's [2001] limited definition [of art] are in media such as ceramic or bone/antler'. Equally, as Joy points out, 'in recent years, the study of the motif has gone out of favour in Celtic Art studies', but those very motifs are vital in terms of how relational object networks were created, connecting different kinds of material culture across space and time (Joy 2020, 111, 112). He goes on to suggest that 'in the words of Küchler, motifs can provide the brain with "a special thing-like tool for thinking" (Küchler 2013, 27)' (Joy 2020, 113); and, we would add, a mechanism by which to approach relations between and across materials.

In conducting research into Celtic Art and Iron Age/Early Roman visual culture across a much broader width of the material spectrum, Downum (2022) investigated pottery, antler/bone, wood, and stone objects, situating them in comparison to the decorated metal items captured within the original 'Technologies of Enchantment' project's Celtic Art database (Garrow & Gosden 2012; subsequently updated as part of the ECAIC project, see Nimura *et al.* 2020). A total of 2014 decorated objects found across southern England were evaluated as part of the study, including 1351 ceramic vessels, 300 antler/bone combs, 13 wood objects, 64 stone objects, and 286 metal objects. Core research questions included: How did artistic expression change during the later Iron Age and into Early Roman Britain? What role did decoration serve, and what can it reveal about various social connections and visual communication? Did different materials present similar forms of visual expression, and were these used to define communities?

Given the complexity of the task envisaged – analysis of decoration across this wide range of materials – a targeted approach was necessarily taken to site selection. The study region was broadly defined as Cunliffe's (2005) three main southern style zones: the Eastern, the Central Southern, and the Southwestern. Within these regions, specific sites were chosen based on the richness of their individual material assemblages, and where possible, because they contained multiple periods of occupation. Ceramic analysis was conducted using material from Danebury in Hampshire, the Meare Lake Villages in Somerset, and Dragonby in Lincolnshire. For the other materials, a wider range of sites was included, since total numbers of relevant objects were far lower (see Downum 2022 for details). Data collection was focused primarily on motifs, but also included information about pattern (*e.g.* combinations of motifs) and decorative techniques. The classificatory scheme adopted was developed initially in relation to the well-known metal corpus, but then adapted to incorporate other materials as well. This methodology drew on a combination of Prieto-Martínez and colleagues' visual approach (Prieto-Martínez *et al.* 2003) and Cunliffe's (1984) coding system established for Danebury and Hengistbury Head (see Downum 2022, 59–63). The total period covered across all sites was approximately 600 BC–AD 100, with the greatest emphasis placed within the latter two centuries of that bracket. Chronological studies were conducted as part of the research, but due to the complexities of dating involved on many sites, these worked most effectively at a site-specific rather than trans-regional level. Our intention in this short chapter is to outline a limited number of key findings identified as a result of this cross-materials data collection and analysis. Our primary focus is, of course, on the relationship between Celtic Art/decoration and materials.

Figure 12.2 outlines which motifs were employed across the entirety of the study period and all materials – the fundamentals of Iron Age/Early Roman visual culture, if you like – by region. As would be expected, certain motifs feature more prominently than others – chevrons, lozenges, arcs, circles, and horizontal, vertical, and diagonal bands. The presentation of the motifs by region allows us possible insight into local traditions and preferences in terms of visual culture that could have *potentially* created spatial differences in terms of the patterning observed. Whilst there is some variation (for example, the Central Southern region has more chevrons and the Southwestern region more arcs), in some ways the most striking aspect of the graph is the *similarity* of motifs employed right across the south of England.

While social connections, traditional styles, and functional roles played a significant part in the selection of decoration, the affordances of different materials and the materiality of the objects themselves would also have influenced the decorative choices made. Differences in surface texture, surface space, and elasticity would have affected what decoration could be employed and how it was applied, while accessibility and preservation of the material would affect what evidence is available for analysis. Pottery is relatively easy to manipulate and shape, the material is relatively easy to acquire, and it allows for a variety of application techniques. Similarly, metal can be manipulated with heat and decoration added using a wide variety of techniques, allowing for both two-dimensional and three-dimensional forms. Although decorated wood has a low representation in the archaeological record, the material itself would have been easy to decorate and manipulate, suggesting that a larger collection probably existed but simply has not been preserved. Antler and bone, on the other hand, would need to be soaked before the material could be decorated, and if this was not completed then the decoration would appear scratched (similar to scratched pottery decoration applied after the material has hardened). In the case of stone, the specific material type would affect the decorative possibilities; overall, objects of shale, chalk, or limestone are most frequently decorated.

Figure 12.3 presents the decorative elements employed on all decorated artefacts by material. The relative amounts of variability are approximately as expected, given the

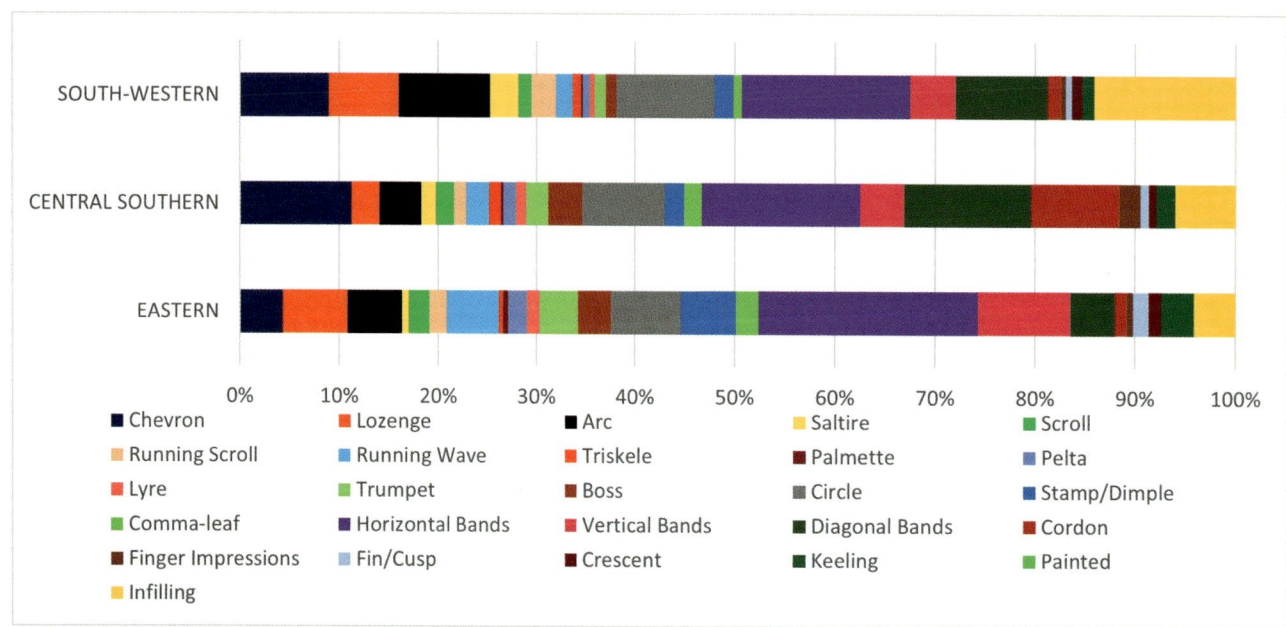

Figure 12.2 Decoration by style zones (n = 843 objects from the Eastern zone, 561 from the Central Southern zone and 610 from the Southwestern zone).

material affordances discussed above. Antler/bone, for example, sees the least variability in terms of motifs. Stone appears roughly similar towards the right-hand side of the graph, but with several additional smaller percentage variables (towards the left-hand side). The sample size for wood is very small, but that material exhibits quite a large amount of variation. Ceramics are similar, with both wood and ceramic materials dominated to a large extent by geometric decoration – horizontal and vertical bands, chevrons, and lozenges. The decorative 'signature' of metal is very different, contrasting significantly with all the other materials. The metals category is characterised by a fairly even spread of decoration types – most of the segments along the bar are fairly evenly weighted. The geometric motifs that dominate non-metals are not there in such numbers. Arguably the most striking aspect of the graph is the overall broad similarity between non-metal objects, and their notable differences as a group when compared with metal objects. The 'swirly' Celtic Art style is much more clearly linked to the latter.

Figure 12.4 depicts the same data from the opposite perspective, so to speak, allowing us to assess how motifs were divided up by material, as opposed to vice versa. The dominant, most clearly visible materials in the chart are ceramics and metal. Ceramics contributed by far the greatest number of motifs to the dataset. Pots were, it seems, decorated in specific ways – notably, as mentioned briefly above, with more geometric motifs (which dominate the left-hand side of the chart). Metals, on the other hand, contributed relatively fewer motifs in numerical terms, but a great *variety* of motif types overall (as seen in Fig. 12.3). As clearly indicated on the right-hand side of Figure 12.4, there were many motifs that *only* feature on metal objects. Antler/bone also features fairly prominently in the chart, but in relation to only a small range of motifs, most notably the saltire and the simple circle. Stone and wood are less prominent overall, but in their cases sample sizes are much smaller.

As discussed above, the plastic medium of ceramics means that all decorative types *could* have been applied to pots, yet in the substantial sample datasets interrogated as part of this study, they were not. By contrast, with one exception (vertical bands), metal objects feature every single motif type, albeit sometimes in small numbers. Taken together with the data presented in Figure 12.3, this patterning suggests that different materials were perceived decoratively in different ways. Ceramics, antler/bone, and metal all have their own distinctive signature that is different to the others. Whilst the material affordances of antler/bone limited their decorative possibilities to some extent, ceramics and metal could have been more similar, but were not. Metals arguably dominate the picture, appearing much more 'promiscuously' across the full spectrum of the decorative range. In this sense, the motif 'language' of metal might be described as more complex, although – as Sharples (2008, 203) reminds us in loosely comparing Celtic Art to Bodi cow hide markings – we must be careful in assuming we fully understand its 'grammar'.

Discussion

The sections above have identified and described clear patterning in relation to the kinds of motif employed on different materials. This identification and description

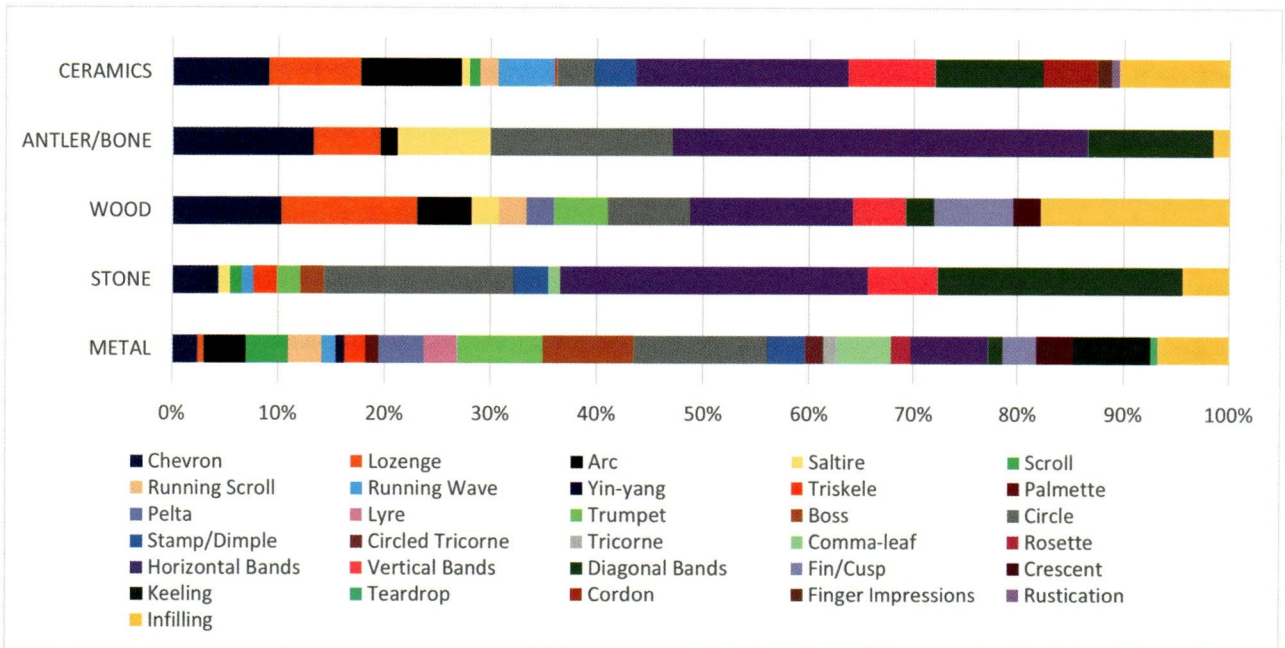

Figure 12.3 Materials and their decoration (n = 1351 ceramic vessels, 300 antler/bone combs, 13 wood objects, 64 stone objects, and 286 metal objects).

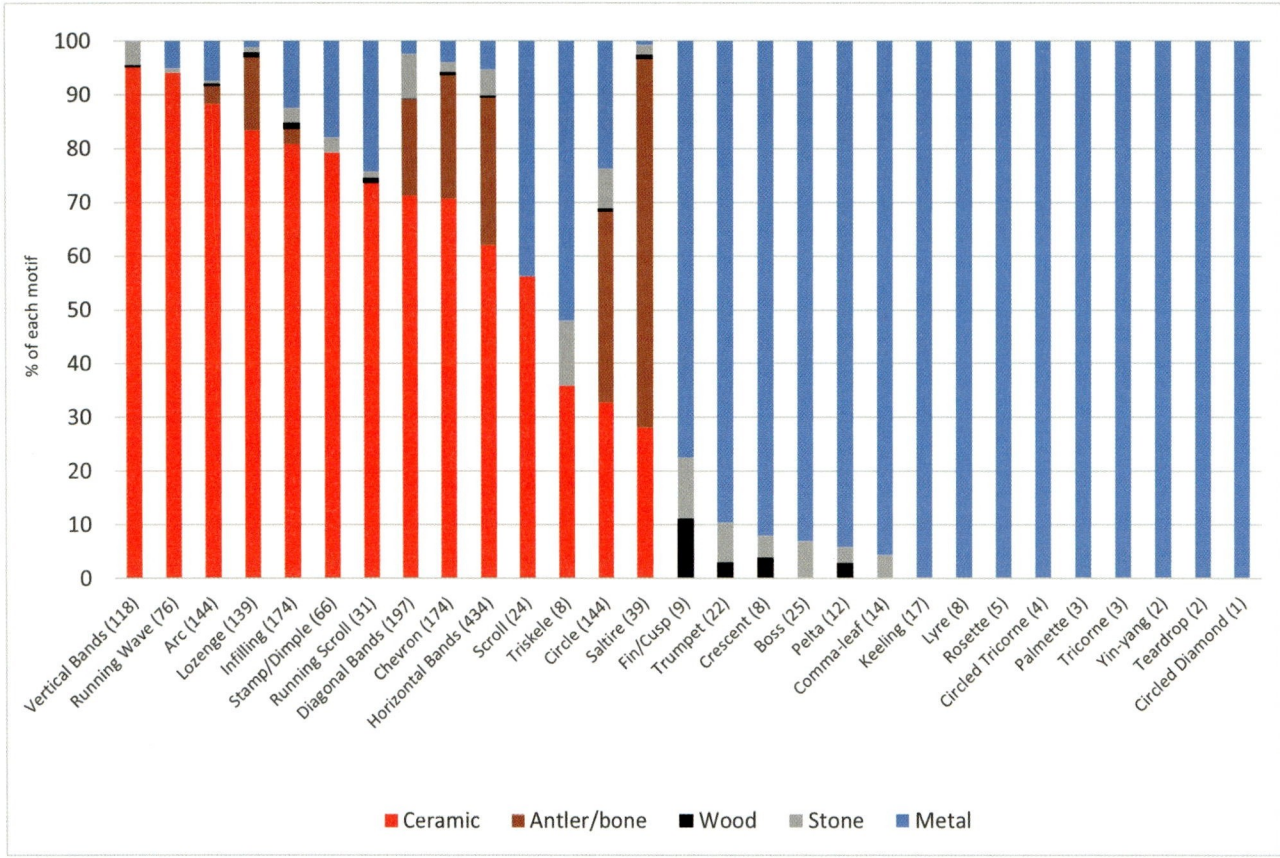

Figure 12.4 The relationship between motifs and materials, ordered by prevalence in ceramics and metal (total numbers of motifs recorded in brackets).

represent a critical first stage in our understanding of the issues we are hoping to explore. Any interpretation of these patterns – the next step in that process – is, however, much trickier to approach. The roles, as well as the meanings, of decoration are, put simply, very difficult to ascertain, as others before us have found as well. Joy (2011, 220), for example, suggested that 'decorated artefacts performed in negotiations of social power and cosmology … decoration was employed for multiple reasons. The decision of what type of artefact to decorate was dependent on specific social and cultural needs', while Chittock (2017, 172) has argued that 'pattern had different purposes in different spheres of activity; it did specific things during specific activities and during deposition, and perhaps affected people in ways that caused them to treat patterned objects in different ways to plain objects' – both necessarily perhaps choosing to remain vague given the interpretive difficulties of the challenge they were facing.

The study of Iron Age and Early Roman visual culture in southern England presented briefly here has added weight to arguments such as these, as well as broader interpretive narratives drawn up by Gosden (2005) and others which have suggested that objects – and their decorative attributes – did important *work* at this time. It has hinted that the complexity of meaning which lay behind decorative decisions was significant and, it must be said, quite impenetrable to us today. At the same time, it has demonstrated that the 'stylistic universe' in which people in southern England were operating within was largely shared – the total range of motifs that people had to draw on were similar in different parts of the country. However, it does seem that there were boundaries (albeit permeable ones) in that stylistic universe between materials. Certain matter was decorated in certain ways. Metal objects were decorated differently. Ceramics and wood could theoretically have been decorated in the same way, but were not. In emphasising differences between metals and non-metals in this way, we do not wish to steer the debate back to tired (and arguably anachronistic) discussions of prestige goods, elite status, and 'fine' metallic art. It is important to remember that those material/decorative 'boundaries' were porous and *could* readily be crossed. Materials are best understood relationally. It is hoped that the small number of core findings of the research presented here, building on other recent studies with similar aims, indicate that a lot is to be learnt by studying metals and other materials in combination. As the recipient of this volume pointed out (Gosden 2005, 208), it is vital that we consider not just the meaning but the *effects* of objects, and 'how those effects emanated in a complicated fashion from things *en masse*'.

Acknowledgements

We would like to thank Chris Gosden, of course, for inspiring Duncan Garrow's and subsequently Sarah Downum's interest in and engagement with Celtic Art; hopefully, Chris, you will feel some comfort in seeing the stylistic universe a bit more complete now. We would also like to acknowledge John Creighton's significant input as co-supervisor (along with Garrow) of Downum's PhD in Reading and thank him for reading a draft of this paper. The intellectual debt owed to, and influence of, many other friends and colleagues working on comparable issues will be evident from the works cited – thank you all.

References

Anon. (1934) Notes: An early British fragment. *Antiquaries Journal* 14, 59–61.

Bradley, R. (2009) *Image and Audience: Rethinking Prehistoric Art*. Oxford, Oxford University Press.

Bulleid, A. & Gray, H. St G. (1917) *The Glastonbury Lake Village (Vol 2)*. Glastonbury, Glastonbury Antiquarian Society.

Bulleid, A. & Gray, H. St G. (1948) *The Meare Lake Village: A Full Description of the Excavations and Relics from the Eastern Half of the West Village, 1910-1933 (Volume 1)*. Privately printed.

Chittock, H. (2014) Arts and crafts in Iron Age Britain: Reconsidering the aesthetic effects of weaving combs. *Oxford Journal of Archaeology* 33, 313–26.

Chittock, H. (2017) *Pattern and purpose in Iron Age East Yorkshire*. Unpublished PhD thesis, University of Southampton.

Conneller, C. (2011) *An Archaeology of Materials: Substantial Transformations in Early Prehistoric Europe*. Abingdon, Routledge.

Cunliffe, B. (1984) *Danebury: An Iron Age hillfort in Hampshire*. London, Council for British Archaeology.

Cunliffe, B. (2005) *Iron Age Communities in Britain*. Abingdon, Routledge.

Deleuze, G. & Guattari, F. (1988) *A Thousand Plateaus: Capitalism and Schizophrenia*. London, Athlone.

Downum, S. (2022) *Visual culture and 'decoration' in Iron Age Britain: Seeing beyond metal*. Unpublished PhD thesis, University of Reading.

Evans, C. (1989) Perishables and worldly goods: Artefact decoration and classification in the light of wetlands research. *Oxford Journal of Archaeology* 8, 179–201.

Fell, C. (1936) The Hunsbury Hill-Fort, Northants. *Archaeological Journal* 93, 57–100.

Garrow, D. (2008) The space and time of Celtic Art: Interrogating the 'Technologies of Enchantment' database. In D. Garrow, C. Gosden & J. D. Hill (eds), *Rethinking Celtic Art*, 15–39. Oxford, Oxbow Books.

Garrow, D. & Gosden, C. (2012) *Technologies of Enchantment? Exploring Celtic Art: 400 BC to AD 100*. Oxford, Oxford University Press.

Giles, M. (2008) Seeing red: The aesthetics of martial objects in the British and Irish Iron Age. In D. Garrow, C. Gosden & J. D. Hill (eds) *Rethinking Celtic Art*, 59–77. Oxford, Oxbow Books.

Gosden, C. (2005) What do objects want? *Journal of Archaeological Method and Theory* 12(3), 193–211.

Gosden, C. (2020) Art, ambiguity and transformation. In Nimura *et al.* 2020, 9–22.

Jope, M. (2000) *Early Celtic Art in the British Isles*. Lincoln, Clarendon Press.

Joy, J. (2011) Fancy objects in the British Iron Age: Why decorate? *Proceedings of the Prehistoric Society* 77, 205–30.

Joy, J. (2020) How can Celtic Art styles and motifs act? A case study from Later Iron Age Norfolk. In Nimura *et al.* 2020, 111–26.

Küchler, S. (2013) Threads of thought: Reflections on art and agency. In L. Chua & M. Elliott (eds), *Distributed Objects: Meaning and Mattering after Alfred Gell*, 25–38. Oxford, Berghahn.

Megaw, R. & Megaw, V. (2001) *Celtic Art: From Its Beginning to The Book of Kells*. London, Thames & Hudson.

Nimura, C., Chittock, H., Hommel, P. & Gosden, C. (eds) (2020) *Art in the Eurasian Iron Age: Context, Connections and Scale*. Oxford, Oxbow Books.

Nimura, C., Gosden, C., Hommel, P., Chittock, H. (2020) Collecting Iron Age art. In Nimura *et al.* 2020, 1–8.

Prieto Martínez, M., Cobas Fernández, I. & Criado Boado, F. (2003) Patterns of spatial regularity in late prehistoric material culture styles of NW Iberian Peninsula. In A. Gibson (ed.), *Prehistoric Pottery: People, Pattern and Purpose*, 147–88. Oxford: British Archaeological Reports S1156.

Sharples, N. (2008) Comment I: Contextualising Iron Age Art. In D. Garrow, C. Gosden & J. D. Hill (eds), *Rethinking Celtic Art*, 203–13. Oxford, Oxbow Books.

Shove, E., Watson, M., Hand, M. & Ingram, J. (2007) *The Design of Everyday Life*. London, Berg.

13

Jet and Gender in Late Roman Britain

Cameron Moffett

This paper considers the gender-specific nature of the various types of jet objects produced in late Roman Britain. The vast majority are items of female jewellery and include amuletic pendants. The main non-jewellery objects relating to women are textile-working implements and equipment for chopping and grinding (e.g. for food, medicine, or magic). There are also a small number of more occult items, potentially used for divining the future. Apotropaic and/or healing magic are probably implied in all the female-gendered jet object types. There are, however, a small number of jet objects from this period that seem to relate specifically to men, and these usually have associations with one of the pagan deities known as the saviour-gods. The sword handle from the Walbrook Mithraeum is the best-known example. There is also evidence for wooden boxes inlaid with jet, and the iconography of the small number of pieces of figured inlay indicates that such boxes were probably male-gendered. Exploration of the significances of jet for both genders in late Roman Britain results in a clear picture of the permeability of the divide between magic and religion.

Keywords: Jet, Late Roman Britain, Gender, Magic, Religion

Background

Jet, a fossilised wood, is a valuable non-metal material that has had intermittent periods of popularity in Britain over the last five millennia. It is easy to carve and polishes up to give an attractive shine. The principal deposits of jet known and used in Britain are around Whitby, North Yorkshire, where it erodes out of the sea bed. Jet has a number of unusual physical properties that come from it being a fossilised organic material, and these must have made it interesting to people in the past. It is electrostatic – rubbing jet can generate a temporary charge sufficient to attract and lift light materials, such as textile fibres or small feathers – and though it looks like a stone, jet floats and can be burned.

There are minor deposits of jet at a number of locations on the Continent (Germany, Spain, Portugal, France, and Russia) (Allason-Jones 1996, 5–6), but the best sources of the hard jet there used for the manufacture of objects are found in southern Germany. The earliest jet artefacts, tiny amulets dating to the end of the Palaeolithic, have been found in Germany. After a long period in which jet does not appear to have been used, there were two episodes in the Iron Age during which jet was a popular material for the manufacture of items of personal adornment. This trend had run its course by the end of the 1st century BC (Muller 1987, 94–5).

Jet is mentioned in a number of key texts of the Classical period. While Pliny, in the 1st century AD (Pliny, Book 36), noted that jet had unusual physical properties, his main focus was on its medicinal uses. It was most commonly prescribed for women's complaints, and it could be ground up and either taken internally, applied externally, or burned and the fumes inhaled. Pliny describes a number of other materials as being suitable for the manufacture of jewellery and amulets, but neither he nor Galen, writing a century later (Galen, Book IX.203), mention jet in this context. However, by the later Roman period, the production of jewellery and amulets had become the principal use for jet. Only Solinus, writing in the 3rd century AD, refers to jet's visual appeal, specifically citing British jet for its quality (Solinus, Book 22.11).

Jet in Roman Britain

Though jet had been used at times for making high-status items of personal adornment in the Neolithic (4000–2300 BC) and Bronze Age (2300–750 BC), its appearance in Iron Age (750 BC–AD 43) contexts in Britain is rare. This may be the result of different burial practices; interment then was uncommon, and jet would not survive the normal cremation process. The origins of the jet jewellery industry are in York (*Eboracum*), the capital city of Britain's Northern Province and only 47 miles from Whitby, where there is evidence for workshops making simple objects such as hairpins from the 2nd century AD (Royal Commission on Historical Monuments England 1962, 142; Allason-Jones 1996, 22–3). From early in the 3rd century AD, jet jewellery became much more common, and the presence at York between AD 208 and 211 of the Emperor Septimius Severus with the Imperial household, including the Empress Julia Domna, presumably accompanied by her retinue, may have been a factor influencing this trend.

From York, the fashion for jet spread quickly through Britain during the 3rd century AD. On the Continent in this period, a group of jet objects extremely similar stylistically to those from Britain is seen in the Rhineland, a region which then had strong trading and military links with Britain. As in Britain, jet had been used intermittently on the Continent (Muller 1987), but in the case of the later Roman period, the presumption in the literature is that, despite the existence of sources of the material nearer to the Rhineland, jet was being traded from Britain to the Rhineland, where the Roman towns of Cologne and Trier have produced large numbers of jet objects (Muller 1987, 95–8). The relationship between jet objects in Britain and the Rhineland in the 3rd and 4th centuries AD is a subject that would benefit from further study.

In Britain, both finished objects and raw jet were traded, and there is evidence of jet working from a number of Yorkshire sites other than York, such as Beadlam Villa, which has produced 11 fragments of jet working waste (HOMS[1]). The Rural Settlement of Roman Britain database (Allen *et al.* 2018), using data from published reports and grey literature, provides a general indication of the distribution of jet objects: they occur mainly in the east but with a noticeable western cluster in the affluent Cotswolds area.

The database also shows the mainly westerly distribution of objects made of shale, the most common of a group of black materials used in the Roman period for the same general purposes as jet. Shale, the principal source of which in Britain is Kimmeridge in Dorset, has its own history (Denford 2000) but is in many ways associated with jet. Its physical properties differ to those of jet, and it is more suited to the making of larger, simpler objects. Shale armlets were produced in the Kimmeridge area from the Iron Age. Two centuries later armlets were still being made of shale there, but in addition, shale was being used to manufacture Roman-style objects like trays and tables, and it was also used, to a lesser extent, to imitate jet in the manufacture of beads and finger rings. Shale could be oiled and polished to enhance its resemblance to jet, but it is likely that consumers could distinguish between the two. Shale was not used for any of the more significant object types that occur in jet, such as amulets.

The inclusion of jet objects in some very rich female inhumation burials is seen particularly in York (Allason-Jones 1996) and London (McKenzie *et al.* 2020), but there are also large assemblages from a small number of civil/secular contexts, principally the centres of towns that were flourishing in the late Roman period: Canterbury (Blockley *et al.* 1995); Silchester (Lawson 1975); Wroxeter (HOMS); and Colchester (Crummy 1983). These groups of jet objects are the principal comparative sources used for this brief study.

The Uses of Jet for Women

By far the most common object type made of jet is the bead, of which there are a number of different shapes. These were used for necklaces, bracelets, and anklets. Hair pins, bangles, armlets, finger rings, and hair rings, the manufacture of which required a larger piece of jet than beads, were relatively common. Spindle whorls, knife handles, palettes and small plaques, and dice (Davis 2018, 74) also occur in jet but in much smaller numbers. More unusual are amulets, in the form of pendants for necklaces, and tiny figurines almost all of which depict animals. The figurines were usually carved incorporating a base plate, allowing them to stand alone, but most had the potential to be suspended from a cord and worn as a pendant. The majority of jet amulets are cameo-style and usually depict either an individual, often a *gorgoneia* (Medusa) or a married/betrothed couple. Magical protection would have been afforded by the amulet for: the people depicted on the amulet, or, in the case of the *gorgoneia*, specifically for the wearer, as the face of the Medusa was believed to deflect evil (Parker 2016).

Medical pendant amulets were also made in jet (Moffett 2019). These are rarer finds, and as there are two from Wroxeter, it may have been the base of an artisan producing them. One is in the form of a breast (Fig. 13.1), presumably intended to promote lactation or treat mastitis, while the other depicts the lower half of a leg with foot. This last amulet is incomplete and prior to a recent reassessment had been provisionally interpreted as depicting an axe or a phallus; however, as Allason-Jones (2005, 124) has noted, there is a noticeable lack of fertility amulets made in jet.

There are many examples of burials of Romano-British women interred wearing jet jewellery and accompanied by high-status accessories made of jet (*e.g.* Find No. Bu90, McKenzie *et al.* 2020, 42); the inclusion of jet in

Figure 13.1 Jet pendant amulet in the form of a breast. Perforated shank for suspending on reverse. English Heritage accession number: 88061214. Diameter: 15 mm (© Historic England/Bob Smith).

male burials is very rare (Allason-Jones 2005, 123). The Catterick *gallus*, however, was buried wearing a number of items of jewellery including jet pieces. This individual has been identified as biologically male from their skeletal remains (Wilson 2002, 41), but presented as female through their personal adornment, providing a rare demonstration from the archaeological record of gender as performative (Butler 1993).

The animal figurines made of jet are usually encountered in child burials. In Nina Crummy's 2010 study of late Roman infant burials, a group of jet figurines depicting bears, which were found in Colchester interred with children, are suggested to relate to the Greek cult of Artemis, goddess of childbirth and childrearing, and are seen as evidence of mothers providing protection for their children into the afterlife.

Jet Objects for Men

While burials of males wearing jet or interred with jet grave goods are very rare, there are a very small number of jet objects from non-burial contexts, the iconography or function of which indicate that they were meant to be used by men. The best-known example is the unparalleled jet sword hilt from the Walbrook Mithraeum, London: a temple of a religious cult restricted to male worshippers. Citing the sword hilt, Allason-Jones (1996, 17) queried the relationship between jet and gender more than two decades ago. Also from the Mithraeum are two probable scabbard fittings in jet, potentially associated with the sword hilt – Finds IX.52 and VIII/IX.8 (Shepherd 1998, 176, figs 206 & 207). An image of the jet hilt can be found on the Museum of London's website (ID No. 18263).

An object type that has not previously been considered to be specific to men is a box inlaid with pieces of jet. It has been observed that in Roman York, jet was not used to decorate the jewellery caskets occasionally found in the inhumation burials of women, in which jet jewellery was otherwise such a notable feature (Allason-Jones 1996, 24). Pieces of jet inlay are not common finds; in general, larger sites have produced only one or two pieces (*e.g.* Wroxeter, York, and Canterbury). These are usually triangular or rectangular in form and often have simple linear decoration: a cross-hatched border along one or two edges is typical, and their reverse sides have often been roughened for better adhesion when glued to the carcass of a wooden box.

There is an excavated example of what may be a complete set of inlay pieces from a single box from Richborough Fort, Kent. Pit 314 was a hastily interred group burial of an adult man, woman, and an adolescent dated *ca.* AD 400 (Bushe-Fox 1949, 36). The burial pit contained two 'trinket' boxes. One was an example of a standard Roman box for valuables made of wooden boards with copper-alloy plating and binding applied to provide reinforcement, a hinged lid, and often a lock plate. This is an object type that occurs fairly frequently in burials. Two good examples of such boxes were found in the cemetery at Butt Road, Colchester (Crummy 1983, 85–9) and another, from a burial in Mansell Street, London, has been analysed and reconstructed (Watson 1997). The other box from Pit 314 was indicated by the presence of 19 pieces of jet inlay, its wooden carcass having decomposed. Within this set of inlay pieces, at least three different patterns of decoration exist, suggesting that the sources of the inlay could lie in the recycling of other flat jet objects, *i.e.* palettes and plaques.

The proposal that these 19 pieces of inlay may derive from the decoration of a box is supported by Paola Pugsley's 2003 work on Roman domestic wood. The evidence from Britain for a small square box carved from a single piece of wood with a drop lid is unusually good. Pugsley (2003, 60; fig. 4.1) identifies this as a traditional Western European form with its antecedents in the Bronze Age. The 19 pieces of jet inlay from Pit 314 at Richborough would provide a continuous linear trim of *ca.* 530 mm: comparable to the total length of four sides of each of the two boxes cited by Pugsley.

The jet-inlaid box from Pit 314 at Richborough was found closest to the female burial and was considered by its excavator Bushe-Fox (1949, 36) to be a woman's trinket box. The other box, from which a number of pieces of jewellery seemed to have spilled, was nearest to the man's body (Cunliffe 1968, 36). Bushe-Fox assumed that both boxes belonged to the woman; 70 years later, this interpretation needs revisiting. Since that publication, the number of excavated pieces of jet inlay has increased, and five figured pieces are now known from Britain, the iconography of which, as a group, provides a context within which jet-inlaid boxes can be considered in greater depth.

Figured Jet Inlay

Two of the five figured pieces are small fragments, making identification of the subject difficult. A piece showing only an arm raised to mouth level was found at Dorchester on Thames (Frere 1984, fig. 30, no. 20), and parallels were observed between this and two other excavated figured inlay pieces from Gloucester and Barton Court Farm (Frere 1984, 139). The fragmentary piece from Barton Court Farm, Oxfordshire (Miles 1984), shows a well-muscled male torso. The more complete Gloucester example depicts a squatting, nude male figure with genitals exposed, possibly wearing a headdress and blowing a horn held in his right hand (Hassall & Rhodes 1974, 79, fig. 29, no. 5 & plate VIb).

The other known pieces of figured inlay are larger and were probably originally trapezoidal in shape. That from the villa at Whitton, Norfolk (Reid Moir & Maynard 1933, 248, fig. 55), is the only one depicting a clearly identifiable deity: Atys, the consort of Cybele, the eastern Great Mother goddess, who is recognisable by his Phrygian hat. The fifth example is from York (Allason-Jones 1996, 46, no. 285) and shows a stylised nude male figure wearing a possible headdress similar to that worn by the Gloucester figure. These larger trapezoidal pieces may have been used differently to the generic rectangular and triangular pieces of jet inlay that occur more frequently.

Henig (1984a, 5:G6–8 & fig. 113.6) interpreted the torso from Barton Court Farm as possibly depicting Hercules or a satyr, and in the same report he compared the Gloucester figure to satyrs seen on gemstones and suggested it to be one of that family. Though so little survives of it, the Dorchester piece, with its upheld hand, could, like the Gloucester example, show a person blowing a horn or possibly calling, and may be tentatively linked with satyrs. Potentially, three cults are indicated here: the probable satyrs (Fig. 13.2), naked and noisy, are likely references to the cult of Bacchus; Allason-Jones (1996, 16) has suggested that jet hairpins with cantharus-shaped heads relate to the cult of Bacchus. The other two cults probably represented in the figured plaques are those of Atys and Hercules. All these are part of the group of later Roman deities called the saviour-gods, who, in their mythologies, either narrowly escaped death or died and were resurrected. Mithras, mentioned above in the context of the jet dagger handle from the Walbrook Mithraeum, is another of the saviour-gods.

A Market for Jet Inlay Pieces

In addition to the box with jet inlay from Pit 314, the larger assemblage of jet from Richborough Fort is of considerable interest in that it is atypical. The site produced a further four fragments of jet inlay, making it the largest group of inlay pieces from a single site in Britain. The jet from Richborough (HOMS), a very large and important late Roman military base, is compared below to that from five major

Figure 13.2 Piece of jet inlay depicting a satyr(?). Yorkshire Museum accession number: SC1987.24. Height: 48 mm; Width (max.): 25 mm (© York Archaeological Trust).

late Roman urban sites with large jet assemblages: Wroxeter; Colchester; York; Silchester; and Canterbury (Table 13.1).

The inlay from Richborough accounts for almost half of the site's jet objects, and the group is sufficiently large that a pattern in the use/reuse of jet inlay can be suggested on its basis. The discrete group of inlay pieces from the pit burial seem to derive from the recycling of potentially at least three different source objects. The other four inlay pieces from the wider site of Richborough are all decorated, all different to each other and to the decoration seen on the inlay from Pit 314. We see here that individual pieces of jet were being routinely reused in this context.

The Significance of Jet as a Material

What the implied market for pieces of jet inlay may show is that a single piece of jet divorced from the larger object from which it derived had the potential to function as a protective amulet. From Crummy's (2010) analysis of child burials in Colchester, there is a similar indication of

Table 13.1 Jet objects from civilian sites compared to jet objects from Richborough Fort.

Site	Jewellery	Pendant amulets	Domestic equipment	Inlay	Misc and unidentified	Total numbers
Wroxeter	98%	0.7%	0.18%	0.18%	0.55%	537
Colchester	99%	0.3%	0	0	0.3%	369
York	84.4%	2.4%	5.7%	2%	5.3%	327
Silchester	90.4%	1.6%	4.7%	0	3%	63
Canterbury	80%	0	0	8%	12%	26
Richborough	52%	0	2%	46%	0	50

a belief in the apotropaic power of jet as a material. These interments contained recurring sets of specific objects, one of which was an item of jet jewellery, which was sometimes represented by a single bead.

The electrostatic property of jet has been described above, but colour, which had great significance in antiquity, must also have been a major factor in jet's popularity. Black was then and is still regarded as chthonic, and Henig (1984b, 185) convincingly suggests that the apotropaic power of jet might lie in an expectation that its blackness could be used to deflect the blackness of death and illness. Hella Eckardt (2014, 115) points out that black pendants of this kind are not known in Italy, suggesting of the Medusa amulets that in Roman Britain, provincials are 'rendering a highly classical apotropaic image in local artistic styles and in what is clearly perceived to be a powerful material'.

Conclusions

Henig (1984a, 5:G8) describes the objects made in jet as a group as 'entirely Graeco-Roman in conception'. Yet there was clearly a strong and specific connection between jet and the women of Roman Britain, which, as Allason-Jones (2005, 124) indicates, the uses of jet described by the authors of antiquity do not seem sufficient to explain. The medical amulets from Wroxeter highlight a close association between jet and good health, including reproductive wellbeing: the offer of a successful birth would have appealed to followers of the cults of the saviour-gods, who may have believed that jet provided a supplementary means of achieving rebirth and resurrection.

Within the group of female jet objects, function is either decorative (jewellery), domestic (*e.g.* spindle whorls), or, occasionally, occult (dice, palettes, and plaques). In the case of men's objects, function is less clear. The jet weapon accessories from the Walbrook Mithraeum are taken to be ceremonial props or votive offerings, and not intended for practical use (Shepherd 1998, 161). Similarly, uses for a box inlaid with a significant material, such as jet, are likely to have been sacred rather than secular.

When compared to the many female-gendered jet objects, the rarity of probable male-gendered items argues for a sequence in which jet and its meanings were adopted by men from women. Women's jewellery had driven the proliferation of jet in Roman Britain, and it was presumably women's requirements that prompted the expansion of the range of objects made in this material, from its initial appearance in the Roman period for making simple hairpins through to its use for complex objects, often imbued with magico-religious significance. The interest in and uptake of mystery religions imported from the east that was a feature of the whole western Empire in the 2nd and 3rd centuries AD is very likely to have been a factor in the adoption of jet by some men (Henig 1984b, 95–127).

The use of jet in Roman Britain was at its peak in the 3rd and 4th centuries AD, but at the end of that period it almost disappears from the archaeological record. Christianity became the official religion of the Roman Empire during the 4th century, and it is suggested here that these contemporary occurrences are likely to be related: the doctrine of the Christian faith forbade the use of magic and amulets, unlike the cults of the other saviour-gods. Changes in women's fashions in later Roman Britain, which may also relate to the transition from pagan beliefs to Christianity, have been observed by Hilary Cool (1990). There is a decrease in the length of hair pins (Cool 1990) as women adopted simpler, more modest, hairstyles – in the Bible shorter hair for women is good but covering the hair is preferred. Running in tandem with these changes is a 4th-century boom in copper alloy bangles (Cool 1983), suggestive of a relocation of zones of the female body acceptable for personal adornment away from the hair to the wrists.

Acknowledgements

Grateful thanks to Lindsay Allason-Jones (University of Newcastle), Adam Parker (Yorkshire Museum), and Sharon Strong (English Heritage), who all generously made time to read this paper in draft, and whose comments and suggestions helped greatly to improve it. Any remaining errors are my own. Thanks to the editors of this volume for involving me in it, as well as to their anonymous reader. The York Archaeological Trust kindly provided an image; thanks to Ellie Drew for her help with that.

Note

1. HOMS: English Heritage Historic Object Management System (internal access only). Accessed by the author 27 January 2022.

Ancient sources

Galen. *De simplicium medicamentorum temperamentis et facultatibus*. Edited by K. G. Kühn (1964), *Claudii Galeni Opera Omnia*. Reprint. Hildesheim, Georg Olms.

Pliny the Elder. *The Natural History*. Translated by J. Bostock (1855). London, Taylor and Francis. Available at: <http://www.perseus.tufts.edu/hopper/text?doc=Perseus:text:1999.02.0137> [accessed 10 October 2022].

Solinus. *Collectanea Rerum Memorabilium*. Edited by T. Mommsen (1895). Leiden and Paris, Weidmann. Available at: <https://archive.org/details/collectanearerum00soliuoft> [accessed 10 October 2022].

References

Allason-Jones, L. (1996) *Roman Jet in the Yorkshire Museum*. York, The Yorkshire Museum.

Allason-Jones, L. (2005) *Women in Roman Britain*. York, Council for British Archaeology.

Allen, M., Blick, N., Brindle, T., Evans, T., Fulford, M. G., Holbrook, N., Lodwick, L., Richards, J. D. & Smith, A. (2018) *The Rural Settlement of Roman Britain: An Online Resource*. York, Archaeology Data Service. Available at: <https://doi.org/10.5284/1030449>.

Blockey, K., Blockley, M., Blockley, P., Frere, S. S. & Stow, S. (1995) *Excavations in the Marlowe Street Car Park and Surrounding Areas*. Canterbury, Canterbury Archaeological Trust.

Bushe-Fox, J. P. (1949) *Fourth Report on the Excavation of the Roman Fort at Richborough, Kent*. London, Society of Antiquaries.

Butler, J. (1993) *Bodies That Matter: On the Discursive Limits of 'Sex'*. London, Routledge.

Cool, H. E. M. (1983) *A study of the Roman personal ornaments made of metal, excluding brooches, from southern Britain*. Unpublished PhD thesis, University of Wales.

Cool, H. E. M. (1990) Roman metal hairpins from southern Britain. *Archaeological Journal* 147, 148–82.

Crummy, N. (1983) *Colchester Archaeological Report 2: The Roman Small Finds from Excavations in Colchester 1971–9*. Colchester, Colchester Archaeological Trust.

Crummy, N. (2010) Bears and coins: The iconography of protection in late Roman infant burials. *Britannia* 41, 37–93.

Cunliffe, B. W. (ed) (1968) *Fifth Report on the Excavations of the Roman Fort at Richborough, Kent*. London, Society of Antiquaries.

Davis, G. (2018) Rubbing and rolling, burning and burying: The magical use of amber in Roman London. In A. Parker & S. McKie (eds), *Material Approaches to Roman Magic: Occult Objects and Supernatural Substances*, 69–83. Oxford, Oxbow Books.

Denford, G. T. (2000) *Prehistoric and Romano-British Kimmeridge Shale* [data set]. York, Archaeology Data Service [distributor]. doi:10.5284/1000090.

Eckardt, H. (2014) *Objects & Identities: Roman Britain and the North-Western Provinces*. Oxford, Oxford University Press.

Frere, S. (1984) Excavations at Dorchester on Thames, 1963. *Archaeological Journal* 141, 91–174.

Hassall, M. & Rhodes, J. (1974) Excavations at the new market hall, Gloucester, 1966-7. *Transactions of the Bristol and Gloucester Archaeological Society* 93, 15–100.

Henig, M. (1984a) A jet plaque. In Miles 1984, Chapter IV.8.4, 5:G6–8.

Henig, M. (1984b) *Religion in Roman Britain*. London, Batsford.

Lawson, A. J. (1975) Shale and jet objects from Silchester. *Archaeologia* 105, 241–75.

McKenzie, M. & Thomas, C., with Powers, N. & Wardle, A. (2020) *In the Northern Cemetery of Roman London; Excavations at Spitalfields Market, London E1, 1991–2007*. London, MOLA.

Miles, D. (ed.) (1984) *Archaeology at Barton Court Farm, Abingdon, Oxon: An Investigation of Late Neolithic, Iron Age, Romano-British, and Saxon Settlements*. York, Council for British Archaeology Research Reports 50. Available at: <https://doi.org/10.5284/1081709>.

Moffett, C. (2019) The amulets of Roman Wroxeter: Evidence for everyday magic. *Transactions of the Shropshire Archaeological and Historical Society* 94, 45–58.

Muller, H. (1987) *Jet*. London, Routledge.

Parker, A. (2016) Staring at death: The jet *gorgoneia* of Roman Britain. In S. Hoss & A. Whitmore (eds), *Small Finds & Ancient Social Practices in the Northwest Provinces of the Roman Empire*, 98–113. Oxford, Oxbow Books.

Pugsley, P. (2003) *Roman Domestic Wood: Analysis of the Morphology, Manufacture and Use of Selected Categories of Domestic Wooden Artefacts with Particular Reference to the Material from Roman Britain*. Oxford, British Archaeological Reports International Series 1118.

Reid Moir, J. & Maynard, G. (1933) The Roman villa at Castle Hill, Witton, Ipswich. *Suffolk Institute of Archaeology Proceedings* 21, Part 3, 240–62.

Royal Commission on Historical Monuments England (1962) *Eburacum, Roman York*. London, Royal Commission on Historical Monuments England.

Shepherd, J. (1998) *The Temple of Mithras, London: Excavations by W F Grimes and A Williams at the Walbrook*. English Heritage Archaeological Report No. 12. London, English Heritage.

Watson, J. (1997) The Reconstruction of a Roman Jewellery Box from Mansell Street, London. *Ancient Monuments Lab Report* 88/97. London, English Heritage.

Wilson, P. (2002) *Cataractonium: Roman Catterick and Its Hinterland. Excavations and Research, 1958-1997. Volume 2*. York, Council for British Archaeology.

14

Using Coinage and Die-studies to Obtain Evidence about Society in the Late Iron Age

John Talbot

This chapter draws on die-studies to reveal information about the nature of coinage, its production, and circulation. It summarises some of the key findings of a complete die-study of Icenian coinage from East Anglia and then compares these findings to preliminary results from an ongoing die-study of Durotrigan coinage from southwest England. By creating 'die-groups', I am able to separate linked batches of coinage from an otherwise homogenous pool and highlight the sub-regional nature of significant quantities of Durotrigan coinage. Die-studies of two hoards from the Isle of Wight point to similarities and differences between the use of Iron Age coinage in the southwest and in East Anglia, which we can use to theorise who was responsible for the production of Iron Age coinage.

Keywords: Iron Age, Coins, Die-Study, Hoards, Durotriges, Iceni, Isle of Wight, East Anglia

Background

The most fortunate thing to have happened to me since I began studying Iron Age coinage was having Chris Gosden as my supervisor for my Oxford DPhil on Icenian coinage. Before I started the DPhil, I had already spent a decade or so in my spare time, alongside my work in business, carrying out a die-study (described below) of the 10,000 known coins of the Iceni (a Late Iron Age 'tribe' based in East Anglia). My first supervision with Chris was incredibly memorable; he thought I should stop all work on the die-study and read for a few months.

My professional career had been focused on rescuing troubled businesses, yet Chris wanted me to use the coinage as a tool to extract a broad sweep of information about the Late Iron Age in East Anglia. To enable me to do this, he needed me to have knowledge about ancient economies and trade, tribal societies, 'Celtic' and other art, the conflicting views about the nature of Iron Age coinage, as well as many other subjects. He wanted me to approach the study in the context of a society whose thought processes and use of coinage may have been radically different from our own. Chris gave me an initial reading list and then saw me fortnightly to debate the various books that I had read.[1] This process lasted almost eight months, and these were some of the most interesting and stimulating learning experiences of my life.

One of the questions Chris wanted me to address was whether Icenian coinage was money in a contemporary sense or whether it had other primary purposes, such as prestige gift. Coin die-studies combined with other analyses can provide many insights into the ways in which coins were produced and used. Icenian coins were produced by striking either an unmarked metal pellet or blank flan between two engraved metal dies, each of which impressed their engraved image onto the newly struck coin. The hammer die (top) provided the coin with its reverse side and the anvil die (bottom) its obverse side. The two dies were physically separate and the hammer dies wore out more quickly and were replaced roughly twice as often as the anvil dies. Several hammer dies were frequently in use at the same time, but it was rare that more than one anvil die was used at the same time. By identifying the dies used to strike specific coins through a die-study, it is possible to prove that particular coins were part of a linked sequence of minting and, by examining die-wear, to demonstrate their relative chronology. Coins that share a common die are 'die-linked'. For types of coin with a significant sample size (usually above an average of four coins per reverse die), I refer to each series of dies which are connected through die-links as a 'die-group' to facilitate further study (Talbot 2017, 5).

In many studies, die-linked sequences of dies are artificially linked to other die-linked sequences based upon a subjective judgement of stylistic development. These studies assumed that the absence of links was caused by a shortage of samples; the simultaneous replacement of all dies in a mint; or intermittent minting with new dies being used after a gap in production. Such subjective links are avoided in my die-studies. By avoiding purely stylistic connections, I have shown that the resultant die-groups can each have distinctive characteristics such as distribution.

The most significant findings in my Icenian die-study came from placing the dies within each die-group into chronological order (Talbot 2017, 4–5) and then using the die-groups to examine both the content of coinage hoards and the distribution maps of single finds. My main conclusions can be summarised here:

- At first, what seemed to be a single issue of a single type of coin was often several die-groups, each sometimes having different and distinctive areas of sub-regional distribution.
- Where a hoard contained many coins from a single die-group, its composition was heavily biased towards coins from later in that sequence.
- Hoards showed that some die-groups of a single type of coin had been issued in parallel and were contemporaneous, as were sometimes different types of coinage.
- Extremely rare die-links between different types of coinage indicated that minting may have taken place in a common facility but with careful separation of dies.

One of the most revealing findings was that the coins in Icenian hoards contained a clear bias towards those that were minted closest to the closure of the hoard. An example is the Dallinghoo hoard of 840 staters found in Suffolk in 2008–9 (de Jersey 2014, hoard 227). The 'Boar Horse B' (BHB) variety of stater (Talbot 2017, 184, and ABC 1441, 1444 and 1447) formed the largest single component of the hoard. BHB staters are straightforward to order into a chronological sequence, from reverse die 1 (the earliest) to reverse die 12 (an example from the sequence is shown in Talbot 2017, 4–5). Table 14.1 shows the quantity of BHB staters in the Dallinghoo hoard divided into three consecutive batches of reverse dies and reveals this strong pattern.

The pattern shown by the analysis of Dallinghoo is not surprising in relation to modern coinage. One would expect a batch of coins taken from circulation in 1940 to contain proportionately more coins from the 1930s than the 1920s and so on. However, Iron Age coinage, particularly gold

Table 14.1 The relative presence of 'Boar Horse B' (BHB) Stater dies in the Dallinghoo hoard (data from Talbot 2015, 233; Talbot & Leins 2010).

BHB stater reverse dies	Quantity in hoard	Coins per die
1–4	49	12.2
5–8	84	21.0
9–12	202	50.5

coinage like these staters, is often assumed to be *unlike* modern coinage. For example, coins can be produced intermittently in large quantities in times of urgent need, such as for military purposes or for buying off a potential aggressor (*e.g.* Sills 2003, 3).

The analysis of Icenian hoards suggested that the various types of local Iron Age coinage were not generally produced in vast batches but were minted steadily over an extended period. This resulted in the age structure seen in the hoards. The age structure further confirms that the coinage also appeared to have undergone some form of circulation, otherwise one would expect to see a much less consistent falling off in the presence of older dies in a minting sequence.

My ultimate conclusions were that whilst some of the earliest forms of local coinage may have been produced to enhance the prestige of local individuals, the coinage always had a 'monetary' use, in a contemporary sense, and that the growth in the production and use of coinage was driven by the increasing sophistication of the local economy and of the traders who participated in it (Talbot 2017, 150). Indeed, I formed the view that it was highly likely that the Iron Age coinage of East Anglia was produced by traders rather than by hierarchical leaders or tribal organisations.

Since concluding my study of Icenian coinage, I have wanted to establish whether East Anglia was exceptional in the Late Iron Age or whether my findings also apply to other contemporary coin-using areas of Britain. I decided to carry out a die-study of another regional coinage and chose the stater coinage of southwest England, commonly attributed to the Durotriges. This is a large coinage, with several well recorded hoards, and it has not been studied in detail since Melinda Mays (1984) completed her doctoral thesis.

The Durotrigan die-study is still in progress after some four years. However, enough analysis has been completed to draw some comparisons between East Anglia and southwest England with regard to the key findings from my earlier study. In this chapter, I will first provide a brief overview of Durotrigan coinage, then subsequently look at some preliminary results from my die-study. I have undertaken a detailed analysis of a few hoards composed of staters where the die-study is largely complete; the resultant findings will be reviewed in respect of two Isle of Wight hoards. Finally, I will summarise what initial conclusions can be drawn from the work I have undertaken to date.

Durotrigan Coinage

The Iron Age coinage of southwest England is commonly called Durotrigan, although its attribution to the 'Durotriges', is a matter of contention (*e.g.* Papworth 2008, 374–5). Its distribution is centred upon Dorset but also encompasses parts of neighbouring counties and the Isle of Wight. There are two main denominations, the modern names for which are 'stater' and 'quarter stater'. The quarter stater was produced less widely and is much less common. The Durotrigan stater coinage includes an issue struck in gold alloy known as the Chute stater, but most were struck in silver of varying degrees of purity or bronze. Many thousands of die-struck Durotrigan staters in silver or bronze have been recorded from the southwest. There was also a series of cast bronze staters which had a sub-regional distribution, and which appear to be from relatively late in the period of coinage issue. This chapter focuses exclusively on these silver and bronze die-struck staters, which is also the main focus of my on-going die-study.

Durotrigan staters are unique as an Iron Age series in that their imagery remained remarkably consistent throughout the century or more of their production. The obverse is an abstracted right-facing head which is similar to other early British staters and is ultimately derived by a process of abstraction from gold coins of Phillip II of Macedon (359–336 BC) (Talbot 2017, 86). Figure 14.1 shows the clear relationship between a Durotrigan stater on the right and an Icenian Norfolk Wolf A stater in the centre. Both are clearly part of the same process of abstraction, an earlier stage of which is represented by Gallo-Belgic A stater ('GBA') shown on the left. GBA is the earliest type of 'Celtic' stater found regularly in England and is stylistically much closer to the Macedonian prototype. All three coins are illustrated by composite images. Composite photographic images provide a fuller indication of the design on the die than any one coin and help in identification. This is because the dies used to strike Durotrigan and Icenian coins often had more than 150% of the surface area of the struck coins.

In Figure 14.1, the curls of hair to the upper left (A), the wreath (B), and the bar through the hair at right angles to the wreath (C) are clearly visible on each of the three coins illustrated. The face to the right is less easily identifiable without seeing the sequence; in the Durotrigan coin it is largely reduced to three crescents. The cloak, below the face, is shown on both later coins as alternate lines and rows of dots. At the top of the cloak on the Icenian example is what seems to be an ornate fibula; the elongated triangle on the Durotrigan coin is probably an abstracted representation of the same object.

The reverse of the Durotrigan staters bears a stylised horse, three examples of which are shown in Figure 14.2. The horse's four rigid vertical legs can be seen at the bottom of each image and to the left, above the first leg, is the horse's head, usually indicated by a large upper pellet with two thin lines dropping to the mouth area. To the right of the head is a single vertical line representing the horse's neck. Above its back are usually 12 circular pellets.

The presence of a horse on the reverse of Iron Age coins is ubiquitous on both sides of the channel. In the southwest, the imagery is unusually static, stylised, and consistent. In most other coinages, the energy of the horse is illustrated through jointed legs and more naturalistic drawing.

The standardisation of the imagery on Durotrigan staters encompasses issues of high silver content through various levels of debasement to coins which are entirely bronze. The poor condition and close similarity of many of the more debased coins has made the die-study difficult. This is compounded by the large numbers of dies. So far, I have separated 520 obverse and 700 reverse dies from a population of some 6100 coins. A small number of sub-types can be separated from the mass of standard Durotrigan staters by additional detail on one or both dies. These sub-types fall within die-groups that are entirely composed of dies with similar features, each of which has a specific sub-regional focus of distribution. An example is shown in Figure 14.3; the 'pelleted' sub-type is distinguished by the two vertical

Figure 14.2 The reverse of Durotrigan staters (Photos: The Celtic Coin Index, School of Archaeology, University of Oxford).

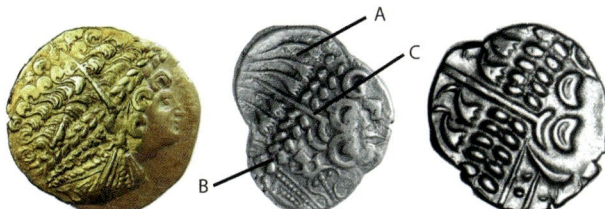

Figure 14.1 From left to right: obverses of Gallo-Belgic A, Icenian, and Durotrigan staters (Photos left and centre: The Celtic Coin Index, School of Archaeology, University of Oxford; right: Classical Numismatic Group, LLC, www.cngcoins.com).

Figure 14.3 Pelleted sub-type (die-group 49) (Photo: The Portable Antiquities Scheme/The Trustees of the British Museum).

Figure 14.4 From left to right: examples of staters from Durotrigan die-groups 1, 2, and 3 (Photos: Classical Numismatic Group, LLC, www.cngcoins.com).

Table 14.2 Indicative phases of Durotrigan coinage.

Phase	Indicative dating	Coinage types
1	Pre-50 BC	Gold; including British A (Lepe)
2	ca. 50–25 BC	Gold (British B or Chute) and silver alloy
3	ca. 25 BC–AD 10	Silver alloy
4	ca. AD 10–Roman	Very debased silver and bronze (including cast bronze)

Table 14.3 Key statistics for three Durotrigan die-groups.

Die group	Known coins	Obv. dies	Rev. dies	Findspots	Likely target weight (in g)
1	911	5	15	11	5.88
2	201	6	9	30	6.01
3	224	6	9	26	6.00

pellets on the obverse above the hair-bar and adjacent to the wreath and, where visible, eyelash-like lines linking the crescent shaped end of the hair-bar to the adjacent crescent. The distribution of the pelleted sub-type is focused in the southwest of the region.

The staters not attributable to sub-types have been the main ongoing focus of the die-study. Those with a higher silver content are generally better preserved and most of these have been allocated to specific dies. In many cases, high levels of coins per die are now identified, which enable accurate chronologically ordered sequences of dies to be constructed. Most silver alloy coins fall into one of 24 die-groups, the largest of these has 27 different dies and several others comprise over 20 dies. The larger die-groups of Icenian staters have similar numbers of dies. The die-groups for the more debased Durotrigan staters are, so far, much shorter – this is probably because many have yet to have their dies identified.

Whilst the die-study is a work in progress, I have started to develop a preliminary working hypothesis for the organisation and chronology of the Durotrigan coinage. This divides the coinage into four main phases of production as shown in Table 14.2. The silver coinage of Phases 2 and 3 have been the principal focus of study to date, and many die-groups have been allocated to the appropriate phase based upon hoard analysis. The dating of the phases is extremely tentative and little statistical work has been undertaken on metal content or weight. Coins from different die-groups are frequently very similar. Figure 14.4 illustrates a coin from each of three common die-groups of Phase 2 silver alloy staters. There are no unusual features on the coins illustrated in Figure 14.4; without a detailed die-study it would not be possible to allocate coins to the appropriate die-group.

A key factor in assessing the likely completeness of a die-study is the sample size relative to the relevant number of dies. As Iron Age coinages usually have more reverse than obverse dies – the latter sometimes being used until the design is practically obliterated – I prefer to assess whether a sample size is likely to be meaningful by the ratio of the total sample to the number of reverse dies. Table 14.3 reveals that the sample sizes for each of these die-groups is very high, ranging from 22 coins per reverse die (CPRD) for die-group 2 to over 60 CPRD for die-group 1. This suggests that most dies and die-links are likely to be known and thus included in the die-charts.

The coins that have been found to belong to die-groups 1, 2, or 3 emanate from many sources. These include well-recorded hoards; single finds recorded with the Portable Antiquities Scheme or the Celtic Coin Index; and un-provenanced coins recorded from dealers' catalogues and other sources, such as eBay. The maps in Figure 14.5 plot the known findspots for each of the die-groups.

The maps in Figure 14.5 reveal that die-group 1 is almost exclusively found on the Isle of Wight and die-group 3 has a similarly restricted distribution in central Dorset focused on the Badbury Rings and Hod Hill hillforts. Die-group 2 is more widespread and has a major focus in central Dorset, but there is a significant scattering of finds to the south of the region, including the Isle of Wight. These findings demonstrate that what appears to be a homogenous coinage is actually composed of multiple die-groups, each often having a distinctive pattern of distribution. Although the Durotrigan die-study is still in progress, it is clear that many other die-groups have a sub-regional focus to their distribution.

14. Using Coinage and Die-studies to Obtain Evidence about Society in the Late Iron Age 113

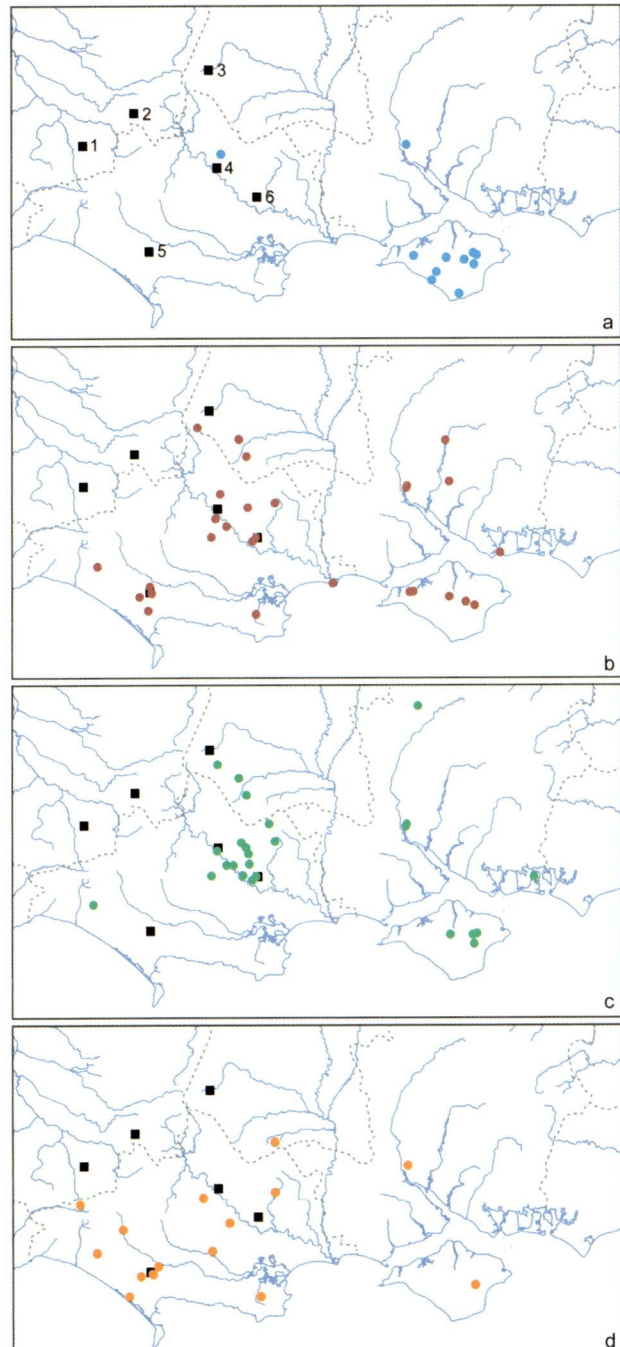

Figure 14.5 Distributions of Durotrigan (a) die-group 1, (b) die-group 2, (c) die-group 3, and (d) pelleted sub-type with rivers, counties (dashed lines), and major hillforts (1: Ham Hill, 2: Cadbury, 3: Cold Kitchen Hill, 4: Hod Hill, 5: Maiden Castle, and 6: Badbury Rings). The maps only include a single find spot for each parish or site to avoid distortion caused by multiple adjacent findspots for dispersed hoards, or in particularly productive sites. More findspots will be added as the study progresses.

Table 14.4 Analysis of the contents of the Brighstone hoard.

Die-group or dies	Number of coins	%	Findspots
1	746	77.1	Predominantly Isle of White (IOW)
2	2	0.2	Dorset & coastal inc. IOW
3	1	0.1	Dorset
6	77	8	Predominantly IOW
23	125	12.9	Mixed IOW & mainland
Dies RR:61	11	1.1	IOW (only 2 findspots)
Singletons or plated	5	0.5	
Total	**967**	**100**	

Table 14.5 Analysis of the Durotrigan staters in the Shorwell hoard.

Die-group	Number of coins	Shorwell %	Brighstone %
1	112	81%	77%
6	10	7%	8%
23	16	12%	13%
	138	100%	98%

example was found by metal-detectorists in 2005 at Brighstone on the Isle of Wight (de Jersey 2014, 219). The hoard contained 967 Durotrigan silver staters, which were photographed and returned to the finders. The British Museum photographs enabled me to carry out a die-study of the hoard, the results of which are shown in Table 14.4.

The composition was remarkably restricted, with 98% of the content from only three die-groups. Over three-quarters of the coins in the hoard came from die-group 1, coins of which are almost exclusively found on the Isle of Wight (see Fig. 14.5). Indeed over 86% of the coins in the hoard belong to die-groups (or dies) predominantly found on the Isle of Wight.

The previous year another hoard had been discovered on the Isle of Wight in the nearby parish of Shorwell (de Jersey 2014, 229). This hoard differed from Brighstone as it contained 18 Chute gold staters, an uninscribed silver unit, and three massive silver ingots. However, the largest numismatic element was 138 Durotrigan silver staters. A die-study of these revealed striking similarities to Brighstone. All were from the same three die-groups that provided the majority of the Brighstone staters and in similar ratios, as shown in Table 14.5.

These two hoards have provided an opportunity for some interim analysis even whilst the overall die-study is incomplete. The contents of both hoards are dated to my Phase 2 of southwestern coinage (Table 14.2). Firstly, I will

Hoards

Many hoards of Durotrigan coinage have been found in the region. One particularly important and well recorded compare the mix of coins from die-group 1 in the hoards to their known chronology based upon die wear. I will then examine some unusual features of die-group 23, which is the only die-group found in volume in these two Isle of Wight hoards as well as on the mainland.

Die-group 1

The large number of die-group 1 staters in the Brighstone and Shorwell hoards reveal the order in which dies were introduced into the minting process. To exemplify this, Figure 14.6 illustrates the gradual deterioration of the most common obverse die (YZ) as different reverse dies were used and replaced. Die YZ was used with reverse dies 3 to 12. The reverse dies are numbered in chronological order. The deterioration can be clearly seen in the pellets forming the left side of the wreath above the hair-bar. The pellets are defined and separate when die YZ was used with die 4 (left); there is significant deterioration when it was used with die 8 (centre); and by the time it was used with die 12, there was widespread deterioration in the pellets of the wreath.

Having ordered the reverse dies into an approximate chronological sequence, Figure 14.7 illustrates their proportional presence in the hoard. There is not the same direct correlation to chronology that is seen in Icenian hoards (Table 14.1). There is a bias towards the latest dies, as 62% of the relevant coins were struck by the final five reverse dies. However, there is an unexpectedly strong presence of coins struck by dies 3 and 4. Figure 14.7 also shows that the die-group 1 coins in the Shorwell hoard have a very similar statistical distribution to those in Brighstone. Possible explanations for the disproportionate level of dies 3 and 4 in the hoards are:

A. Dies 3 and 4 were exceptionally productive dies, which distorts the analysis.
B. The hoards are each composed of two batches, one of which predated the introduction of the later dies.
C. Minting of die-group 1 was conducted rapidly; the relative presence of dies in the hoards simply reflects the varying production volumes achieved from each die before it was abandoned.

The similarity between the content of Shorwell and Brighstone hoards raises interesting questions. Despite the hoards being found in succeeding years and only a few miles apart, it appears clear from the records summarised in de Jersey (2014, hoards 126 & 132) that they are separate finds.

Figure 14.6 From left to right: the ageing of die YZ paired with reverse dies 4, 8, and 12, with the least worn on the left to the most worn on the right (Photos: Classical Numismatic Group, LLC, www.cngcoins.com).

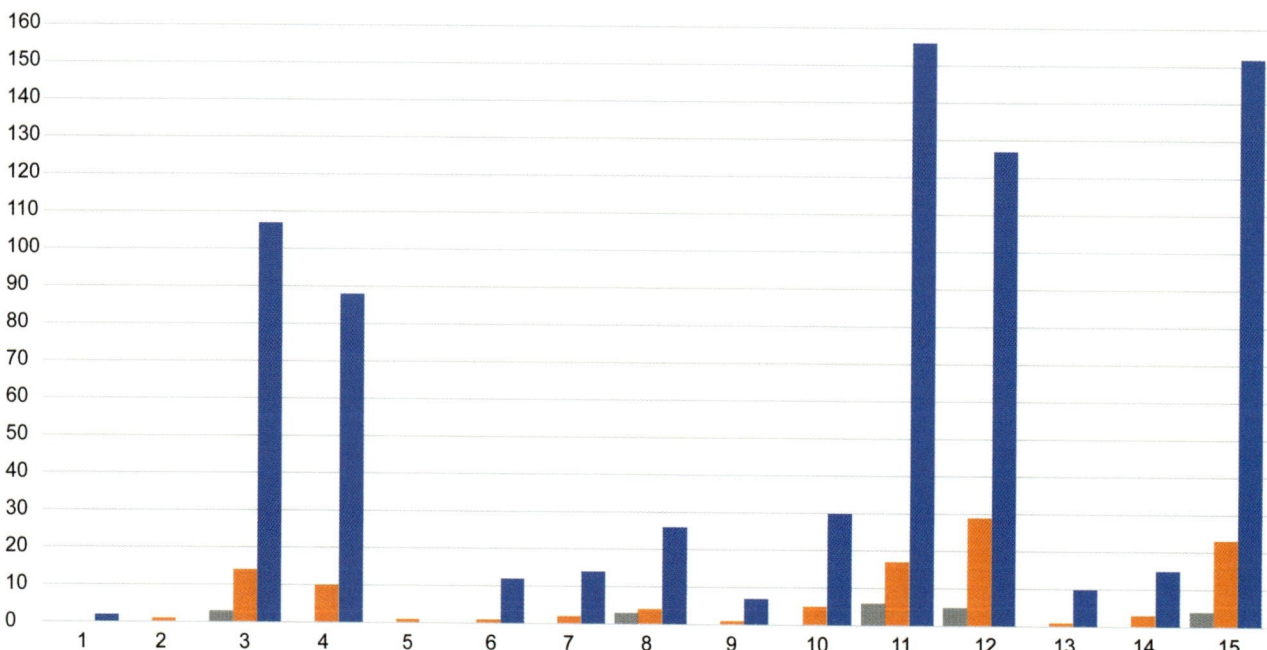

Figure 14.7 (blue) The presence of the 15 die-group 1 reverse dies in the Brighstone hoard; (orange) the numbers of die-group 1 reverse dies in the Shorwell hoard; and (grey) die-group 1 reverse dies from neither Brighstone nor Shorwell.

This leaves open the possibility that the Durotrigan silver staters in the hoard may have emanated from a common pool before deposition. There was evidence that something similar occurred in respect of Icenian hoards deposited at the time of the Boudiccan revolt (Talbot 2017, 117–9).

An analysis of die-group 1 coins recorded prior to the discovery of the two hoards or with a clear alternative provenance are also shown in Figure 14.7. With a couple of exceptions, this reveals a broadly similar pattern. This does not definitively point to explanation C, as many of these coins may emanate from other hoards with a similar profile to Brighstone and Shorwell.

Die-group 23

Die-group 23 is smaller, with only two obverse and seven reverse dies. The chronology within the die-group is clearly established by the reverse die that provides the link between the two obverses. All provenanced die-group 23 coins are shown in Table 14.6. The known die combinations are listed in chronological order with A:1 being the earliest. The two obverse dies in this die-group show clear signs of damage and deterioration with use, making the relative chronology of the reverse dies easy to assess, although certain reverse dies, such as 6 and 7, overlapped in their use.

The presence of die-group 23 coins in the Brighstone and Shorwell hoards is heavily biased towards the dies that were most recent when the hoards were closed. No coins in either hoard were struck from the first obverse die in the die-group, despite other finds on the Isle of Wight including four coins struck from this die. Conversely the Nursling hoard found in Hampshire includes only coins struck from the first obverse die in the group and none from the second.

I have analysed many hoards which contain multiple examples of coins struck by the later dies in a particular die-group, and these invariably contain at least a scattering of coins from the earlier dies in that die-group. This, and the absence of later die-group 23 coins from Nursling hoard,

Table 14.6 Finds of die-group 23 coins with a provenance in chronological order.

Dies	Brighstone hoard	Shorwell hoard	Nursling hoard	Other mainland	IOW
A:1			12	2	3
A:2			4	3	1
A:3			2		
B:3	3			1	
B:4	8	1			
B:5	3				
B:6	61		7		
B:7	50		8	1	

raises the question of whether the location of this workshop moved from the mainland to the Isle of Wight at about the time a new obverse die was introduced. I suspect that this was the case notwithstanding earlier coins from the die-group having been found on the island and two late coins having been found in mainland hoards (Table 14.6).

The key observations from the die-study to date can be summarised as follows:

- Durotrigan coinage includes parallel die-groups with differing distributions, some of which appear to have been produced simultaneously.
- The dies of certain die-groups were designed to be identifiable by the inclusion of specific iconography, but most die-groups were not so separable.
- The composition of Durotrigan hoards appears to be correlated to the internal chronology of the die-groups they contain. However, so far, the correlation is weaker than with Icenian hoards.
- There is some evidence to suggest that the minting of die-group 23 was moved from the mainland to the Isle of Wight during production.

Discussion: Icenian and Durotrigan Coinage Compared

Although Durotrigan stater coinage initially gives the impression of being an amorphous mass of identical material, the die-study shows that it breaks down into numerous die-groups, each of which often has a sub-regional distribution. I have found little evidence of efforts being made to enable these different die-groups to be identified by users. There are a few exceptions, including some sub-types that can be easily separated, one of which is discussed above.

Icenian coinage had a similar pattern of die-groups, but there were several parallel coinages in use at once as the coinage developed. These parallel coinages were readily separable. Icenian coinage in hoards had a clear chronological bias towards the most recently produced coins, and weight and metal content were tightly controlled. Taken together these factors strongly suggested that the coinage had a monetary usage, was circulating, and was likely to have been issued by a number of traders, who were probably concerned with 'brand awareness' to facilitate trade.

At this stage of the Durotrigan study, the evidence is less compelling than for East Anglia. The pattern of local distribution suggests sub-regional issuance of the coinage, and the uniformity of design indicates a desire to make the coins known and acceptable and lends weight to their having a monetary usage. The absence of a clear chronology in the samples of coinage in the two hoards so far examined in depth precludes certainty over whether the coins were taken from a revolving pool of coinage. This was a major element in the ultimate conclusions about the usage of Icenian coinage. The similarity of most Durotrigan staters implies that they were not designed to enable the ready identification of different issuers. The few sub-types so far identified are exceptions.

On balance it appears likely that Durotrigan coinage was issued as a form of money (in a contemporary sense) to facilitate trade, which resulted in it being produced and used locally, but also following regionally acceptable criteria as to weight, probably metal content, and imagery.

The differences noted between the structure of Isle of Wight and East Anglian hoards are extremely interesting. These could be the result of one or several factors, such as differences in the nature of hoarding and in the use of coinage or of die-groups being produced over a shorter timeframe in the southwest. These attempts to draw interim conclusions from the research to date have highlighted both similarities and differences with East Anglian Iron Age coinage. The continuing work on the die-study should eventually provide a clearer understanding of Durotrigan coinage.

The work raises fascinating questions about the relationship between the Isle of Wight and the main 'Durotrigan' centre in Dorset. This will be examined further as the Durotrigan die-study is completed, as it challenges the current tribal model of Late Iron Age Britain, which links the Isle of Wight with 'the southern region'. During Phase 2, the staters used in both locations cannot be separated without a die-study; in later phases this changed, and Isle of Wight die-groups contained additional detail that makes them distinguishable. Despite the commonality of coinage during Phase 2, it is remarkable that of the 967 coins in the Brighstone hoard, only three emanate from die-groups focused primarily on the mainland, and none of the silver staters in the Shorwell hoard are from such mainland die-groups. More is likely to be learned about the relationship between the Isle of Wight and Dorset as the die-study is completed.

Note

1. A few of the books that Chris suggested I read, which I found particularly interesting, include: Munn 1986; Zanker 1988; Bloch & Parry 1989; Freedberg 1989; Cohen 1992; Gell 1998; and Seaford 2004.

References

Bloch, M. & Parry, J. (1989) *Money and the Morality of Exchange*. Cambridge, Cambridge University Press.

Cohen, E.E. (1992) *Athenian Economy and Society: A Banking Perspective*. Princeton, NJ, Princeton University Press.

de Jersey, P. (2014) *Coin Hoards in Iron Age Britain*. The British Numismatic Society Special Publications 12. London, Spink.

Freedberg, D. (1989) *The Power of Images: Studies in the History and Theory of Response*. Chicago, IL, The University of Chicago Press.

Gell, A. (1998) *Art and Agency: An Anthropological Theory*. Oxford, Oxford University Press.

Mays, M. (1984). *A social and economic study of the Durotriges, from 150 B.C. to A.D. 150, with particular reference to coinage*. Unpublished Doctoral thesis, University of Oxford.

Munn, N. D. (1986) *The Fame of Gawa: A Symbolic Study of Value Transformation in a Massim (Papua New Guinea) Society*. Durham, NC, Duke University Press.

Papworth, M. (2008) *Deconstructing the Durotriges: A Definition of Iron Age Communities within the Dorset Environs*. Oxford, British Archaeological Reports British Series 462.

Seaford, R. (2004) *Money and the Early Greek Mind*. Cambridge, Cambridge University Press.

Sills, J. (2003) *Divided Kingdoms: The Iron Age Gold Coinage of Southern England*. Aylsham, Chris Rudd Ltd.

Talbot, J. (2015) *What is Icenian coinage?* Unpublished Doctoral thesis, University of Oxford.

Talbot, J. (2017) *Made for Trade: A New View of Icenian Coinage*. Oxford, Oxbow Books.

Talbot, J. & Leins, I. (2010) Before Boudicca: the Wickham Market hoard and the middle phase coinage of East Anglia. *British Numismatic Journal* 80, 1–23.

Zanker, P. (1988) *The Power of Images in the Age of Augustus*. Ann Arbor, MI, The University of Michigan Press.

15

'Keep on Truckin' – Thoughts from the Back of a Bus

A. M. Pollard

This contribution pays tribute to the initiative devised by Jessica Rawson, in which a group of scholars from Oxford and Peking University (Beijing) met on several occasions to travel around China, Siberia, Mongolia, Kazakhstan, and Turkmenistan, in order to discuss the relationship between Chinese archaeology and the broader Eurasian context. In addition to these specific objectives, spending many hours travelling across the backroads of East Asia with an exceptional group of scholars allowed more philosophical and fanciful discussions to develop. Topics addressed included 'is the Universe sentient?' and 'what is the Bronze Age?', but also the role of magic in technological change. Whilst crossing Siberia, we also discussed whether the School of Archaeology in Oxford would be better served by including a Shaman in the administrative structures, but this did not meet with much enthusiasm back in the cold rain of Oxford. This contribution is a tribute to many such discussions with Chris Gosden, Jessica Rawson, and others on these journeys.
Keywords: Bronze Age China, Sentient Universe, Technology, Magic, Alchemy

Introduction

The aim of this article is to pay particular tribute to two people. One is Jessica Rawson, who made our journeys possible, and the other is Chris Gosden, who, amongst several others, provided much mental stimulation whilst travelling around some of the more remote places in East Asia. The impetus for such travels came from Jessica. She had access to funding over several years to promote collaboration between scholars in Oxford and Peking University. Rather than take the obvious route of arranging an annual conference, she made the inspired suggestion of an annual joint 'journey' over a couple of weeks, a minibus of maybe 8–10 staff and students from both Oxford and Peking to see the archaeology, visit the museums, and meet the local archaeologists. Initially this was to various parts of China – the Central Plains, the Gansu Corridor, Qinghai, and Inner Mongolia – but it also included southern Siberia, Kazakhstan, and, just before Covid made such travel impossible, Turkmenistan. Consequently, each year, Jessica, Chris, myself, and some postdocs and students from Oxford (including Peter Hommel, Ruiliang Liu, Yiu-Kang Hsu, Rebecca O'Sullivan, and Xiaojia Tang) would meet with a group of scholars from Peking University (Xu Tianjin, Zhang Chi, Wu Xiaohong, and Chen Jianli) in some city, and then travel by plane, train, and minibus around the archaeological sites and museums. The overarching theme was understanding the relationship of China with her steppe neighbours.

The direct outputs of these travels have been documented in a number of collaborative papers regarding the flow of metal in China and beyond (*e.g.* Pollard *et al.* 2014; 2015; 2017; 2019; Bray *et al.* 2015; Cuénod *et al.* 2015; Pollard & Bray 2015; Hsu *et al.* 2016; Jin *et al.* 2017; Liu *et al.* 2018; 2019; 2021; Zhang *et al.* 2019; Sainsbury *et al.* 2021). Beyond the specifics of Chinese archaeology, however, these journeys also facilitated extensive discussion about other more speculative archaeological issues that normal academic commitments would not easily allow. These ranged from questions like 'Is the universe sentient?' or 'What is the Bronze Age?' to broader issues, such as the role of magic in technological change. I particularly remember writing the first draft of a paper on the flow of metal through ancient societies (Bray *et al.* 2015) in a bus travelling along a dirt road in the Hexi Corridor in China to visit a jade mine near the Mongolian border. The road was so rough that the author's bottom spent very little of an eight-hour journey in comfortable contact with the seat, and several times the laptop had to be grabbed in mid-air as it made a bid for freedom. The result, however, was eventually a paper that took the concept of materiality as applied to an individual archaeological metal object and generalised it to consider the materiality of the underlying metal. Thus, an object is an instantiation of a larger flow of metal. Analogies with a 'river of metal' are partially helpful here, with the object being a portion of the metal abstracted from the river and potentially returned to it. The chemical (and isotopic) composition of the river changes over time, depending on the nature of the tributary streams (inputs from metal sources) and the ways in which humans interact with the river. A river, however, flows from a small source stream to become a large estuary, whereas our hypothetical river of metal starts big (close to the source of the metal) and shrinks as metals are abstracted. The Okavango Delta, perhaps, provides a better model, but even this is unsatisfactory. I would, however, never have embraced the concept of materiality at all if it had not been for extensive conversations with Chris Gosden.

A Sentient Universe?

Our discussions on sentience lead to a joint paper in John Barrett's festschrift (Gosden & Pollard 2021), in which we started from the position that 'people are enmeshed in a series of relationships with other people, plants, animals, earth, air, fire and water. All of these need serious consideration and should be thought about within a single conceptual framework'. My own exploration of this started from a Gaia-type perspective: the theory that living organisms interact with their inorganic surroundings to form a mutually beneficial and self-regulating system. To me, this position seems self-evident, although it is disputed in several quarters. I would wish to go further and suggest that the distinction between 'living organism' and 'inorganic surroundings' is largely a construct of the viewpoint of the observer. We all know what a living organism is: an entity that can reproduce, respond to external stimuli and some of which can move under their own volition. I would argue that inorganic materials – minerals, rocks, even landscapes – can also replicate themselves, and change in response to external stimuli, but that the timescale is often much longer than a human lifetime, and so to a human observer these materials appear inert.

I think minerals provide a useful example in this context. Minerals can grow and replicate themselves via crystallisation processes from aqueous media – mostly slowly, although noticeable on the scale of a human lifetime, hence the old miner's perception of minerals re-growing underground. They can disappear or grow in response to local chemical conditions. In this sense, minerals respond to their environment, which is one interpretation of 'sentience'. At the larger scale, rocks and mountains, as well as rivers, lakes, and oceans, also respond to changes in their local environment, but usually slowly, unless in response to catastrophic processes such as earthquakes, volcanism, or flooding. If we take sentience to mean responding to environment, then minerals and landscapes might be seen as no less sentient than plants but respond over a longer timescale. Like plants, they can reproduce themselves and move as a species in response to changes in environmental conditions.

Technology and Technological Change

What significance do these thoughts have for archaeology? Perhaps little, if any, in themselves, but they do lead into another line of thought (and sets of conversations) about how people of the past viewed their world, and, in particular, how their world view impacted on the processes of technological change. 'Technology' of itself is a tricky concept. Older definitions usually emphasised the role of machinery, thereby differentiating practical things from purer thought or 'science' – no doubt reflecting an underlying, perceived class distinction between aristocratic scientists and less socially distinguished engineers. The history of technology, which, unlike the histories of science and medicine, only became a recognised academic discipline during the late 20th century, thus began largely with studies of the powered looms and steam engines of the 'Industrial Revolution', followed by a growing interest in such things as water wheels, siege engines, and transport systems of the more distant past. This definition and approach leads to the idea that the history of technology is synonymous with the history of inventions. However, the definition of technology has broadened more recently, to include: 'the sum of any techniques, skills, methods, and processes used in the production of goods or services or in the accomplishment of objectives, such as scientific investigation' (Technology). Even this, in my opinion, is still too restrictive. It is, for example, a one-way definition. Humans produce 'goods or services', but this does not recognise that it is in fact a two-way process – we make things, but things also make us. The recent phenomenon of smart phones shows how quickly a manufactured object can change the way we live. More broadly, I would argue that the technological packages referred to as agriculture and animal domestication exercised a decisive impact on the way many humans have lived over the last four or five thousand years.

Consequently, in a recent volume introducing a new series on the history of technology (Pollard & Gosden 2022), we have adopted an even broader interpretation of the term 'technology', stating that it encompasses the totality of the interaction between humans and their environment.

Clearly, this potentially opens up the scope of technology to encompass all of human life. For example, religious beliefs and practices could fall into such a definition, since religion might be argued to be a mechanism whereby humans mediate with other worlds. In order to contain the scope of the series of volumes, we agreed to focus it on those interactions that involve material culture. Even this, of course, does not necessarily exclude activities such as religion, since many religions are affected through specialised paraphernalia, statues, buildings, and the like.

However it is defined, it is clear that the technological capacity of humans has changed over time, and varies by place and culture. It has been conventional in recent European and American literature to talk about 'technological progress', but we have argued that, in a global archaeological context, this is a somewhat exceptional circumstance applicable mainly to Europe during the 18th and 19th centuries. That is not to say that technology has not changed in the rest of the world in the past – clearly it has – but the rate of change has been relatively slow, and, crucially, the impetus for change may not have been a striving for 'progress'. If we dismiss the pressure to improve the efficacy of a particular tool or process, then what was the driving force? In reality, of course, as now, there are probably many different drivers (including improvement, but also including appearance, social status signalling, spiritual forces, etc.) and the question is more one of determining if there is a *dominant* motive.

So, here we might unite the questions of 'is the universe sentient?' and 'can we understand the motivation(s) for technological change?' From our Western scientific perspective, grounded in Newton and Dalton, and articulated over the last 800 years by various authors from Roger Bacon to Immanuel Kant, these two are not obviously related. However, for much of the past, and even for parts of the present, the overarching natural philosophy – the view of how the world works – has embraced the living and the inanimate, the human and the spiritual. Gods or supernatural beings intervened in the lives of humans, and many inanimate objects housed spirits that needed to be avoided or placated in some way. It is in the context of this perceived world that we need to consider technological processes and change. This is not to deny the value and power of modern scientific studies, but it is to point out that our frame of reference is not the same as their frame of reference (the parallels with the reverse contrast between Newtonian mechanics and Einstein's relativistic world are striking, at least superficially).

It has long been recognised that technological processes in the past often included 'irrational' elements – actions that, from a modern Western scientific perspective, have no conceivable effect – but that nevertheless were seen as important from the original perspective. Examples of sacrifices being necessary for effective smelting or the need for ritual purity are numerous in the archaeological and anthropological records. By considering these examples, we must ask ourselves the question 'by what frame of reference should we seek to explain technological change?' If we seek to do it from a purely Western scientific perspective, then we may be able to observe certain physical changes (*e.g.* higher temperature, greater reduction), and these would naturally lead us to assume that such changes were effected in order to improve the process in some physical way; in other words, lead us to an assumption that the driving force was the quest for some form of technical improvement. But what if it were not? What if, to imagine a fictitious scenario, it was felt important to blow harder on the bellows in order to drive away pernicious demons? The *consequence* might be the same – achieving a higher temperature – but the *motivation* may have been completely different, and one that leaves no visible trace other than the fortuitously achieved higher temperature. If such a scenario were true, then does it really matter that the motivation was 'irrational', since the outcome was the same? Well, yes. It both imbues the actors with thoughts they did not have, and simultaneously deprives them of thoughts and beliefs that they did. As such, it is a distortion of reality, a simple example of projecting our world system back into the past. Naturally, the necessity for understanding their world system also begs the question of how might we reconstruct such systems? That is undeniably difficult to do, especially in worlds without writing to give us glimpses of thought patterns. But the first step of a journey starts by recognising that there is a journey to be made!

Alchemical Magic

There are other sources of information about past world systems apart from scarce ancient technical or philosophical tracts. For the last three millennia, this has included texts on alchemy from Egypt and the Classical Mediterranean, many of which were subsequently translated and added to by Arabic scholars, and also from India and China. Often seen as confused and obscure, these texts in fact embody a record of human engagement with the material and spiritual world. The aims were predominantly twofold: the conversion of base metal into gold using the 'philosopher's stone' and the related quest for immortality via the 'elixir of life', but the ramifications extended far beyond these specific aims into medicine and the general relationship with the material world. When seen from the viewpoint of Western science, they are often irrational and deemed likely to be ineffective, but this is not the only frame of reference from which to view these works, nor, perhaps, the best. When seen from within the contemporary understanding of the world in Western and Eastern philosophies, one based on a four- or five-fold component material world, with transformation from one form to another possible via human addition or subtraction of these component properties, they make much more sense. As has been pointed out by Cyril Stanley Smith (1968, 639):

> Transmutation was a thoroughly valid aim, a natural outgrowth of Aristotle's combinable qualities, and its truth was demonstrated by every child growing from the food he ate, by every smelter who turned green earth into red copper or black galena into base lead or virgin-hued silver, by every founder who turned copper into gleaming yellow brass, by every potter who glazed his ware, by every goldsmith who produced niello, by every maker of stained glass windows, and by every smith who controlled the metamorphosis of iron during its smelting, conversion to steel, and hardening. Such changes of properties, seen physically, are transmutations, but they are not chemical in the purified modern sense.

Thus, many technical processes are transformational – ore to metal, sand to glass – and, to the non-adept, most easily explained by magic. This might explain the power of new materials when they first appear. The holder of a copper-alloy axe in the Late Neolithic, or one who sports personal adornment made from iron in the Late Bronze Age, self-identifies not necessarily as an embracer of new technology, but someone who has – or is close to someone who has – magical powers. After all, if he or she can transform a green rock into a red liquid, then might they not with equal facility transform an uncooperative observer into a frog? Thus, the long history of magic carries within it, as well as charlatanry and prestidigitation, a core of the record of human relationships with the material and spiritual worlds (Gosden 2020). It is perhaps not unreasonable to suggest that any serious study of technology and technological change in the past should take cognisance of the contemporary world view, however irrational that might appear from a Western scientific perspective.

Medicine for the Material World

Thoughts of the ambiguity between the animate and inanimate worlds have also triggered a very specific thought about the development of technology. This is to make the connection between the use of medicinal compounds to cure human ills and the use of similar compounds in technological processes to cure perceived 'ills' in the inanimate world. The majority of *materia medica* in the ancient world were based on plant materials – only a relatively small minority were based on mineral compounds, but this still encompasses a wide range of materials, such as mercury, arsenic, and antimony compounds. There are many likely examples of this transfer, such as the use of antimony or its sulphide, stibnite Sb_2S_3, which was used in Egypt for black eye makeup (kohl), and also as a cure for headache and conjunctivitis. Technologically, traces of antimony were used to clarify glass or harden copper from the Roman period onwards. Perhaps even more significantly, zinc oxide was used in Antiquity to treat diseases and diseases of the eye and skin. Compounds of zinc were used extensively in metallurgical process of later Antiquity to make brass and was perhaps one of the candidates for the philosopher's stone, turning red copper into golden brass.

More research is needed into the potential links between *materia medica* and their use as technological agents. This could be very significant for studies of the potential drivers of technological change. It should be remembered, however, that most *materia medica* were plant-based, and it is highly unlikely that any evidence of their use in pyrotechnological processes would survive to be discovered by modern science. Thus, we may suspect that there were a whole range of plant-based material used technologically, of which we are completely unaware.

How Would You Know If You Were Wrong? The Value of Enemies

A recurring theme in our speculative conversations over the years has been the question of 'how would you know if you were wrong?' in archaeological research. The nature of archaeology means that human thoughts, beliefs, and actions in the past are reconstructed based (primarily) on interpretations made from an all-too-fragmentary material record. In this respect, archaeology can be thought of as the past tense of forensic science, since the latter also focuses on the reconstruction of thought and deed from material evidence. Forensic science is, however, carried out within a legal framework, in which such evidence is ultimately tested in a courtroom context, with the associated need for proof 'beyond reasonable doubt'. Archaeology lacks such a rigorous framework, and all too often, it is reduced to telling plausible stories based on the currently available evidence. In such circumstances, it becomes important for the self-reflexive process of asking 'how would we know if we were wrong?' This is second-nature within the modern philosophy of science, but is perhaps less well-developed in archaeology.

The normal response in 'the hard sciences' is the requirement for the replication of experimental or observational data, firstly by the researcher herself or himself, and eventually by colleagues in the same discipline. Hence the over-riding importance of including carefully reported experimental details in the publications, and, increasingly, the publication of all the necessary data in open-source data repositories. As an aside, certain areas of archaeology have historically been averse to sharing archaeological materials and data, precluding simple replication and re-evaluation of the data. Thankfully, this tendency is now substantially reduced.

It is in this context that we may reflect on the thoughts of Georg von Békésy, Nobel Laureate in Physiology and Medicine in 1961 'for his discoveries of the physical mechanism of stimulation within the cochlea'. He extolled the value of having scientific enemies (von Békésy 1960, 8):

One way of dealing with errors is to have friends who are willing to spend the time necessary to carry out a critical examination of the experimental design beforehand and the results after the experiments have been completed. An even better way is to have an enemy. An enemy is willing to devote a vast amount of time and brain power to ferreting out errors both large and small, and this without any compensation. The trouble is that really capable enemies are scarce; most of them are only ordinary. Another trouble with enemies is that they sometimes develop into friends and lose a great deal of their zeal. It was in this way the writer lost his three best enemies. Everyone, not just scientists, needs a good few enemies.

As can be seen from the above, friendships are essential for creative discussions. However, as von Békésy urges, we should also cherish competent academic opponents. Although archaeologists by nature tend to be conciliatory, and generally seek consensus – which is admirable but tends to result in conclusions of the lowest common denominator – it has also been blessed with a good number of reliable enmities.

Friendships are obviously more comfortable than enmities, so I am extremely grateful to Chris and Jessica, and the whole jolly crew, for many happy hours spent bouncing around in the back of a bus along the backroads of East Asia.

References

Bray, P., Cuénod, A., Gosden, C., Hommel, P., Liu, R. & Pollard, A. M. (2015) Form and flow: the 'karmic cycle' of copper. *Journal of Archaeological Science* 56, 202–9.

Cuénod, A., Bray, P. & Pollard, A. M. (2015) The 'tin problem' in the Near East – further insights from a study of chemical datasets on copper alloys from Iran and Mesopotamia. *Iran* 53, 29–48.

Gosden, C. (2020) *The History of Magic: From Alchemy to Witchcraft, from the Ice Age to the Present*. London, Viking.

Gosden, C. & Pollard, A. M. (2021) Is the universe sentient? What implications might this have for archaeology? In M. J. Boyd & R. C. P. Doonan (eds), *Far from Equilibrium: An Archaeology of Energy, Life and Humanity: A Response to the Archaeology of John C. Barrett*, 313–24. Oxford, Oxbow Books.

Hsu, Y.-K., Bray, P. J., Hommel, P., Pollard, A. M. & Rawson, J. (2016) Tracing the flows of copper and copper alloys in the Early Iron Age societies of the eastern Eurasian steppe. *Antiquity* 90, 357–75.

Jin, Z. Y., Liu, R., Rawson, J. & Pollard, A. M. (2017) Revisiting lead isotope data in Shang and Western Zhou bronzes. *Antiquity* 91, 1574–87.

Liu, R., Pollard, A. M. & Rawson, J. (2018) Beyond ritual bronzes: multiple sources of radiogenic lead used across Chinese history. *Scientific Reports* 8, article number: 11770.

Liu, R., Pollard, A. M., Rawson, J., Tang, X. & Zhang, C. (2019) Panlongcheng, Zhengzhou and the movement of metal in Early Bronze Age China. *Journal of World Prehistory* 32, 393–428.

Liu, R., Hsu, Y.-K., Pollard, A. M. & Chen, G. (2021) A new perspective towards the debate on highly radiogenic lead in Chinese archaeometallurgy. *Archaeological and Anthropological Sciences* 13, 33.

Pollard, A. M. & Bray, P. J. (2015) A new method for combining lead isotope and lead abundance data to characterise archaeological copper alloys. *Archaeometry* 57, 996–1008.

Pollard, A. M. & Gosden, C. (2023) *An Archaeological Perspective on the History of Technology*. CUP Elements. Cambridge, Cambridge University Press.

Pollard, A. M., Bray, P. J. & Gosden, C. (2014) Is there something missing in scientific provenance studies of prehistoric artefacts? *Antiquity* 88, 625–31.

Pollard, A. M., Bray, P., Gosden, C., Wilson, A. & Hamerow, H. (2015) Characterising copper-based metals in Britain in the first millennium AD: A preliminary quantification of metal flow and recycling. *Antiquity* 89, 697–713.

Pollard, A. M., Bray, P., Hommel, P., Hsu, Y.-K., Liu, R. & Rawson, J. (2017) Bronze Age metal circulation in China. *Antiquity* 91, 674–87.

Pollard, A. M., Liu, R., Rawson, J. & Tang, X. (2019) From alloy composition to alloying practice: Chinese bronzes. *Archaeometry* 61, 70–82.

Sainsbury, V. A., Bray, P., Gosden, C. & Pollard, A. M. (2021). Mutable objects, places and chronologies. *Antiquity* 95, 215–27.

Smith, C. S. (1968) Matter versus material: A historical view. *Science* 162, 637–44.

Technology. *Wikiversity*. [online] Available at: <https://en.wikiversity.org/wiki/Technology> [accessed 10/06/2022].

von Békésy, G. (1960) *Experiments in Hearing*. New York, McGraw-Hill.

Zhang, C., Pollard, A. M., Rawson, J., Huan, L., Liu, R. & Tang, X. (2019) China's major Late Neolithic centres and the rise of Erlitou. *Antiquity* 93, 588–603.

16

Biography and Technology: A Bronze *Ding* Vessel of the Iron Age in China

Xiuzhen Li

The life history of objects can provide rich insights into technology and culture. As a way of revealing the variety of relationships between people and things in different cultural contexts, scientific analysis can also augment object biographies. The bronze ding *vessel examined for this paper has an informative life history, from its production techniques to the inscriptions added in a variety of social and cultural contexts. These inscriptions concern China's Iron Age, particularly the Warring States period, the Qin Empire, and the Han Dynasty. Such palaeographic evidence provides a wealth of information about the vessel's experiences, including the people and events with which it was involved from the pre-imperial to early imperial periods. Using SEM techniques, it is possible to characterise tool marks and thus identify changes in technology used to inscribe the vessel in different periods and contexts. These include Warring States and Qin Empire chiselling techniques and possible Han Dynasty wheel incising techniques. This bronze* ding *vessel not only accumulated meaning within different cultural contexts and witnessed the unification of early imperial China, it also bears marks that demonstrate technological changes over the Iron Age.*

Keywords: Bronze *Ding* Vessel, Biography, SEM Imaging, Technological Changes

Introduction

Large quantities of bronze *ding* vessels served as symbols of civilisation in the Chinese Bronze Age, particularly in the Shang (1600–1050 BC) and Zhou (1050–221 BC) dynasties (*e.g.* Chang 1982; Falkenhausen 2006). They were mainly used for ritual practices and marked social status in burials (*e.g.* Bagley 1990; Rawson 1996). Some of them bear a variety of inscriptions, including long narratives, production information, and the names of clans or manufacturers (*e.g.* Shaughnessy 1991; H. Wang 2006). The inscriptions can be divided technologically into those cast or incised onto the vessels (*e.g.* Li *et al.* 2011; C. Zhang 2012; F. Li 2015). However, the study of these bronze objects has mainly focused on the casting technology, artistic decoration, and ritual practices involving them, whereas far less attention has been paid to the objects' biographies, including the social and technological changes to which they bear witness as a result of their involvement with people and events throughout their life history.

The theory of cultural biographies of objects understands the way that the histories of human and objects inform each other. 'As people and objects gather time, movement and change, they are constantly transformed, and these transformations of person and object are tied up with each other' (Gosden & Marshall 1999, 169). However, it is not just simple use-life (Tringham 1994), nor just a snapshot of one point in their existence of the processes and cycles of production, exchange, and consumption (Kopytoff 1986). Sometimes, the life of the object becomes a metaphor for those lives of humans (Dant 2001). A study of the biographies of objects is one way of understanding details in the changing patterns of social lives, including cultural and technological changes. However, the objects present their accumulated histories in a variety of ways. It is interesting to consider that the changes in the life of the object are sometimes reflected in the material itself, sometimes in the re-contextualisation process, and sometimes both.

Technology is broadly defined as 'a system of practices interrelating transformation of material resources, abstract and practical knowledge, social and political relationships, and cultural beliefs' (Brezine 2011, 82), while the term 'technology' itself 'can be used effectively and meaningfully at several different levels of abstraction' (Flad 2018), with the result that its very definition arguably varies according to different perspectives (*e.g.* Dobres 2000; 2010). Technology can be seen as an intermediary for interaction between objects and people as they form the objects' biographies. For this particular case study of a bronze *ding* vessel (Fig. 16.1), the technological practices involved include the initial bronze casting and the later addition of inscriptions using other tools or devices. Scientific methods can help to investigate such technological traits (Sáenz-Samper & Martinón-Torres 2017; Legarra & Martinón-Torres 2021; Plaza & Martinón-Torres 2021) and produce richer accounts of the technological and cultural contexts in which the objects were made, moved, modified, exchanged, and valued.

The aim of this paper is to understand the biography of the *ding* vessel, and how it was involved with people and events in China's Iron Age, particularly during the Warring States period (475–221 BC) and Qin (221–206 BC) and Han (202 BC–AD 220) dynasties. This paper also examines the inscriptions on the vessel, added at different stages, that bear witness to social and technological changes during these periods. This bronze *ding* vessel was even smuggled and shipped to France sometime in the late Qing Dynasty (1644–1911) before being returned to China in 2006 (Wu 2007). The dynamic history of this particular object is not only about its production, usage, exchange, burial, and collection, but also the meanings and value accumulated in different cultural and technological contexts.

Biography as an Analytical Tool

The bronze *ding* vessel was cast using piece moulds. It is 3.06 kg in weight, 17 cm in height, and 24.5 cm in diameter

Figure 16.1 The bronze ding *vessel. Emperor Qinshihuang's Mausoleum Site Museum collection, Shaanxi Province.*

between the two handles. It was dated to the Warring States period by its style and inscriptions (H. Wang 2007; M. Zhang 2007). There are no decorative patterns on its surface, but it bears 50 incised characters, including the names of cities, as well as the *ding*'s weight, volume, and function.

Altogether, four city names were incised on the surface of the bronze vessel, Yiyang (宜陽), Mian (黽), Xian (咸), and Linjin (臨晉) (Jiang & Liu 2007; H. Wang 2007), and these correlate with fighting among the Warring States and the Qin unification, as well as the royal palace of the Han (漢) Dynasty. Yiyang and Mian were cities of one of the seven warring states, Han (韓) State, which was later conquered by the Qin. Xian refers to Xianyang, which was the capital of Qin State and later the Qin Empire. Linjin was a county name in both the Qin and Han dynasties, and a temporary royal palace was built in the county city because of its crucial location. Thus, the vessel's movement between these cities seems to relate to regional conquests and war booty rather than trade and exchange, as observed for stone vessels found in the Bronze Age Mediterranean (Bevan 2007, 124–5) and Mesopotamia (Potts 1989).

Yiyang was located southeast of Luoyang in the Central Plain, in present-day Henan Province. The Han State was a small neighbour of Qin, whose mountainous territory directly blocked Qin's passage to the east. Although Han attempted to reform its government to improve state administration and strengthen its military, these reforms were not enough to defend itself, and it was the first of the seven warring states to be conquered by Qin. In 307 BC, the Qin occupied Yiyang, and in 260 BC they invaded Han's Shangdang Commandery, leading to the Battle of Changping with Zhao State, allegedly the bloodiest battle of the Warring States period in which up to 400,000 soldiers died (Loewe & Shaughnessy 1999; Lewis 2007). In 230 BC, the Qin had completely occupied Han territory. The characters for Yiyang were carefully chiselled near the vessel's upper edge, and it was probably created after the vessel had been cast. The chiselled marks clearly form curves to create the characters, which are similar to those carved on bronze weapons for the Qin Terracotta Army (Li *et al.* 2011; X. Li 2020). Details are discussed in the following section on technology. It remains uncertain, however, whether this *ding* vessel had a ritual or practical purpose when it was originally produced.

Xianyang was the capital of Qin State from 350 BC. Later, after 221 BC, it served as capital of the Qin Empire. There is one character, *xian* (an abbreviation of Xianyang), providing further information about the *ding* vessel's life history. The *xian* character is written in a different style from those for Yiyang, and calling it Xian rather than the formal Xianyang gives the inscription a more casual style. Next to *xian* are inscriptions detailing the volume of the vessel, which relates to the Qin's policy after their unification of the warring states. The First Emperor of Qin standardised characters, laws, coins, weights, and measures to consolidate the newly established empire. The inscription reads *yi dou si sheng* (一斗四升), meaning '1 *dou* [and] 4 *sheng*', which equals approximately 2.8 l. This is roughly the same as the vessel's actual volume, which is 2.89 l (Dang 2007). It was most probably taken to the Qin capital after the unification as war booty and its volume double checked before being chiselled onto it. This also indicates that it probably served as a measure for a certain period of time during the Qin.

A long inscription of 34 characters was added to the *ding* vessel in the Han Dynasty, which shows that it was later used for cooking. The inscription includes a county name (Linjin), the vessel's function, volume, and weight, as well as a sequential number. This shows that this *ding* vessel was passed down from the Qin to Han Dynasty, perhaps, again, as one of the spoils of war. Linjin was a Han Dynasty county that has been known by its present-day name of Dali, Shaanxi Province, since AD 266. Due to its location on the way from the Han Dynasty capital to the regions east of the Yellow River, a temporary palace for the royal family or a base for governmental officials was built here, according to written documents (H. Wang 2007). At this time, the vessel was now emphasised as a cooking vessel in the temporary palace and no longer served any ritual or measurement purpose. The inscription lists it as the 49th object in the royal kitchen. There is a record of another cooking utensil with a similar inscription used in this temporary palace, which was listed as the 35th object (Jiang & Liu 2007). This shows that a series of bronze vessels were used in the palace kitchen.

The inscriptions on the bronze *ding* vessel mentioned above cover part of its life history from the Warring States period to the Han Dynasty, but there is a huge gap between this period and the moment it was found in the private collection of a French family before being returned to China in 2006, specifically to Emperor Qinshihuang's Mausoleum Site Museum in Xi'an. It was supposedly excavated or looted in Shaanxi Province at the end of the Qing Dynasty then smuggled to Europe. Its return from France bore witness to the international relationship between China and France, reflecting its role in informing modern Chinese cultural identity and signifying another chapter in its life history.

In addition to observing the object's features and interpreting the inscriptions added in different periods to understand its general historical background, this chapter further investigates the early Iron Age technologies and people involved in creating these inscriptions. What kinds of tools and devices were available and practical for use? What was the intention of the people involved? Why did they put inscriptions on such a vessel?

Investigating Technological Changes

Technology can be seen as a medium that negotiates interactions between objects and people in the formation of object biographies. At the same time, the object's biography is closely related to their cultural (Thomas 1991) and technological contexts. For the bronze *ding* vessel, traditional casting technology was employed for its production; however, the inscriptions were not cast directly onto the bronze vessel as they were on many other early bronzes of the Shang and Zhou dynasties. The different techniques used to create the inscriptions, such as chiselling or wheel cutting, can be identified by making impressions using silicone rubber that are observed under Scanning Electron Microscope (SEM) (for details of the method, see Li *et al.* 2011; 2012). The results provide evidence of technological changes from the late Bronze Age to the Iron Age in ancient China, particularly from the Warring States to the Qin unification, and then to the Han Dynasty.

As mentioned above, bronze *ding* vessels were mainly used for ritual purposes in the Shang and Zhou dynasties, and inscriptions were mainly cast directly onto the surfaces of bronze objects. Cast inscriptions can also be divided into positive and negative styles. As bronzes were cast using ceramic piece moulds (Bagley 1990), a negative intaglio inscription was incised into the ceramic mould to yield a positive relieve inscription on the cast vessel (Škrabal 2021). While the relievo inscriptions were easy to produce, they were also more susceptible to mechanical damage during post casting finishing and regular cleaning when the vessels were used for ritual purposes. However, creating intaglio inscriptions on the ritual bronzes would have required ancient craftspeople to create positive lines on the ceramic moulds, because it was very difficult to engrave negative lines on the surface of hard bronze before the emergence of iron and steel (Wagner 1993; 2008; Li *et al.* 2011). The casting method for intaglio inscriptions on ritual bronze objects has been discussed over the past decades, with a modelling technique on the clay piece moulds argued as being a method of producing such inscriptions (Nickel 2006; Bagley 2009; C. Zhang 2010; F. Li 2015; Škrabal 2021).

It has also been argued that the engraved/carved negative inscriptions on the bronze were not solely the result of the emergence of iron and steel tools. The earliest attempts, simple incised lines or simple characters, come from Yinxu, the capital of the Bronze Age Shang Dynasty (Yang 2000, 100; Yue *et al.* 2012; Liu 2019, 106–9). Furthermore, engraving was occasionally used from the 9th century BC to produce longer inscriptions (Yang & Yang 2020), as seen on a set of bells belonging to Marquis Su of Jin State. These are dated to the middle of the 9th century BC and bear 355 engraved characters (C. Li 1998; Guan *et al.* 2002). From the 4th century BC, with iron tools coming into widespread usage, carved inscriptions gradually came to dominate the epigraphic landscape of ancient China (Škrabal 2021), with the techniques employed varying slightly from object to object, such as carved lines as opposed to chiselled marks (Li *et al.* 2011; C. Zhang 2012).

The tools (knives or chisels) used to create carved inscriptions would have to have been made of a material both harder and tougher than the bronze surface of the vessel. This could be hard stone, high-tin bronze, cast iron, or quenched steel. Both stone and high-tin bronze are relatively brittle and thus not suitable for carving or chiselling long inscriptions and marks (Li *et al.* 2011). The earlier, simpler characters carved on Shang bronzes could have been created with tools made of hard stone or high-tin bronze; however, the long inscriptions on the bells of Marquis Su were probably made using tools of high tin, siderite, or possibly iron (Guan *et al.* 2002). The chiselled inscriptions on the bronze weapons of the Terracotta Army, meanwhile, seem more likely to have been made of either cast iron or quenched steel (Li *et al.* 2011). During the Warring States period, iron tools and implements were already employed, and evidence for quenched steel and cast iron have been discovered at many contemporary archaeological sites in Central China (Han & Ke 2007; Wagner 2008, 98–128). Some even show evidence of earlier iron/steel production activities, which can be traced back to the 8th century BC in Jin State (Han & Duan 2009, 99). The inscriptions on the bell of Marquis Su of Jin may thus be the earliest examples created by iron tools.

Some inscriptions on this particular bronze *ding* vessel, including Yiyang, Xian(yang), and '1 *dou* [and] 4 *sheng*', were chiselled. The strokes of the characters were formed with elongated triangle marks that were clearly created by a chisel (Fig. 16.2), similar to that on the bronze weapons of the Terracotta Army (Li *et al.* 2011). The SEM micrographs on the left of Figure 16.2, taken from silicone impressions, show the character *yi* in reverse, so that the positive features in the micrographs represent the negative on the bronze. The chisels used for these marks were probably made of cast iron or quenched steel.

The long inscriptions created in the Han Dynasty are different from these chiselled marks (Fig. 16.3). The SEM micrograph on the left of Figure 16.3 was similarly taken of a silicone impression and shows the inscribed character in reverse. The strokes of the character are relatively straight and smooth. The positive curvature in the micrograph represents the negative curvature in depth on the bronze. In comparison with the wheel engraving on Mesopotamian cylinder seals (Sax *et al.* 2000), the stroke curvature suggests that these inscriptions were most probably produced using rotary wheels.

Although evidence for metal cutting can be traced back to the Shang and Zhou dynasties (Dong 2006, 44–5), rotary wheels were in use for jade carving in the Neolithic period (*ca.* 10,000–2000 BC), long before they were used on bronzes (Dong 2006, 104; Sax *et al.* 2007, 25). The square holes on the forehead of the bronze masks from Sanxingdui (2800–1100 BC; Chen 2000) are thought to have been cut

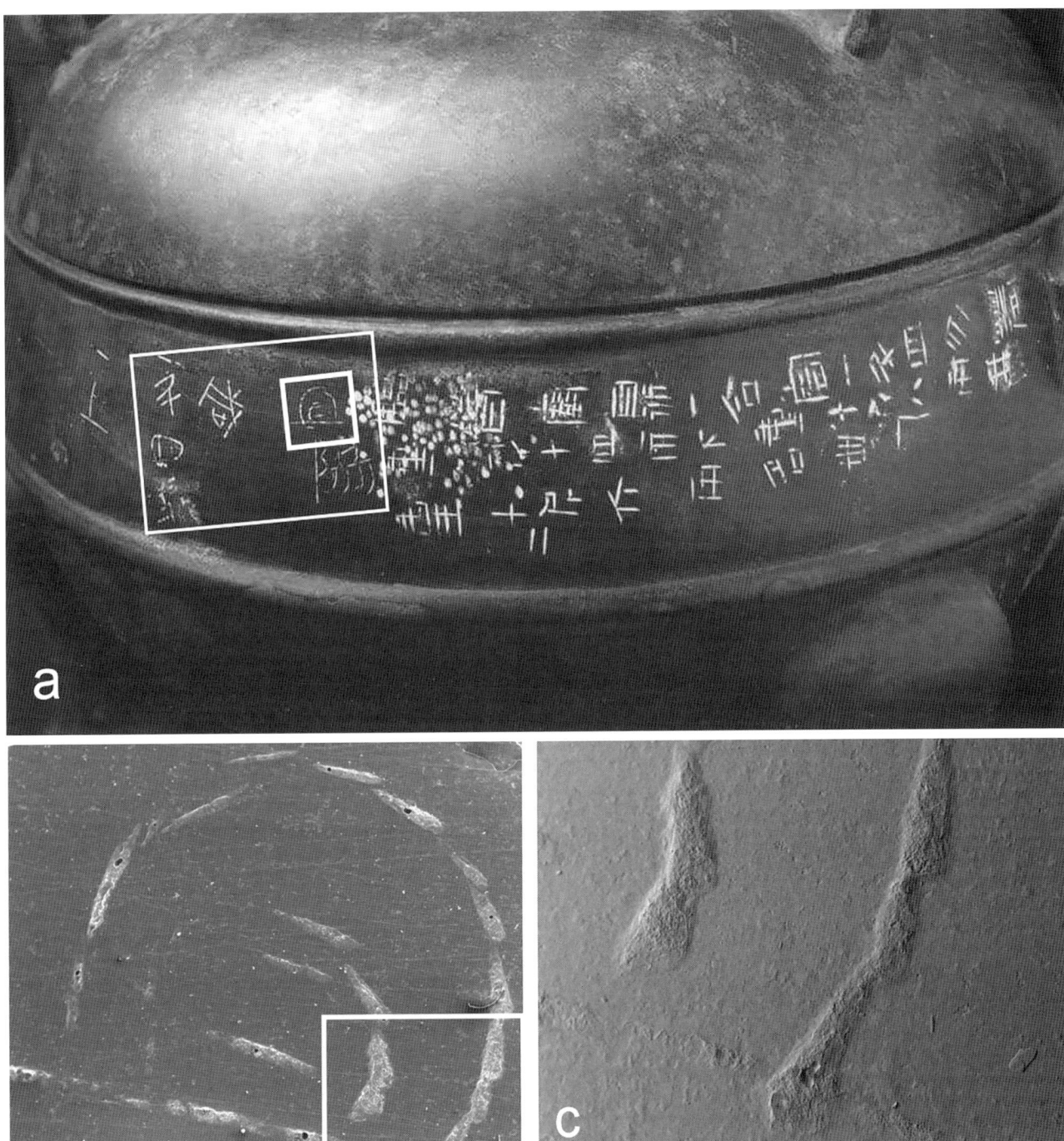

Figure 16.2 (a) Chiselled inscriptions of the Qin period, including the phrase yi dou si sheng *and characters for Xian and Yiyang; (b) Magnification of the character* yi *showing the use of a combination of chisel marks; (c) SEM micrograph showing detail of overlapping chisel marks used for the character* yi.

using a rotary wheel, because the four corners display the intrusive crossed lines that are typical of wheel cutting (Dong 2006, 44–5). If accurate, this would constitute the earliest example of the use of rotary mechanical means on Chinese bronzes known to date. In the Qin Dynasty, positive identification of a rotary mechanical device used on a systemic and industrial scale is provided in the form of grinding marks on the tens of thousands of bronze weapons made to equip the Terracotta Army (Li *et al.* 2011; X. Li 2020). The inscriptions on top of these grinding marks were, however, chiselled. The rotary incising wheel was supposedly slightly different from the rotary grinding wheel, but they both involved the same mechanical principles. As part of such a mechanical device, a cogwheel and

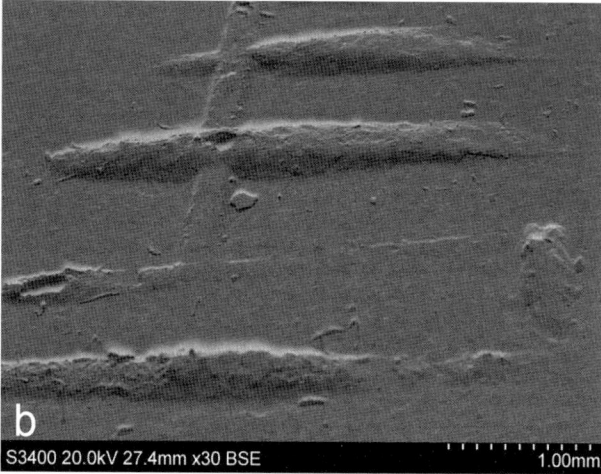

Figure 16.3 (a) Wheel-incised inscription created in the Han Dynasty; (b) SEM micrograph showing the curvature of lines used in the Han inscription.

a related mould have also been found at an archaeological site dating to the Han Dynasty (Loewe 2005).

However, the use of the rotary wheel for creating inscriptions on Chinese bronzes has been discussed only infrequently before. As mentioned above, use of the rotary wheel can be seen on jade from as early as the Neolithic period, and evidence for rotary lathes on jade objects has been identified from the Western Zhou (1045–771 BC) period (Sax & Ji 2013). With this long-term development, wheel incising would have been a mature technology by the Han Dynasty, the period to which the bronze *ding* vessel examined here dates, and combining iron blades with this technology would have been an easy way to create longer inscriptions on bronze. This vessel can tentatively be considered the earliest evidence for wheel-incised inscriptions on bronze vessels, though this interpretation needs further evidence to confirm this.

This bronze *ding* vessel embodies the technological changes that occurred over China's Iron Age. Technology, as an agent, bound people and objects together. The piece-mould casting technique, a tradition of the Bronze Age, was still used by craftspeople to produce bronze vessels during the Warring States period. Then, the inscriptions were added later by skilled workers using iron or steel chisels after its production and after the vessel had shifted various hands throughout the Qin period, from Qin State in the Warring States period to the Qin Empire. In addition to the craftspeople, many other people will have interacted with this vessel, imparting their feelings and concepts of self-identity onto it and simultaneously having these reinforced or even altered by the vessel itself, though none of this has left any marks behind. In the Han Dynasty, rotary wheels were most likely employed instead of iron chisels to mark it, which were an easy and efficient way to create inscriptions on bronze objects, and particularly those that formed part of a set (at least 49 altogether) used in kitchens of the temporary royal palace.

Discussion and Conclusion

The dynamic history and technological traits of this bronze *ding* vessel do not just simply reflect the variety of people involved, but also their social identities and a specific style of material culture (Stein 1999). In addition to the craftspeople and skilled workers who contributed to its production and inscriptions, this particular vessel interacted with many other people, including conquerors, owners, users, and organisers in the early stage of its life, as well as smugglers, collectors, donors, and museum curators in the later stage. Objects and people enrich both their life histories during their mutual interactions (Hoskens 1998). As a symbolic vessel in ancient China, the *ding* accumulated significance and meaning from the people who interacted with and used it, and some people's lives were enriched from being tied to it within specific social contexts, such as the donors who acquired the vessel. Even as its function and status transformed in different stages, its potential to emanate social and cultural meaning was embedded deep within it.

If the inscriptions on the bronze vessels are regarded as a type of social memory (M. Wang 1999), it was a selective memory related to specific locations, people, and events that can be compared to historical documents in the way that such memory exists within different cultural scenarios. The inscriptions were added with the aim of covering the previous memories in a new social and cultural context, and it could be presented to new people and be involved with new events. Yiyang is believed to be the earliest inscription that shows the vessel's original production location. The characters for Yiyang are written in the style of Qin (State) rather than the local Han State style, similar to those found within the First Emperor's mausoleum complex (W. Wang 2000; H. Wang 2007; Yuan & Liu 2009, 246) and on the Qin bronze weapons (Li *et al.* 2011), meaning these characters were not made immediately after production. The inscription Yiyang was probably added by Qin craftspeople after the Qin occupied the city in 307 BC or by local craftspeople who were following the Qin style. The inscription of Xian(yang) and the vessel's volume in standardised Qin measurements were added using the same technique and demonstrate that it was taken into the Qin system as war plunder after the unification. In the following Han Dynasty, the inscription's emphasis on the fact that the *ding* was a cooking vessel in the royal palace adds another new memory that changed its identity. Adding new inscriptions was also a way of changing the bronze vessel's role and perceptions of it in a dynamic context.

However, when the *ding* vessel was unearthed and smuggled to Europe at the end of the Qing Dynasty, this represented a sharp break from its earlier social and cultural significance, and it reflected a specific social identity. The vessel passed through several collectors in France. In 2005, Mr Bernard Gomez, president of the Association for the Protection of Chinese Art in Europe (APACE), and Ms Huang Xinlan, president of the Meidu Company of Real Estate in Xi'an, worked together to buy the *ding* vessel at auction in France and donated it to the Emperor Qinshihuang's Mausoleum Site Museum. A ceremony was held when the vessel arrived at the museum in 2006, and many people were involved. Thus, the *ding* took on a new social role as a symbol of modern Chinese identity, reflecting the nation's participation in international networks of cooperation.

Biographies of objects can be approached in a variety of ways (Gosden & Marshall 1999). One can recover the social contexts in which the object emerged through a reconstructive narrative (Dant 2001). This bronze *ding* vessel did not simply accumulate physical changes in the form of incised inscriptions on its surface, but it was also involved with a variety people and events within dynamic social and technological contexts. It witnessed the fighting during the Warring States period, the Qin unification, and the following Han Dynasty. By reconstructing a narrative, scientific approaches can help to build rich biographies that in turn inform the technological and cultural contexts of where artefacts were made, moved, modified, exchanged, and valued. Though the *ding* vessel was smuggled abroad, it also forged another story of social and cultural identity in France and later when it was returned to China. The life-history of this bronze *ding* vessel can thus be narrated from a multitude of perspectives in a variety of particular cultural and technological contexts.

Acknowledgements

I would like to thank Professor Marcos Martinón-Torres and Professor Andrew Bevan for their comments and suggestions. Many thanks are due to Shiona Airlie for editing the draft version.

References

Bagley, R. W. (1990) Shang ritual bronzes: Casting technique and vessel design. *Archives of Asian Art* 43, 6–20.

Bagley, R. W. (2009) Anyang mold-making and the decorated model. *Artibus Asiae* 69(1), 39–90.

Bevan, A. (2007) *Stone Vessels and Values in the Bronze Age Mediterranean*. Cambridge, Cambridge University Press.

Brezine, C. J. (2011) *Dress, technology and identity in colonial Peru*. Unpublished PhD thesis, Harvard University.

Chang, K. C. (1982) *Studies of Shang Archaeology*. New Haven, Yale University Press.

Chen, D. (2000) *Sanxingdui*. Chengdu, Sichuan People's Press (in Chinese).

Dang, S. (2007) A study on the Yiyang bronze tripod. In Wu 2007, 19–21.

Dant, T. (2001) Fruitbox/toolbox: Biography and objects. *Auto/Biography* 9(1–2), 11–20.

Dobres, M. (2000) *Technology and Social Agency: Outlining a Practice Framework for Archaeology*. Oxford, Blackwell Publishers.

Dobres, M. (2010) Archaeologies of technology. *Cambridge Journal of Economics* 34(1), 103–14.

Dong, Y. (2006) *Mould-Cast Bronzes*. Beijing, Beijing Art and Science Electronic Publishing House (in Chinese).

Falkenhausen, L. von (2006) *Chinese Society in the Age of Confucius (1000-250 BC)*. Los Angeles, University of California.

Flad, R. (2018) Urbanism as technology in early China. *Archaeological Research in Asia* 14 (2018), 121–34.

Gosden, C. & Marshall, Y. (1999) The cultural biography of objects. *World Archaeology* 31, 169–78.

Guan, X., Lian, H., Bai, R., Liu, X. & Hua, J. (2002) An investigation of how the carved inscriptions were created on the bell of Marquis Su of the Jin State. In Shang Museum (eds), *International Symposium on the Bronze Objects Found in the Cemetery of Marquis Jin*, 331–345. Shanghai, Shanghai Shuhua Press (in Chinese).

Han, R. & Duan, H. (2009) An early iron-using centre in the ancient Jin State region (8th-3rd century BC). In J. Mei & T. Rehren (eds), *Metallurgy and Civilisation: Eurasia and Beyond*, 99–106. London, Archetype Publications.

Han, R. & Ke, J. (eds) (2007) *A History of Science and Technology in China – Mining and Metallurgy*. Beijing, Sciences Press (in Chinese).

Hoskens, J. (1998) *Biographical Objects: How Things Tell the Story of People's Lives*. London, Routledge.

Jiang, W. & Liu, Z. (2007) Interpretation of the inscriptions on the Qin Yiyang tripod. In Wu 2007, 26–35.

Kopytoff, I. (1986) The cultural biography of things. In A. Appadurai (ed.), *The Social Life of Things: Commodities in Cultural Perspective*. Cambridge, Cambridge University Press.

Legarra Herrero, B. & Martinón-Torres, M. (2021) Heterogeneous production and enchained consumption: Minoan gold in a changing world (ca. 2000 BCE). *American Journal of Archaeology* 125, 333–60.

Lewis, E. M. (2007) *The Early Chinese Empires: Qin and Han*. Cambridge, Harvard University Press.

Li, C. (1998) Engraved inscriptions on the bronze bell of Marquis Su of Jin and the use of iron in the Western Zhou. In Department of History of Sichuan University (ed.), *A Festschrift in Honour of the 100 Birthday of Xu Zhongshu,* 116–21. Chengdu, Bashu Press (in Chinese).

Li, F. (2015) Solving puzzles about the casting method of bronze inscriptions of the Western Zhou Dynasty. *Chinese Archaeology* 15(1), 140–52.

Li, X. (2020) *Bronze Weapons of the Qin Terracotta Warriors: Standardisation, Craft Specialisation and Labour Organisation.* Oxford, Bar Publishing.

Li, X., Martinón-Torres, M., Meeks, N. D., Xia, Y. & Zhao, K. (2011) Inscriptions, filing, grinding and polishing marks on the bronze weapons from the Qin Terracotta Army in China. *Journal of Archaeological Science* 38, 492–501.

Li, X. J., Martinón-Torres, M., Meeks N. & Xia, Y. (2012) Scanning electron microscopy imaging of tool marks on Qin bronze weapons using silicone rubber impressions. In N. Meeks, C. Cartwright, A. Meek & A. Mongiatti (eds), *Historical Technology, Materials and Conservation: SEM and Microanalysis,* 62–8. London, Archetype and The British Museum.

Liu, Y. (2019) *Casting Techniques of Ritual Bronzes Unearthed from Yinxu.* Guangzhou, Guangdong Renmin Press (in Chinese).

Loewe, M. (2005) *Everyday Life in Early Imperial China.* London, Batsford.

Loewe, M. & Shaughnessy E. L. (eds). (1999) *The Cambridge History of Ancient China: From the Origins of Civilization to 221 BC.* Cambridge, Cambridge University Press.

Nickel, L. (2006) Imperfect symmetry: Re-thinking bronze casting technology in ancient China. *Artibus Asiae* 66(1), 5–39.

Plaza, M. T. & Martinón-Torres, M. (2021) Technology, use and reuse of gold during the Middle Period: The case of Casa Parroquial, Atacama Desert, Chile. *Cambridge Archaeological Journal* 31, 613–37.

Potts, T. F. (1989) Foreign stone vessels of the late third millennium B.C. from southern Mesopotamia: Their origins and mechanisms of exchange. *Iraq* 51, 123 –64.

Rawson, J. (1996) The ritual bronze vessels of the Shang and Zhou. In J. Rawson (ed.) *Mysteries of Ancient China: New Discoveries from the Early Dynasties.* London, British Museum Press.

Sáenz-Samper, J. & Martinón-Torres, M. (2017) Depletion gilding, innovation and life-histories: the changing colours of Nahuange metalwork (Colombia). *Antiquity* 91(359), 1254–67.

Sax, M. & Ji, K. (2013) The technology of jades excavated at the Western Zhou, Jin Marquis cemetery, Tianma-Qucun, Beizhao, Shanxi Province: Recognition of tools and techniques. *Journal of Archaeological Science* 40(2), 1067–79.

Sax, M., Meeks, N. D. & Collon, D. (2000) The introduction of the lapidary wheel in Mesopotamia. *Antiquity* 74(284), 380–7.

Sax, M., Meeks, N. D., Ambers, J. & Michaelson, C. (2007) The introduction of rotary incising wheels for working jade in China. In J. G. Douglas, P. Jett & J. Winter (eds), *Scientific Research on the Sculptural Arts of Asia. Proceedings of the Third Forbes Symposium at the Freer Gallery of Art.* London, Archetype Publication.

Shaughnessy, E. L. (1991) *Sources of Western Zhou History: Inscribed Bronze Vessels.* Berkeley, University of California Press.

Škrabal, O. (2021) How were bronze inscriptions cast in ancient China? New answers to old questions. In J. B. Quenzer (ed.), *Exploring Written Artefacts, Objects, Methods, and Concepts.* Berlin, De Gruyter.

Stein, G. J. (1999) Material culture and social identity: The evidence for a 4th millennium BC Mesopotamian Uruk colony at Hacinebi, Turkey. *Paléorient* 25(1), 11–22.

Thomas, N. (1991) *Entangled Objects: Exchange, Material Culture and Colonialism in the Pacific.* Cambridge, MA, Harvard University Press.

Tringham, R. (1994) Engendered places in prehistory. *Gender, Place and Culture* 1(2), 169–203.

Wagner, D. B. (1993) *Iron and Steel in Ancient China.* Leiden, Brill.

Wagner, D. B. (2008) *Science and Civilisation in China. Volume 5: Chemistry and Chemical Technology. Part 11: Ferrous Metallurgy.* Cambridge, Cambridge University Press.

Wang, H. (2006) *Inscriptions on the Bronzes of the Shang and Zhou Dynasties.* Beijing: Cultural Relics Press (in Chinese).

Wang, H. (2007) Research on the newly discovered Qin Yiyang tripod. In Wu 2007, 10–15.

Wang, M. (1999) Western Zhou remembering and forgetting. *Journal of East Asian Archaeology* 1, 231–50.

Wang, W. (2000) Pottery inscriptions found at the Qin site of Nandu in Xinfeng, Lintong, Xi'an. *Archaeology and Cultural Relics* 1, 7–8 (in Chinese).

Wu, Y. (ed.) (2007) *A Bronze Treasure Returns Home: Record of an Important Donation.* Xi'an, Chuangwei Printing (in Chinese).

Yang, H. & Yang, J. (2020) Examination of the inscribing techniques applied on cast bronze of the Shang and Zhou dynasties. *Jianghan Archaeology* 3, 100–106 (in Chinese).

Yang, X. (2000) *Reflections of Early China: Decor, Pictographs, and Pictorial Inscriptions.* Seattle & London, The Nelson-Atkins Museum of Art in association with the University of Washington Press.

Yuan, Z. & Liu, Y. (2009) *New Edition of Qin Pottery Inscriptions.* Beijing, Cultural Relics Press (in Chinese).

Yue, Z., Yue, H. & Liu, Y. (2012) Techniques for creating the inscriptions on the Yinxu bronzes. *Cultural Relics of Central China* 4, 62–8 (in Chinese).

Zhang, C. (2010) The bronze inscription techniques used during the Shang and Zhou dynasties: A case study focusing on the bronzes of Zheng State. *Cultural Relics* 8, 61–70 (in Chinese).

Zhang, C. (2012) Some considerations of the bronze inscription techniques used during the Shang and Zhou synasties. Translated by L. Lu. In N. Pearce & J. Steuber (eds), *Original Intentions: Essays on Production, Reproduction, and Interpretation in the Arts of China,* 264–81. Gainesville, University Press of Florida.

Zhang, M. (2007) Discussion of the date of the newly discovered Yiyang tripod. In Wu 2007, 16–18.

17

Rewriting Global Histories of Human–Material Relations in Different Cultural Contexts

Shadreck Chirikure

Humans have always entangled themselves with material and immaterial worlds in stunningly dissimilar ways. Conditioned by varying cosmologies and epistemologies (ways of knowing and doing), this diversity of experiences stimulated different trajectories of how materials and their materialities shaped human histories in various places. For instance, based on materials that preserved well, large parts of Africa initially used stone and directly transitioned to iron (ca. 1000–800 BC) before bronze emerged 1000 years later (ca. AD 1000). In the Americas, the reduction of iron was unknown before the Columbian exchange (beginning AD 1492); stone and wood were dominant materials in Australasia before European colonisation. Eurasia experienced the transition from stone to the steam engine via bronze and iron. Inspired by Chris Gosden's works on colonialism and post-colonialism, this contribution speaks to the variety of human–material relations across cultures and geographies and argues for inclusive epistemologies to make studies of materials in archaeology less colonial and less Eurocentric.

Keywords: Human–Material Relations, Cosmologies, Epistemologies, Three-Age System, Post-Colonial

Introduction

Generally, the way humans of different societies reflected on their universe to a large extent shaped how they interacted with materials around them (Gosden 2004). This prompted some theorists to argue that technologies, and by extension human–material relations, are all materialised worldviews (Tresch 2014). Cosmologies or worldviews encapsulate philosophies about the world and the heavens. Not surprisingly, various cultures in different places across the broad spectrum of the human past developed culture-specific origin myths (Renfrew & Bahn 2012). For example, the book of Genesis in the Bible puts forward the version that the Christian God created the world and its elements in seven days. Meanwhile, some among the Dorobo of East Africa believe that *Torooret*, or God who lives high in the sky, created the world and let their founding ancestor glide down to earth using a leather cord (Kenny 1981). Dissimilar cosmologies often translated into different ways of knowing and thinking about the world and the materials in it (Chirikure 2017; 2020). Worldviews also shape conceptions and understandings of time, and therefore societies the world over have varied ways of thinking about it (Fabian 1983; Gosden 1994; Lucas 2004). Amongst groups now known as Shona in southern Africa, time is known as *nguva*. In the Shona world, *nguva* is important but complex, as it speaks to how long ago, and how long after, things occur. The past (*pasichigare*), present (*parinhasi*), and future (*ramangwana*) are all different periods that, depending on context, are interwoven into each other (Chirikure 2020). One of the most important observations is that events are not ordered by materials: the ordering is shaped by this cosmology where the past, present, and future co-exist. For example, when people die, they transition into the living dead (*vadzimu*, plural; ancestors) who guide the present and future, which become the past as days succeed each other (Chirikure 2020). This Shona worldview and conception of time is one of the multiple ways of knowing and doing currently marginalised by dominant ways of interpreting human–material relations through time. The Shona past is not ordered based on materials but is considered in a more integrated and holistic sense, speaking to the totality of human experience – the material, immaterial, and the tangible and intangible.

Like humans living in Africa or Oceania, those occupying regions now known as Europe had their own cosmologies and ways of thinking about what came before them in the long ago (Gosden 1994). For example, ancient Greeks and Romans philosophised about the ages that antecedent societies had passed through, leading to their own time (Rowley-Conwy 2004; 2007). According to Renfrew and Bahn (2012), Hesiod, the ancient Greek philosopher, delineated five ages of human existence, beginning with the original, long-vanished stage in which humans enjoyed near divine existence – the age of gold – through the ages of silver, bronze, and epic heroes to his time, which he called the age of iron and dreadful sorrow, characterised by pain and suffering. The Roman poet Ovid outlined the ages of gold, silver, bronze, and iron – each symbolically linked to decreasing values attached to the metals/alloys (Gräslund 1987). These ideas were sustained in various shades into the 19th century, when curiosity about human existence (from the point of view of Western cosmology) gained momentum during the Enlightenment period in Europe. The theories of stratification and uniformitarianism, combined with the recovery of stone tools in river gravel in France, prompted the realisation that humans had great antiquity on earth (Rowley-Conwy 2007; Renfrew & Bahn 2012). C. J. Thomsen developed the three-age system to order collections in the Royal Museum in Copenhagen. Later, Worsaae matched this classification with stratigraphic information to conclude that humans first used stone technologies (Stone Age), before moving to bronze (Bronze Age) and iron (Iron Age). Emerging in Denmark and southern Sweden in the years 1835–43, the Stone-Bronze-Iron system provided a system of relative ordering of archaeological artefacts and epochs with doses of Eurocentric explanatory potential (Rowley-Conwy 2007). However, this ordering of past events was evidently biased in favour of materials that preserved well and/or left durable traces.

From this Eurocentric origin and worldview, the system of assigning relative chronology to past events based on ages that correspond to selected materials became heavily embedded in archaeological thought and practice. Between the 19th and mid-20th centuries, the three-age system was coupled with ideas about social evolutionism (Gosden 1994; Lucas 2004). These advanced the flawed thinking that society evolved linearly from the simple savagery stage through the in-between barbarism level on the way to the highest stage of civilisations (see Trigger 1989). Civilisations like those of Mesopotamia and Egypt were identified by distinctive traits, such as sophisticated technology, literacy, long distance trade, monumental architecture, and bureaucracy, to mention just a few (Childe 1936; 1950). Comparatively, some populations resident in different parts of the world that did not fit this framework were assigned to lower levels.

These include the Kalahari hunters and gatherers who mostly used stone to make tools and were thus pigeonholed as belonging to the savagery stage (Fabian 1983; Stahl 1993).

While the discovery of radiocarbon and other absolute dating techniques added more certainty to when phenomena occurred in the past, the three-age system remained the major temporal structural unit in European and, with minor exceptions, world archaeology (Gräslund 1987; Rowley-Conwy 2004). With European colonial expansion, the three-age system was exported to different regions that obviously had different histories of engagement with materials and the universe. In the process, local cosmologies and ways of assigning events to the past were side-lined (Chirikure 2017; 2020). The European way, which developed specific to a given cultural context and cosmology, became the universal in regions like Africa and elsewhere. Considered from this perspective, the imprinting of the three-age system's logics onto other areas of the world might represent some form of colonialism (Gosden 2004). Is it possible to bring into the discussion other ways of thinking about human–material relations in ways rooted to worldviews of the Global South? This contribution grapples with this and other matters in its discussion of different histories of material relations across cultures and geographies. The next section presents the evidence from Africa as a backdrop to segue into other parts of the world.

Materials and the Past in Africa

A significant body of research has, with limitations, enabled us to understand the history of human–material relations in Africa through different time periods (Phillipson 2005). However, retrospective discussions are always easier than predictions that may or may not be validated. Synthesising what we know about Africa, especially for the period AD 1500–1884/5 (before European colonialism), there was a great deal of variability regarding the broad spectrum of materials used by humans living in vast stretches of land from the Cape of Good Hope in the south to the Mediterranean in the north (Chirikure 2015). In southern Africa, an assessment of materials used by different communities between AD 1500 and the 1900s demonstrates interesting similarity and variety. Materials were used to make permanent and temporary shelter, permanent and improvised tools, and other objects for symbolic and utilitarian purposes. Examples of these include clay, both fired and unfired, wood, metals and alloys, stones, and organic materials like resin, egg and seashells, and leather.

These were used by various communities with different lifeways (Mitchell 2002); some communities had livelihoods based on hunting and gathering, while others were transhumant pastoralists, moving from place to place with cattle, sheep, and goats (Deacon & Deacon 1999). The Khoe herders in the Cape were pastoralists, while the San lived on hunting and gathering. Amidst these were sedentary farming communities that were organised as states, chiefdoms, and acephalous communities. The farmers made metal, stone tools, and pottery; they worked leather and built permanent shelter of clay, wood, and stones. Conditioned by their ways of life, the San and Khoe lived in less permanent settlements, often using stone as a construction material. These groups – farmer, pastoralist, and hunter-gatherer – were not mutually exclusive as the categories might imply. Often, humans pigeonholed into these groups through archaeological classifications exchanged materials and intermarried, creating a mix of genes, worldviews, and material and visual culture.

This picture of admixture also prevailed to humans in central and eastern Africa, where different groups of sedentary farmers were making metal, working wood, moulding pottery, and keeping cattle and small livestock. The pastoralists in East Africa were transhumant but also made iron (Iles & Martinon-Torres 2009), unlike pastoral groups in the south, such as the Khoe. The Pygmies of Central Africa also co-existed alongside farmers responsible for some of the great empires, such as the Luba or Lunda. In West Africa, different communities also engaged with the world around them in both similar and different ways. The farmers worked iron, copper, and other alloys, as well as gold. They also used stone for anvils in making tools and other objects. Pastoralism was a developed way of life amongst groups such as the Fulani. In North Africa and Ethiopia, the same materials – metals, stone, clay, and wood – were worked but in culturally specific ways.

The conclusion that emerges is that humans in different parts of Africa engaged with materials around them in culturally specific ways shaped by different worldviews. Often, exchange resulted in the incorporation of traits from other systems, as well as in the production of new innovations. The spread of religion, especially Islam in West Africa and East Africa, introduced new human–material relations, including the construction of places of worship such as mosques and a new ordering of space. Even then, not everybody was converted. From AD 1500 onwards, the Portuguese interacted with the Shona in northern Zimbabwe and central Mozambique but did not make many converts (Schmidt & Pikirayi 2018).

When we add the picture from Africa's deep past, we see very interesting pictures emerging. For a very long time, stone tools dominate the archaeological record, perhaps owing more to their high rate of preservation and less to the fact that they were the only material in use. There is also a preservation bias, in that perishables like wood rarely survive in deep time deposits similar to other materials, such as unfired clay. Nevertheless, from a wider usage of stone to make tools, different regions of Africa followed varied trajectories related to the appearance of other materials in the material record. Regions such as the West African Sahelian regions of Mali developed pottery making around 10,000 cal BC (Huysecom *et al.* 2009). Metal initially appeared in the form of iron and iron and/or copper with gold, with tin bronze appearing around AD 700 (Chirikure 2015). However, sophisticated craftsmanship involving clay and copper-based metals was well established across West Africa from AD 900 onwards, as shown by terracotta from the Inland Niger Delta of Mali and remarkable finds like those from Igbo Ukwu. The same material engagement record applies to eastern and southern Africa. North Africa is, however, very different: in ancient Egypt, copper was followed by bronze and eventually iron. However, in places like Carthage (Tunisia), Ethiopia, and many others, tin bronze, gold, and glass were worked before iron. What is therefore interesting is that, within a single landmass such as Africa and its islands, humans entangled themselves with similar materials and at different times, but the relationship was mediated by different worldviews. Given the different contexts, the values attached to different materials also differed from place to place and time to time.

The situation that emerges is that, once available, different materials were worked and used during periods that overlapped chronologically by people in different locations. Some humans, however, chose to stick to stone, *e.g.* the Kalahari hunter-gatherers, while the East African pastoralists opted to make iron, which was not the material of choice for Khoe herders. Indeed, even in excavations of some archaeological sites dating to the last 2000 years, all materials are present, showing that assigning major period names based on one material, such as stone, is applicable to part of but not the whole picture. As such, as a chronological label, the three-age system is not fit for purpose in Africa south of the Sahara. This motivates concept revision (Chirikure *et al.* 2017) to rid archaeology of some of its Western prejudices (Schmidt & Pikirayi 2018). What, therefore, are the implications of this history of material engagement for global histories of human–material relations? This is the focus of the section below.

Discussion: Recasting Human–Material Relations Across Geographies and Cultures

Human entanglements with materials extend far back in time (Hodder 2012). Available evidence suggests that these were shaped by, and in turn shaped, different worldviews. A short overview of human–material relations in Africa shows varied and context-specific responses that saw some materials being used by humans all over the continent at chronologically overlapping times (*e.g.* stone), but in some situations some humans (*e.g.* Kalahari hunter-gatherers) preferred stone to metals, even when the latter were available (Chirikure 2015). On the face of it, this choice may appear to defy logic, particularly from the perspective of worldviews conditioned to consider the properties of iron, such as its malleability and ductility, to be superior to stone. But this is also a humbling observation, because materials are materials and are often selected based on culturally specific motivations. Owing to its origins in a Eurocentric worldview, archaeology borrows heavily from the three-age system, and interprets the value attached to different materials, as well as the ease with which they could be worked compared to others (Rowley-Conwy 2004), from this system.

Within the three-age-system logic, stone was considered an easier material to work compared to native copper, which requires hot and cold hammering. The smelting of copper was also considered easier compared to that of iron, which appeared last of the three ages (Daniel 1943; Gräslund 1987). In this categorisation, stone constitutes a preliminary stage, because it can be used for tools without much working. By comparison, a bronze knife implies that somebody controlled the heat, temperature, and atmospheric conditions. Iron is also considered to be very advanced, with its smelting requiring control of furnace temperature and gases to reduce ores to metal, while forming a slag to mop away the impurities (Phillipson 2005; Chirikure 2015). While the identification of certain technologies or materials and their diversification or sophistication might as well be a sign for cultural change, continuity is less emphasised in the three-age system. At the heart of such a hierarchical ordering of materials – stone-bronze-iron – stands a rather biased and unrefined understanding of technology and human–material relations. For instance, the working of stone to produce tools was not simple; it required complex organisation and a skillset to select suitable stones and shape them to produce desired outcomes. This contradicts the view that some of the earliest use of stone by humans was based on trial-and-error or chance usage (Deacon & Deacon 1999).

While useful in some respects, the hierarchical ordering of materials based on processing difficulty and production organisation implied in the three-age system has consigned much of what happens in Africa to levels of stagnation and stasis (Chirikure 2015; 2017). Therefore, the 19th-century Kalahari hunter-gatherers were erroneously considered as remnant societies from the deepest Stone Age. Such careless classifications are oblivious of the fact that the same people were marrying into other groups that were sedentary and iron-making, nevertheless they consistently chose not to make iron or copper, shunning a sedentary life (Mitchell 2002; Phillipson 2005). In any case, stone as a material continues to be worked by humans even in the present, demonstrating remarkable continuity in the adaptation of stone as a material in varied contexts (Chirikure *et al.* 2018). Clearly, human–material relations were not as clear cut as traditionally implied by the linear evolutionary logic set in the three-age system. In any case, sub-Saharan Africa inverts the human–material relations predicted by the three-age system because of transitions from stone to iron to bronze.

The situation in Africa is somewhat similar but also different to that in other parts of the world, such as the Americas. In North and South America, various materials, such as stone, copper, and silver, were in use before Columbus's arrival (Lechtman 1988; Hosler 1994; Schultze *et al.* 2009). Iron was only transferred from Europe as part of the Columbian exchange. Archaeologists working in the Americas recognised that it was not possible to transplant the template for exploring human–material relations developed for Europe (*i.e.* the three-age system) (Pauketat & Sassaman 2020). Rather than periods defined by materials, the main stages in the Americas are the Lithic, Archaic, Formative, Classic, and Post-Classic stages (Cobb 1991; Sassaman 2010). This system has been modified to suit different regions of North and South America, taking into account regional diversity across this vast continent. As with Africa, different humans had varying worldviews to the extent that a bottom-up approach using local cosmologies is critical for building a comparison of human–material practices in other parts of the world (Romain 2009; Pauketat *et al.* 2015).

China is yet another part of the world with a very interesting history of human–material relations. While in some contexts China mirrors the situation in Eurasia, one of the most iconic materials worked in the region was jade. Jade was mostly popular in the period 4500 to 2000 BC. According to Dematté (2006), some Chinese scholars have previously argued for a Jade Age, which would occupy the intermediate place between the Stone and Bronze ages. Had this been accepted, the Jade Age would have differentiated the situation in China from the history of material succession depicted in the three-age system in the rest of Eurasia. The issue is not, however, about creating ages to mimic the situation in Europe. Rather, the main point is that different humans living in different parts of the world had varied histories of working and assigning values to different materials (Smith & Wobst 2004; Smith 2005). This reinforces the point that perhaps what is required is a bottom-up approach, where the situations in different regions are compared to provide global histories of human–material relations.

The application of the three-age system erases and marginalises other cosmologies and ways of relating to and understanding materials (Mavhunga 2014; Pauketat *et al.* 2015). According to Gosden (2004), the imposition of European ways of thinking on regions of the world like Oceania has, in some ways, colonised and suppressed ways of knowing in those places. In line with post-colonial theories, there is no universal theory or understanding of human–material relations (Gosden 2004). Rather, it is essential to factor in different epistemologies and worldviews (Mavhunga 2014). For example, as mentioned earlier, communities like the Shona in southern Africa do not order the past based on materials (Chirikure 2020). In some cases, it is the task at hand and social context that determines the appropriate material within a canvas provided by worldviews and epistemologies. The big question then becomes how do we order past events free from the shackles of the three-age system?

Absolute methods of dating, such as radiocarbon, may be of greater use in producing age estimates that can be moulded into a chronology free from the 'ages of materials'. This will allow an understanding of the full range of materials and not just the signature material giving its name to an age. This will avoid the dangers of transposing culture-bound concepts, such as the three-age-system, to other cultural areas, thereby offering an opportunity to develop inclusive theoretical frameworks (Demattè 2006). Across the story of human–material relations, some materials were discovered only to disappear and be rediscovered much later. Others like stone continued through space and time. Any comparative understanding of human–material relations across cultures and geographies must be guided by local epistemologies and cosmologies to produce less colonial versions of the past, as championed by Gosden (2004).

Conclusion: Recasting Human–Material Relations

In conclusion, the human story is one in which people were continuously entangled with the material world. However, the expansion of Europe resulting in the colonisation of different parts of the world imposed outside worldviews on other cultures. Consequently, it is vital to adopt a bottom-up approach whereby the situation in different areas is combined to build a more global picture. This will enrich the comparison because there are similarities between Africa, the Americas, and Eurasia, but there are also massive differences. This bringing together of different perspectives is central to what Chris Gosden worked on during his time at the Pitt Rivers and continues to work on now.

References

Childe, V. G. (1950) The urban revolution. *The Town Planning Review* 21(1), 3–17.

Childe, V. G. (1936) *Man Makes Himself*. London, Watts & Co.

Chirikure, S. (2015) *Metals in Past Societies: A Global Perspective on Indigenous African Metallurgy*. New York, Springer.

Chirikure, S. (2017) The metalworker, the potter, and the pre-European African 'laboratory'. In C. Mavhunga (ed.), *What Do Science, Technology, and Innovation Mean from Africa*, 63–77. Cambridge MA, MIT Press.

Chirikure, S. (2020) *Great Zimbabwe: Reclaiming a 'Confiscated' Past*. Abingdon, Routledge.

Chirikure, S., Nyamushosho, R. T., Chimhundu, H., Dandara, C., Pamburai, H. H. & Manyanga, M. (2017) Concept and knowledge revision in the post-colony: Mukwerera, the practice of asking for rain amongst the Shona of southern Africa. In M. Manyanga & S. Chirikure (eds), *Archives, Objects, Places and Landscapes: Multidisciplinary Approaches to Decolonised Zimbabwean Pasts*, 14–55. Bamenda, Langaa.

Chirikure, S., Fredriksen, P. D. & Manyanga, M. (2018) Adaptation, craftscapes and knowledge networks: Introductory remarks on historical ecology and state formation in southern Africa. *Azania: Archaeological Research in Africa* 53(4), 425–38.

Cobb, C. (1991) Social reproduction and the long durée in the prehistory of the Midcontinental United States. In R. Preucel (ed.), *Processual and Postprocessual Archaeologies: Multiple Ways of Knowing the Past*, 168–82. Carbondale IL, Southern Illinois University Press.

Daniel, G. (1943) *The Three Ages. An Essay on the Archaeological Method*. Cambridge, Cambridge University Press.

Deacon, H. J. & Deacon, J. (1999) *Human Beginnings in South Africa: Uncovering the Secrets of the Stone Age*. Cape Town, New Africa Books.

Demattè, P. (2006) The Chinese Jade Age: Between antiquarianism and archaeology. *Journal of Social Archaeology* 6(2), 202–26.

Fabian, J. (1983) *Time and the Other*. New York, Columbia University Press.

Gosden, C. (1994) *Social Being and Time*. London, Blackwell.

Gosden, C. (2004) *Archaeology and Colonialism: Cultural Contact from 5000 BC to the Present*. Cambridge, Cambridge University Press.

Gräslund, B. (1987) *The Birth of Prehistoric Chronology*. Cambridge, Cambridge University Press.

Hodder, I. (2012) *Entangled: An Archaeology of the Relationships Between Humans and Things*. London, John Wiley and Son.

Hosler, D. (1994). *The Sounds and Colors of Power*. Cambridge MA, MIT Press.

Huysecom, E., Rasse, M., Lespez, L., Neumann, K., Fahmy, A., Ballouche, A., Ozainne, S., Maggetti, M., Tribolo, C. & Soriano, S. (2009) The emergence of pottery in Africa during the tenth millennium cal BC: New evidence from Ounjougou (Mali). *Antiquity* 83(322), 905–17.

Iles, L. & Martinón-Torres, M. (2009) Pastoralist iron production on the Laikipia Plateau, Kenya: Wider implications for archaeometallurgical studies. *Journal of Archaeological Science* 36(10), 2314–26.

Kenny, M. G. (1981) Mirror in the forest: The Dorobo hunter-gatherers as an image of the other. *Africa* 51(1), 477–95.

Lechtman, H. N. (1988) Traditions and styles in Central Andean metalworking. In R. Maddin (ed.), *The Beginning of the Use of Metals and Alloys*, 344–78. Cambridge MA, MIT Press.

Lucas, G. (2004) *The Archaeology of Time*. London, Routledge.

Mavhunga, C. C. (2014) *Transient Workspaces: Technologies of Everyday Innovation in Zimbabwe*. Cambridge MA, MIT Press.

Mitchell, P. (2002) *The Archaeology of Southern Africa*. Cambridge, Cambridge University Press.

Pauketat, T. R. & Sassaman, K. E. (2020) *The Archaeology of Ancient North America*. Cambridge, Cambridge University Press.

Pauketat, T. R., Alt, S. M. & Krutchten, J. D. (2015) City of earth and wood: Cahokia and its material-historical implications. In N. Yoffee (ed.), *Early Cities in Comparative Perspective, 4000 BCE–1200 CE*, 437–54. Cambridge, Cambridge University Press.

Phillipson, D. W. (2005) *African Archaeology*. Cambridge, Cambridge University Press.

Renfrew, C. & Bahn, P. (2012) *Archaeology: Theories, Methods and Practice*. London, Thames & Hudson.

Romain, W. F. (2009) *Shamans of the Lost World: A Cognitive Approach to the Prehistoric Religion of the Ohio Hopewell.* Lanham MD, AltaMira Press.

Rowley-Conwy, P. (2004) The three age system in English: New translations of the founding documents. *Bulletin of the History of Archaeology* 14(1), 4–15.

Rowley-Conwy, P. (2007) *From Genesis to Prehistory: The Archaeological Three Age System and Its Contested Reception in Denmark, Britain, and Ireland.* Oxford, Oxford University Press.

Sassaman, K. E. (2010) *The Eastern Archaic: Historicized.* Lanham MD, AltaMira Press.

Schmidt, P. R. & Pikirayi, I. (2018) Will historical archaeology escape its western prejudices to become relevant to Africa? *Archaeologies* 14(3), 443–71.

Schultze, C. A., Stanish, C., Scott, D. A., Rehren, T., Kuehner, S. & Feathers, J. K. (2009) Direct evidence of 1,900 years of indigenous silver production in the Lake Titicaca Basin of southern Peru. *Proceedings of the National Academy of Sciences of the United States of America* 106(41), 17280–3.

Smith, L. T. (2005) Building a research agenda for indigenous epistemologies and education. *Anthropology & Education Quarterly* 36(1), 93–5.

Smith, C. & Wobst, H. M. (eds) (2004) *Indigenous Archaeologies: Decolonising Theory and Practice.* London/New York, Routledge.

Stahl, A. (1993) Concepts of time and approaches to analogical reasoning in historical perspective. *American Antiquity* 58(2), 235–60.

Tresch, J. (2014) Cosmologies materialized: History of science and history of ideas. *Rethinking Modern European Intellectual History*, 153–72. Oxford, Oxford University Press.

Trigger, B. G. (1989) *A History of Archaeological Thought.* Cambridge, Cambridge University Press.

18

Collections of Aboriginal Ground Stone Tools from the Murray Darling Basin: Function, Temporality, and Social Context

Richard Fullagar, Elspeth Hayes, and Colin Pardoe

In this essay we examine the links between people and objects and the structuring agency of space, including artefact distributions and their use-lives. We highlight form, function, life history, and chronology of ground stone tools: investigating their place in the lives of Aboriginal peoples and their distribution across country. Examples presented here come from the Murray Darling Basin, the largest river system in Australia. Although Chris Gosden (Gozza as we know him) did not work so much on Australian archaeology during his time here, his interest in life histories of artefacts and object mutability (e.g. Sainsbury et al. 2020) finds fertile ground. Aboriginal grinding implements occur throughout archaeological contexts in Australia, becoming more common in the Holocene. A passing glance might show a ground-edge axe or seed grinding dish, but we now know that they had long lives intertwined with generations of people dependent upon them. They had many uses in fibrecraft, ceremonies, and specialised tasks (Hayes et al. 2020), as well as multi-purpose domestic furniture indicative of residential patterning. Surface concentrations of ground stone implements are essential for understanding the subsurface archaeological record, in part because of their more substantial sample size, and are of increasing interest to Aboriginal people, in part because of their distributions, tool function, and artefact associations.
Keywords: Axes, Quandong Stone, Basket Stones, Grinding Stones, Artefact Biography, Usewear

Introduction

Australian Aboriginal ground stone implements consist of a range of artefact forms serving a range of functions, but there have been few attempts at a comprehensive typology of stones that have been modified by grinding through use or manufacture. Stone technologists commonly make a broad distinction between knapping (flaked stone) and grinding (ground stone) technologies, recognising that some artefacts bear signs of both. Jenny Adams (2014, 3) provides a broadly useful definition of what archaeologists think to be ground stone tools: 'any stone item that is primarily manufactured through mechanisms of abrasion, polish, or impaction or is itself used to grind, abrade, polish, or impact'.

McCarthy (1976) distinguished nine major groups of Australian Aboriginal stone implements that variously have evidence of grinding, pounding, abrading, smoothing, polishing, and/or engraving and sculpting. Implements were assigned to these major groups based on a classification outlined previously by McCarthy and colleagues (McCarthy *et al.* 1946), who determined function and shaping processes as key criteria for determining implement types. However, function may not be constant. Ground stone implements, whether multi-functional, manufacture-ground, or use-ground, can have complicated use-lives. We discuss below what a life span of some ground stone implements might be like, and special relationships between objects and people, including within Aboriginal communities, as family heirlooms (see Fig. 18.3d).

The sources of data presented here include ground stone artefacts in Murray Darling Basin (MDB) collections and a selection of available excavated contexts (Fig. 18.1). The ideas and data that we describe draw on analyses of regional collections from the MDB, where understanding of the grinding technology has been enhanced by long-term collaboration with Aboriginal and farming communities, quantitative study of surface collections, and selective application of use-wear/residue analysis to determine tool function.

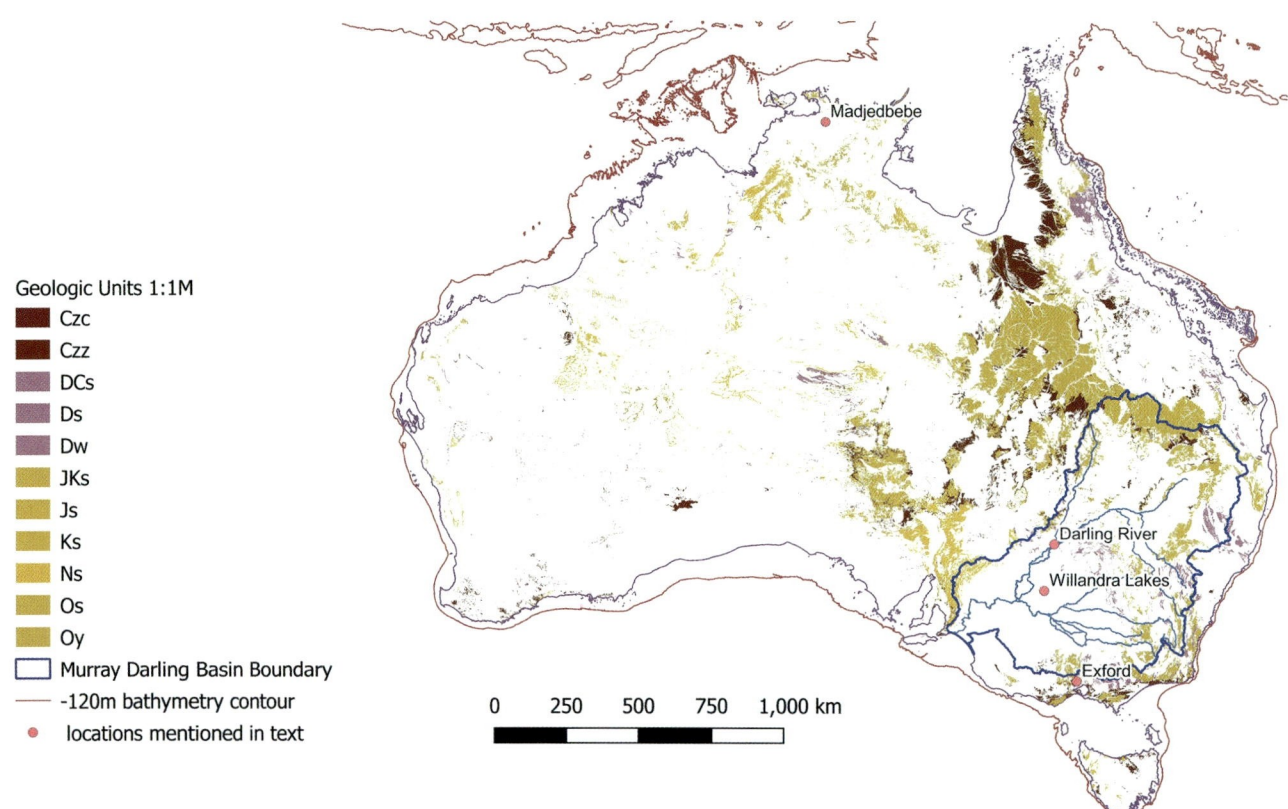

Figure 18.1 Location of Murray Darling Basin (blue rivers), locations mentioned in the text, and the distribution of potential grinding stone materials. Purple indicates quartzite, either in the form of cobbles or tabular slabs. Dark red indicates hard sandstone and silcrete. Buff indicates various sandstones. Stone axe sources (e.g. dolerite & greenstone) and less common tool stone materials used for pounding and grinding (e.g. granophyre, granite & banded ironstone) are not included on this map.

We provide a brief (spatial, ecological, social) overview of the context of grinding implements in the MDB and suggest how use-wear/residues can provide evidence to illuminate the social lives of these artefacts, especially in conjunction with contemporary Aboriginal interpretations and connections. We present four specific examples, each of which has different uses, temporalities, and social lives: axes, quandong stones, basket stones, and seed grinding stones.

Murray Darling Basin Collections and Taxonomy

The MDB is a large region, part of which has been the subject of research by Colin Pardoe and Penny Taylor over the past 40 years, with a focus on Aboriginal occupation, village sites, and artefact collections held by Aboriginal Land Councils or in private and public museums. Of particular interest has been the farmer's shed, where substantial collections of worked stones are the result of clearing paddocks for pasture and ploughing (Fig. 18.2). Many collections appear to be a fair sampling of the variety of objects, while others may show some bias to larger or more complete pieces. The point of collection is sometimes

Figure 18.2 A farmer's shed with grinding stones collected from paddocks.

precisely known and other times that knowledge may be lost. While the main theme for original study was to examine trade across the Basin, we have expanded our goals to functional, use-wear, and residue studies within an updated taxonomy. Currently, we have examined and measured 1893 ground stone implements, still held in dozens of relatively small public and private collections across the traditional territories of Barapa Barapa, Nari Nari, Wiradjuri, Ngiyampaa, Barkandji, and Gunu.

Most of these come from the four main rivers of the Basin: Barka or Darling (31%), Calare or Lachlan (43%), Murrumbidgee (8%), and Murray (13%). Some of these figures (the composition and proportion of collections documented in riverine regions) are partly fortuitous, reflecting travels and interest from friends and colleagues. There is, however, a rough relation to the location of suitable rock, mainly quartzite and sandstone for grinding stones and metamorphics for axes, as is likely for the whole continent (see Fig. 18.1). Sandstone and quartzite were selected for their properties that form distinct grinding and use-wear patterns (Hayes *et al.* 2018).

As Gozza maintained an eye for our perceptions of chronology and how time served to structure our thoughts, we thought to provide a baseline for such views by maintaining a largely non-chronological view. His interest in a spatially oriented view of the archaeological record, solidly based on ecological and environmental underpinnings, is more similar to our own perspective (Gosden *et al.* 2021). Historical large-scale erosion of a slowly aggrading surface across the Murray Darling Basin has produced an archaeological landscape that fulfils the promise of Processualism, with its focus on distribution and ecology rather than chronology and history. Although dating of items in collections is not currently possible, what items have been found at or near the surface by collectors likely represent a very good sampling of the variety, number, and distribution of implements used by Aboriginal people. In the following sections we discuss several categories of implement.

Axes

Hafted edge-ground hatchets are ubiquitous among people who interact with and use forests. The heads are generally made from tough metamorphic stone and are universally known as axe heads or simply axes. Greenstone stone axe heads from one particular set of quarries north of Melbourne were the subject of sourcing, distribution, and ethnohistorical analysis of trade across southeastern Australia (McBryde 1984); relatively few complete axes, however, have ever been excavated anywhere in Australia.

Axe heads are quite common in the MDB collections, making up just over 14% of all items studied. The variety of forms are dependent on many factors, including function, the type of stone, and the extent of maintenance and re-sharpening. Some cobbles are retouched to varying degrees, ground, and polished, whereas others have been minimally altered with short ground cutting edges on cobbles (Fig. 18.3). Many specimens in collections are broken. The edge angles are high (>60°) compared with most steel axes (<30°) and were commonly hafted with a wrap-around wooden handle, with various gums or resin compounds to secure the haft (Dickson 1981). In some parts of the MDB, tough sedimentary rocks like quartzite are readily accessible and more common in collections than basalts and fine-grained volcanic rocks. The taxonomy of all the MDB collections is a work in progress, and so it is with the axes. Use-wear and residue analysis will help to establish the function(s) of these tools. As an example, heavier axes with higher edge angles are occasionally seen. These clearly would have been used in ways different from the more common, smaller axes (mean weight 464 g). We refer to a class of heavier axes (mean weight 1134 g) as 'block splitters', as they would have been used for heavy duty work. Other axe forms are suggestive of functional differences because their hafting and therefore heft and handling would have varied. Grooved and waisted axes are rare throughout most of the Basin, as for the rest of the country (Akerman 2014), until one arrived at the banks of the Barka or Darling River in western NSW. Although the Barkandji had less use for axes than their neighbours in more forested country, they favoured grooved axe heads.

The design and function of axes was likely intended for chopping wood. Like so many aspects of the material culture of peoples across the country, axes also served a range of functions, with evidence of the multi-functional uses from surface collections in eastern NSW (*e.g.* Attenbrow & Kononenko 2019). The six complete axe specimens excavated at Madjedbebe (northern Australia) include four distinct forms (grooved, oval, waisted, and stemmed), and their chronology suggests more variety *ca.* 30,000–50,000 years ago (Clarkson *et al.* 2017). The combination of large-scale spatial collections with the smaller numbers found in excavations starts to flesh out a picture of pattern through time and across large areas.

Quandong Stones

Quandong stones were recently identified as a specialised tool type in the MDB collections with evidence that they are more widespread and track a southern Australia-wide distribution of Quandong (*Santalum acuminatum*) (Pardoe *et al.* 2019). Quandong is a valued species, whose fruit is edible, highly nutritious, and with medicinal qualities. The nuts are used by Aboriginal people today for ornaments, and the kernels can be ground with a top stone to prepare a paste with analgesic and cosmetic properties for nourishing skin (D. Doyle and D. Stephenson, Barkandji friends, and colleagues [pers. comm.]). The cobble-sized ground stone implements have at least one deep, manufactured pit and a concave mortar surface on the opposing surface (Fig. 18.3b–c). Metric analyses show a range of tool forms with a similar cross-section of the manufactured pits. Several specimens have been found within established groves of quandong. Preliminary analysis of use-wear and residues indicates traces of manufacture, nut cracking in the deep pits, and grinding oily nuts on the mortar side. Residues recovered included cellulose and patches of crushed fibrous plant tissue, with the same appearance as the inner fibrous tissue of quandong nutshells, cracked experimentally.

No excavated specimens had been reported until after our initial study had been published in 2019. However, a recent report describes a pitted stone that was excavated near Exford in Victoria on the southernmost edge of known quandong tree distribution (Fullagar 2019). It has characteristics similar to the MDB quandong stones (Fig. 18.3). Although the primary function of quandong stones was to crack the nuts and grind kernels to a paste, they may have been used for other tasks. The stone material is a very tough quartzite, and we have seen few broken specimens. We suggest that quandong stones had longer use-lives than edge-ground hatchet heads and were probably cached near quandong groves rather than carried around day to day.

Basket Stones

Basket stones were mentioned by Brough Smyth (1878) but then disappeared from archaeological sight. A leaflet describing basket stones in some MDB collections was prepared for Barkandji and other colleagues and friends (Pardoe 2019a). This leaflet has sparked considerable interest and further study is underway. Basket stones are waterworn quartzite cobbles found in streambeds where Devonian sediments have been exposed. They are egg shaped and weigh a little less than 3 kg on average [$\bar{x} = 2576 \pm 938$ g]. If noticed, they might be identified as a mortar as they have a fairly flat facet parallel to their longest and widest axes (Fig 18.3e). More importantly, there is a distinctive asymmetrical pattern of smoothing or polish around the circumference near the base, which can be detected with fingers and observed under appropriate lighting conditions. Preliminary study suggests the polish is a combination of handling and dragging plant fibres (including siliceous taxa) in a diagonal direction up from the base. Further experiments are needed to understand the wear on these tough stones from natural agency at the source, human handling and plant fibres (with and without phytoliths), and on-site weathering. We have only just begun. Basket stones are perhaps a weight to hold steady and fix the base of coil-and-stitch baskets, as strands of fibre were drawn tightly up. They might also be used for string bags. Many basket stones are to be seen in place, usually close to reed and rush wetland, and on the edge of larger residential areas.

Potential stone sources have been found in Barkandji country, where cobbles are eroding from ancient riverbeds. The stone is a very tough quartzite which is highly resistant to wear, and use-wear from the basket weaving process would have hardly any impact on the tool's efficiency or use-life. The wear probably develops very slowly on such tough stone material, and those with the most extensive wear could have been used and passed down from mother to daughter for generations. Current experimental studies are underway with Aboriginal communities to assess wear patterns and residues on source cobbles and artefacts.

Grinding Stones

McCarthy (1976) included six types of abrading stones which most archaeologists would recognise as grinding stones, including millstones, mullers, whetstones, bone shaping stones, two kinds of smoothing/polishing stones, and shell fishhook files. Millstones are large seed grinding dishes that are usually paired with mullers – a faceted upper stone that fits the grooves formed on the lower millstones. Percussion and abrading implements comprise mortars and rock engraving stones. Within the MDB collections, a range of grinding stones are recognised, several of which are paired (upper and lower) and multi-purpose – designed for portability and travel. Millstones are among the heaviest of all Aboriginal artefacts and weigh up to 30 kg. They are quarried, shaped, and transported to locations where they would remain for extended periods, rather than part of the daily travel kit. Sandstone millstones used for grinding soft seeds and the tougher quartzite dishes used for crushing hard seeds can wear down rapidly and break frequently. The broken pieces could be recycled as top stones or re-worked into smaller dishes.

One of the regularly shaped, smaller sandstone grinding dishes found within the MDB collections is the travelling plate (Pardoe 2019b), which is used in pairs and can be held in each hand (Fig. 18.3f & g). These implements were used for grinding soft seeds and one of their other functions was to sharpen the digging stick, as can often be seen in the grooves on the outer surface. Another regularly shaped pair of small multi-purpose implements designed for daily travel include the Kulki and the Portable Mortar & Pestle (PMP) – often called the little and big brother respectively (and respectfully). The outside Kulki edge is used as a pestle, and the faces used for light duty crushing and pounding of seeds or ochre on the PMP. The pits are used as anvils for bipolar flaking, as a stable holder to crack nuts, and possibly for holding a drill. The PMP has one or two mortar faces often with pitting from use.

Some grinding stones in the MDB collections have been examined for use-wear and residues and have distinctive traces of use indicating plant and animal resources (Pardoe & Stephenson 2013). Here and in other parts of Australia, functional studies of grinding stones indicate the use of specific resources including hard seeds like acacia and nardoo fern sporocarp, soft grass seeds, bone and animal flesh, insects, wood, tubers, tree nuts, ochre, and fibre for making string (*e.g.* Fullagar & Field 1997; Fullagar *et al.* 2008; 2015; Pardoe & Stephenson 2013; Hayes 2015;

Smith *et al.* 2015; Hayes *et al.* 2016; Stephenson *et al.* 2020; Wallis *et al.* 2020). Analyses of excavated grinding stone function indicate an emphasis on seed grinding during times of stress in the Pleistocene (Hayes 2015; Hayes *et al.* 2022). Grinding stones, mortars, and pestles are not expected to have long use-lives but are likely to be re-used, broken, recycled, and worn until they are not recognisable as artefacts, particularly in parts of the MDB where stone is scarce.

Discussion and Conclusion

Although Gozza's time in Australia was short, archaeologists and other researchers have benefited from his insights about how people relate to objects, and particularly about how we conceive time when we interpret function, use, and longevity of implements (Gosden 1994; Gosden & Marshall 1999). While the concept of landscape is nothing new in archaeology, particularly for those of us with a spatial or ecological view, taking that word from a physical description to an interpretive framework in archaeology has been infused in his work (*e.g.* Gosden & Head 1994).

Scale of landscape and its relation to social groupings is plainly in view in the big sky country of the Murray Darling Basin. We could not do more than use this as a backdrop for discussion of items that are at once necessary to make a living there, but the examples of axe, quandong stone, and various grinding implements also serve as markers of the social links between people. In particular, we wanted to pick up a thread on the life of an artefact. In our examples, we chose some items that would be used frequently by their owner and which might outlive them. The variety of grinding items, both portable and permanently placed, are at the centre of the kitchen. The mortar and the seed grinding dishes would be used each day at home and their portable counterparts during the daily round. These items have much to teach us of residential patterning and mobility; of the menu via residues; of hard use via use-wear; and of trade across tribal areas. Of all the stone artefacts, flaked or ground, it is these that wear out and are recycled or repurposed (Gosden & Marshall 1999; Pardoe 2012; Sainsbury *et al.* 2021).

Axes also must endure hard work, as can be seen in excavations where chips from re-tooling abound. They would be in use by foresters, and many are likely to break. We do not know their lifespan, but one could imagine having several during a working life. Quandong stones would find heavy use each season, as people would knuckle down to producing the year's supply of preserved fruit and unguents. They might be expected to last a very long time, perhaps millennia, but are ultimately doomed to breakage, even if hit only by a wooden mallet. The basket stones are likely to have use-lives beyond our imagining. When seeing how they are held by the descendants of the women who used them, how they are cradled, and the reluctance to let them go once picked up, it is no wonder the past 'elbow grease' of their ancestors has polished smooth one of the hardest stones – metamorphosed Devonian quartzite cobbles. There is much more to this story, but for the moment it is sufficient to consider an item that might be in use for thousands of years – 50 or 500 generations. Such a time scale places an inanimate object in a special position in its society, just as we encounter them today in special positions around the village, camp, or settlement.

The studies and objects discussed here have been enhanced by Aboriginal knowledge about resources such as quandongs, what parts were used, and how they were processed. The shed collections, made available by farmers, provide detailed information about artefact forms, functions, and distributions. Quantitative study of the MDB surface collections has helped organise our understanding of the variety of objects that are distinguished on functional attributes and morphology and go a long way to constructing a taxonomy for multi-functional tools. Studies of ground stone tools, and indeed all the archaeology of the Murray Darling Basin, are encouraged by Aboriginal local communities, who express considerable knowledge and interest in ground stone implements. The application of use-wear and residue analysis is helping to evaluate hypotheses about tool function and the use of specific resources. While a festschrift is some kind of looking back over a career, and one that we hope has made a bit of sense of the impact of Gozza's thoughts and words here in Australia, we also want to look forward.

Artefacts in farm sheds and local museums, like those in big faraway public museum collections (so ably run by the likes of Gozza), are extremely valuable for understanding Australian ground stone tools. Having remained on the Aboriginal country where they were found, they also are particularly significant for local communities, who have retained knowledge about their function and significance. Artefacts in these local collections can often be assigned to a paddock or landscape if not a specific location, and although they have poor temporal resolution, they can be combined to understand distributions related to functional variability, resources, geology, and social significance.

Acknowledgements

We are grateful to the editors, referees, and Lesley Head for comments. We hope to have shown aspects of a living typology where ground stone artefacts are informed by concepts of lifespan and life history elaborated by Gozza and increasingly more by Aboriginal participation in the study of these implements. The image of the Exford stone (Fig. 18.3b) is published with the support and permission of Wurundjeri Elders Ron Jones, Allan Wandin, and Bobby Mullins; Bill Green (the Sponsor of the Cultural Heritage Management Plan, during the implementation of which the

Figure 18.3 (a) MDB axe heads; (b) Excavated quandong stone from Exford, Victoria; (c) Quandong stone from the MDB; (d) Family heirloom. Picture (from the 1960s) of Harold and Evelyn Bates, on a portable grinding stone (Pardoe 2019c); (e) MDB basket stones; (f) MDB sandstone travelling plate pair; (g) Using travelling plates to grind seeds.

artefact was identified); and Ochre Imprints. We particularly thank Barapa Barapa, Nari Nari, Wiradjuri, Ngiyampaa, Barkandji, Gunu and Mirarr, who have shaped our ideas.

References

Adams, J. (2014) *Ground Stone Analysis*. Salt Lake City, UT, The University of Utah Press.

Akerman, K. (2014) Observations on edge-ground stone hatchets with hafting modifications in Western Australia. *Australian Archaeology* 79, 137–45.

Attenbrow, V. & Kononenko, N. (2019) Microscopic revelations: The form and multiple uses of ground-edged artefacts of the New South Wales, Central Coast, Australia. *Technical Reports of the Australian Museum* (Online) 29, 1–100.

Brough Smyth, R. (1878) *The Aborigines of Victoria*. Melbourne, John Currey, O'Neil.

Clarkson, C., Jacobs, Z., Marwick, B., Fullagar, R., Wallis, L. A., Smith, M., Roberts, R. G., Hayes, E. H., Lowe, K., Carah, X., Florin, S. A., McNeil, J., Cox, D., Arnold, L. J., Hua, Q., Huntley, J., Brand, H. E. A., Manne, T., Fairbairn, A., Shulmeister, J., Lyle, L., Salinas, M., Page, M., Connell, K.,

Park, G., Norman, K., Murphy, T. & Pardoe, C. (2017) Human occupation of northern Australia by 65,000 years ago. *Nature* 547(7663), 306–10.

Dickson, F. P. (1981) *Australian Stone Hatchets: A Study in Design and Dynamics*. London, Academic Press.

Fullagar, R. (2019) *Functional study of the Exford stone artefact (4A, #3008)*. Unpublished report for Exford Waters P/L.

Fullagar, R. & Field, J. (1997) Pleistocene seed grinding implements from the Australian arid zone. *Antiquity* 71(272), 300–7.

Fullagar, R., Field, J., & Kealhofer, L. (2008) Grinding stones and seeds of change: Starch and phytoliths as evidence of plant food processing. In Y. M. Rowan & J. R. Ebeling (eds), *New Approaches to Old Stones: Recent Studies of Ground Stone Artifacts*, 159–72. London, Equinox Publishing P/L.

Fullagar, R., Hayes, E., Stephenson, B., Field, J., Matheson, C., Stern, N. & Fitzsimmons, K. (2015) Evidence for Pleistocene seed grinding at Lake Mungo, south-eastern Australia. *Archaeology in Oceania* 50(1), 3–19.

Gosden, C. (1994) *Social Being and Time*. Oxford, Blackwell.

Gosden, C. & Head, L. (1994) Landscape: A usefully ambiguous concept. *Archaeology in Oceania* 29, 113–16.

Gosden, C. & Marshall, Y. (1999) The cultural biography of objects. *World Archaeology* 31, 169–78.

Gosden, C., Green, C. & the EngLaId team (2021) *English Landscapes and Identities: Investigating Landscape Change from 1500 BC to AD 1086*. Oxford, Oxford University Press.

Hayes, E. (2015) *What was ground? A functional analysis of grinding stones from Madjedbebe and Lake Mungo, Australia*. Unpublished PhD thesis, University of Wollongong.

Hayes, E., Fullagar, R., Mulvaney K. & Connell. K. (2016) Food or fibercraft? Grinding stones and *Triodia* grass (spinifex). *Quaternary International* 468(B), 271–83.

Hayes, E., Pardoe, C. & Fullagar, R. (2018) Sandstone grinding/pounding tools: Use-trace reference libraries and Australian archaeological applications. *Journal of Archaeological Science: Reports* 20, 97–114.

Hayes, E. H., Fullagar, R. & Marwick, B. (2020) Australian use-wear/residue studies, artefact design and multi-purpose tools. In J. Gibaja, I. Clemente, N. Mazzucco & J. Marreiros (eds), *Hunter-Gatherers' Tool-Kit: A functional perspective*, 256–86. Newcastle upon Tyne, Cambridge Scholars Publishing.

Hayes, E., Fullagar, R., Field, J., Coster, A., Matheson, C., Nango, M., Djandjomerr, D., Marwick, B., Wallis, L. & Smith, M. (2022) 65,000-years of continuous grinding stone use at Madjedbebe, Northern Australia. *Scientific Reports* 12(11747), 1–17.

McBryde, I. (1984) Kulin greenstone quarries: The social contexts of production and distribution for the Mount William site. *World Archaeology* 16, 267–85.

McCarthy, F. (1976) *Australian Aboriginal Stone Implements: Including Bone, Shell and Tooth Implements*. Sydney, The Australian Museum Trust.

McCarthy, F., Bramwell, E. & Noone, H. (1946) The stone implements of Australia. *Australian Museum Memoir* 9, 1–94.

Pardoe, C. (2012) *Telkuk tirr – Good or Beautiful Axe: Ground-edge Axe Heads from the Koondrook State Forest, Murray River, NSW*. Information Pamphlet for The Joint Indigenous Group of the Koondrook Perricoota Forests Flood Enhancement Works Project, Barapa Barapa Nation Aboriginal Corporation, Yorta Yorta Nation Aboriginal Corporation, Forests NSW, and State Water Corporation NSW.

Pardoe, C. (2019a) *Travelling Plates: An Important Rediscovery*. Leaflet for Ngiyampaa, Barkandji and Wiradjuri people. Available at: <https://www.academia.edu/40724890/TRAVELLING_PLATES_AN_IMPORTANT_REDISCOVERY>

Pardoe, C. (2019b) *Aboriginal Basket Stones in the Murray-Darling Basin*. Leaflet for Barkandji and other groups of the Murray-Darling Basin. Available at: <https://www.academia.edu/40724895/ABORIGINAL_BASKET_STONES_IN_THE_MURRAY_DARLING_BASIN>

Pardoe, C. (2019c) *Harold and Evelyn Bates Ground Stone implement Collection*. Unpublished report. Available at: <https://www.academia.edu/40724893/Harold_and_Evelyn_Bates_Ground_Stone_implement_Collection>

Pardoe, C. & Stephenson, B. (2013) *The World's First Hamburger and Chips*. Leaflet for West Wyalong Local Aboriginal Land Council.

Pardoe, C., Fullagar, R. & Hayes, E. (2019) Quandong stones: A specialised Australian nut-cracking tool. *PloS One* 14(10), p.e0222680.

Sainsbury, V., Bray, P., Gosden, C. & Pollard, M. (2021) Mutable objects, places and chronologies. *Antiquity* 95, 215–27.

Smith, M., Hayes, E. & Stephenson, B. (2015) Mapping a millstone: The dynamics of use-wear and residues on a Central Australian seed-grinding implement. *Australian Archaeology* 80(1), 70–9.

Stephenson, B., Bruno, D., Fresløv, J., Arnold, L., GunaiKurnai Land and Waters Aboriginal Corporation, Delannoy, J., Petchey, F., Urwin, C., Wong, V., Fullagar, R., Green, H., Mialanes. J., McDowell, M., Wood, R. & Hellstrom, J. (2020) 2000-Year-old Bogong moth (*Agrotis infusa*) Aboriginal food remains, Australia. *Scientific Reports* 10(22151), 1–10.

Wallis, L., Stephenson. B. & Yinhawangka Aboriginal Corporation (2020) A nardoo processing grinding stone from a rockshelter in the Pilbara, Western Australia. *Australian Archaeology* 86, 112–17.

19

Cultural and Landscape Change in Australia's World Heritage Wet Tropics Bioregion, Northeast Queensland

Richard Cosgrove

Recent mtDNA analysis has suggested that Aboriginal Australia has a deep antiquity of residential attachment to country (Tobler et al. *2017). Within a relatively short time of landfall, strong residential links to landscape were established by Aboriginal populations. These earliest ties to country indicate a permanent footprint of peoples occupying territory across multiple generations. Based on mtDNA results, it has been argued that populations remained unaffected by environmental change, since there is little genetic evidence of bottlenecking. However, the Australian archaeological record provides extensive evidence of cultural transformations in the social, economic, and ritual fabric of people's lives, demonstrating flexible behavioural responses to environmental fluctuations that appear to have mitigated bottlenecking. Since mtDNA results suggest continuity of residential occupancy, these cultural transformations allowed generations to sustain links to country during significant climatic perturbations over the Late Glacial Maximum and Holocene periods. One such area that illustrates this is the central area of the Wet Tropics World Heritage Bioregion, northeast Queensland. It provides an example of Aboriginal behavioural variability and flexibility over the longer term. Principally it focuses on the distinctive Aboriginal rainforest culture developed there during the late Holocene, that arose at the end of the Last Glacial Maximum when the environmental conditions were very different.*

Keywords: Australia, Rainforest Stone Technology, Plant Exploitation, Trade, Exchange

Introduction

The Wet Tropics World Heritage Bioregion (WTWHB) extends from Townville to Cooktown in northeast Queensland. These rainforests have developed over thousands of years to create one of the richest and most biodiverse natural systems on the planet. Although this chapter focuses on the central portion of the WTWHB (Fig. 19.1), the WTWHB rainforests developed only within the last 10,000 years at the end of the Ice Age, with increasing temperatures and humidity. The central region's western edge has remained remarkably stable over this period based on pollen evidence from Witherspoon Swamp (Moss *et al.* 2012). Before this, the vegetation consisted of open sclerophyll woodland (Haberle *et al.* 2010) with rainforest localised in refugia in the north, central, and southern zones (Hilbert *et al.* 2007; VanDerWal *et al.* 2009). Human occupation began about 8000 years ago at very low levels, only intensifying from about 2000 years on an economy of toxic nut exploitation and other plant and animal use (Horsfall 1996; Cosgrove & Raymont 2002; Cosgrove 2005; Tuechler *et al.* 2014). Sites across the Atherton Tablelands and the coast show a pattern of early, less intense occupation with a more intensive phase of permanent rainforest settlement later (Cosgrove *et al.* 2007). This is evident at the sites of Urumbal Pocket, Goddard Creek, Jiyer Cave, and Murubun Shelter (Fig. 19.1). These patterns align with evidence from numerous archaeological sites depicting an intensification of use from the middle to late Holocene period (Ulm 2013; Williams *et al.* 2015).

Despite the evidence for permanent settlement within the last 2000–1800 years (Cosgrove *et al.* 2007), the mtDNA suggest many generations of Aboriginal people had long residential ties to what is now the Wet Tropics World Heritage Area. Oral histories tell stories associated with recent

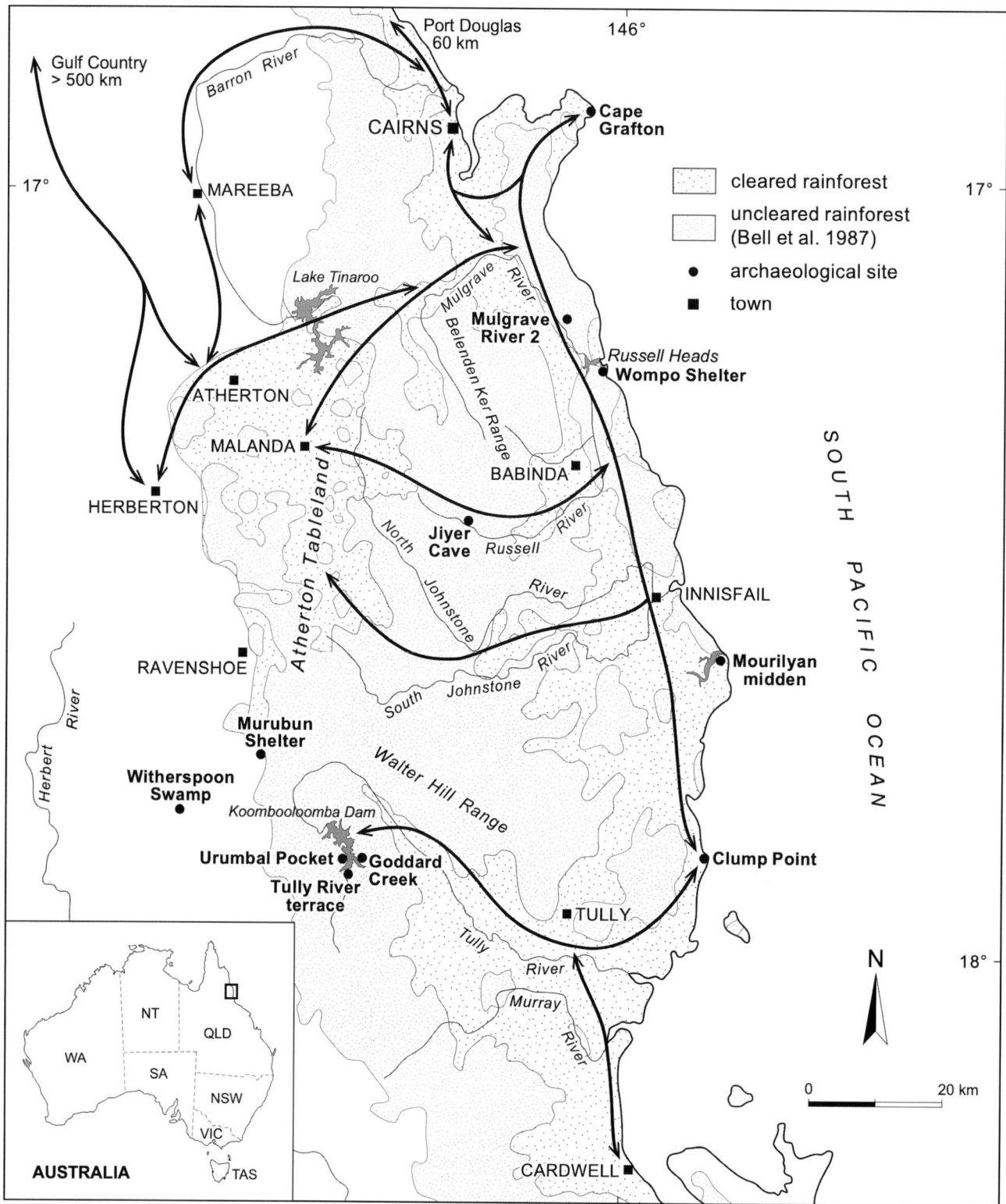

Figure 19.1 Map of the study area showing the extent of current rainforest, archaeological sites mentioned in the text, ooyurka artefact localities and number found, directions of trade, and exchange items described by Walter Roth (1910).

volcanism on the Atherton Tableland, geological events that occurred *ca.* 20,000 years ago, as well as sea level rises and past open vegetation conditions where rainforest now exists (Dixon 1972; 1976; Steinberger 2014, 78). Mythologies that have been handed down over at least 1200 generations support a long cultural and mtDNA association with the region.

Successive changes in palaeoclimate indicate behavioural variability and flexibility within rainforest Aboriginal society. This is reflected in the material culture of the central rainforest area, as well as the appearance there of two distinctive language groups and tribal boundaries within a few millennia of permanent rainforest settlement (Dixon 1972; Buhrich *et al.* 2016). It suggests that Indigenous cultures significantly altered their social and economic structures at the end of the late Pleistocene to accommodate vegetation and faunal resource fluctuations, as open sclerophyll forest gave way to the spread of dense tropical rainforest across the region, where anthropogenic fires tempered its advance (Haberle *et al.* 2010; Roberts *et al.* 2021).

It was argued previously that this late occupation was caused by intensifying ENSO conditions that made living in the semiarid area more unpredictable (Cosgrove *et al.* 2007). However, similar patterns of intensification are now also recorded for areas outside the rainforest zone, making this explanation less likely (Lourandos *et al.* 2012), although increasing competition for resources in these drier areas may have made the rainforest more attractive after 2000 years. On the western edge and to the north in the semiarid regions, occupation extends back at least 37,000 years (David *et al.* 1999; Cosgrove *et al.* 2007). Why there are no earlier sites from the Wet Tropics Bioregion is still unclear but may be a product of limited sampling, the difficulty of access into the more impenetrable terrain, or taphonomic factors (*e.g.* Anderson and Robbins 1988). It can also be speculated that the archaeological sites that have been analysed lie mainly in the modelled rainforest Late Glacial Maximum refugia (VanDerWal *et al.* 2009) and only reflect this later adaptation, where earlier, more mobile occupation was limited. It may be that areas outside these modelled vegetation zones have a greater antiquity.

Evidence suggests that Aboriginal management of these rainforests had a profound impact on their composition and diversity since their settlement by Aboriginal people in the early Holocene (Stanton *et al.* 2014; Roberts *et al.* 2021). The creation of small open pockets of grassland maintained by fire and mechanical clearing allowed the construction of villages deep inside the rainforest that were occupied on a seasonal basis (Cosgrove 2005; Steinberger 2014, 63, 68 & 71). This increased the diversity and produced unevenly aged vegetation complexes, with more ecotonal boundaries supporting open vegetation containing *Xanthorrhoea johnsonii*, *Corymbia intermedia*, *Eucalyptus tereticornis*, *Allocasuarina torulosa*, and vine forest (Steinberger 2014). This in turn provided habitat for a wide range of marsupials, birds, and reptiles (Bird *et al.* 2005; Bliege Bird *et al.* 2012). Such a sophisticated approach to land management, involving deliberate niche construction, determined both Aboriginal settlement and economic exploitation in the Holocene (Field *et al.* 2016).

The early descriptions of the central rainforest Aboriginal material culture evince technologies developed to efficiently exploit the surrounding environment as well as those that played a specific role in social and economic activities (Roth 1910). The ceremonial gatherings at the start of the early wet season provided an opportunity to trade, exchange, and settle disputes using the large, one-handled, wooden swords and elaborately decorated shields made from the buttress roots of the native fig tree (Lumholtz 1889, 263). Body adornment included multi-coloured ochre paints and accoutrements, such as cockatoo feather headdresses, nasal septum piercings with macropod fibulae, shell necklaces, and shell headbands. Much of the material culture was made from wood and vines necessitating the use of ground-edge stone implements, such as axes, wedges, and sharp flakes, to acquire and manufacture them. The importance of plant-based food collection is reflected in the elaborate variety of baskets constructed from lawyer cane (*Calamus australis*), plant fibre, and bark. They were used to collect rainforest fruits, nuts, honey, and water, and many were painted and decorated with detailed geometric designs (Aaberge *et al.* 2014). These items were exchanged both within the rainforest and between communities further up the coast and into the drier inland. Figure 19.1 illustrates the movement of people and items exchanged as recorded by Roth between 1887 and 1904 in the study area (1910).

Despite the presence of two different languages within the central rainforest zone, Yidjin in the north and Dyribal in the south (Dixon 1972; 1976), fluid movement of tribal groups was frequent, wide-ranging, and associated with trade and exchange between rainforest people. This is compelling since both languages contain a number of different dialects but are in no way grammatically similar (Dixon 1972, 27). Nevertheless, groups from Cairns moved between Port Douglas and the Mulgrave River, while the Barron River people moved between Port Douglas, Kuranda, and Mareeba in the drier country. The Russell River groups, who lived deep in these incised valleys, went to Pyramid Mountain just south of Cairns, onto the Mulgrave and Johnston Rivers. The people from the latter area moved between Clump Point and Liverpool Creek south of Innisfail.

Trade and exchange were widespread across the region, with crescent-shaped woven baskets (or bi-cornual) being traded to Port Douglas, the Mulgrave and Barron Rivers, Marbeeba, and Herberton. These were only manufactured in the area between Cairns and the Tully River, and on the Atherton Tableland (Khan 1996, 23). Fish spears, straight spear throwers, as well as grass necklaces were traded from Cape Grafton to the valleys of the Mulgrave, Russell, Johnston Rivers, and to Clump Point. In addition, bent spear throwers, large fighting shields and long single-handled swords were traded to the Barron River and further northwards.

Goods imported into Cape Grafton could come from hundreds of kilometres away and included a range of decorative items, such as large oval cut pearl shell chest ornaments from the Gulf Country, 650 km to the west, that travelled via Atherton and Herberton down the Mulgrave

River to Cape Grafton. Other imported items from the Port Douglas and Barron River regions were hourglass dilly bags, beeswax necklaces, varieties of bamboo spears, square-cut nautilus shell necklaces and cockatoo topknot headdresses. From the Mulgrave River, they obtained long swords, boomerangs, shields, and possum fur string amulets (Roth 1910).

It is clear from these interactions that rainforest people were economically interconnected with neighbouring tribal groups and the physical landscapes around them, as well as long-distance cultural networks beyond the rainforest. It suggests these people were aware of a world beyond their rainforest borders but saw themselves as intrinsically rainforest people symbolised in their material culture. Unfortunately, the organic items discussed above seem not to survive archaeologically, so their antiquity cannot be traced and, although rock art motifs apparently also show connections wider afield, they remain undated (Buhrich et al. 2016). Thus, it is the stone tool technology that assists us to understand the emblematic rainforest culture that has adapted to climatic changes since the terminal Pleistocene. Flaked stone tools are abundant in archaeological sites (Cosgrove et al. 2007), but these can be indistinguishable from the technology found in the drier regions and periods prior to 10,000 years ago, when rainforest was restricted to upper montane regions (VaDerWal 2009). However, two such items that are characteristically rainforest in origin, the ooyurka and incised grinding stones, highlight the unique rainforest lithic material culture and economic focus (Fig. 19.2a & c).

Distribution

Spatially, the ooyurka and incised grinding stones occur within the central WTWHB, defined as Complex Mesophyll Evergreen Vine Forest. They are found nowhere else in Australia, and ooyurkas appear to be concentrated in the region of lowland rainforest.

The ooyurka is concentrated in the coastal Babinda-Innisfail-Tully region and the valleys of the South and North Johnston Rivers and Russell River. They have also been found in the uplands about 800 m.a.s.l., but in very small numbers. A total of 123 have been analysed, 29 recovered from Waughs Pocket, 14 from the collections of the Queensland Museum, 30 from the James Cook University Museum, and the remainder from private collections. The locations of 85 undated surface ooyurka specimens with map coordinates found during earlier surveys are shown in Figure 19.1 (Cosgrove 1984). Two specimens were identified outside this area. One 50 km north of Waughs Pocket and another 60 km south of Tully at Lucinda, recovered from a coastal shell midden (Cosgrove 1984). The concentration within this geographical area could be a reflection of the intensive land-use practices carried out on the coastal lowlands. Ninety-six percent of all artefacts examined came from the coastal zone, whilst 4% were found in montane rainforest at elevations of between 750 and 800 m.a.s.l. However, their concentration within this relatively small area, surrounded by equally intensive and widespread cane farming both north and south, would suggest that this pattern is robust, as visibility of all ploughed ground would have been similar. Therefore, the chance of discovery appears relatively equal across the whole zone. Nevertheless, it is within the Mamu and Djiru traditional owner groups that the highest numbers appear: 34% (n=40) and 38% (n=47) respectively of the total analysed. Both groups have their residence in the core lowland rainforest zone (Cosgrove 1996). The ooyurka has not been found north of Cairns, south of Tully, or beyond the Atherton Tablelands. As speculated above, there may be several reasons for this: taphonomic factors, limited survey, and visibility. Nevertheless, their restricted distribution may represent a material culture province, defined by a polythetic set of attributes that include ooyurkas and incised grinders, setting it apart from the rest of the WTWHB cultural landscape.

The incised grinders are found throughout the central rainforest region from the coastal lowlands to the Atherton Tablelands. Principally, they are located in open-air and sometimes rock shelter sites, where the starchy plants were processed. Some have even been found embedded in nut tree trunks, cached there, and forgotten while the tree grew around them (C. Mansfield, pers. comm.) They are associated with the exploitation of at least eight species of noxious rainforest food (Field et al. 2016) and played a significant role in providing valuable starchy food sources by grinding, steaming, and leaching techniques (Cosgrove 2005). The access to higher starch yields probably led to a much larger aggregation of people and population increases in the late Holocene (Cosgrove 2005). The use of toxic foods is first recognised in the archaeological record by at least 1474±40 cal BP (OZJ–718, Ferrier & Cosgrove 2012, 107) and aligns with similar increases in artefact discard rates in widely spaced sites about the same time (Cosgrove et al. 2007; Lourandos et al. 2012).

Most ooyurkas and incised grinding stones are recovered as undated surface finds associated with large slate axes and pitted nut stones, the latter being used to crack open the hard endocarps of the tree nuts. One incised grinder fragment excavated at Urumbal Pocket associated with thousands of burnt nutshells was dated to ca. 560 BP. These implements all reflect the distinctive Aboriginal stone technology of rainforest Aboriginal people used to process the toxic nuts that provided an essential source of carbohydrates, oil, and nutrition (Cosgrove 2005; Ferrier & Cosgrove 2012; Field et al. 2016).

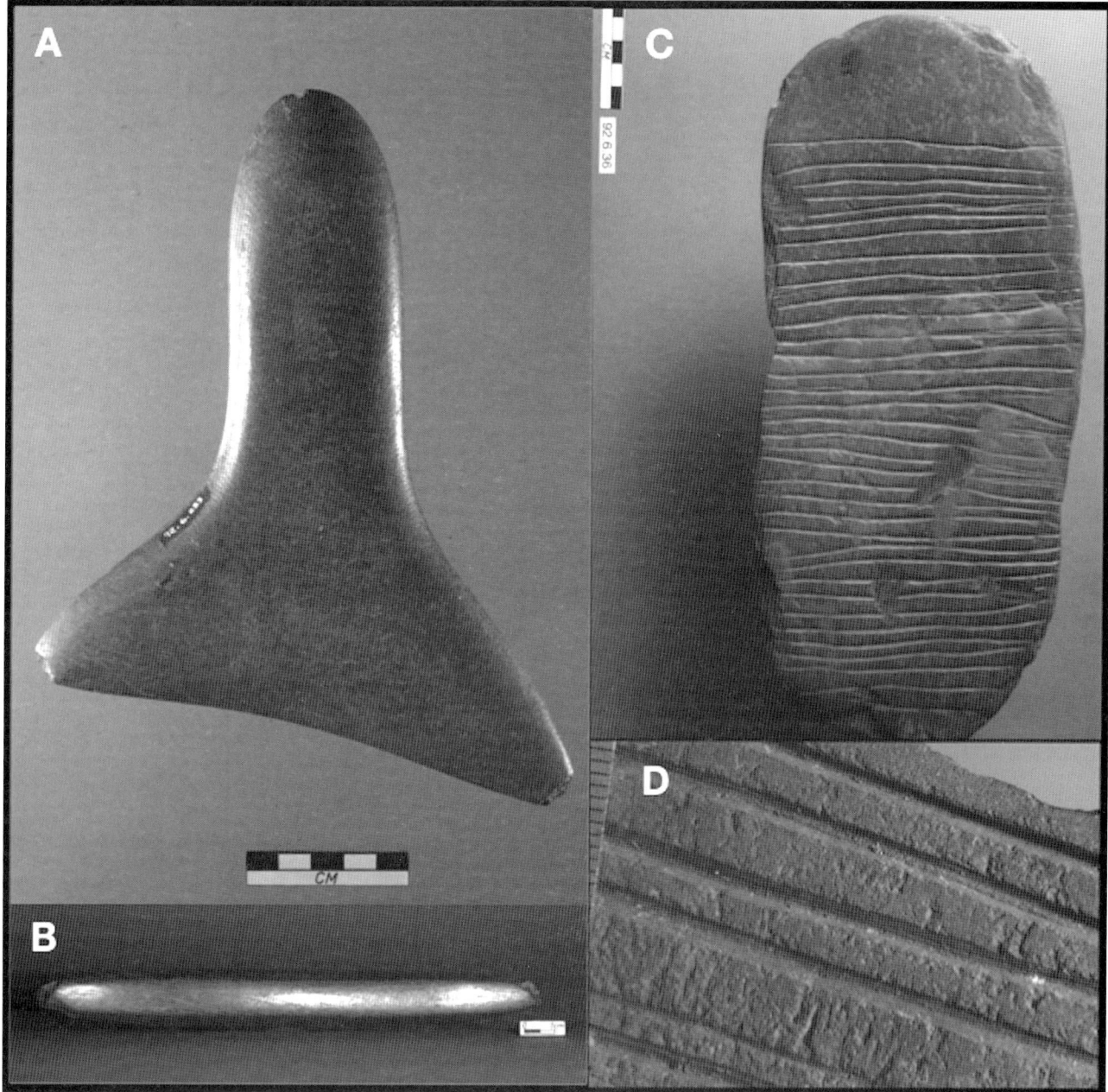

Figure 19.2 A and B show the ooyurka artefact (accession #92.6.283) and its flat polished working edge. Note the lustrous polish on the 'handle'. C and D are the slate incised grindstone (accession #92.6.36) with incisions made across the body of the tool. D is a close-up of the grooves from which starch grains were retrieved.

Form

Ooyurka

Most of these tools are made from soft slate and micaceous schist, with the thin lamellar bonded by cordierite crystals derived from the Barron River Metamorphics (Cosgrove 1984; Willmott & Stephenson 1989). The material is difficult to work, especially to shape by hammer dressing or flaking (Cosgrove 1984). Extensive striations across the body, handle, and working edge suggest grinding with sandstone or similar natural files to achieve the unusual shape, which includes a working face that is ground down and perpendicular to the medium plane rather than oblique, as is found on axe edges. Some have a well-shaped 'tang' or 'handle' forming a 'T-shape', whilst others are simply triangular with a smoothed body and flat working edge (Fig. 19.3a–h). These characteristics posed questions for functional and stylistic classification.

The ooyurka can be defined according to its limited geographical range, the degree of shoulder wasting, the major influence on its ultimate shape, the orientation of the

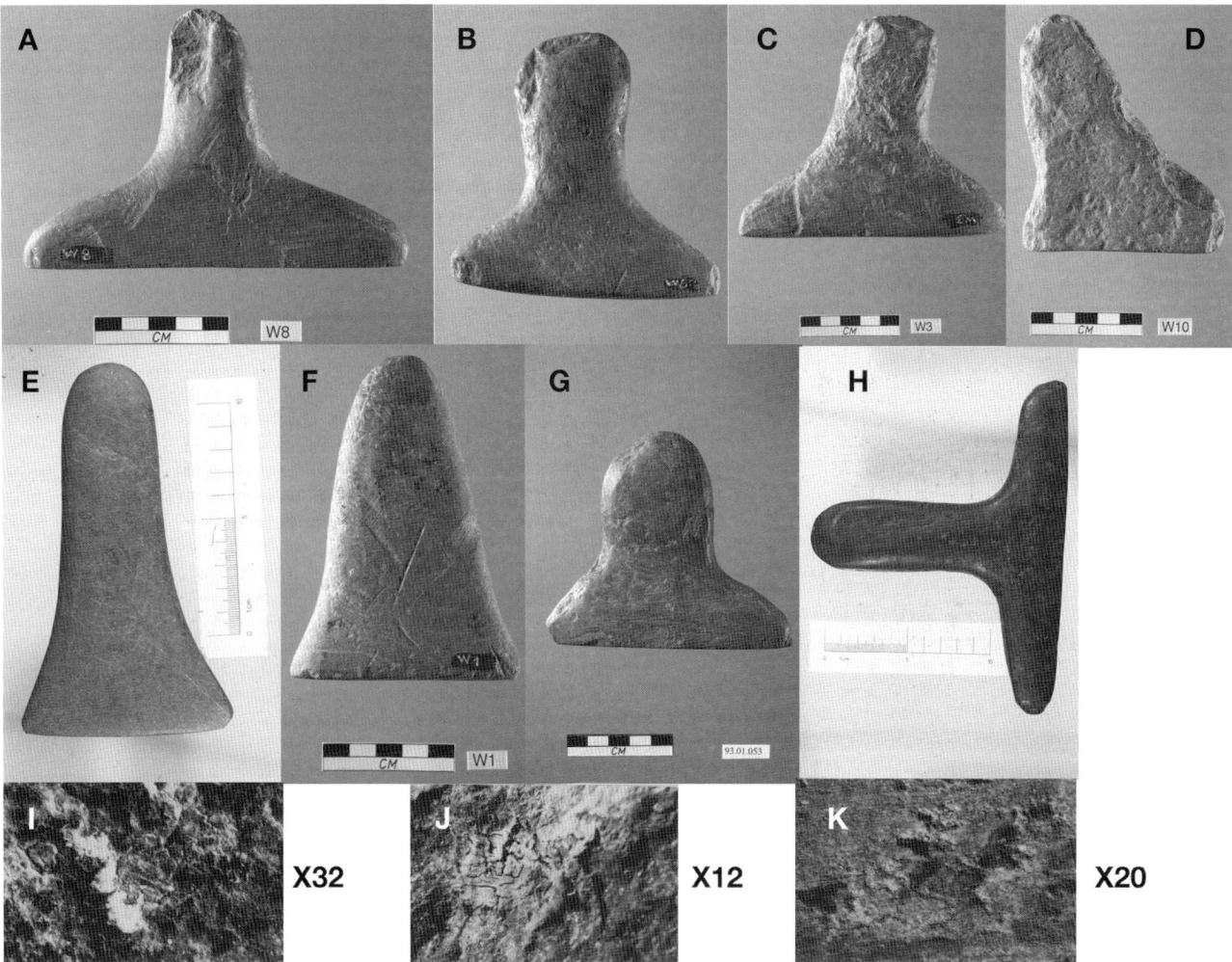

Figure 19.3 Eight ooyurkas (A–H) demonstrate the variety of shapes and sizes of these tools: E and F are triangular, whilst A, B, G, and H have a pronounced 'tang' or 'handle'. Artefacts C and D are off-set and asymmetrical. Images I–K display the organic residue adhering to the working surfaces of three ooyurkas identified as beeswax, Callophyllum *oil, and an unidentified resin.*

working face, which is perpendicular to the median plane, the occurrence of polish, striation, or organic residue on the working face, the narrowness of the working face, the selection of slate as the major raw material, and the degree of concavity, convexity, or flatness of the working face (Fig. 19.2b).

Slate Incised Grinders

The slate incised grinders are flat pieces of grey slate, probably retrieved from the cobble beds of rivers flowing through the granite contact zones of the Barron River Metamorphics (Fig. 19.2c; Willmott & Stephenson 1989). Larger examples are up to *ca.* 50 cm in length and weigh up to 10 kg. Smaller, seemingly more portable ones can be *ca.* 20–30 cm in length, but all have similar parallel, incised grooves that appear to have been made by sharp-edged stone flakes (Fig. 19.2d). The walls of the incisions are angular, straight, and narrow, about 3–5 mm in width. On the reverse side of some specimens, groups of cupules accommodated nuts that were cracked to obtain the kernels. One specimen from the Curatto collection (Cosgrove 2005, 51) had curved designs at one end reminiscent of the outline of an ooyurka.

Function

Ooyurka

Use-wear, residue, and starch grain analysis on these tools have revealed different functions. Small deposits of waxy resin adhered to the body and working face of the implements. Although *Xanthorroea* sp. (grass tree) resin has been identified from deposits at Early Man Shelter, Laura, dated to 3000 BP (Bowden & Reynolds 1982), the survival of such residue in a wet, humid, tropical environment would suggest that the tool type belonged to the late Holocene period and probably reflects the general trend of stone tool development and specialisation across mainland Australia in the last *ca.* 5000 years. Several specimens had surviving residue on their working surfaces as well as furrows and sleeks covered with a lustrous gloss. The manufacturing marks on the working edge are overlain in many examples by residue and polish. The results of thin-layer gas liquid chromatography of purple and brown substances sampled from the edges of three implements identified a range of residue types adhering to the working edges of specimens M4, AAB.79.1.16, L78.3.168, and V15 (Fig. 19.3i, j & k; Cosgrove 1984). These included beeswax, *Callophyllum* sp. tree oil, and an unidentified resin. *Callophyllum* grows on the tropical coast and has oily seeds contained in a small nut. These are poisonous. The presence of polish and residue suggests that they have been used in a grit-free environment, as slate is very soft, and any abrasive material like sand would destroy any polish or residue when rubbed across wooden or stone surfaces. Use-wear replication experiments and the survival of polish and residue on the working surfaces suggests some were used on soft fibres or plant materials containing a high silica content, such as palms or grasses (Cosgrove 1984).

Slate Incised Grinders

Research by Field *et al.* (2016) have used comparative starch grain analysis to determine the function of these tools. On several specimens, small deposits of vegetable residue were adhered deep within the interstices of the fine parallel incisions on the slate grinding stones (Fig. 19.2d). These were collected. Many contained starch grains that were compared using geometric morphometric analysis to contemporary, known starch from various species of nut. Several species were identified, including *Lasjia whelanii* (Whelan's macadamia), *Endiandra palmerstonii* (black walnut), and *Beilschmiedia bancroftii* (yellow walnut), whilst *Endiandra insignis* (hairy walnut), *Cycas media* (cycad), *Lepidozamia hopei* (Hope's cycad), *Tacca leontopetaloides* (Polynesian arrowroot), *Sundacarpus amarus* (black pine), and *Prunus turneriana* (almondbark) were tentatively identified. It nevertheless showed that these implements were used to process a range of rainforest nuts and tubers. In addition, sediment samples from two archaeological sites, Urumbal Pocket and Goddard Creek contained high numbers of starch grains, especially in the upper layers dated to less than 1000 years (Field *et al.* 2009; 2016). The earliest, surviving burnt endocarps and nutshell of the yellow and black walnut are found in the mid-layers of Urumbal Pocket site, dated to 2910–2750 cal BP (Wk-13573) and corroborates the late Holocene uptake of tree nuts and their processing (Ferrier & Cosgrove 2012, 116).

Conclusion

Just as agriculture took a considerable time to fully develop in the Middle East and elsewhere (Graeber & Wengrow 2021, 234), so too did it take time for rainforest people to incorporate the exploitation of toxic foods by adopting and developing a specialised technology. Between 8000 and 2000 years, manipulation of rainforests by fire and exploitation of its food resources was low intensity (Haberle *et al.* 2010) up to 2500 years ago (Ferrier & Cosgrove 2012, 116), with people moving in and out of the rainforest to acquire resources. During these forays, people would have been familiar with the range of rainforest animals, as well as noxious and non-toxic vegetable produce. However, it is suggested that exploiting noxious tree nut resources with high yields would have been time consuming without appropriate technologies. Apparently, the permanent settlement of the rainforest only began when resources like abundant starchy nuts were fully exploited, beginning at least 2500 years ago, which provided conditions for population increase and the intensification of economic activity. Despite the lag between mobile rainforest forays to permanent settlement, at least 1200 generations of Aboriginal people continued to have a residential footprint in this area. *In situ* transformations in the social and economic lives of people permitted them to culturally adapt to long term climatic forces, demonstrating remarkable behavioural flexibility, a hallmark of our species.

Acknowledgements

There are many to whom thanks are due. Many private collectors generously lent their artefacts for analysis. Special thanks to Mr and Mrs Veivers, Waugh's Pocket, as well as R. Stager, C. White, I. Strowd, Davies, R. Taylor, A. Curatto, N. Maxwell, C. Mansfield, and the Aboriginal people of Murray Upper, Yarrabah, Palm Island, and Mossman. I would like to thank Masie Barlow, Ernie Raymont, Pompy Langdon, Robert Paterson, and Frank Davidson. For help in material analysis: Dr C. Cuff, Dr M. Rubenach, and Professor J. Stephenson at James Cook University; D. Wright and N. Chadwick of the Material Culture Unit, J.C.U.; J. Darley of the Scanning Electron Microscope Unit, J.C.U.; W. Jablonski, University of Tasmania; P. Finch of the Photography Department; and Dr B. Bowden of the Chemistry Department, J.C.U. Dr Judith Field, Dr Å. Ferrier, and Professor Richard Fullagar are thanked for advice and assistance with fieldwork. I am grateful to the Queensland Museum for photographs and information on their ooyurka collections. I would like to acknowledge the assistance of M. Quinnell and R. Robins of the Anthropology and Anthropology Department. I am grateful to: Wei Ming and Rudy Frank for the maps, photographs, and diagrams; my brother Michael for help with field experiments; and Fran Cosgrove for reading the final drafts. I would like to thank Dr J. Kamminga for advice on use-wear, Professor Jim O'Connell, and Professor Chris Gosden for compelling discussions on rainforest archaeology.

The ARC Discovery Project (DP0210363) funded part of the project, whilst initial fieldwork support was provided by the Australian Institute of Aboriginal and Torres Strait Islander Studies. Financial assistance was given by James Cook University for fieldwork and the loan and analysis of specimens. Any errors or omissions are of my own making.

References

Aaberge, B., Barnard, T., Greer, S. & Henry, R. (2014) Designs on the future: Aboriginal painted shields and baskets of tropical North Queensland, Australia. *eTropic* 13(2): Special Issue: Value, Transvaluation and Globalization, 56–74.

Anderson, C. & Robins, R. (1988) Dismissed due to lack of evidence? Contemporary Kuku- Yalangi campsites and the archaeological record. In B. Mehan & R. Jones (eds), *Archaeology with Ethnography: An Australian Perspective*, 182–205. Canberra, Australian National University.

Bird, D. W., Bird R. B. & Parker, C. H. (2005) Aboriginal burning regimes and hunting strategies in Australia's Western Desert. *Human Ecology* 33, 443–64.

Bliege Bird, R., Codding, B. F., Kauhanen, P. G. & Bird, D. W. (2012) Aboriginal hunting buffers climate-driven fire-size variability in Australia's spinifex grasslands. *PNAS* 109(26), 10287–92.

Bowden, B. F. & Reynolds, B. (1982) The chromatographic analysis of ethnographic resins. *Australian Institute of Aboriginal Studies Newsletter* 17, 41–2.

Buhrich, A., Goldfinch, F. & Greer, S. (2016) Connections, transactions and rock art within and beyond the Wet Tropics of North Queensland. In S. Greer, R. Henry, R. McGregor & M. Wood (eds), *Transactions and Transformations: Artefacts of the Wet Tropics, Queensland*. Memoirs of the Queensland Museum – Culture 10, 23–42. Brisbane, Queensland Museum.

Cosgrove, R. (1984) *A stylistic use-wear study of the Ooyurka*. Unpublished MA thesis, James Cook University.

Cosgrove, R. (1996) Past human use of rainforests: an Australasian perspective. *Antiquity* 70, 900–12.

Cosgrove, R. (2005) Coping with noxious nuts. *Nature Australia* 28(6), 47–53.

Cosgrove, R. & Raymont, E. (2002) Jiyer Cave revisited: Preliminary results from northeast Queensland rainforest. *Australian Archaeology* 54, 29–36.

Cosgrove, R., Field, J. & Ferrier, Å. (2007) The archaeology of Australia's tropical rainforests. *Palaeogeography, Palaeoclimatology, Palaeoecology* 251, 150–73.

David, B., McNiven, I., Bekessy, L., Bultitude, B., Clarkson, C., Lawson, E., Murray, C. & Tuniz, C. (1999) More than 37,000 years of Aboriginal occupation. In B. David (ed.), *Ngarrabullgan: Geographical Investigations in Djungan Country, Cape York Peninsula*, 157–78. Clayton, Department of Geography and Environmental Science, Monash University.

Dixon, R. M. W. (1972) T*he Dyirbal Language of North Queensland*. Cambridge, Cambridge University Press.

Dixon, R. M. W. (1976) Tribes, languages and other boundaries in northeast Queensland. In N. Peterson (ed.), *Tribes and Boundaries in Australia*, 207–38. Canberra, Australian Institute of Aboriginal Studies.

Ferrier, Å. & Cosgrove, R. (2012) Aboriginal exploitation of toxic nuts as a late- Holocene subsistence strategy in Australia's tropical rainforests. In S. Haberle & B. David (eds), *Peopled Landscapes: Archaeological and Biogeographic Approaches to Landscapes*, 104–20. Terra Australis 34. Canberra, Australian National University Press.

Field, J., Cosgrove, R., Fullagar, R. & Lance, B. (2009) Starch residues on grinding stones in private collections: A study of morahs from the tropical rainforests of NE Queensland. In S. Nugent, L. Kirkwood, A. Crowther & M. Haslam (eds), *Archaeological Science Under a Microscope: Studies in Residue and Ancient DNA Analysis in Honour of Tom Loy*, 218–28. Terra Australis 30. Canberra, Australian National University Press.

Field, J. H, Kealhofer, L., Cosgrove, R. & Coster, A. C. F. (2016) Human-environment dynamics during the Holocene in the Australian Wet Tropics of NE Queensland: A starch and phytolith study. *Journal of Anthropological Archaeology* 44(part b), 216–34.

Graeber, D. & Wengrow, D. (2021) *The Dawn of Everything: A New History of Humanity*. London, Allen Lane.

Haberle, S. G., Rule, S., Roberts, P., Heijnis, H., Jacobsen, G., Turney, C., Cosgrove, R., Ferrier, A., Moss, P., Mooney, S. & Kershaw, P. (2010) Paleofire in the wet tropics of northeast Queensland, Australia. *PAGES* 18(2), 78–80.

Hilbert, D. W., Graham, A. & Hopkins, M. S. (2007) Glacial and interglacial refugia within a long-term rainforest refugium: The Wet Tropics Bioregion of NE Queensland, Australia. *Palaeogeography, Palaeoclimatology, Palaeoecology* 251, 104–18.

Horsfall, N. (1996) Holocene occupation of the tropical rainforests of North Queensland. *Tempus* 4, 175–90.

Khan, K. (1996) *Catalogue of the Roth Collection of Aboriginal artefacts from North Queensland. Volume 2*. Technical Reports of the Australian Museum 12, 1–189.

Lourandos, H., David, B., Roche, N., Rowe, C., Holden, A. & Clarke, S. J. (2012) Hay Cave: A 30,000- year cultural sequence from the Mitchell-Palmer limestone zone, north Queensland, Australia. In S. G. Haberle & B. David (eds), *Peopled Landscapes: Archaeological and Biogeographic Approaches to Landscapes*, 27–63. Terra Australis 34. Canberra, Australian National University Press.

Lumholtz, C. (1889) *Among Cannibals: An Account of Four Years' Travels in Australia and of Camp Life with the Aborigines of Queensland*. John Murray, London.

Moss, P. T., Cosgrove, R., Haberle, S. G. & Ferrier, A. (2012) Holocene landscape change in the sclerophyll woodlands of the Wet Tropics of northeastern Australia. In S. Haberle & B. David (eds), *Peopled Landscapes: Archaeological and Biogeographic Approaches to Landscapes*, 329–341. Terra Australis 34. Canberra, Australian National University Press.

Roberts, P., Buhrich, A., Caetano-Andrade, C., Cosgrove, R., Fairbairn, A., Florin, S. A., Vanwezer, N., Boivin, N., Hunter, B., Mosquito, D., Turpin, G. & Ferrier, A. (2021) Reimagining the relationship between Gondwanan forests and Aboriginal land management in Australia's 'Wet Tropics'. *iScience* 24(3), 102190.

Roth, W. E. (1910) North Queensland ethnography. Bulletin no.14. Transport and trade. *Records of the Australian Museum* 8(1), 1–19.

Stanton, P., Parsons, M., Stanton, D. & Stott, M. (2014) Fire exclusion and the changing landscape of Queensland's Wet Tropics Bioregion 2. The dynamics of transition forests and implications for management. *Australian Forestry* 77(1), 58–68.

Steinberger, L. M. M. (2014) *Hands in pockets: Cultural environments of the Atherton Tablelands of the past 1,500 years*. Unpublished PhD thesis, University of Queensland.

Tobler, R., Rohrlach, A., Soubrier, J., Bover, P., Llamas, B., Tuke, J., Bean, N., Abdullah-Highfold, A., Agius, S., O'Donoghue, A., O'Loughlin, I., Sutton, P., Zilio, F., Walshe, K., Williams, A. N., Turney, C. S. M., Williams, M., Richards, S. M., Mitchell, R. J., Kowal, E., Stephen, J. R., Williams, L., Haak, W. & Cooper, A. (2017) Aboriginal mitogenomes reveal 50,000 years of regionalism in Australia. *Nature* 544, 180–4.

Tuechler, A., Ferrier, A. & Cosgrove, R. (2014) Transforming the inedible to the edible: An analysis of the nutritional returns from Aboriginal nut processing in Queensland's Wet Tropics. *Australian Archaeology* 79, 26–33.

Ulm, S. (2013) 'Complexity' and the Australian continental narrative: Themes in the archaeology of Holocene Australia. *Quaternary International* 285, 182–92.

VanDerWal, J., Shoo, L. P. & Williams, S. E. (2009) New approaches to understanding late Quaternary climate fluctuations and refugial dynamics in Australian wet tropical rain forests. *Journal of Biogeography* 36, 291–301.

Williams, A. N., Ulm, S., Turney, C. S. M., Rohde, D. & White, G. (2015) Holocene demographic changes and the emergence of complex societies in prehistoric Australia. *PLoS One* 10, e0128661.

Willmott, W. F. & Stephenson, P. J. (1989) *Rocks and Landscapes of the Cairns District*. Brisbane, Queensland Department of Mines.

20

What's Involved in Technological Change? Aboriginal Marine Hunting in Tropical North Australia

Harry Allen

Marshall Sahlins defined hunter gatherers as the 'original affluent society', but also criticised them for their lack of productivity. Others have asserted that Aboriginal totemic philosophy prevented them from taking up technological innovations. These propositions are tested against a review of changes in Aboriginal marine hunting technology in northern Australia over the past 400 years. There has been a shift from rafts and bark canoes to sailing dugouts, outrigger canoes, and now to aluminium dinghies. These innovations have arisen through contact with other Aboriginal people and with a succession of outsiders, Torres Strait Islanders, Macassans, and Europeans. Aboriginal agents have been selective about which items served Aboriginal purposes. In some cases, Aboriginal traditions include the new items, but in other cases they lie outside mythological authority. This does not seem to have inhibited Aboriginal people from making pragmatic choices about what was useful to them. Marine hunting can produce large returns, and the sharing of dugong and turtle reinforces relationships between people, places, and mythology. Marine hunting is highly productive, economically, socially, and in terms of the way Aboriginal people see themselves relative to others.

Keywords: Northern Australia, Marine Hunting Technology, Aboriginal Foragers, Technological Change, Relations of Production

Introduction

Marshall Sahlins (1968; 1974) praised Aboriginal hunter gatherers for their work/life balance, but he also criticised them for their lack of productivity: '[they] seem not to realize their own economic capabilities. Labor power is underutilized, technological means are not fully engaged, natural resources left untapped' (Sahlins 1974, 41). In addition, Sahlins (1974, 63–9) saw ritual as alleviating the need for greater foraging efficiency by diverting labour away from the food quest and taking pressure off the environment. Similarly, Meggitt (1964, 31) thought that Aboriginal totemic ideology worked as a 'quasi technology', one that took the place of practical innovation. While these are early views, they are not so distant from more recent ones. Sutton (in Sutton & Walshe 2021, 35) notes that Aboriginal society was hyper-conservative, allowing change only when it was sanctioned in mythology; he speaks of 'spiritual gardening', concluding that when innovations took place they represented variations in response to local ecological circumstances.

MacKenzie and Wajcman (1985) note the term *technology* has multiple layers of meaning; it is a set of objects (hardware), a set of activities, and also the knowledge and skills necessary to manufacture, maintain, and use equipment. Specific modes of work coordination also come into play (Pfaffenberger 1992, 498). The classic study of technological change in an Australian Aboriginal population is Sharp's 'Steel Axes for Stone-Age Australians' (1952). Sharp argued that the missionaries' distribution of steel axes to women and young males created dependency on the mission and negatively impacted on Aboriginal totemic and status relationships, predicting that Yir Yoront society would collapse as a result. In a follow-up study, Taylor (1995) found that Sharp's predictions had not eventuated. Aboriginal society proved to be flexible, seniors remained seniors, women still borrowed axes from adult males, and myths had been adjusted to accommodate the changes brought about by the missionaries. Sharp was right in seeing the impact of change in terms of its impact on social relations; however, he underestimated Aboriginal resilience and flexibility in dealing with change.

Reviewing the technological means that Aboriginal marine hunters had available to them through the 20th century provides an opportunity to test the proposition that Aboriginal hunter gatherers were not fully engaged in the economic quest and that opportunities for change were hindered by totemic philosophy. In addition, there will be an exploration of the manner in which Aboriginal hunting continues to be embedded within the social and totemic relationships that define Australian Aboriginal being.

Marine Hunting Technology in Tropical Northern Australia

Interest in Australian Aboriginal watercraft dates from the time of the first European visitors to Australia, with Dampier's observation in 1688 that Aboriginal people on the northwest coast swam from island to island without the aid of 'boats, canoes, or bark-logs', and Cook's observation in 1770 of an outrigger canoe in the Whitsunday Islands (Rowland 1986, 74; Akerman 2015, 83). Anthropological interest was stimulated by Haddon's Cambridge Expedition to Torres Strait (1898–9) which documented the use of outrigger canoes on both sides of the straits (Haddon 1935). This was followed by general surveys of Aboriginal watercraft by Thomas (1905), Davidson (1935), and Thomson (1952). Payne (2016) provides an excellent technical discussion of Australian Aboriginal rafts and canoes in the collection at the Australia National Maritime Museum.

Lloyd Warner (1937, 453–68) conducted anthropological fieldwork in northeastern Arnhem Land between 1926 and 1929, documenting Indonesian/Macassan influences on Aboriginal culture including sailing dugout canoes and harpoons. Finally, between 1929 and 1932, Donald Thomson spent two seasons undertaking ethnographic research on eastern Cape York giving a detailed account of the technology and social relationships involved in dugong hunting, the use of dugout canoes with outriggers, and Torres Strait Island influences on Aboriginal social and ceremonial life (Thomson 1933; 1934a; Fig. 20.1).

Northwestern Western Australia

In northwestern Western Australia, dugong and marine turtle hunting was largely shore-based, where a communal workforce drove animals onto reefs where they were killed by hand (Akerman 1976; Rouja 1998, 124, 131). Less important was the use of a single or double raft and a simple pointed spear (Akerman 1975a). When the pearling industry was initiated in the 1860s, the wooden spear was tipped with a barbed iron spearhead (English harpoon) and attached to the raft with a rope. Later, with Japanese involvement in the pearling industry, this was replaced by a small detachable iron harpoon on a wooden shaft (Akerman 1975a, 21; Rouja 1998, 124). In terms of archaeology, a shelter on the offshore High Cliffy Island had shellfish, fish, dugong, turtles, and turtle eggs in layers dated 2700–3000 yrs, with turtle predominant in all layers. The open site (HC-3) contained turtle, dugong, fish, and shellfish dated *ca.* 650 years ago to the recent past (O'Connor 1999, 100, 108–12). Middens on the Dampier Peninsula, however, showed turtle and dugong remains as surface features only, probably a result of the manner in which dugong and turtle were shared (Akerman 1975b; Litster *et al.* 2020, 126, 130).

Northern Territory

Elsewhere in Northern Australia, dugong and turtle hunting was carried out using one, two, or three-piece sewn bark canoes, wood or bone-tipped harpoons, ropes, and floats (Davidson 1935, 81–3, 137–9). Bradley (1991; 1997) provides a detailed description of Yanyuwa dugong hunting in the Sir Edward Pellew Islands in the southern Gulf of Carpentaria. By custom, dugongs had to be speared twice, with the float allowing paddlers to follow as the animal swam away. Once it had tired, it was not brought near the bark canoe for fear the canoe would be upset; instead one of the hunters jumped overboard and plugged the dugong's nostrils to drown it. It could then be lashed to the side and towed to shore. Sewn bark canoes, the largest of which might carry eight people, were used between the Northern Territory and the east coast of Queensland. They might be fitted with a variety of pole gunwales, ribs, cross-ties, bark reinforcing, rope ties, and used with carved paddles (Davidson 1935, 137–9).

From sometime in the 16th or 17th centuries, 'Macassan' fisherman voyaged from Southeast Asia to locations across Northern Australia to gather trepang (bêche-de-mer or sea slugs). Staying through the Wet Season, they established close, sometime conflicting, relationships with Aboriginal people through employment, trade, and cultural exchanges (Clark & May 2013b). Among numerous items of Indonesian origin, some of the most important – *e.g.* dugout canoes, rigging, sails, and iron harpoon heads – originated as gifts or were left when the Macassans departed but were manufactured subsequently by Aboriginal people from discarded iron materials. Proof of origin comes from the many Macassan loan words – *e.g.* boat, canoe, anchor, mast, sails, ropes, rudder, paddle, harpoon – in the languages of northern Arnhem Land (Evans 1992).

Dugouts could be large (7 m), carry between eight and 12 persons, and could cover distances up to 50 km with sail, enabling people to hunt and gather from distant islands and reefs (Tindale 1925, 104–11; Thomson 1948–9, 58–60). On the Coburg Peninsula in western Arnhem Land, Mitchell (1996) noted a marked increase in the number of middens

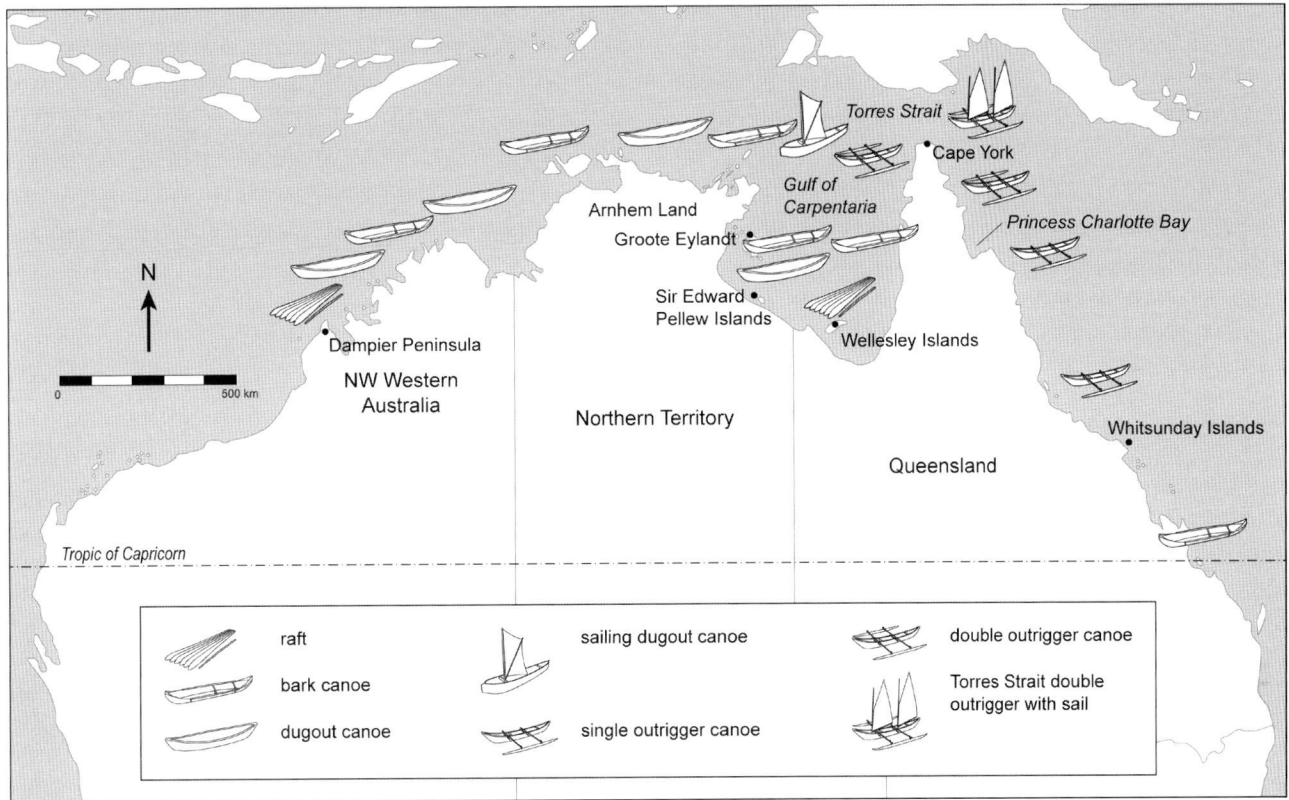

Fig. 20.1 Distribution of Aboriginal watercraft types across northern Australia (Illustration: Seline McNamee, University of Auckland).

containing turtle and dugong in the period after contact with Macassan fisherman, a change he puts down to the adoption of dugout canoes and iron harpoon heads. An archaeological survey of Vanderlin Island in the Sir Edward Pellew Group, also visited by Macassans, however, failed to find evidence of intensification in the relevant time period (Sim & Wallis 2008). In these islands the shift from bark canoes to dugouts occurred during the lifetime of informants (Bradley 1997, 285), and this may have been too late to register in the middens.

Queensland

McNiven (2015, 130) sees some form of sea-going canoe in use in the Torres Strait Islands *ca.* 7000 years ago, followed by intensified occupation across western Torres Strait *ca.* 4000 BP, probably associated with outrigger canoes. About 2600 years ago, however, there was a further increase in the intensity of marine exploitation, which McNiven associates with the westward expansion of Austronesian peoples and the arrival of double canoes. Finally, multiple constructed mounds of dugong bones across the Torres Strait Islands from *ca.* 500 years ago indicate the emergence of a cult associated with dugongs (David *et al.* 2009; McNiven *et al.* 2009; Rowland 2018).

Davidson (1935) and Thomson (1952) document the distribution of single and double outrigger dugout canoes on Cape York Peninsula, with double outriggers being used as far south as the Stewart River and single outriggers between Princess Charlotte Bay and the Whitsunday Islands. Thomas (1905, 73) states 'everything points to a Papuan origin for this type of canoe', suggesting a long period of interaction between northern Cape York and the Torres Strait Islands. A variety of canoe forms and outrigger attachments were present, along with platforms to assist the carrying of gear and the spearing of marine animals (Davidson 1935, 69–71). Using both archaeological, ethnographic, and linguistic evidence, Wood (2018) documents the complexity of technological exchanges between Papua and Cape York, suggesting that time, deliberation, and multiple personnel were involved. Similarly, Dousett and Di Piazza (2021) demonstrate the possibility of multiple contacts between Papua and the Queensland coast.

Thomson (1934a) provides details of dugong hunting with double outrigger canoes at Princess Charlotte Bay, including a description of the canoe, outrigger, spearing platform, harpoon and ropes, the positioning of the crew, the distribution and cooking of dugongs, and a discussion of taboos, magic, and rituals associated with dugong hunting. At Princess Charlotte Bay, mounds of dugong bones were used to mark the graves of renowned dugong hunters (Thomson 1934a, 254). In a survey of the area, Minnegal (1984, 70) recorded primary and secondary dugong consumption sites, as well as a mound of dugong bones similar to that recorded by Thomson, noting that dugongs were likely to be underrepresented in middens as dugong were widely shared after being cooked.

Off-shore islands, such as the Whitsunday and Keppel Islands on the Great Barrier Reef to the south, show middens with marine animals suggesting the availability of watercraft from ca. 3000 BP. Bark canoes in coastal Queensland enabled Aboriginal people to exploit offshore island resources up to 50 km from the mainland with the pattern of regular off-shore island exploitation being in place for the last 500 years (Barker 2004, 148; Rowland et al. 2015, 159).

An exception to bark canoes in Queensland was a shore-based method practised in the Wellesley Islands in the southeast of the Gulf of Carpentaria, where people on land and hunters on rafts herded dugong towards nets placed at the end of tidal channels, where the dugong could then be speared or drowned (Memmott 1979; Dalley 2012, 263).

Changes in Marine Hunting Since 1930

Over the past 90 years, Aboriginal societies in northern Australia have experienced multiple impacts ranging from punitive police expeditions, Christian missions, mining, plantations, cattle stations, military installations, tourism, and conservationism. While the legal tenure of many areas in northern Australia is now in Aboriginal hands, these often incorporate townships, mining reserves, or national parks (Povinelli 1993). To a greater or lesser extent Aboriginal marine hunting has continued, with Aboriginal people now using aluminium dinghies and outboard motors. Additional substitutions include nylon ropes, polystyrene floats, and Toyota tappet rods as harpoon heads (Bradley 1991, 94).

The course of change from rafts and bark canoes to dugouts and dinghies has been different in each area. In northwest Western Australia, Crawford (1969, 332) argues the Macassans had little impact on Aboriginal material culture. Dugout canoes became prevalent only in the 1930s when the missionaries arrived and the Worora copied the use of canoes from their Aboriginal neighbours (Davidson 1935, 77). Similarly, Aboriginal people in the Wellesley Islands adopted dugout canoes after 1914, following contact with Aboriginal missionary workers from Cape York (Dalley 2012, 15–16). On Groote Eylandt, use of a mission motorboat was initiated by Fred Gray, the superintendent, who sent Aboriginal men out on a weekly basis for turtle, turtle eggs, and dugong to supplement the mission diet. Worsley (1961, 169) calculated about 550 sea turtles were caught in this manner during 1951. Dalley (2012, 268) notes one of the first outboards in the Wellesley Islands was purchased by the artist Dick Roughsey, using cash from art sales.

Motorboats, aluminium dinghies, and outboards have changed hunting patterns. They allowed travel over greater distance and had the power to cope with difficult tides, but travel also became necessary as dugong and turtle graze further away. Dugongs are sensitive to noise, so oars or paddles have to be used to get close enough to spear the animals (Rouja 1998, 125–7). In traditional times, mature men were the most skilled hunters. Today, elders criticise the younger men, who do most of the hunting, for their lack of skill and traditional knowledge (Bradley 1997, 418; Rouja 1998, 136).

Spiritual and Social Relations in Aboriginal Marine Hunting

Aboriginal territorial estates extended beyond the land to include islands, the seabed, banks, and reefs, where ancestral heroes travelled, hunted, and left significant sites above and below the water (Smith 1984; Bradley 1997, 106, 161, 196; Peterson & Rigsby 1998). Bradley (1997, 269–81) documents how the Sea Turtle Spirit Ancestor and the Dugong Hunter Spirit Ancestor travelled from east to west, across Limmen Bight, to the Sir Edward Pellew Islands, creating places as they journeyed. Their songs were sung at ceremonies for the dead of their associated kin. On visiting a distant island, Bradley (1997, 172) records a man calling,

> I have come back, I have not been here for a long time, hear me I am a kinsman of this country… I stand here, I who know the song cycle which is running here, I am kin to the Dugong Hunter Spirit Ancestors which came here from the east and continued on into the west, I have not forgotten… I can name this place, I can call the names of this place…

Present day hunters retrace the steps of these ancestral heroes, re-enacting the drama of creation as they move through a seascape filled with cultural meaning (see Gosden & Head 1994; Tamisari 1998). The Dugong Hunter Spirit Ancestor used harpoons, ropes, and a bark canoe, all of which have moiety and totemic names; dugout canoes do not (Bradley 1997, 153).

Relationships between people and the totemic land/seascape regulated all aspects of a dugong hunt, the manufacture and ownership of equipment, skills acquisition, magic, the crew, the territory hunted over, behaviour towards the dugong or turtle, cooking, butchery, distribution, and final disposal of the animal (Thomson 1934a, 247–53; Bradley 1997, 322; Rouja 1998, 234–9; Dalley 2012, 265). These relationships were on display in the sharing of the catch, where cooking, butchery, and distribution were often done by the hunter's older male relatives, with the hunter receiving the least desirable parts of the animal (Bradley 1997, 318; Rouja 1998, 235).

In northeastern Arnhem Land, McIntosh (2011; 2013) documents the impact of successive waves of materially wealthy peoples (Bayini, Macassans, and Europeans) on Yolngu culture and their reaction to this. While the new technologies were attractive, their adoption endangered Yolngu definitions of themselves as Aboriginal. Acceptance of Macassan canoes, harpoon, sails, and knives etc., may have been made easier by the fact that stories of an earlier group of Asian voyagers, the Bayini, had been incorporated into Yolngu mythology. Hidden within the Bayini narrative are stories of Birrinydji (the Dreaming Macassan) whose desire (*marr*) drew these seafarers to the Yolngu, introducing a world of material riches, but also a sense of inequality and dispossession (McIntosh 2011, 349). Bayini's law makes no mention of the Macassans and reinforces the use of traditional material culture – bark canoes and stone axes – as aspects of Yolngu resistance (McIntosh 2013, 99–100). Nonetheless, items of Macassan origin have been incorporated into Yolngu ritual, and Macassan subjects figure prominently in both the rock and portable art of Arnhem Land (Warner 1937, 466–7; McIntosh 2011, 341; Taçon & May 2013). On Cape York, hero cult stories and rituals, similar to those observed in the Torres Strait Islands, are associated with drums and outrigger canoes which Thomson (1933; 1934b) suggests represent a recent stratum of Aboriginal tradition.

Conclusions

Marine hunting technology used by Aboriginal hunters has greatly changed over the past two hundred years, from rafts and bark canoes to dugout canoes to dinghies with outboard motors. There have been comparable changes in harpoon technology. Many of these items of technology have been incorporated into Aboriginal foundational myths, but others have not. Aboriginal people appear to have been pragmatic in their selection of items to take up or to discard. The lack of mythological authority does not appear to have inhibited Aboriginal usage, while at the same time people acknowledge the foreignness of later inclusions. Aboriginal people were not passive recipients of the materials brought by successive waves of foreigners, but active agents in selecting what they might use from the assortment of goods introduced by the newcomers and deciding what they might be willing to give up in exchange.

Viewing the shift from bark canoes to dugouts to aluminium dinghies in evolutionary terms would be misguided. Each set of hunting components has their own costs and benefits. These have to be seen within an Aboriginal accounting system, where relationships of tradition, people, and place might stand alongside questions of efficiency and return. Aluminium dinghies and outboards are expensive to buy, run, and maintain. It is often individuals within family groups which own and share these among their family members with the ownership of hunting gear giving rights to part of the catch (Rouja 1998, 125).

Dugong and marine turtles are now on International Lists of endangered species and Aboriginal hunters have been criticised in the national press for their use of 'non-traditional' technology. Responses include the provision of conservation plans and the employment of Aboriginal Rangers to implement them. Aboriginal rangers expressed concern about the use of new technologies (boats, outboards, storage freezers) and the potential for the catch to be privately consumed. On the other hand, rangers were reluctant to impose conditions on hunters in the interest of maintaining relationships within the community (Dalley 2012, 275–6).

Povinelli (1992, 178, 181) notes that Aboriginal people on the Cox Peninsula (near Darwin) ranked their activities in terms of their ability to hunt 'in the old way', where hunting was not just hunting, but an activity that ensured the reproduction of the material, ecological, and mythic landscapes and seascapes. Reductions in animal numbers, or seagrass patches, might be put down to a reduction in Aboriginal hunting or to the unknowing intrusions of outsiders, who have destroyed places of significance and wantonly killed dugong (Bradley 1997, 394–5).

Sahlin's 1968 description of hunter gatherers as 'the original affluent society' and his argument that they are economically underproductive is to impose a set of non-Aboriginal values onto Aboriginal people. Povinelli (1993, 26) argues that Aboriginal hunting is both an activity and a discourse, reaffirming relationships between people, the land/sea, and the mythological world in a manner that defines their Aboriginality. A young man's decision to hunt dugong is to embark on a dangerous enterprise making use of the technology available. A successful hunt gives prominence, allows the sharing of the catch, and the right to tell the story – personal returns that rank highly within the community.

Acknowledgements

Seline McNamee, School of Social Sciences, University of Auckland, drew Figure 20.1.

References

Akerman, K. (1975a) The double raft or Kalwa of the west Kimberley. *Mankind* 10, 20–3.

Akerman, K. (1975b) Aboriginal camp sites on the western coast of Dampier Land, Western Australia. *Occasional Papers in Anthropology* 4, 93–104.

Akerman, K. (1976) Fishing with stone traps on the Dampierland Peninsula, Western Australia. *Mankind* 10, 182.

Akerman, K. (2015). A review of the Indigenous watercraft of the Kimberley region, Western Australia. *Journal of the Australian Association for Maritime History* 37, 82–111.

Barker, B. C. (2004) *The Sea People: Late Holocene Maritime Specialisation in the Whitsunday Islands, Central Queensland*. Canberra, Pandanus Books, Australian National University.

Bradley, J. J. (1991) 'Li-Maramaranja': Yanyuwa hunters of marine animals in the Sir Edward Pellew Group, Northern Territory. *Records of the South Australian Museum* 25, 91–110.

Bradley, J. J. (1997) *Li-Anthawirryarra, people of the sea: Yanyuwa relations with their maritime environment*. Unpublished thesis, Charles Darwin University.

Clark, M. & May, S. K. (eds) (2013a) *Macassan History and Heritage: Journeys, Encounters and Influences*. Canberra, ANU ePress.

Clark, M. & May, S. K. (2013b) Understanding the Macassans: A regional approach. In Clark & May 2013a, 1–19.

Crawford, I. M. (1969) *Late prehistoric changes in Aboriginal cultures in Kimberley northwestern Australia*. Unpublished thesis, University of London.

Dalley, C. (2012) Dugong hunting as changing practice: Economic engagement and an Aboriginal ranger program on Mornington Island, southern Gulf of Carpentaria. In N. Fijn, I. Keen, C. Lloyd & M. Pickering (eds), *Indigenous Participation in Australian Economies II: Historical Engagements and Current Enterprises*, 261–86. Canberra, ANU ePress.

David, B., McNiven I. J., Crouch, J., Mura Badulgal Corporation Committee, Skelly, R., Barker, B., Courtney, K. & Hewitt, G. (2009) Koey Ngurtai: The emergence of a ritual domain in Western Torres Strait. *Archaeology in Oceania* 44, 1–17.

Davidson, D. S. (1935) The chronology of Australian watercraft. *Journal of the Polynesian Society* 44, 1–16, 69–84, 137–52, 193–207.

Dousset, L. & Di Piazza, A. (2021) Mapping prehistoric sailing routes to Lizard Island and beyond. *Journal of Pacific Archaeology* 12, 16–31.

Evans, N. (1992) Macassan loanwords in top end languages. *Australian Journal of Linguistics* 12, 45–91.

Gosden, C. & Head, L. (1994) Landscape – a usefully ambiguous concept. *Archaeology in Oceania* 29, 113–16.

Haddon, A.C. (1935) *Reports of the Cambridge Anthropological Expedition to the Torres Straits: Volume 1, General Ethnography*. Cambridge, Cambridge University Press.

Litster, M., Barham, A., Meyer, J., Maloney, T. R., Shipton, C., Fallon, S., Willan, R. C. & O'Connor, S. (2020) Late Holocene coastal land-use, site formation and site survival: Insights from five middens at Cape Leveque and Lombadina, Dampier Peninsula, Kimberley, Australia. *Australian Archaeology* 86, 118–36.

McIntosh, I. S. (2011) Missing the revolution! Negotiating disclosure on the pre-Macassans (Bayini) in North-East Arnhem Land. In M. Thomas & M. Neale (eds), *Exploring the Legacy of the 1948 Arnhem Land Expedition*, 337–54. Canberra, ANU ePress.

McIntosh, I. S. (2013) Unbirri's pre-Macassan legacy, or how the Yolngu became black. In Clark & May 2013, 95–106.

MacKenzie, D. & Wajcman, J. (1985) *The Social Shaping of Technology: How the Refrigerator got its hum*. Milton Keynes, Open University Press.

McNiven, I. J. (2015) Canoes of Mabuyag and Torres Strait. *Memoirs of the Queensland Museum – Culture* 8, 127–207.

McNiven, I. J., David, B., Kod, G. & Fitzpatrick, J. (2009) The great *Kod* of Pulu: Mutual historical emergence of ceremonial sites and social groups in Torres Strait, Northeast Australia. *Cambridge Archaeological Journal* 19, 291–317.

Meggitt, M. (1964) Aboriginal food-gatherers of Tropical Australia. Ninth Technical Meeting of the International Union for the Conservation of Nature, Nairobi. *IUCN Publications* 4, 30–7.

Memmott, P. (1979) *Lardil properties of place: An ethnological study in man–environment relations*. Unpublished thesis, University of Queensland.

Minnegal, M. (1984) Dugong bones from Princess Charlotte Bay. *Australian Archaeology* 18, 63–71.

Mitchell, S. (1996) Dugongs and dugouts, sharptacks and shellbacks: Macassan contact and Aboriginal marine hunting on the Cobourg Peninsula, northwestern Arnhem Land. *Bulletin of the Indo-Pacific Prehistory Association* 15, 181–91.

O'Connor, S. (1999) 30,000 years of Aboriginal occupation, Kimberley, Western Australia. *Terra Australia* 14. Canberra, ANH Publications and Centre for Archaeological Research, Australian National University.

Payne, D. (2016) *Australia's First Watercraft* [online]. Available at: <https://www.sea.museum/2016/12/15/australias-first-watercraft> [accessed March 2022].

Peterson, N. & Rigsby, B. (1998) *Customary Marine Tenure in Australia*. Sydney, Oceania Publications.

Pfaffenberger, B. (1992) Social anthropology of technology. *Annual Review of Anthropology* 21, 491–516.

Povinelli, E. (1992) 'Where we gana go now': Foraging practices and their meanings among the Belyuen Australian Aborigines. *Human Ecology* 20, 169–201.

Povinelli, E. (1993) *Labor's Lot: The Power, History and Culture of Aboriginal Action*. Chicago, IL, University of Chicago Press.

Rouja, P. M. (1998) *Fishing for culture: Toward an Aboriginal theory of marine resource use among the Bardi Aborigines of One Arm Point, Western Australia*. Unpublished thesis, University of Durham.

Rowland, M. J. (1986) The Whitsunday Islands: Initial historical and archaeological observations and implications for future work. *Australian Archaeology* 3, 72–87.

Rowland, M. J. (2018) 65,000 years of isolation in Aboriginal Australia or continuity and external contacts? An assessment of the evidence with an emphasis on the Queensland Coast. *Journal of the Anthropological Society of South Australia* 42, 211–40.

Rowland, M. J., Wright, S. & Baker, R. (2015) The timing and use of offshore islands in the Great Barrier Reef Marine Province, Queensland. *Quaternary International* 385, 154–65.

Sahlins, M. (1968) Notes on the original affluent society. In R. B. Lee & I. DeVore (eds), *Man the Hunter*, 85–89. Chicago, IL, Aldine.

Sahlins, M. (1974) *Stone Age Economics*. London, Tavistock.

Sharp, R. L. (1952) Steel axes for stone age Australians. *Human Organization* 1, 1–69.

Sim, R. & Wallis, L. A. (2008) Northern Australian offshore island use during the Holocene: The archaeology of Vanderlin Island, Sir Edward Pellew Group, Gulf of Carpentaria. *Australian Archaeology* 67, 95–106.

Smith, M. (1984) Bardi relationships with sea. *Anthropological Forum* 5, 443–47.

Sutton, P. & Walshe, K. (2021) *Farmers or Hunter Gatherers? The Dark Emu Debate*. Melbourne, Melbourne University Press.

Taçon, P. S. C. & May, S. K. (2013) Rock art evidence for Macassan–Aboriginal contact in northwestern Arnhem Land. In Clark & May 2013a, 127–40.

Tamisari, F. (1998) Body, vision and movement: In the footprints of the Ancestors. *Oceania* 68, 249–70.

Taylor, J. (1995) Goods and Gods: A follow-up study of 'Steel Axes for Stone-Age Australians'. In J.-L. Chodjiewicz (ed.), *Peoples of the Past and Present: Readings in Anthropology*, 364–73. Toronto, Harcourt Brace.

Thomas, N. W. (1905) Australian canoes and rafts. *Journal of the Anthropological Institute of Great Britain and Ireland* 35, 56–79.

Thomson, D. F. (1933) The hero cult, initiation and totemism on Cape York. *Journal of the Anthropological Institute* 63, 453–537.

Thomson, D. F. (1934a) Dugong hunters of Cape York. *Journal of the Anthropological Institute* 64, 237–64.

Thomson, D. F. (1934b) Notes on a hero cult from the Gulf of Carpentaria, North Australia. *Journal of the Anthropological Institute* 64, 217–34.

Thomson, D. F. (1948–9) Arnhem Land: Explorations among an unknown people. *Geographical Journal* 112(4/6), 146–64.

Thomson, D. F. (1952) Notes on some primitive watercraft in Northern Australia. *Man* 52, 1–5.

Tindale, N. B. (1925). Natives of Groote Eylandt and the west coast of the Gulf of Carpentaria. *Records of the South Australian Museum* 3, 61–102, 103–34.

Warner, W. L. (1937) *A Black Civilization: A Social Study of an Australian Tribe*. New York, NY, Harper and Brothers.

Wood, R. (2018) Wangga: The linguistic and typological evidence for the sources of the outrigger canoes of Torres Strait and Cape York Peninsula. *Oceania* 88, 202–31.

Worsley, P. (1961) The utilisation of natural food resources by an Australian Aboriginal tribe. *Acta Ethnographia* 10, 153–90.

21

The Yolŋu System as a Regional Polity

Howard Morphy with Frances Morphy

This chapter focuses on the society of the Yolŋu people of northeast Arnhem Land in northern Australia. Recent research combining archaeological, linguistic, and anthropological data has shown that a set of toponyms ending in -tjpi have likely been attached to the same places for at least 3000 years. The distribution of these place names opens up the possibility of reconstructing the regional trajectory of Yolŋu society over the longue durée. *Ancestral precedent as 'cultural' memory and regional ecological systems work together to produce a landscape that constantly brings the past into the present. It is argued that, over time, the Yolŋu marriage system has expanded in tandem with the expansion of the Yolŋu languages.*

Keywords: Australian Aboriginal, Toponyms, Memory, Holocene High Stand, Yolŋu, Regional Systems

Introduction

From the 1993–4 annual report of the Pitt Rivers Museum, we learn that Chris Gosden 'together with Howard Morphy developed and co-taught a new advanced course on the material culture of regional systems'.[1] The course came out of the collaborative and cross-disciplinary environment that developed for anthropology and archaeology in Oxford at that time. Chris and I were both interested in regional systems and looking at long-term socio-cultural trajectories. I had just finished writing a chapter on cultural adaptation for a volume edited by Geoffrey Harrison, then Professor of biological anthropology (H. Morphy 1993). Although the chapter was general in its scope, the main case study was based on research with the society of the Yolŋu people of eastern Arnhem Land in Australia's Northern Territory. Regional systems in that region and their duration have continued to be one focus of my research in the company of Frances Morphy.[2] Thirty years on, Chris's collaborative research on the *longue durée* in Britain and ours in Australia show surprising synergies, though they are apparently focused on quite different cultural systems and durations.

Working with Frances (a linguistic anthropologist), archaeologist Pat Faulkner, and environmental anthropologist Marcus Barber, I have been engaged in research that has the potential to show the trajectory of Yolŋu society over at least 3000 years. Chris Gosden and the EngLaId team's historical present is a little earlier than ours, but the start date roughly corresponds – they are looking at trajectories over a period of 2500 years from the Middle Bronze Age (1500 BC) to the Domesday Book of 1086 (Gosden *et al.* 2021, 2). Their inspiration comes partly from landscape historian W. G. Hoskins' dictum that 'everything is older than we think' (Hoskins 1977, 212).

The rationale for setting the Domesday Book as the end point is that the history of Britain after that date is well served by historical documents and 'histories taught in schools or written about in popular books generally focus on the millennium since the Norman Conquest in 1066' (Gosden *et al.* 2021, 1). The EngLaId team argue that 'there is a case to be made for a longer set of histories, combining archaeology and historical records' (Gosden *et al.* 2021, 1). In Australia the absence of precolonial written sources and the erasure of Indigenous presence in the minds of settler Australians as consequence of colonisation has meant that for much of Australia's recent past both the Indigenous past and the Indigenous present were absent from school curricula.[3] However, the absence of precolonial written records also means that more recent accounts of Indigenous

Australian societies are absolutely essential to understanding past histories. In the case of eastern Arnhem Land new data that enables one to stretch back into the past and uncover the trajectories of regional cultures will not come from historical records, but through archaeology, linguistics, and anthropology. While the material evidence of archaeology arguably provides the most direct evidence from the past, the temporal processes sedimented in linguistic and anthropological data are, to echo Hoskins, 'older than we think'.

Country

I will argue that in northeast Arnhem Land the landscape or 'Country' provides a major entry point into the past.[4] The regional geography and the affordances of the environment provide a framework for understanding the ways in which people might have lived in it as hunters and gatherers over time – for example, in the distribution of resources, the seasonal cycle, the logic of the terrain, and the routes across Country that potentially influence patterns of production, trade, and exchange. And from the perspective of Yolŋu ontology, as we shall see, the Country as an ancestral (Waŋarr) presence is an eternal determinant of relationships between people – a pattern that is infinitely reproduced through time (H. Morphy 1991; Keen 1994). Ancestral precedent as 'cultural' memory (Connerton 2009) and regional geography work together to produce a landscape that constantly brings the past into the present. Hoskins' starting point, summarised by Gosden and colleagues, that 'the landscape is the greatest historical document we posses' resonates well with a Yolŋu perspective (Gosden et al. 2021, 4).

Yolŋu Rom (Ancestral law) governs all aspects of human lives and has structural features that ensure it remains in place over time. This does not mean that Yolŋu relationships with the environment are unchanging or that the environment itself does not change over time. Their way of life, which is adapted to the dynamic environment in which they live, enables them to respond to seasonal change, to demographic vicissitudes, and to the contingencies of everyday life (Peterson 1972). But for most of Yolŋu precolonial history, since at least the mid-Holocene, the process of larger-scale environmental change was slow and changes in technology did not result in a major rapid transformation of their hunter-gatherer economy or in significant population growth.

3000 Years Ago at Malitjpi

I begin with a bold statement. On a peninsula in the north of Blue Mud Bay – as Matthew Flinders named it in 1802 – is a place named Malitjpi. Today it is close to the site of a huge ground sculpture in the form of a stingray. The body of the stingray covers around a hectare – the size of a 400 m running track – and the tail extends behind it for at least a further 100 m. Two large hollows for eyes are scooped out of its head. Before hunting for fish or crabs people will sometimes place the remains of an earlier meal in each eye, or scoop out a little of the sandy soil and cast it into the air – for luck in hunting. The ridges of sandy soil that mark the outline of the stingray's body are periodically renewed. We know that this sculpture was in place nearly 100 years ago; Yolŋu oral tradition records the anthropologist Donald Thomson visiting the place (in 1937). But it could have been there very much longer. Our research has shown that if we could transport ourselves back in time some 3000 years and ask to be taken to Malitjpi, it is highly probable that we would be taken to this place.

The remainder of this chapter addresses the question of why we are reasonably confident that we have established the minimum age for this and some other place names and looks at the implications for the deep history of the society which today continues to use that name. The answer to the latter is entangled with the former because it involves delineating the possible social and cultural factors that have allowed the name to remain where it is for so long.

The Stratigraphy of Language

The region occupied by the Yolŋu Matha languages (Fig. 21.1) extends from the Walker River in the south to the Arafura Sea in the north, and from the Gulf of Carpentaria in the east to the Blyth River in the west (F. Morphy 1983). In 2000 Frances Morphy and I, along with Marcus Barber and archaeologists Pat Faulkner and Annie Clarke, began to conduct research on what became known as the Blue Mud Bay case. The claim area in the northern half of Blue Mud Bay off the Gulf of Carpentaria is the region whence, linguistic analysis suggests, the Yolŋu languages originally spread to their current locations. The project was funded by an Australian Research Council linkage grant, with the Northern Land Council as industry partners, to undertake background research for a Yolŋu claim to rights in the sea under the Commonwealth's *Native Title Act 1993* and the *Aboriginal Land Rights (Northern Territory) Act 1976*. The details of the project can be found elsewhere (F. Morphy & Morphy 2009). As far as the court case was concerned it was necessary for us to research Yolŋu systems of land and sea tenure, the evidence for continuity of occupation of the land since the time of European sovereignty,[5] the nature of people's rights in land and sea, and their marine economy. Central to the case was the making of a detailed map of the area under claim, including the names of places, the location of different clan estates, and the tracks of major ancestral beings connecting different places. Archaeological research was conducted to provide evidence for continuity of occupation and evidence of economic activity over time. It was the combination of the mapping of place names

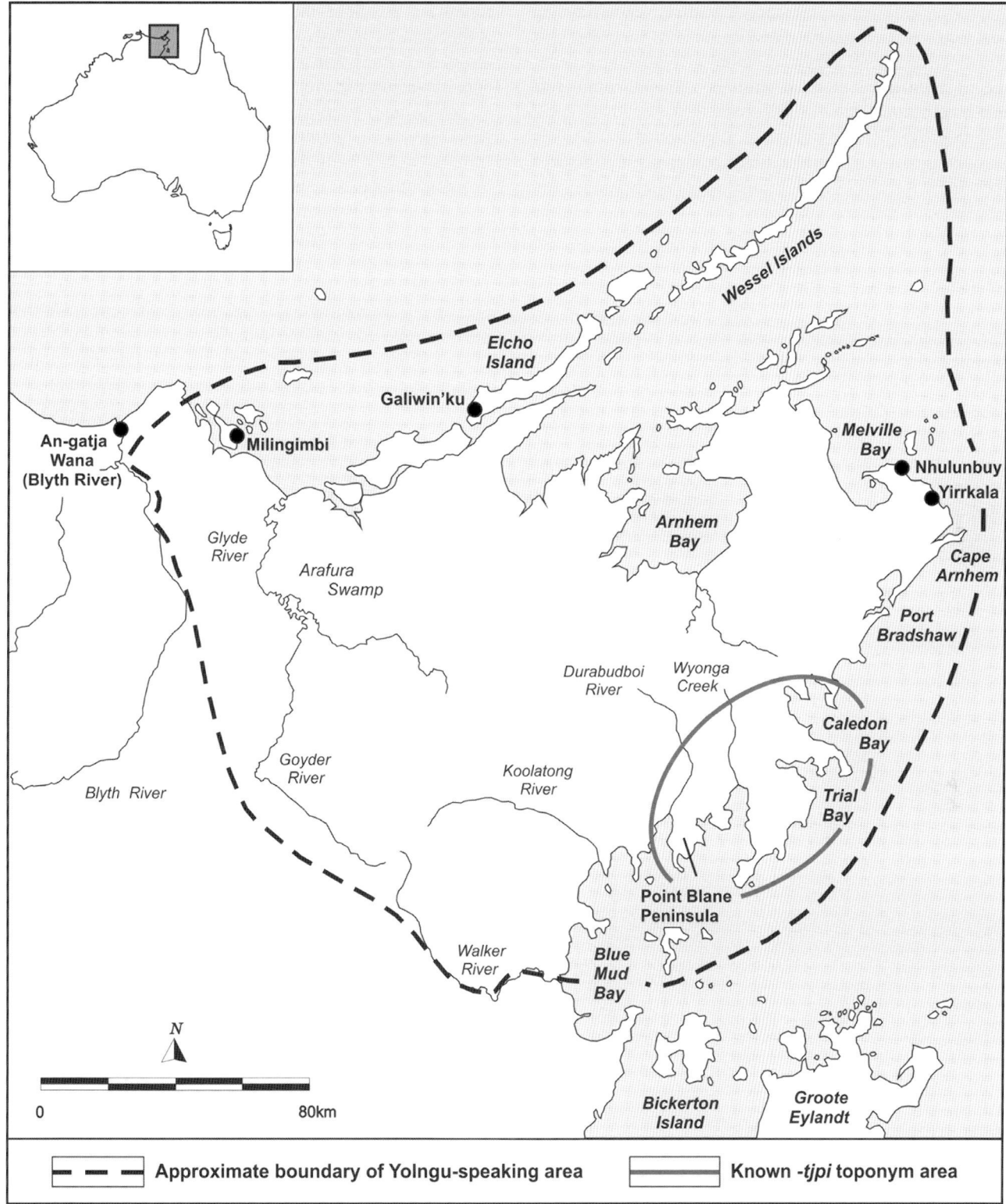

Figure 21.1 Location of Yolŋu Matha languages and of known -tjpi toponyms (area circled).

with archaeological and geomorphological evidence that provided an unanticipated breakthrough.

In previous research in the region Frances had encountered a series of names that puzzled her as a linguist – a set of names ending in *-tjpi*, which are treated today by Yolŋu as monomorphemic words. Frances had an intuition that these might be of considerable age, and that once in some distant time, possibly in the proto-Yolŋu Matha (pYM) language, -*tjpi* had been a productive suffix. As the mapping continued the number of -*tjpi* place names increased and a distinctive pattern began to emerge. The 28 -*tjpi* names within the claim area, mapped using GPS, formed a line inland and along part of the coast. The Eureka moment happened when Pat Faulkner produced his map showing the contour of the coastline at the time of the Holocene high stand, around 3000 years ago (Faulkner 2013). The mapped -*tjpi* names occur almost without exception on the old Holocene high stand coastline (Fig. 21.2). Since that time the coastline has prograded at the outlets of the major rivers in the region, creating mangrove-fringed floodplains and wetlands, and today some of the -*tjpi* places are 12 km or so inland. Subsequent research has identified another 50 -*tjpi* place names that fit the same pattern and are located in the same region or very close to it. The inland -*tjpi* places in the claim area are situated on or near a low laterite cliff that, at the time of the high stand, was on the seashore. Several (*e.g.* Gumurryaṉutjpi and Ḏilmitjpi) are the sites of large, early shell mound and midden complexes. The earliest of these complexes are dated to around 2900 years ago – the time when intensive shellfish exploitation became a major component of the local economy. This focus on shellfish lasted for a considerable time – the latest midden deposits of this type date to around 500 years ago.

Are these -*tjpi* names survivors from pYM? Frances's analysis suggests that the -*pi* part of -*tj-pi* is a contracted form of –*p/wuy*, a still-productive Yolŋu Matha suffix which is found as the final element of many Yolŋu toponyms (F. Morphy *et al.* 2020, 168). It can be translated, approximately, as 'place associated with'. The segment -*tj-* is harder to pin down. It bears no relation to any productive suffix found in the YM languages of today. However, there is evidence to suggest that it may once have been a productive derivational suffix, meaning, approximately, something like English -ed in 'gifted' or -y in 'stony'. Removing these suffixes yields lexemes of which approximately 30% are still words in today's Yolŋu Matha languages of the region. For example Gumurryaṉutjpi, a place on the laterite ridge now 12 km inland, is analysable as *gumurr-yäṉu-tj-pi '*shore-gully-ed-place' (Fig. 21.2). The topography of the site is where the lateritic ridge is bisected by a gully.

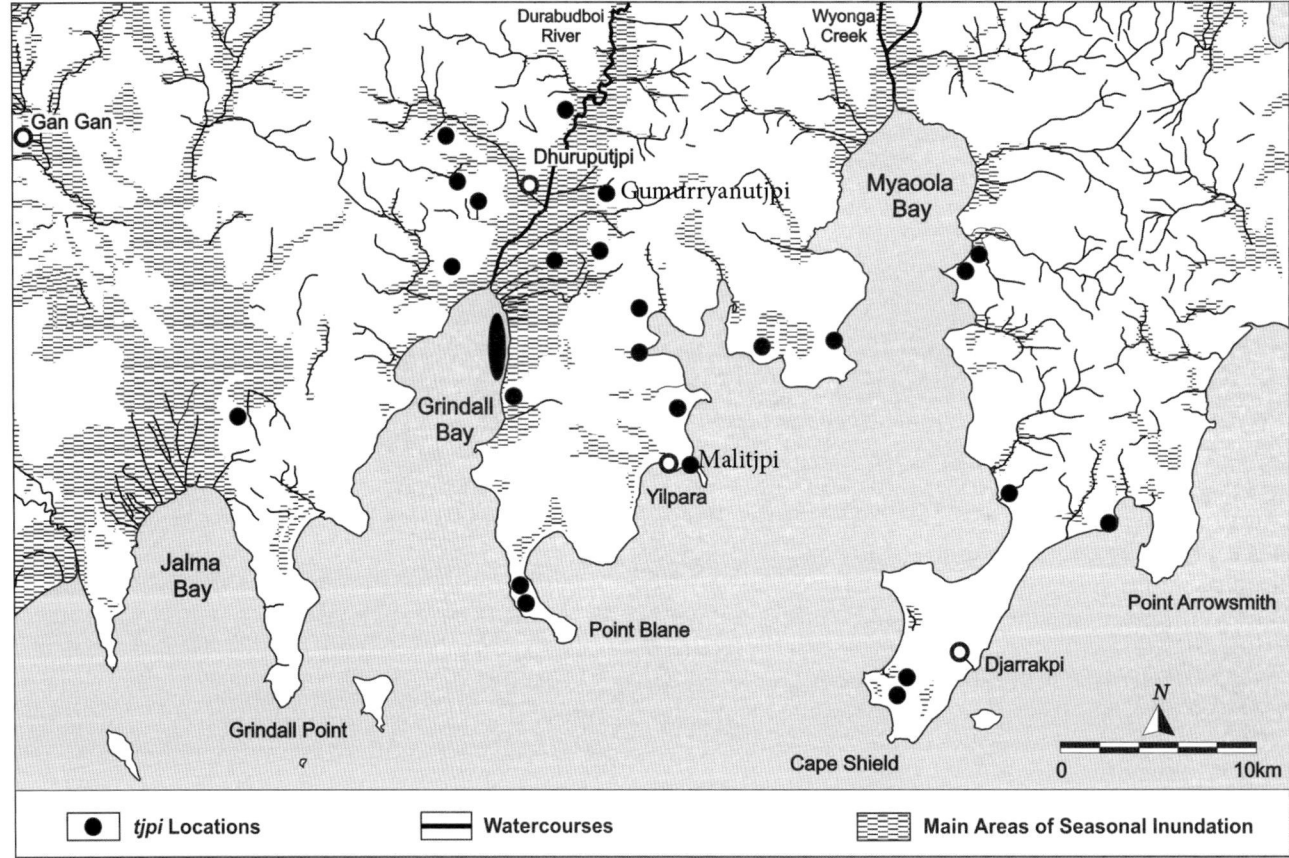

Figure 21.2 Location of the mapped -tjpi *toponyms. The map shows the distribution of* -tjpi *toponyms and locates Gumurryaṉutjpi and Malitjpi. The names are located on the Holocene high stand which today in many places is inland, separated from the present day prograded coastline by mangroves and seasonally inundated wetlands. The only* -tjpi *name located away from the Holocene high stand is a regional name that includes an area of wetlands. The map was originally published in F. Morphy* et al. *(2020), which provides a more detailed account of the data.*

Three thousand years ago this ridge was indeed on the shore overlooking the tidal mudflats.

The aforementioned Malitjpi can plausibly be associated with the ancestral Yirritja moiety stingray Lulumu. *Mali* is the word for shade or ancestral spirit and, with the stingray ground sculpture close by, it is possible that this is a place named for the shade of Lulumu. The story as told today is that the stingray was speared by the ancestral Dhuwa moiety hunter Liyawaday. Wounded, it leapt out of the sea and rushed inland, biting the ground as it went on to form a series of billabongs. Then, turning, it raced back towards the sea and entered it again off the peninsula where the ground sculpture is situated. Its body now lies under the sea as a series of rocks and reefs. The ordinary word for this species of stingray is *gurrtjpi* – one of the few *-tjpi* words that is not a place name. So the hypothesis that this place has been associated with an Ancestral stingray for a very long time is not implausible. And in a nearby coastal estate of a Dhuwa moiety clan is a place called Garatjpi (*gara-tj-pi *spear-ed-place), which is explicitly linked to Liyawaday, who speared fish there.

Why Retain Names in Place?

In this section I address two broad questions: is it plausible to suggest that Yolŋu have 'archival' mechanisms that have maintained names in place for over 3000 years, and if so, what would those mechanisms imply about the nature of Yolŋu society over time? The starting point has to be what we know about Yolŋu society from current research and what mechanisms are in place today for passing knowledge of toponyms from generation to generation. How secure are those mechanisms? What kind of evidence do we have for the stability of toponyms in the present?

Names for places are often also bestowed as personal names and these names have their origins in the actions of ancestral beings in the Waŋarr. Of those that are analysable and therefore translatable, some refer to the actions of ancestral beings or events that befell them in place: the place where the sun burnt the ancestral women's skin, the place of the shark's liver, or the sand-crab beach. But the untranslatable *-tjpi* names are equally ancestral in their reference to place. Gunmurrutjpi, for example, refers to the general area where Bäru the ancestral crocodile transformed from human being to crocodile. Yolŋu place names are organised in a hierarchy of 'big' named and 'small' named places. *Yindi yäku* – big names – reference the primary ancestral beings associated with a particular clan estate. Under their umbrella cluster sets of names marking the places where particular events occurred, or referencing particular attributes of the ancestor and or the topography. In the case of Gunmurrutjpi in Madarrpa country, it is the place where Bäru was set on fire by his wife Blue Tongued Lizard, the place of the crocodile's nest, the place where his arm was transformed into a rock, the place of his tail and so on. Each place has more than one name, only one of which is in general use at a particular time. The other names, however, circulate in a number of ways. A near full set of names exists in the words of the songs that are sung and the sacred names that are intoned by ritual specialists during ceremonial performance. The most powerful names occur in the more restricted contexts but are nevertheless widely known. The small names are not simply referred to in songs; they are the frame for the narratives carried by songs. They index the presence and actions of Waŋarr in place and much of the meaning of many songs is carried in the sequencing of names.

It is important to note at this stage that just as knowledge of place names is distributed beyond the boundaries of a particular clan's estate so too are the songs associated with each clan's estate. Songs are distributed in two ways: by the ancestral connections between places and through the kinship relations between people. Ancestral beings travelled from one estate to another and the performance of a song cycle connects one estate to another along the path taken. People from one clan learn and join in with the songs of other clans that are connected by the same ancestral journey, forming a virtual memory bank across a wide region. And people related to a clan – in particular *wäku*, children of women of the clan, and *gutharra*, women's daughter's children, will also learn the songs.

Like places, people have multiple names. Their names reflect both place and kinship. People acquire names at birth from their own clan's estate, connected to the ancestral spirit that initiated their conception. And as they grow older they acquire more names associated with life events, status, and achievements. Some names can be shared by men and women but the majority of names are associated with one or other gender. In addition to names from a person's own clan estate men and women may have names from another clan of the same moiety, and these too may be bestowed at birth. People are frequently gifted at least one name from their *märi* or mother's clan. And in certain circumstances they may be given the name of a closely related 'sister' clan (*yapa*) of the same moiety, in particular if that clan is demographically weak. The inheritance of names is linked closely to the politics of marriage and to the politics of succession.

Yolŋu marriage involves an asymmetric exchange relationship between a minimum of six patrilineal groups. The society is divided into two exogamous moieties called Dhuwa and Yirritja. Women clan members are bestowed in marriage not by members of their own clan but through an arrangement (*milmarra*) involving their mother's clan. In effect a man bestows his daughter as mother-in-law, and his daughter's daughter as wife, to his sister's daughter's son, who belongs to the same moiety as he does. The bestower calls the bestowee *gutharra* ([sister's] daughter's child – a term used both by the grandmother and by her brothers), the reciprocal term being *märi* (mother's mother, mother's mother's brother). These relationships are not just between individuals, but are projected to the level of the clan as a whole. A woman's mother's clan, acting as *märi,* in effect bestow her in marriage to an individual from another clan of the same moiety which they call *gutharra*. The bestowal relationships are not reciprocal, since a woman's clan cannot bestow daughter's daughters in marriage back to their *märi* clan. Those women will be bestowed to a member of a clan categorised as *gutharra*. Hence an alliance or sequence of negotiations is required between at least six clan groups of both moieties before a connubial marriage cycle involving opposite sex siblings can be completed. Frequently more than six groups are involved. While the system seems complex, our evidence shows those *milmarra* relationships between *märi* and *gutharra* clans to be enduring over many generations, and they are understood by Yolŋu as central to the structure or ancestrally determined pattern of their society. People refer to the relationship between *märi* and *gutharra* as the 'backbone' of their society (H. Morphy 1997).[6]

The relationships between groups and the transmission of knowledge about places is reinforced by social practices and associated emotional ties to Country that develop over the life cycle. On marriage a man will spend some time living in his wife's Country learning about its places, and when he returns with his wife to his own clan's Country his wife's younger brother is likely to accompany her (Peterson 1974). As both men and women reach the end of their lives there is a tendency to return to their clan's ancestral Country – their spiritual home (Peterson 1972).

The pattern of marriages is understood to follow ancestral Waŋarr precedents. Names are central to this process of cultural memory. Yolŋu speak as if the marriage of people is the marriage of Waŋarr ancestral beings and Country (H. Morphy & Morphy 2006). In ceremonies the connections between Countries along the matriline are emphasised – rivers flow from mother's mother's brother's clan, to daughter's clan, to daughter's daughter's clan and so on – from one big name place to another. And the bestowal of names follows the direction of the marriages, since a man gets one or more of his names bestowed on him by his *märi* clan. Almost literally the movement of names is a means of accounting for the movement of women in marriage. It is a sign of the boundaries of the connubium from whence a return can be anticipated. And indeed, the movement of names from their place of origin is a signal not of alienation, but rather of a relationship.

The process of bestowal in turn is linked to the politics of succession. The ancestral footprint of Country cannot be changed. Each Country has belonged from the beginning of time to the same moiety. Hence groups cannot succeed to their mothers' Country, since this belongs to the opposite moiety. The primary claim to succession is for *gutharra* to take over their *märi* country and, in effect, for the names they bear to be returned to their Country of origin. The process of succession involves a reallocation of people to Countries that remain the same. The group taking over will sing the songs that were already there, will take on the ritual responsibilities associated with that Country, and will speak the dialect associated with it. So one can see the importance of names to the politics of everyday life yet at the same time begin to apprehend how, in an oral landscape-based culture, they have a role in masking the particularities of historical processes and restoring the appearance of the eternal.

But How Long?

The stability of the names is a key mechanism whereby the system of social organisation of connubial relationships between sets of clans is reproduced over time. The names mark the imprint (*djalkiri*) of Waŋarr in the clan estates and Yolŋu ideology and ontology stress the generative nature of the ancestral beings. The linking of clans in overlapping connubial sets enables the ancestral order to be continually re-imposed even if the exigencies of demography and the effects of clan warfare mean on occasions the actual human descendants have to reorder their relationships to place. Clans do sometimes become too large and others fail to reproduce, but the system allows the reallocation of people to named places (H. Morphy 1997).

So far, we have produced a model of how a system operates at a particular moment in time and the ideological rationale for its maintenance over time, with the past viewed from the present. Can we look forward from the past? We are not going to find a Domesday Book that provides information on the stability of names over a period of 1000 years or more. We do have the evidence of the *-tjpi* names, but to rely on them alone would be to create a somewhat circular argument. However, we do have very good evidence for the continuity of components of the system over quite a long period of time. The ethnographic record for Yolŋu (Murrngin) society is impressive and stretches back 100 years to the beginning of effective European control of their region in the 1920s. The record includes systematic collections of paintings and documentation of ceremonial objects, songs, and ritual performances across the region.[7] The documentation confirms the continuity of names in place over that time.

Over 100 years earlier, beginning in 1801, Matthew Flinders was circumnavigating the continent, mapping and naming Australia as it was being colonised. In 1802 he encountered Yolŋu people along the coast of the Gulf of Carpentaria. His visit was short, but the vocabularies and names recorded by Robert Brown, the ship's surgeon, suggest that the same clan dialects were being spoken in the same places as today. And the name of the Yolŋu person who provided the vocabulary was Wäka. This in turn is the name of one of the leading applicants in the Blue Mud Bay case, 200 years on.

However, my final argument brings me back to the -*tjpi* names. I want to link them to what we are beginning to understand to be a distinctive and enduring regional system, an expanding polity which arguably has emerged over a period of 3000 years. On present evidence the language first appeared in the north of Blue Mud Bay (in the area where the -*tjpi* names are found) and subsequently spread to the north, south, and west. How did the Yolŋu Matha languages spread? To understand the problem it is necessary to backtrack a little to provide a wider regional context. The Yolŋu Matha languages belong to the Pama-Nyuŋan language family, the most recent language family on the continent. It has a time-depth of some 6000 years, and today covers most of the Australian mainland, having displaced the languages previously spoken there.[8] The non-Pama-Nyuŋan language families of more ancient origin, of which there are several, remain only in the Kimberley region of Western Australia and in the Top End of the Northern Territory. The languages that surround Yolŋu Matha landwards to the west and south are all non-Pama-Nyuŋan; Yolŋu Matha is an enclave of Pama-Nyuŋan separated from the rest of its language family by unrelated languages. The Yolŋu system of kinship and marriage is also very different from the more symmetrical Arandic systems that surround it (Keen 2006). There is intermarriage at the boundaries and people have developed ways of adjusting their kinship systems to allow marriage to occur on the border zones. Over the long term it is possible to construct a scenario whereby, around 3000 years ago, speakers of proto-Yolŋu Matha arrived on the coast in the Blue Mud Bay area. Arguably the language then gradually moved inland as the asymmetrical connubial marriage system brought neighbouring groups into the Yolŋu connubial structure, and new connubia developed. Over time, the Yolŋu marriage system has expanded in tandem with the expansion of the language group. The Yolŋu languages on the margins of the region show evidence of long-term contact with neighbouring non-Pama-Nyuŋan languages (Heath 1978), and certainly there is evidence today of the continuing movement of the marriage system into neighbouring non-Pama-Nyuŋan-speaking groups.

Taken overall the evidence does confirm that the place names with -*tjpi* endings are distributed along the Holocene high stand coastline that can be dated to 3000 years BP. The distribution of Yolŋu place names is densest along the coast, the edges of the coastal wetlands and river valleys. No -*tjpi* names occur along the more recent prograded coastline. Malitjpi is situated in a place where the coastline has remained relatively stable since the time of the Holocene high stand (Fig. 21.2). Three thousand years ago proto-Yolŋu Matha speakers were hunting stingray as an important food resource and today their descendants do the same. We cannot say for certain that 3000 years ago Malitjpi was associated with an ancestral stingray, but Yolŋu today say that it is. And the evidence we have makes it at least plausible that it always has been.

A Connubial Polity

Gosden has argued strongly for an articulation between anthropology and archaeology (Gosden 1999), but in stretching back deeper in time he argues for the primacy of material evidence and archaeological method (Gosden *et al.* 2021, 2). There are good reasons for this in the case of the Bronze Age in Britain, and the naïve use of ethnographic parallels in interpreting archaeological data across time and place is acknowledged to be problematic (Gosden 1999, 9). However, boldness and a strong comparative perspective is also something that characterises his research – 'information is part of what is needed to write such long-term histories, but we also need ideas concerning how past societies worked together, formed internal solidarities and differentiated one from another' (Gosden *et al.* 2021, 2).

One of the most difficult problems people have had in interpreting hunter-gatherer societies over the *longue durée* is to simultaneously acknowledge the continuity and relative stability of such societies over time and yet recognise the dynamic features of their social and cultural systems. On the whole archaeological research on hunter-gatherer societies has found it difficult to recognise their dynamism and diversity, while pointing to revolutionary moments in human history – the development of polished stone axes, the use of ochre in mortuary practices, and evidence of successful adaptations to climate change. Certainly, the stereotype of Indigenous Australians as unchanging people in an unchanging land has long been overturned. But arguably we have also failed to recognise the contemporaneous diversity of Australian societies in the archaeological record except at the most general level (Keen 2006; Ulm 2013). Our research on Yolŋu society is an argument both for the recognition of this diversity and an argument that the different systems may have been very durable. Recognising this means that in the Australian context a deep understanding of the recent systems in place across the continent is essential when approaching evidence from archaeological contexts.

The Yolŋu region at the time of European colonisation was about the size of Wales. The Yolŋu system was arguably a polity: not one with a centralised system of governance but one that recognised the same Rom (law) operating across its region. Today Yolŋu recognise the albeit somewhat fuzzy boundaries of the polity in many different ways. They have a concept of *gurrutu* (kinship – relatedness) that differentiates people like themselves from outside groups. They have a set of ceremonies that characterise the region as whole. They have a sense of their dialects being components of a systematically related whole that is linked to an emic theory of dialectal variation. They have a system of permission that enables them to move across the region and find a place in the most distant Yolŋu groups. Their regional polity and sense of conceptual unity is also strongly present in their response to the colonial state. Their ceremonial system and its structural properties has formed the basis of a theory of education and the cross-cultural articulation of systems of knowledge. In fighting for land rights, they have been able to act on a regional basis and gained recognition for their legal system in the Federal courts. They have developed forms of contemporary music that have had a major impact on Australian cultural life and have done so by drawing on performers on a regional basis. In all of these cases the structured naming system and its articulation with the networks of connubial relationships that extend across the region continue to play a central role both as evidence of connections, markers of identity, and agents in political process.

Notes

1. https://england.prm.ox.ac.uk/englishness-PRM-teaching-Part-6.html
2. When we were in Oxford, Frances worked for OUP as an Assistant and later Commissioning Editor. On our return to Australia, she rejoined academia. Much of the research and many of the ideas that I report on here have been jointly produced since then. But while we were in Oxford, Chris and Howard were close colleagues, and Frances was more of an 'affinal' connection. So, our joint decision is that this essay should have Howard as the primary author and Frances should be listed as 'with' rather than as a full co-author. The reader may be assured, however, that Frances has rigorously fact-checked the essay and that any remaining infelicities and errors may be laid at the door of both Morphys!
3. Prior to the establishment of the colony of New South Wales in 1788, with the arrival of the First Fleet, Australia was visited on a number of occasions by British, French, Dutch, Chinese, and Portuguese voyagers as well as by people from neighbouring Islands of Eastern Indonesia (Veth *et al.* 2008), but very few records exist from these times.
4. I capitalise Country when using it in the sense that Yolŋu do when speaking English, to refer to the ancestrally transformed clan estates where humans have the role of *Wäŋa-waṯaŋu* ('Country-holder'). Country encompasses both land and sea.
5. The date of declaration of sovereignty varies across Australia, but Yolŋu country was just included in the original 1788 boundary of the colony of New South Wales.
6. Our evidence so far suggests that there are a number of focal, loosely bounded but enduring connubia across the Yolŋu region that are connected by marriages outside the focal sets.
7. The foundational ethnography is Lloyd Warner's *A Black Civilisation* (1958) followed by Donald Thomson's *Economic Structure and the Ceremonial Exchange Cycle in Arnhem Land* (1949). But the richest data are provided by the archival and material culture collections of Thomson held in the Melbourne Museum and of Ronald and Catherine Berndt in the Berndt Museum in the University of Western Australia.
8. The precise dating and location of the origin of the Pama-Nyuŋan language family remains a matter of debate, as do the reasons for its spread across most of the Australian continent. In a recent paper (Bouckaert *et al.* 2018), combining 'basic vocabulary data…with Bayesian phylogeographhoc methods', it is hypothesised that the language family originated in the Gulf Plains (south of the Gulf of Carpentaria) in the mid-Holocene.

References

Bouckaert, R. R., Bowern, C. & Atkinson, Q. D. (2018) The origin and expansion of Pama-Nyungan languages across Australia. *Natural Ecology & Evolution* 2, 741–9.

Connerton, P. (2009) *How Modernity Forgets*. Cambridge, Cambridge University Press.

Faulkner, P. (2013) *Life on the Margins: An Archaeological Investigation of Late Holocene Economic Variability, Blue Mud Bay, Northern Australia*. Terra Australis 38. Canberra, ANU Press.

Gosden, C. (1999) *Anthropology and Archaeology: A Changing Relationship*. London, Routledge.

Gosden, C., Franconi, T. & ten Harkel, L. (2021) Introduction. In C. Gosden, C. Green & the EngLaId team, *English Landscape and Identities: Investigating Landscape Change 1500BC to 1086*, 1–26. Oxford, Oxford University Press.

Heath, J. (1978) *Linguistic Diffusion in Arnhem Land*. Canberra, Australian Institute of Aboriginal Studies.

Hoskins, W. G. (1977) *The Making of the English Landscape*. New edition. London, Hodder and Stoughton.

Keen, I. (1994) *Knowledge and Secrecy in Aboriginal Religion*. Oxford, Clarendon Press.

Keen, I. (2006). Constraints on the development of enduring inequalities in Late Holocene Australia. *Current Anthropology* 47(1), 7–38.

Morphy, F. (1983) Djapu, a Yolngu dialect. In R. M. W. Dixon & B. J. Blake (eds), *Handbook of Australian Languages*, 1–188. Vol. 3. ANU Press, Canberra.

Morphy, F. & Morphy, H. (2009) The Blue Mud Bay case: Refractions through saltwater country. *Dialogue* 28, 15–25.

Morphy, F., Morphy H., Faulkner P. & Barber, M. (2020) Toponyms from 3000 years ago? Implications for the history and structure of the Yolŋu social formation in north-east Arnhem Land. *Archaeology in Oceania* 55, 153–67.

Morphy, H. (1991) *Ancestral Connections, Art and an Aboriginal System of Knowledge*. Chicago, University of Chicago Press.

Morphy, H. (1993) Cultural adaptation. In G. A. Harrison (ed.), *Human Adaptation*, 98–150. Oxford, Oxford University Press.

Morphy, H. (1997) Death, exchange and the reproduction of Yolngu society. In F. Merlan, J. Morton & A. Rumsey (eds), *Scholar and Sceptic: Australian Aboriginal Studies in Honour of L.R. Hiatt*, 123–50. Canberra, Aboriginal Studies Press.

Morphy, H. & Morphy, F. (2006) Tasting the waters: Discriminating identities in the waters of Blue Mud Bay. *Journal of Material Culture* 11(1–2), 67–85.

Peterson, N. (1972) Totemism yesterday: Sentiment and social organization among Australian Aborigines. *Man* 7, 12–30.

Peterson, N. (1974) The importance of women in determining the composition of residential groups in Aboriginal Australia. In F. Gale (ed.), *Women's Role in Aboriginal society*, 17–27. Canberra, Australian Institute of Aboriginal Studies.

Thomson, D. (1949) *Economic Structure and the Ceremonial Exchange Cycle in Arnhem Land*. Melbourne, Macmillan.

Ulm, S. (2013) 'Complexity' and the Australian continental narrative: Themes in the archaeology of Holocene Australia. *Quaternary International* 285, 182–92.

Veth, P., Sutton, R. & Neale, M. (2008) *Strangers on the Shore: Early Coastal Contacts in Australia*. Canberra, National Museum of Australia Press.

Warner, L. (1958) *A Black Civilization*. Chicago, Harper and Row.

22

Anthropology and Archaeology: A Necessary Unity

Lambros Malafouris

This chapter looks at the current relation between anthropology and archaeology. I revisit some key arguments about the interrelationship of the two subjects, focusing specifically on some emerging concepts and trends in the study of biosocial, temporal, and material processes that present novel challenges both for archaeology and anthropology and call for an extended synthesis. I argue that such an extended synthesis is critical to the future of both disciplines.

Keywords: Materiality, Temporality, Ethnographic Analogy, Becoming

Introduction

This paper looks at the current relation between anthropology and archaeology. I revisit some key arguments about the interrelationship of the two subjects as proposed by Chris Gosden in his *Anthropology and Archaeology* (1999), and explore what has changed in the last 20 years since this book was published. Needless to say, both disciplines, archaeology and anthropology, would have been very different without the other. Their joint heritage, which Gosden (1999, 9) describes as 'a double helix with their histories linked, but distinct', provided fertile ground where 'ideas and evidence from one can feed into the other'. This exchange of ideas and viewpoints between archaeology and anthropology is ongoing, but it has not been without problems. Still, when properly realised (and by 'properly' I mean both critically and symmetrically), it has benefited both disciplines and has the potential to transform their contribution relevant to new and old themes in the comparative and situated study of human becoming.

The major aim of this paper is to call for such an extended synthesis between anthropology and archaeology. I argue that this unity is needed now more than ever. The necessary background for such an extended synthesis is very much in place, as can be seen in current theoretical trends that are replacing traditional polarities between processual and post-processual thinking in archaeology and foregrounding issues of ontology and materiality in anthropology. Questions of creativity and skill, technique and technology, art and aesthetics, memory and personhood, space and landscape, body and gender, or globalisation and locality are only some of the areas that the interaction between anthropology and archaeology has been especially productive and where new insights and critical perspectives have been introduced as a result. Perhaps more than anything else, the developing field of material culture studies (Hicks 2010; Ingold 2012) provided a unifying ground for true dialogue between archaeology and anthropology, especially relevant to their relationship to time, the making of meaning, and the study of 'otherness'. Beyond those traditional areas of exchange, there are also new, pressing issues that can benefit from the extended synthesis. Examples include the Anthropocene and climate change, postgenomic[1] research and the embodied mind, as well as the links between materiality and mental health. These are all issues that cross the boundary between archaeology and anthropology and open the fields up to other disciplines. They also relate to indigenous ontologies and ways of thinking. The impetus for an extended synthesis between anthropology and archaeology is undoubtedly complex and multifaceted. Here, I focus on insights derived from three research areas – temporality, materiality, and human becoming – that, as I describe below, reveal convergent themes that present novel challenges and offer unique opportunities for integrating archaeological and anthropological thinking. The list of topics and questions I choose to discuss in this paper are indicative of this emerging trend and by no means are complete.

I will structure my argument as follows: First, I clarify the meaning of the term anthropological archaeology and explain what I mean by a necessary union between archaeology and anthropology, how it relates to previous conceptions of anthropological archaeology, and what it actually implies in practice. Then, I focus on the issue of analogy and unpack the vision of the proposed extended synthesis relevant to contemporary debates and critiques. In the final part, I focus on the themes of temporality, materiality, and human becoming.

Anthropological Archaeology

Before I proceed to present my argument, I would like to offer some clarifications about the term 'anthropological archaeology', which I will use to denote, for the sake of simplicity, the proposed necessary union and extended synthesis between anthropology and archaeology. Different attempts at unifying archaeological and anthropological perspectives have been variously expressed and practised. Understandably then, the use of any specific label to describe the common ground between anthropology and archaeology is bound to be problematic and, in some cases, misleading. The term anthropological archaeology has taken a variety of meanings over time and space (*e.g.* in North America and Britain) (Gosden 1999). Many of those meanings of anthropological archaeology, especially those preoccupied with the production of middle-range theory (*e.g.* Binford 1962; 1981) and associated with the ways in which analogy could be employed to interpret past phenomena with similar phenomena seen in the present of a given society (see also next section), are often seen as problematic. These problems, especially relevant to the utility and contemporary nature of ethnoarchaeology (see González-Ruibal 2016; Gosselain 2016; Lane 2016), have been widely discussed and debated in recent years. New descriptive labels, such as contemporary archaeology (Buchli & Lucas 2001a; 2001b; Garrow & Yarrow 2010; Graves-Brown *et al.* 2013), postcolonial ethnoarchaeology (González-Ruibal 2016), archaeological ethnography (Meskell 2012), and ethnographic archaeology (Hamilakis 2011),[2] have been proposed by scholars who wish to distance and dissociate themselves from traditional meanings of anthropological archaeology and 'standard' ethnoarchaeological practices.

My proposal draws on these contemporary debates and discussions over the meaning of contemporary anthropological archaeology and its various manifestations. However, it does not map exactly with any of the mentioned trends. Rather, my view of anthropological archaeology stems from my work in cognitive archaeology (Malafouris & Renfrew 2010), material engagement theory (Malafouris 2013; 2019; 2020), and the field of process archaeology (Gosden & Malafouris 2015; Gosden & Pollard 2021; Malafouris *et al.* 2021). In particular, my argument for the necessary unity of archaeology and anthropology is grounded on the continuity of mind and matter and the primacy of becoming over being, which both demand and stimulate new ways of understanding the relationship between humans and the material world (past or present). Once those assumptions are made explicit, I see no reason why the term anthropological archaeology should be abandoned or replaced with a new label. Anthropological archaeology is broad enough to encompass the full range of 'contemporary archaeologies' concerned with issues of materiality, ontology and temporality, the evolving field of experimental archaeology with its emphasis on skill and technique, but also those ethnoarchaeological practices explicitly concerned with the development and refinement of analogical modes of archaeological inference. These ways of practicing anthropological archaeology have different epistemic and theoretical orientations, but they are part of a continuum (González-Ruibal 2016)[3] and thus provide a good basis for an extended synthesis between anthropology and archaeology.

The Senses of 'Unity'

In this section, I want to clarify what I mean by a 'necessary union' between archaeology and anthropology. Important to note first is that when I use the word 'necessary' to predicate their union, I do not mean it in a derivative sense but in a primary originary sense. To explain, the union between archaeology and anthropology is not 'necessary' as the natural product of their joint history and overlapping subject matter. Rather the opposite is the truth: it is the necessary union between archaeology and anthropology that is responsible for their joint history and overlapping subject matter. The 'necessity' I talk about has deeper ontological and epistemological dimensions that relate to the exact ways of producing knowledge (in the sense of discourse) about what it means to be human. The necessary union between archaeology and anthropology expresses the foundational assumption that the study of the human condition can only be situated and dependent, or else impossible, without a proper understanding both of its long term and cross-cultural dimensions. In this connection, the joint heritage between archaeology and anthropology can be a blessing as much as a curse. We have examples where the intersection between archaeology and anthropology has provided the foundation for critical rethinking and unthinking, but also examples where it led to unproductive power dynamics and asymmetries, as when one discipline claims ownership over a specific knowledge domain or aspect of human life.

Consider the example of 'analogy' mentioned before and discussed in the next section. Traditionally the relationship between archaeology and anthropology was primarily conceptualised as an asymmetrical use of ethnographic analogies to aid with the interpretation of past material evidence. Anthropology could provide rich descriptions of living cultures (their categories and practices) that could give meaning to what was generally perceived as a fragmented archaeological record based on lifeless material traces and remnants of the past. We know now that none of that is true. The proposed unity between archaeology and anthropology is largely the product of the gradual realisation of the significance of things (in the broadest sense of material relations, signs, practices, and objects) in human social and cognitive life.

Another key epistemological feature that archaeology and anthropology share is the creative tension between theory and practice. Although these days both anthropology and archaeology are multi-method, multi-modal, and multi-disciplinary, they can still be defined and differentiated on the basis of their primary methodologies. Anthropologists conduct participant observation and archaeologists excavate. Ethnography is to anthropology as excavation is to archaeology. Still, anthropology is not ethnography (Ingold 2014), and archaeology is not excavation. One argument of this paper is that it is in that common space of integration between theory and practice where anthropology and archaeology meet and exchange information and ways of seeing. What brings together archaeology and anthropology is not just their emphasis on the study of human becoming (past and present), but also their commitment, at the methodological level, to 'concerned' participant observation. When I say that they share a commitment to concerned participant observation, I do not mean that in the narrow, conventional, ethnoarchaeological sense revolving around the use or abuse of analogy. Rather, I refer to their shared commitment to the situatedness of the human condition and the value of comparison, often in the face of radical alterity and ontological pluralism. Putting it more simply, a good archaeologist is one who can also be described as an ethnographer of the socio-cognitive dimensions of material processes that she excavates, and a good anthropologist is one who can also be described as an excavator of the material dimensions of socio-cognitive processes revealed in the course of ethnographic research. The archaeologist is immersed in the life of the materials she studies, and the anthropologist is often surveying and temporalising the traces of observed human action.

Proposing an extended synthesis or 'union' between archaeology and anthropology should not be seen in the soft interdisciplinary sense where neighbour disciplines borrow from each other in an attempt to tackle different aspects of the questions that they share (see, for instance, the discussion of time-perspectivism in the following section).

Rather, the proposed vision about the union between archaeology and anthropology denotes the 'strong' critical sense, where ideas and methods are not just exchanged but also transformed during that process (reconceptualised and recontextualised) leading to new questions, observations, and understandings. Archaeology and anthropology offer different but complementary perspectives in an array of important, related themes and questions. This provides the possibility of a genuine dialogue. The aim of this dialogue is not simply to inform each other about new findings and exchange knowledge and insights but above all to use that knowledge to question fundamental assumptions about issues that both disciplines, working in isolation, had more or less taken for granted. Anthropological archaeology is not a synthetic or integrative but primarily an emergentist project, meaning that the union of anthropology and archaeology allows and gives rise to emergent properties and insights that none of the two disciplines can have in isolation. As an emergentist project, it raises some fundamental challenges for conventional ways of doing archaeology and anthropology.

The Use and Abuse of Analogy

A common source of fallacious reasoning in anthropological archaeology, with regard to the relationship between past and present and the study of human–material entanglements, can be found in two opposing assumptions that inevitably constrain any discussion about the prospects and scope of this field. The first general premise, characteristic of that strand of ethnoarchaeology linked to the New Archaeology of the 1960s and 70s (Wylie 1985; 2002; Skibo 1992), state that the past can be understood using systematic analogies with the present (*e.g.* Binford 1962; 1967; 1981). The second premise is that the past can be understood (which here means described or measured) in some unmediated, objective sense that cancels any need for analogical inference. This paper argues that, as generally stated, both contrasting positions are misleading and need radical reconfiguration.

I begin with the first premise: are we justified in thinking that the study of contemporary communities, objects, and practices could act as proxies for the people in the past? Introducing his approach to the special relationship between anthropology and archaeology, Gosden describes ethnoarchaeology as 'immoral'. 'Ethnoarchaeology is immoral', he writes, 'in that we have no justification for using the present of one society simply to interpret the past of another … Societies ought to be studied as interesting in their own right or not at all' (Gosden 1999, 9).

Our instinctive reaction is to agree that any society deserves to be understood in its own terms, rather than in the terms of another. But what does that really mean? What is it to know or describe a society, a process, or a thing in its own terms? Is such an epistemological objective achievable or even possible? It is important that we consider carefully what this presumably uncontested statement means or implies in practice relevant to the value of analogical inference and comparison. Cases like those described by Gosden (1999, 9) where the present is 'seen as a latter-day survival of stage [sic] passed elsewhere in the world, for instance where hunter-gatherer groups from Africa or Australia are used to throw light on the European Palaeolithic' have long been criticised, and any such use of analogy must be clearly rejected. The development of contemporary archaeology and archaeological ethnography (see, for a good review, Hamilakis 2011; Harrison & Breithoff 2017) has shown that there are more productive ways in using the present than as a means to fill in the missing details of the past (that applies to the ethnographic present as well). However, this abuse of analogy, common as it may be, should not lead us to condemn all modes of analogical inference as fallacious or to think that once we get rid of analogical inference the past will be revealed to us in some pure and uncontaminated sense.

For one thing, both archaeological and anthropological thinking (inferential or affective) is thoroughly analogical. As González-Ruibal (2016, 688) points out '[a]nalogy is not exclusive to ethnoarchaeology ... Archaeology, like anthropology and history, is based on analogical reasoning: it builds knowledge through comparison'. For another, human understanding (and archaeologists or anthropologists are no exception) is always mediated by concepts that belong to our present and have their own histories and intellectual baggage. There is no universal language, and all we can say and think is mediated by local concepts that often describe reality from their own unique perspective. Every concept we use is a small theory, a specific way of classifying and making sense of the world. Meanings may vary from one context to another or emerge in unexpected ways and forms. Familiar concepts may be absent or manifest through unfamiliar practices. Familiar practices may refer to unknown concepts. We should not be thinking about those issues only as language-related problems. Bodily practices, material agencies, and skills are also situated and thus dependent. Those dependencies apply to our measures, models, practices, and experimental devices as well. They also construct or allow worldviews that could provide controlled but necessarily partial descriptions and representations (or mis-representations) of unknown realities (past or present). Thus, in a way, all archaeological and anthropological interpretation, even the most descriptive or scientific, is inherently analogical. What differs is that often the source domains of our analogies are hidden in the form of well-entrenched facts of our situation and our learned (primarily Western) ways of knowing and thinking.

What should we do? The solution that this paper proposes is not to abandon analogies but to complicate their use and meaning by developing new ways of engaging with the materiality of the present. No single ethnographic analogy will help you understand why and how something happened in the past but having a good anthropological comparative understanding of a given phenomenon will allow us to make far better sense of a comparable instance from the past. In particular, on the one hand, comparative analogical thinking can protect us from our own biases of how the world works (the more the better), and, on the other, it can allow us possibilities for juxtapositions in the sense of cross-domain mappings and conceptual blends from which new interpretations can emerge. Every potential comparison between past and present includes as much sameness as difference. The key challenge here is how to choose our questions and analytical units, as well as set the boundaries and conditions of comparability.

So, returning to Gosden's critique, my argument will be that a prehistorian studying how hunter-gatherers exchange information about the world would benefit from working with the Kalahari Bushmen or the Australian Aborigines (Gosden 1999, 5) as much as he/she would benefit from studying the nature of information exchange in a contemporary digital data mining setting. In other words, the value of comparison is not based on historical, geographical, and functional continuities/similarities/homologies or proximities but instead on the principles of abductive comparative resonance (this is essentially a form of abductive relational analogy) and structural attunement. These two principles provide the basis for metaphorical mappings and emerging conceptual blends that make use of 'material anchors' (Hutchins 2005; Malafouris 2013) and allow for analogical inferences that could work in bidirectional fashion, changing our understanding both of the source and the target domain (past or present).

There are good reasons to worry that too much reliance on anthropological analogies will limit our ability to enter the 'strangeness' and think about the different nature of the past. But at the same time, one may suggest that the only way to avoid reducing that strangeness to our own familiar ontology, as scientific archaeology often does, is through the use of analogies that could help us conceptualise alternative relations and meanings outside our own familiar experiences. These alternative analogies are not necessarily closer to the past than our own, but they can protect us from the usual illusions of objectivity.

The kind of analogical thinking I propose here is based on the principles of estrangement and metaphor. Given that comparison is unavoidable, as we cannot but make sense of difference in a relational manner – bear in mind that in most general terms, information can be defined as a 'difference that makes a difference' (Bateson 1979) – it seems to me that analogy in itself, as an analytical means for drawing distinctions, was never the real problem. The problem is created by misusing and misinterpreting analogical inference either a) by pretending, as with standard ethnoarchaeology, that analogies are doing the kind of objectivist explanatory work they cannot do, namely, using the present to explain the past, or, b) by thinking, as is often the case in archaeological science and ethnographic description, that because you observe and record without explicitly using a specific analogy you are not implicitly thinking analogically. As for instance, when an archaeologist using the Munsell colour chart to classify dirt or the exact colour of a sherd thinks that her perception is now objectively structured rather than the emergent product of specific analogical practices, instruments, and material signs that constitute her situated field of professional vision (Goodwin 1994). It is those 'looping effects', to extend a term Ian Hacking (1999) uses to describe the effects of classification on those classified and classifying, that define among anything else the epistemological challenge of anthropological archaeology. The study of material and temporal relations that I discuss below offer a good example of that. More important than describing those relations in objective terms (necessary and sufficient conditions) is to change how we see those relations, 'raising consciousness' using analogies and comparisons.

The Temporality of Material Relations

The field of material relations is where archaeology and anthropology meet most closely (Gosden 1999, 152; 2005; 2008). The meaning of the term 'material relations' should not be confused with the modernist one that sees matter as passive and relations as quantities and physical properties. Rather, material relations here retain their primary sentient and social character. They denote the material dimension of human social and cognitive life, which means they are also thoroughly temporal, in the Bergsonian durational sense (Bergson 1922), blurring any real distinction between past and present. As Gosden (1994, 17) points out in his *Social Being and Time*, '[a]ll action is timed action', it 'uses the imprint of the past to create an anticipation of the future'. Time is also multi-scalar, it 'flows on a number of levels' and operates at different speeds.

Do not think only of linear sequential time that archaeologists use to distinguish what came 'before' and 'after' through the refinement of past chronologies. Time relates to calendars and clocks, as much as it relates to the ways people think about their past, experience their present, and imagine their future. The most powerful links with material culture are not with objective time-measurements and the various material devices and techniques that humans invented to represent it but with situated, intersubjective time-consciousness. This intimate experiential link between time and material culture can be easily extended to incorporate various aspects of time-consciousness. For instance, conceptualisations of selfhood and social memory.

The importance of time and materiality and the need to think deeper about their linkages is perhaps more evident than ever before in archaeology and anthropology. Time and material culture are probably the two themes that both unite and differentiate, more than anything else, current archaeological and anthropological thinking. Archaeology has traditionally been seen as the discipline responsible for making sense of the human past[4] by improving dating methods and techniques (both *relative* and *absolute*), for instance, through the development of Bayesian modelling of radiocarbon dates (Bayliss 2009; Whittle *et al.* 2011; Whittle 2018). However, as mentioned, time relates to chronology as much as it relates to human consciousness, memory, and action. Time is objectified and measured but also experienced and remembered.

The extended synthesis of archaeology and anthropology takes us beyond the mere study of dating techniques and chronologies and allows us to approach the issue of time as a phenomenon that is multi-dimensional and operates at different scales contributing to our understanding of the significance and meaning of time in human becoming. This is not to argue that anthropological archaeology should abandon the use of dates or the traditional view of time as chronological sequence, rather the claim is that chronologies, and the linear conception of time that they embody, can only be a partial view of time (Lucas 2015, 5). Time can no longer be seen simply as continuous, linear, and uniform. Instead, a new, non-linear dynamical conception of time is needed, one that can accommodate the temporal variability of the archaeological record and be more attentive to the constant tension between continuity and change.

There are two main challenges for the anthropological archaeology of time. The first relates to how we gain access to the temporal experience of people in the past and in different cultures. The way we go about to meet this challenge depends on whether people's fundamental understanding of time is something we learn or is an innate universal capacity. One important distinction that is worth emphasising in this context is the one between abstract or representational and experiential time (*e.g.* Shanks & Tilley 1987). The former relates to the ways time is conceptualised and measured (for instance using calendars and clocks). The latter is the time of human experience, that is, time embodied in the rhythms and tempos of human practices. One important question here is about what connects the varieties of time-consciousness operating over different scales, durations, and rhythms. If, as we said, time flows on a number of different levels, how does the unity of human consciousness emerge? This question often leads to a conflation between real time (as experienced in human life) and chronological time (as represented and measured) (*cf.* the A series/B series of Gell 1992). While it makes good sense to think, as with many other anthropological notions, that there must be some basic sense of time that all societies and people share, it is unhelpful and rather misleading to assume that this basic sense can be strictly separated from the culture-specific representations of time, and that some of those representations or concepts of time are more objective or real than others, especially when they derive from Western scientific measures of reality.

The second challenge for anthropological archaeology is how to reconcile long-term archaeological phenomena with the anthropological emphasis on small-scale events of individual action, perception, and interaction. Does this difference in chronological resolution also imply an incommensurability of temporal perspectives? One methodological implication of time perspectivism (Bailey 2007, 200) seems to be that anthropological questions and theories of time derived from ethnographic timescales may not be relevant to, or suitable for, understanding and interpreting the 'deep time' of the archaeological record. However, without denying the importance of timescales, or their differences, it would be misleading to suggest that the slow timescales of long-term archaeological change and the fast, ethnographic timescale of bodily and socio-material practices do not meet or interact. Archaeological and anthropological timescales may differ, but they essentially describe different aspects of the same phenomenon, namely, human becoming. The use of the term 'becoming' signifies an important shift in perspective from entities, types, and objects into processes, events, and actions (Gosden & Malafouris 2015; Malafouris et al. 2021). Becoming is always becoming *with* and *through* (Malafouris 2016; 2019). It speaks of the human condition as world-involving and relational. Becoming signifies a process ontology able to accommodate the different modes of being and forms of human participation.

This description of human becoming is not exactly the one we see depicted in the familiar picture of 'descent with modification' by means of natural selection. For one thing, the role of humans in Darwinian selection is passive, adaptation is asymmetrical; humans, like any other organism, adapt to their environment, never the other way around. In stark contrast, human becoming brings forth the process of change by actively altering the 'landscape of affordances' (Rietveld & Kiverstein 2014). Humans undergo situated ontogenetic histories leading creative, self-aware, and social lives. Notice also the use of present tense: evolution is not of the past but ongoing. As I have pointed out elsewhere (Malafouris 2016; 2021), this vision of humanity as 'complete' and 'fixed', and the concomitant neo-Darwinian assumptions about the origins of human 'modernity' (anatomical, behavioural, or cognitive) that it entails can be criticised. The ontogenetic change at the developmental timescale of life history, and phylogenetic and historical change over the long term, intersect in the context of situated material practices. History and culture are not separate from evolution. This basic point has been forcefully advanced in the work of Tim Ingold in the past three decades:

[I]t is pointless to seek the moment when history 'began', or to attribute certain embodied capacities – such as speech or bipedal locomotion – to evolution while relegating others – as the ability to read and write, or to ride a bicycle – to history. For human history is but the continuation of the evolutionary process by another name (Ingold 1994, 9).

As the anthropologist Nancy Munn observes relevant to the issue of time in the context of her ethnography of the Kula ceremonial exchange system: the nature of time embodies the paradox of being both 'something we live in' and something we are constantly constructing. We are 'in some sense always "in" time ... and yet we make, through our acts, the time we are in' (Munn 1992, 94). This raises a challenging question both for archaeology and anthropology: if human practices shape time, and if we make time as much as we live in time, how can we ever truly understand and compare different time-experiences? Taken-for-granted assumptions about the value of linear chronology (based on sequence and succession) that form the basis for many grand archaeological narratives of change, such as those associated with human origins or the emergence of agriculture, are now challenged (Lucas 2005; 2021). It is increasingly recognised that the meaning of temporality in human life and evolution has more than a single dimension. Human becoming cannot be expressed accurately as a unidirectional sequence of events following this linear order of time. Indeed, one possible way to meet those challenges that has recently gained momentum is to look at the human engagement with the material world (in the broadest sense of artefacts, landscapes, skills, and material practices).

The realm of material engagement provides a point of intersection for three interconnected time-scales: evolution, development, and situated action. When humans engage the material world by using a simple artefact, they establish a bridge with the larger-scale processes at work beyond their awareness or control that are embodied in the objects at hand. What the human brain and body can experience on a scale of seconds or minutes can now be extended through the physical persistence of material artefacts to time experiences on the scale of years, decades, centuries, or even millennia. Our experience of things highlights the 'pastness' of the present (see also Olivier 2004; Ingold 2010). By engaging an object in the present, we simultaneously share something of the object's past. With things, the past becomes present. Thus, the temporal structures they embody influence and partially constitute the temporal experience of our shared present. This central idea that objects are capable of accumulating time through their use and associations has been expressed in archaeology through the notions of 'social life', 'life-history', and 'biography' (Appadurai 1986; Kopytoff 1986; Hoskins 1998; 2006; Gosden & Marshall 1999; Joy 2009). Objects and landscapes alike are making time and are made of time (Ingold 1993; Gosden 1994; Harrison & Schofield 2010; Malafouris & Gosden 2020). At any one time, at any given moment, multiple temporalities are present (Lucas 2021). Time is multi-layered; change happens at different scales; different aspects of material culture persist over varied periods of time. In short, the human engagement with the material world provides temporal anchoring and binding that helps us to move and think across the scales of time, thus, escaping the temporal limits and rhythms of individual human life and experience (Malafouris 2013, 246–7).

Conclusions

It is well established by now, in archaeology and anthropology, that proper understanding of social relations is impossible without some attempt to ground it in the material conditions of people's lives. This was not always the case. Traditionally, anthropologists have been very good at dealing with social relations and archaeologists with material relations. For the biggest part of their histories, anthropologists felt that they could study social relations without the need to understand material relations, and conversely archaeologists felt they could study material relations without necessarily having to understand social relations.

The so-called material culture-turn of the last three decades (Hicks 2010), with its new emphasis on socio-material practices and the human (also non-human) modes of material engagement, have played a major role in reconfiguring the importance of the unity between anthropology and archaeology for understanding the material conditions of human social and cognitive life. In this paper, I have argued that this necessary unity is needed now more than ever, and I propose that the study of the temporality of the biosocial and material processes are key areas that present new challenges where anthropology and archaeology can learn from one another.

Archaeology can help anthropology to establish new connections with postgenomic research and contemporary, embodied cognitive sciences, as well as to embrace more fully new developments in the study of material culture, the nature of change, and the meaning of temporality. Archaeology, on the other hand, needs help from anthropology, in order to re-establish its critical consciousness through more intimate contact with topics related to personhood, creativity, skill, art, craft, aesthetics, and the nature of socio-material relations. These are topics to which anthropological archaeology may have a lot to contribute. They are also topics that, regrettably, are increasingly pushed outside the main purview of archaeological science, due to the dominance of 'impactful scientism' these days in archaeology. Re-approaching anthropology and archaeology as 'forming a necessary unity' will help archaeological science to escape 'cognitive capitalism' that condemns matter to an inert substance and to reconnect with its main objective, which is: to rethink the concept of matter in order to account for its sentient and vital capacities and use it to understand the meaning of difference and continuity in human becoming (Gosden & Malafouris 2015; Gosden & Pollard 2021; Malafouris et al. 2021).

Archaeological emphasis on temporality and materiality could help to intensify the descriptive powers of anthropology. Anthropological emphasis on 'otherness' and ontological pluralism could help to intensify the comparative powers of archaeology. The archaeological grounding in materiality could protect anthropology from the threat of radical ontological irreducibility. The anthropological grounding in alterity could protect archaeology from representationalism and epistemological universalism. Anthropological archaeology is also characterised by its unique ability to operate across the scales of time. Tim Ingold (2018) has described anthropology as philosophy with people. We may extend his logic to describe anthropological archaeology as philosophy with people and things.

Another important recognition that unites archaeology and anthropology relates to what is actually present or absent, accessible or inaccessible, complete or fragmentary in the archaeological and anthropological record. Archaeologists came to see that there is more in matter than they initially thought when they talked about what can be studied with the material remains of the past. Anthropologists have also realised that language is not everything and that they have been missing more than they traditionally recognised when they thought of artefacts as secondary, neglecting the value and complexity of material relations. This extended synthesis has proven mutually beneficial and is critical to the future of both disciplines and the intellectual contribution and impact they can have in the broader academic scene, as well as in response to a host of contemporary issues.

Notes

1. The term postgenomic here refers to the epigenetic challenge of positivist genetics. In particular, the discovery of mechanisms *beyond the genome* that challenge traditional views of biology as something 'innate' and 'fixed', emphasising (through mechanisms like methylation) the entanglement of biological and socio-material processes (Keller 2000; Dupré 2012; Ingold & Palsson 2013; Lock 2015).
2. Defined by Hamilakis (2011, 405) as 'a transdisciplinary and transcultural space, a locality and a ground that allows for multiple meetings, conversations, and interventions to take place … Materiality and temporality are the two defining features of this novel space; archaeological ethnography is constituted at the intersection of these two features'.
3. According to González-Ruibal (2016, 689–90), this continuum 'is determined by two variables: context and analogy. If one places the stress on analogy, then we are speaking of ethnoarchaeology, if on the contrary, context is stressed, then it is archaeology that we have. Or to say it in another way, it is possible to put the local context at the service of archaeology or archaeology at the service of the local context'.
4. Archaeology can be seen as a form of *chronesthesis,* that is, a form of historical consciousness of the past in the present (Malafouris 2013).

References

Appadurai, A. (ed.) (1986) *The Social Life of Things: Commodities in Cultural Perspective*. Cambridge, Cambridge University Press.

Bailey, G. (2007). Time perspectives, palimpsests and the archaeology of time. *Journal of Anthropological Archaeology* 26(2), 198–223.

Bateson, G. (1979). *Mind and Nature*. York New, Dutton.

Bayliss, A. (2009) Rolling out revolution: Using radiocarbon dating in archaeology. *Radiocarbon* 51, 123–47. doi:10.1017/S0033822200033750.

Bergson, H. (1922) *Duration and Simultaneity*. Translated by L. Jacobson (1965). Indianapolis IN, Bobbs-Merrill.

Binford, L. R. (1962) Archaeology as anthropology. *American Antiquity* 28(2), 217–25.

Binford, L. R. (1967) Smudge pits and hide smoking: The use of analogy in archaeological reasoning. *American Antiquity* 32(1), 1–12.

Binford, L. R. (1981) Behavioral archaeology and the "Pompeii Premise". *Journal of Anthropological Research* 37, 195–208.

Buchli, V. & Lucas, G. (eds) (2001a) *Archaeologies of the Contemporary Past*. London/New York, Routledge.

Buchli, V. & Lucas, G. (2001b) The absent present. Archaeologies of the contemporary past. In Buchli & Lucas (2001a), 3–18.

Dupré, J. (2012) *Processes of Life. Essays in the Philosophy of Biology*. Oxford, Oxford University Press.

Garrow, D. & Yarrow, T. (eds) (2010) *Archaeology and Anthropology: Understanding Similarity, Exploring Difference*. Oxford, Oxbow Books.

Gell, A. (1992) *The Anthropology of Time: Cultural Constructions of Temporal Maps and Images*. Oxford, Berg.

González-Ruibal, A. (2016) Ethnoarchaeology or simply archaeology? *World Archaeology* 48(5), 687–92.

Goodwin, C. (1994) Professional vision. *American Anthropologist* 96(3), 606–33.

Gosden, C. (1994) *Social Being and Time*. Oxford, Blackwell.

Gosden, C. (1999). *Anthropology and Archaeology: A Changing Relationship*. London, Routledge.

Gosden, C. (2005) What do objects want? *Journal of Archaeological Method and Theory* 12(3), 193–211. doi:10.1007/s10816-005-6928-x.

Gosden, C. (2008) Social ontologies. *Philosophical Transactions of the Royal Society B: Biological Sciences* 363(1499), 2003–10. doi:10.1098/rstb.2008.0013.

Gosden, C. & Malafouris, L. (2015) Process archaeology (p-arch). *World Archaeology* 47(5), 1–17. doi:10.1080/00438243.2015.1078741.

Gosden, C. & Marshall, Y. (1999) The cultural biography of objects. *World Archaeology* 31(2), 169–78.

Gosden, C. & Pollard, M. (2021) Is the universe sentient? What implications might this have for archaeology? In M. Boyd & R. Doonan (eds), *Far from Equilibrium: An Archaeology of Energy, Life and Humanity: A Response to the Archaeology of John C. Barrett*, 313–26. Oxford, Oxbow Books.

Gosselain, O. P. (2016) To hell with ethnoarchaeology! *Archaeological Dialogues* 23, 215–28. doi:10.1017/S1380203816000234.

Graves-Brown, P., Harrison, R. & Piccini, A. (eds) (2013) *The Oxford Handbook of the Archaeology of the Contemporary World*. Oxford, Oxford University Press.

Hacking, I. (1999) *The Social Construction of What?* Cambridge MA, Harvard University Press.

Hamilakis, Y. (2011) Archaeological ethnography: A multitemporal meeting ground for archaeology and anthropology. *Annual Review of Anthropology* 40(1), 399–414.

Harrison, R. & Breithoff, E. (2017) Archaeologies of the contemporary world. *Annual Review of Anthropology* 46, 203–21.

Harrison., R. & Schofield, J. (2010) *After Modernity: Archaeological Approaches to the Contemporary Past*. Oxford and New York, Oxford University Press.

Hicks, D. (2010) The material-cultural turn: Event and effect. In D. Hicks & M. C. Beaudry (eds), *The Oxford Handbook of Material Culture Studies*, 25–98. Oxford and New York, Oxford University Press.

Hoskins, J. (1998) *Biographical Objects: How Things Tell the Stories of People's Lives*. London, Routledge.

Hoskins, J. (2006) Agency, biography and objects. In C. Tilley, W. Keane, S. Küchler, M. Rowlands & P. Spyer, P (eds), *Handbook of Material Culture*, 74–85. Sage Publishing.

Hutchins, E. (2005). Material anchors for conceptual blends. *Journal of Pragmatics* 37(10), 1555–77.

Ingold, T. (1993) The temporality of the landscape. *World Archaeology* 25(2), 152–74. doi:10.1080/00438243.1993.9980235.

Ingold, T. (1994) Introduction to humanity. In T. Ingold (ed.), *Companion Encyclopedia of Anthropology: Humanity, Culture and Social Life*, 3–13. London, Routledge.

Ingold, T. (2010) No more ancient; no more human: The future past of archaeology and anthropology. In Garrow & Yarrow 2010, 160–70.

Ingold, T. (2012) Toward an ecology of materials. *Annual Review of Anthropology* 41, 427–42. doi:10.1146/annurev-anthro-081309-145920.

Ingold, T. (2014) That's enough about ethnography! *HAU: Journal of Ethnographic Theory* 4(1), 383–95.

Ingold, T. (2018) *Anthropology: Why It Matters*. New York, John Wiley & Sons.

Ingold, T. & Palsson, G. (eds) (2013) *Biosocial Becomings: Integrating Social and Biological Anthropology*. Cambridge, Cambridge University Press.

Joy, J. (2009) Reinvigorating object biography: Reproducing the drama of object lives. *World Archaeology* 41(4), 540–56.

Keller, E. F. (2000) *The Century of the Gene*. Cambridge MA, Harvard University Press.

Kopytoff, I. (1986) The cultural biography of things: Commoditization as process. In Appadurai 1986, 64–92.

Lane, P. (2016) Editorial. *World Archaeology* 48(5), 605–8.

Lock, M. (2015) Comprehending the body in the era of the epigenome. *Current Anthropology* 56(2), 151–77.

Lucas, G. (2005) *The Archaeology of Time*. London, Routledge.

Lucas, G. (2015) Archaeology and contemporaneity. *Archaeological Dialogues* 22(1), 1–15.

Lucas, G. (2021) *Making Time: The Archaeology of Time Revisited*. London, Routledge.

Malafouris, L. (2013) *How Things Shape the Mind: A Theory of Material Engagement*. Cambridge MA, MIT Press.

Malafouris, L. (2016) On human becoming and incompleteness: A material engagement approach to the study of embodiment in evolution and culture. In G. Etzelmüller & C. Tewes (eds), *Embodiment in Evolution and Culture*, 289–306. Tübingen, Mohr Siebeck.

Malafouris, L. (2019) Mind and material engagement. *Phenomenology and the Cognitive Sciences* 18(1), 1–17. doi:10.1007/s11097-018-9606-7.735.

Malafouris, L. (2020) Thinking as 'thinging': Psychology with things. *Current Directions in Psychological Science* 29(1), 3–8. doi:10.1177/0963721419873349.

Malafouris, L. (2021) Making hands and tools: Steps to a process archaeology of mind. *World Archaeology* 53(1), 38–55.

Malafouris, L. & Gosden, C. (2020) Mind, time, and material engagement. In I. Gaskell & S. A. Carter (eds), *The Oxford Handbook of History and Material Culture*, 105–20. Oxford, Oxford University Press.

Malafouris, L. & Renfrew, C. (eds) (2010) *The Cognitive Life of Things: Recasting the Boundaries of the Mind*. Cambridge, McDonald Institute for Archaeological Research.

Malafouris, L., Gosden, C. & Bogaard, A. (2021) Process archaeology. *World Archaeology* 53(1), 1–14.

Meskell, L. 2012. Archaeological ethnography: Materiality, heritage and hybrid methodologies. In D. Shankland (ed.), *Archaeology and Anthropology: Past, Present and Future*, 133–44. London, Berg.

Munn, N. D. (1992) The cultural anthropology of time: A critical essay. *Annual Review of Anthropology* 21, 93–123.

Olivier, L. (2004) The past of the present. Archaeological Memory and Time. *Archaeological Dialogues* 10, 204–13.

Rietveld, E. & Kiverstein, J. (2014) A rich landscape of affordances. *Ecological Psychology* 26(4), 325–52.

Shanks, M. & Tilley, C. (1987) *Social Theory and Archaeology*. Cambridge, Polity.

Skibo, J. M. (1992) Ethnoarchaeology and experimental archaeology defined. In J. M. Skibo (ed.), *Pottery Function*, 9–30. Boston MA, Springer.

Whittle, A. (2018) *The Times of Their Lives: Hunting History in the Archaeology of Neolithic Europe*. Oxford, Oxbow Books.

Whittle, A., Healy, F. & Bayliss, A. (2011) *Gathering Time: Dating the Early Neolithic Enclosures of Southern Britain and Ireland*. Oxford, Oxbow Books.

Wylie, A. (1985) The reaction against analogy. *Advances in Archaeological Method and Theory* 8, 63–111.

Wylie, A. (2002) *Thinking from Things: Essays in the Philosophy of Archaeology*. Berkeley, University of California Press.

23

On Ontological Impurity: Conceptualising Time in Archaeology

John Robb

This article explores the nature of ontologies, taking the concept of time as an example. Anthropologists have tended to think of time in terms of 'our' time, a linear, capitalist form, and 'other' times, often cyclical in nature. This draws upon a tradition of ontological analysis which contrasts our world and the worlds of alterity inhabited by the people we study. But, as this paper argues, a logically consistent, pure ontology of time is an illusion created by the method of analysis, which involves ontological purification. Instead, people in all groups experience time by drawing upon multiple ways of understanding time in ongoing bricolage. Ontologies thus are not so much like crystalline, logical worlds of alterity, or even different modes of operation like car gears; they are more flexible palettes of experiential possibilities.

Keywords: Ontology, Alterity, Time, Multi-Modality, Purification, Bricolage

Introduction: What are Ontologies Like?

The recent 'ontological turn' in social theory is immensely stimulating. It makes us think seriously about alterity, and it shows the ontological consistency underlying disparate moments of social life. But, like any concept, it comes with baggage. Anthropology uses a concept of ontology imported from philosophy, which sees ontologies as logically consistent, self-contained, and totalising. Moreover, it employs this concept in a surprisingly traditional anthropological mission, reducing the diversity of human experience to a small number of internally homogeneous, mutually exclusive worlds – often reducing to 'Us' and 'The Other'. These are useful heuristic tactics, but, taken too literally, they can result in ontological purification. The result prevents us from understanding some of the most important features of ontologies themselves.

The goal of the paper is to provoke discussion more than to resolve it. In it, I explore these issues through the example of how archaeologists have thought about time – building on Gosden's (1994) pioneering work. Discussions of time in anthropology and archaeology have tended to view time through two polarised oppositions: 'linear' vs. 'cyclical' time and 'chronometric' vs. 'experiential' time; these are sometimes aligned with 'our' time and 'indigenous' time. But such simplified schemes are achieved through a kind of ontological purification that distorts our understanding of how people interact with ontologies.

Ontologies in Philosophy and Anthropology

The concept of ontology originated in analytical philosophy as 'metaphysics', part of a project to define the basic parameters of existence working from first principles. A recent addition, supplementing 'metaphysics' only from 1930 onwards (van Inwagen & Sullivan 2015), ontology was inherently a totalising concept: the goal was to understand reality as a whole. An ontology was expected *a priori* to be internally homogeneous, governed by stable logical rules, and exemplified in particular, privileged examples. In anthropology, the concept of ontology has been applied to ethnographic theorisations of non-Western ways of understanding the world, particularly in studies of animistic world-experiences (Viveiros de Castro 1998; Holbraad 2007; Willerslev 2007; Descola 2013). Such analyses both describe non-Western understandings, and also assert its irreducible difference and reality. Arguments such as Viveiros de Castro's 'multinaturalism' show us worlds that defy what seem to us the axiomatic rules of life (Hallowell 1960; Alberti & Marshall 2009; Holbraad 2009; Zedeño 2009). Hence the oft-repeated assertion that such difference is ontological rather than epistemological (Alberti *et al.* 2011).

The 'ontological turn' has expanded our anthropological imagination, but it smuggles in a particular intellectual construction. Methodologically, to demonstrate ontological coherence, it follows the classic anthropological tactic of assembling diverse moments of ethnographic observation to evoke a patterned world. In this sense, it presumes a wholeness in advance of finding it, justifying this retroactively by the triumphant unveiling of the intellectual edifice it allows. Moreover, although virtually every ontological analysis begins with a ritual disclaimer that ontologies are multiple rather than totalising, this tends to vanish during the actual analysis. Instead, what emerges is surprisingly traditional heuristic ideal types, sometimes mapped on to an implicit colonialist concept of an unbridgeable gulf between theorist and subject: the Other vs. 'Us'. It thus anticipates a logically consistent wholeness and evokes a crystalline logic from the muddy chaos of life.

Ontologies: Purity and Promiscuity

Ontology is an important concept, but no concept fits all intellectual uses. We leave aside here the realist question of engagement with a world which is fundamentally material and not infinitely malleable. The question at hand is what the consequences of implementing this vision of ontology are. Human experience includes wayward impulses and divergent models of reality. It includes moments of indeterminacy whose significance may be impossible to separate from the act of construal itself. It includes both free speculation and quite channelled direct practical action, vagueness as well as definition, obscurity as well as illumination, contradiction as well as coherence. How does representing this experiential world in terms of ontology silently transform it?

Trying to isolate a pure logic of ontologies renders them more crystalline, hard-edged, and exclusive of other logics than beliefs and actions may ordinarily be. (This may particularly be the case for societies that do not have professional reality specialists such as priests, scientists, scholars, and political authorities, who often partially accomplish the act of ontological purification in advance of the anthropologist.) In extracting an underlying system and logic, we overstate systematicity and ignore encumbering qualifications and frame-shifting contexts.

To take an example, we readily (often unconsciously) make the jump from:

> 'the Ojibwa sometimes attribute spirits to stones and landscape features' (Hallowell 1960)

to

> 'the Ojibwa inhabit an animated world'

and onwards to

> 'the Ojibwa are animists'.

We rarely ask about how the Ojibwa world might at the same time be animated and inanimate, or whether sometimes it is actually unclear which is the case and the Ojibwa (unlike anthropologists) are sensible enough not to actually ask the question (Harris & Robb 2012). We also assume that 'the Ojibwa' exist univocally. As Willerslev's (2013) study of the Yukaghir hunters of Siberia shows, rather than simply 'believing in' a spirit world, people may be situational, practical animists with complex, subtle relations to spirits which may include ironic distance, humour, and ritual adherence.

In other words, people certainly have ontological principles for constituting reality, but these may not automatically translate into 'their ontology'. The methodology of ontological critique is to privilege selective moments of experience as the key to logical paradigms of ontological difference. But human life may have multiple, partial patterns, often latent but in tension. Rather than purity, we live in promiscuous regimes of partial order, and ambiguity and slippage between ontological modes is not incidental but fundamental. If we acknowledge this fact cursorily and then go on to present one order as a self-enclosed totality, we commit an act of ontological purification. 'Purification' here, much as in Latour's (1993) use of the term to describe the 17th-century construction of a nature/culture split, involves isolating selected elements and disregarding their contexts and relationships to create a representation of wholeness according to a global cosmological scheme. All societies have multiple ways of understanding and experiencing the world, among which they alternate situationally (Veyne 1982; Tarlow 2011; Robb & Harris 2013). These 'multimodalities of belief' allow us to switch contextually, often seamlessly, between modes of operating with different reality criteria: between seeing the world as fundamentally rational or irrational, flatly self-evident or pregnant with mystical significance, evaluable empirically or veiled in transformation and deception. Multimodality allows people to operate in inconsistent realities. Indeed, considering human cognition in evolutionary terms, a rigidly consistent unimodality provides us with poor tools. The world is complex and to negotiate different situations successfully, we need to be able to understand it in contradictory ways – as mechanistic and also spiritual, as rule-bound and also full of exceptions, as linear and also cyclical.

Like its intellectual parent structuralism, ontological purification may be a useful servant but a bad master. It lets us find system, but only after assuming systematicity. Ontological purification is not a neutral act of clarifying an observational muddle through some inevitable selection and omission. Close-up, it misrepresents the native's experience and obscures the process of bricolage through which people think and act; variations in systemness create space for dealing with contingencies and communicating across cultures. On a larger scale (Robb & Harris 2013), it is the existence not of one ontology but of multiple partial ontologies interacting with each other which give shape and dynamism to historical process.

'Linear Time' and 'Cyclical Time' as Ontological Models

The anthropology of time is a cumulative masterpiece of ontological dichotomisation. Since the 1930s, anthropologists have seen time and memory as embedded in social structures (Evans-Pritchard 1940), ritual process (Turner 1969), daily routines (Hagerstrand 1977), habitus (Bourdieu 1977), experience (Gosden 1994), and spatialised time and taskscapes (Ingold 2000). As Gell (1996) points out, from Durkheim through Lévi-Strauss, Leach, Geertz, Bourdieu, and other writers, it has been assumed that 'traditional' societies live in a static round of cyclical time. (This image of timelessness may result both from focusing selectively upon 'traditional' activities, such as recurrent rites, and from excising the societies studied from their colonial histories and placing them within the bubble of the 'ethnographic present'.) Throughout the 20th century, an important strand of anthropological thought contrasted people who lived in experience-based, static, or cyclical time and people who lived in linear, moving, or pragmatic time (Munn 1992; Gell 1996). More recent work has both consolidated a notion of quantifiable, linear 'capitalist time' and revealed the complex relations such an abstract model has with actual practices (Bear 2014).

Transferred into archaeology, this grounding authorised Shanks and Tilley's (1987) claim that modern, Western people think in terms of abstract, linear, chronometric time, while indigenous people live in terms of experiential, often cyclical time. Following this, archaeological theorisations of time have been cleanly split into two non-interacting branches. The default common-sense model embedded in our working practices has always been that of linear time, value-free, uniform, infinitely subdivisible into equal units, and objectively measurable, the focus of scientific and historic dating. This is built into the fabric of the discipline; not even the most vociferous post-processualist critics of 'chronometric time' ever stopped using radiocarbon dates. Explicit theoretical discussion in this line has focused around Braudelian themes of multi-scalar history (Bintliff 1991; Iannone 2002; Bailey 1983; 2007). In contrast, Shanks and Tilley (1987) argued that we need to understand time as culturally constructed and experienced. These claims were followed by more in-depth considerations in phenomenology and in other directions (Gosden 1994; Thomas 1996; Lucas 2005). Parallel research upon culturally constructed or experiential time focuses upon the 'past in the past' (Bradley 2002) and the 'archaeology of memory' (Mills & Walker 2008; Borić 2010).

Thus, while we modern Westerners exist in the 'natural', linear, chronometric time of science, the people we study exist in 'cultural', experiential, socially embedded time, often cyclical. Intentionally or not, we have transformed the nature:culture underlying the science/humanities split into ontological models and mapped them onto a colonialist chronotypical geography. This mapping to separate 'our' time and 'other' time into separate ontological worlds is accomplished only through an act of ontological purification carried out both upon our subjects and ourselves. How successful is this purification? Not very: it happens mostly through an act of faith.

To take one example, the ancient Maya provide a classic example of 'cyclical time' (Milbreth 2000). The Maya conceptualised time in terms of grand recurrent cycles of worldly existence and recreation. Multiple smaller calendrical cycles were nested within these, defining repeating moments which were characterised by predestined events and characters. Humans made use of this cyclicity, for instance aligning important events with propitious dates to ride the waves of calendrical fortune (Marcus 1993). Yet in many ways, Maya understandings of time were also clearly linear. Genealogies of historical development recorded unique events in sequence. Moreover, in fields as diverse as farming, urban design, political strategy, and the technological processing of materials, the Maya engaged in goal-directed practical action, which presumed complex, ordered linear sequences of cause and effect. In this sense, defying a simple dichotomisation, they understood time as both cyclical and linear, depending upon what scale we are considering and whether it was grand narrative and cosmological framework, practical action, or some blending of the two.

What about we people who supposedly believe in 'linear time'? 'Chronometric time' has three basic features:

1. Events unfold within a unique, linear sequence extending from the beginning of time to its end.
2. Time can be objectively measured as a natural phenomenon which simply happens.
3. The units of time, like other units of measurement, are of a standard and uniform duration.

These ideas are enshrined in scientific practice, in capitalist economics (for instance in the time-money equivalencies by which labour is bought and sold) and in common-sense technologies for coordinating people such as timetables. Hence its characterisation as 'capitalist time' (Bear 2014).

Yet this model has little to do with how we actually experience the passage of time psychologically. Experiential time diverges radically from chronometric time; compare five minutes of undergoing dental work with five minutes of having an engaging conversation. Moreover, experiential time may be distinctly non-linear, with Proustian loops and zig-zags triggered by things, places, tastes, smells, music, memories, and gestures. Whether or not we *think* that time is linear, regular, and chronometric, we certainly experience and remember it as (in the words of Doctor Who) 'a big ball of wibbly wobbly timey wimey stuff'.

Moreover, our own time understandings and practices integrate cyclicities of many kinds. These range from formally defined recurrences, including astronomically-based ones – days, seasons, years – as formally chronometric as any linear time. Cyclicities such as years may be composed of linear units such as minutes and hours and built into bigger linear units such as decades and centuries, making a formal distinction between linear and cyclical structures of time rather a moot point. We also have cycles of ritual time. Ritual time uses a heightened sensorium and ostentatiously repeated experience to create alternative modes of time (or timelessness) bracketed off from daily life (Turner 1969; van Gennep 1977; Bradley 1991). As part of its power comes from the evocation of exceptional precedents, both sensorily and liturgically, ritual time is often more explicitly cyclical than everyday time. Similar tactics are used to construct political time (Gosden 1994). Political rituals such as coronations, inaugurations, and state funerals often evoke cyclical precedent. Biographical times of the body provide another set of cyclicities which help create social time (Robb 2002; see below). Other recurrent timemarks derive from repeated experiences, for instance seasonal timemarks; at the informal end of the spectrum, the recognisability of experience always draws upon some recurring precedent.

Thus, we certainly live in linear or 'capitalist' time in many contexts, but declaring it to be 'our ontology of time' is an act of ontological purification. Both we and other peoples intimately interweave linearity and cyclicity. Recognising historical particularity and carrying out a sequence of goal-directed action involves linearity in some way; recognition and repetition of precedent involves cyclicity in some way. People mix and match these two principles as needed to cope with the needs of the situation. To develop an overall sense of time, we bootstrap various measures, ranging from astronomical orbits and rotations to calendrical systems, into a generally coherent framework of reasonable precision that works for most purposes. This may be phrased within an overall concept of linearity or cyclicity, but that never captures the complexity of practice which can implement quite different scales and characters of time to understand particular forms of experience. In this sense, ontological understandings of time are not monolithic ontologies. In contrast to concepts of distinct modalities of time (such as 'ritual time'), they are not really even like gears on a car, which you switch between but are clearly distinct. Instead, they are more like shades or colours on the time palette.

Material Time and Bricolage with the Ontological Palette

What would an alternative approach to time which recognises the inherent impurity of ontology-making look like? Space here permits exploring only some initial ideas.

A key starting point is 'time perspectivism', the idea that history consists of a palimpsest of processes of different kinds which unfold at different paces (Bailey 2007). This provides the basis for a multi-scalar archaeology, one which combines (without determinism or reductionism) scale-dependent narratives of entities from the sub-human to humanistic accounts of agency and experience to ones on far greater scales – planetary climate change, for instance, or the development of long-term historical ontologies (Robb & Harris 2013). Yet conventional time perspectivism retains a concept of time which is prior to consideration of the processes unfolding within it. In other words, uniform, linear time exists absolutely; it provides a framework for meshing and coordinating different processes. This is a first-principles ontological assumption about time which transforms archaeological problems into epistemological issues: absolute time exists; the limitation is how we earth-bound mortals can find a measure of it. I would argue the contrary ontological assumption: time is fundamentally a material process. It is definable and understandable only through the material events and changes happening in it (Robb 2020). This is implicitly recognised in chronometry and dating methods, which involve an ongoing quest to find a process so regular that it can provide a proxy of time – the earth's rotation, a pendulum swinging, radiocarbon decay. However, in a realist philosophical position, time emerges from the interaction between the observer and the world: one cannot have chronometric time without clocks, or geological time without strata. But, if time is an emergent phenomenon which derives from our observation of a process, then measured time is also experiential time. History, therefore, is not only a palimpsest of processes unfolding at different speeds. Instead, it is a palimpsest of interacting material processes which define their own time.

Let us take a simple example. The temporality of the body has been best explored with the deceptively simple question of how old a person is (*cf.* Appleby 2011; Sofaer 2011). We live in a record-keeping bureaucratic society, and our reflexive answer is to give a simple chronometric age; establishing such chronometric ages has been the traditional, universally assumed goal of bioarchaeological analysis. But fixing such an age is challenging because there are no bodily processes so metronomically regular as to provide good proxies for the passage of time. We patch together bodily development and degeneration, and both reflect multiple factors besides time, making their relationship to chronometric time approximate at best. If you are living in a society in which chronometric age is not bureaucratically important, for both archaeologists and the people we study, the real bottom line on how old you are is the truth of your body, what it looks like, and what it can do. This leads to the question: what we actually want an age estimate to tell us. Biological-process age, after all, provides a better measure of the capacity to reproduce, to work, and to accumulate knowledge. Someone who has lived for 50 years may have an aching back, minimal teeth, and grandchildren in one culture, and be running marathons pushing their toddler in a stroller in another. And if we are interested in how people identified their own age, chronometric age is of little use; before the 20th century, many, perhaps most, people had little precise idea of their chronological age. Biological-process age provides the observable affordances for the social categorisation of age: size, the development of secondary sexual features, greying or vanishing hair, wrinkles, lost teeth, and so on. Such evidence provides the basis for categorisations into a third measure of biographical duration, social age, the assigning of people to intervals structuring social interaction and adding up to a normative social biography (Robb 2002).

Generalising this example, there are three kinds of time of interest to archaeologists:

1. Process time – duration measured in terms of the specific process unfolding. In this example, it might be the accumulation of biological experience which fixes a body in a locally specific sequence of development and decay.
2. Social time – how the people enmeshed in a process categorised and experienced it – for instance, how they categorised bodies into ages based upon the body's accumulation of biological experience, and how this categorisation fed back into the flow of social action.
3. Chronometric time – our traditional measure of time in absolute chronology. This is an extrapolation to an idealised quantity never directly observable and instead assumed to exist. Chronometric time is useful to us mostly as a general framework for calibrating how different processes coincided, but it does not necessarily tell us much about how any specific process unfolded, and a fixation on chronometric time as a unique baseline might actually prevent us from understanding the past – for instance, understanding how old someone was in real social-biological terms.

We have yet to trace out the implications of this approach. If there are no units of time which can be absolutely defined independently of the processes which happen in them, is a year, a century, or a millennium the same duration regardless of when it occurs? Perhaps not. Whether or not a year is a long time in politics obviously depends on what kind of politics you have. A style or technology remaining unchanged for 100 years would be a rock of stability in the 20th century. It would be entirely unremarkable in medieval or Classical times. In later prehistoric times, the periodicity of change is much longer, but a style or technology remaining unchanging for 1000 years might require explanation of some kind (the society involved was particularly insular, the technology was exceptionally resilient, the style was retro). But we routinely see this lack of change, and over far longer terms, in the Palaeolithic. So either Palaeolithic people lacked agency by modern human standards, or relations between humans, technology, and time were different then. Was a year in the Palaeolithic really the same as a year now? Not if we measure it by process time.

Concluding by Beginning

The paper has two main themes, both of which really require lengthier discussion. The first is about ontological purity. I argue that it is an error to assume a prior state of ontological purity, such as asserting that in any society there is a single master kind of time. This assumption arises from the conceptual baggage of ontology, which implies a single, logically consistent established truth about the basic nature of the world, and from the anthropological tactics of pulling unified order from the chaotic fragments of experience and of characterising difference through heuristic ideal types. Moreover, one feature of Western society for several centuries has been a meta-narrative of science as a single, exclusionary ontology, a mechanistic, rule-bound Newtonian clockwork universe – an act of ontological purification resulting from the coalescence of science and capitalism – and clockwork universes, of course, have clockwork time. And this in turn generates counter-narratives of Other ontological universes which run by other, wilder kinds of time. But if we are interested in ontology ethnographically rather than philosophically, we need to see it not as monolithic, crystalline logical systems but as experientially varying messes resulting from the collision of situationally useful partial orders. People work fluidly with palettes of ontological possibilities.

The second point is what this implies for thinking about time. Archaeologists traditionally, indeed unthinkingly, regard linear, chronometric time as the only way to measure and understand time. But as we develop multi-scalar approaches to the past, we need to work with multiple scales of time whose duration and rhythm are generated by material processes. In such an approach, chronometric time really has only one use, that of helping to establish synchrony and connections among different processes; and even such connections simply mark the fact that the linked processes are part of some larger system which may have its own temporality, much as the village church bells tie varied processes into the temporality of a landscape (Ingold 2000). Considering time as a material process means that we may need to define time locally with reference to specific processes: a century in Palaeolithic social time was not the same as a century in Roman time, and perhaps the body aged differently in Neolithic times than in modern times. This is a project we are not intellectually prepared for: imagine the conceptual shifts required for an analysis which made no mention of conventional chronology but worked exclusively in Neolithic years or medieval body time. Yet painting the past in multiple, free-form local times may be the only way to understand the nature of history.

Acknowledgements

I am grateful to Tim Flohr Sørensen, Rane Willerslev, and Oliver Harris for comments that have improved the argument.

References

Alberti, B., Fowles, S., Holbraad, M., Marshall, Y. & Witmore, C. (2011) 'Worlds Otherwise': Archaeology, anthropology, and ontological difference. *Current Anthropology* 52 (6), 896–912.

Alberti, B. & Marshall, Y. (2009) Animating archaeology: Local theories and conceptually open-ended methodologies. *Cambridge Archaeological Journal* 19 (3), 344–56.

Appleby, J. (2011) Why we need an archaeology of old age, and a suggested approach. *Norwegian Archaeological Review* 43, 145–68.

Bailey, G. (1983) Concepts of time in quaternary prehistory. *Annual Review of Anthropology* 12, 165–92.

Bailey, G. (2007) Time perspectives, palimpsests and the archaeology of time. *Journal of Anthropological Archaeology* 26, 198–223.

Bear, L. (2014) Doubt, conflict, mediation: The anthropology of modern time. *Journal of the Royal Anthropological Institute* 20, 3–30.

Bintliff, J. (ed.) (1991) *The Annales School and Archaeology*. London, Leicester University Press.

Borić, D. (ed.) (2010) *Archaeology and Memory*. Oxford, Oxbow Books.

Bourdieu, P. (1977) *Outline of a Theory of Practice*. Cambridge, Cambridge University Press.

Bradley, R. (1991) Ritual, time and history. *World Archaeology* 23, 209–19.

Bradley, R. (2002) *The Past in Prehistoric Societies*. London, Routledge.

Descola, P. (2013) *Beyond Nature and Culture*. Chicago, IL, University of Chicago Press.

Evans-Pritchard, E. (1940) *The Nuer*. Oxford, Clarendon.

Gell, A. (1996) *The Anthropology of Time: Cultural Constructions of Temporal Maps and Images*. Oxford, Berg.

Gosden, C. (1994). *Social Being and Time*. London, Blackwell.

Hagerstrand, T. (1977) *Culture and Ecology: Four Time-Geographic Essays*. Lund, Lund Universitets Kulturgeografiska Institutionen.

Hallowell, A. I. (1960) Ojibwa ontology, behaviour and world view. In S. Diamond (ed.), *Culture in History: Essays in Honor of Paul Radin*, 49–82. New York, NY, Columbia University Press.

Harris, O. J. T. & Robb, J. E. (2012) Multiple ontologies and the problem of the body in history. *American Anthropologist* 114(4), 668–79.

Holbraad, M. (2007) The power of powder: Multiplicity and motion in the divinatory cosmology of Cuban Ifá (or mana, again). In A. Henare, M. Holbraad & S. Wastell (eds), *Thinking Through Things: Theorising Artefacts Ethnographically*, 189–225. London, Routledge.

Holbraad, M. (2009) Ontology, ethnography, archaeology: An afterword on the ontography of things. *Cambridge Archaeological Journal* 19(3), 431–41.

Iannone, G. (2002) Annales history and the ancient Maya state: Some observations on the 'dynamic model'. *American Anthropologist* 104, 195–207.

Ingold, T. (2000) *The Perception of the Environment: Essays on Livelihood, Dwelling, and Skill*. London, Routledge.

Latour, B. (1993) *We Have Never Been Modern*. Cambridge, MA, Harvard University Press.

Lucas, G. (2005) *The Archaeology of Time*. London, Routledge.

Marcus, J. (1993) *Mesoamerican Writing Systems: Propaganda, Myth, and History in Four Ancient Civilizations*. Princeton, NJ, Princeton University Press.

Milbreth, S. (2000) *Star Gods of the Maya: Astronomy in Art, Folklore, and Calendars*. Austin, TX, University of Texas Press.

Mills, B. J. & Walker, W. H. (2008) *Memory Work: Archaeologies of Material Practices*. Santa Fe, NM, School for Advanced Research Press.

Munn, N. D. (1992) The cultural anthropology of time: A critical essay. *Annual Review of Anthropology* 21, 93–123.

Robb, J. (2002) Time and biography. In Y. Hamilakis, M. Pluciennik & S. Tarlow (eds), *Thinking Through the Body: Archaeologies of Corporeality*, 153–71. London, Kluwer/Academic.

Robb, J. (2020) Material time. In I. Gaskell & S. A. Carter (eds), *The Oxford Handbook of History and Material Culture*. Oxford, Oxford University Press. doi:10.1093/oxfordhb/9780199341764.013.19

Robb, J. & Harris, O. (2013) *The Body in History: Europe from the Palaeolithic to the Future*. Cambridge, Cambridge University Press.

Shanks, M. & Tilley, C. (1987) *Social Theory and Archaeology*. Cambridge, Polity in association with Blackwell.

Sofaer, J. (2011) Towards a social bioarchaeology of age. In S. Agarwal & B. Glencross (eds), *Social Bioarchaeology*, 283–311. Oxford, Blackwell.

Tarlow, S. (2011) *Ritual Belief and the Dead Body in Early Modern Britain and Ireland*. Cambridge, Cambridge University Press.

Thomas, J. (1996) *Time, Culture and Identity: An Interpretive Archaeology*. London, Routledge.

Turner, V. (1969) *The Ritual Process: Structure and Anti-Structure*. Chicago, IL, Aldine.

van Gennep, A. (1977) *The Rites of Passage*. London, Routledge / Kegan Paul.

van Inwagen, P. & Sullivan, M. (2015) Metaphysics. In E. N. Zalta (ed.), *The Stanford Encyclopedia of Philosophy*. Stanford, CA. Available at: <https://plato.stanford.edu/entries/metaphysics/>.

Veyne, P. (1982) *Did the Greeks Believe in Their Myths? Essay on the Constitutive Imagination*. Chicago, IL, University of Chicago Press.

Viveiros de Castro, E. (1998) Cosmological deixis and Amerindian perspectivism. *Journal of the Royal Anthropological Institute* 4, 469–88.

Willerslev, R. (2007) *Soul Hunters: Hunting, Animism, and Personhood among the Siberian Yukaghirs*. Berkeley, CA, University of California Press.

Willerslev, R. (2013) Taking animism seriously, but perhaps not too seriously. *Religion and Society: Advances in Research* 4, 41–57.

Zedeño, M. N. (2009) Animating by association: Index objects and relational taxonomies. *Cambridge Archaeological Journal* 19(3), 407–17.

24

Archaeology, Heritage, and the Heritage of Archaeology

Ian Lilley

Archaeology seems to be prospering with the ascent of the heritage industry. Driven by factors such as the expansion of World Heritage, corporate social responsibility programmes, and the influence of World Bank heritage policies, the future for 'archaeology-as-heritage' looks bright. But what is this doing to the discipline of archaeology? Our claims to scholarly authority and integrity are continually tested by planners and developers, descendants and locals, looters and collectors, and politicians and diplomats, few of whom understand our craft. What should we do? Just carry on and hope for the best? Bury ourselves in technical work and ignore the noise? Replace science with spiritualism? Colleagues try all those things, but I cannot agree with any of them. We certainly need to accommodate diverse circumstances and plural perspectives. Yet we also need to embrace our own disciplinary heritage, in which archaeology and heritage management have always been entwined, so we have a clear vision of what we do and why. Without such a vision, how can we convince others to value what we do, rather than see it as a frivolous waste of money, an impediment to progress, a colonialist imposition, or the plaything of diplomats?

Keywords: Disciplinary Heritage, The 'Australian Turn', Ucko, Mcbryde, Childe, Lubbock

Chris Gosden and I first worked together in 1984 when he was in Australia as a Leverhulme Postdoctoral Fellow. He joined me on my, until-then, solo fieldwork in the remote Siassi Islands in Papua New Guinea (PNG). My wife Cathy arrived shortly after Chris, and, to my lasting shame, my advice saw them both get malaria. It hit Chris when he returned home to the UK and landed him in an isolation ward for Christmas. I am still not entirely sure he's forgiven me (and Cathy has never come on fieldwork with me again!). Chris and I later worked together on the Lapita Homeland Project in PNG, with much less medical drama, and I have since spent several sabbaticals and other sojourns with him in Oxford. On one such visit in 2010–11, when Chris won a Leverhulme Visiting Professorship on my behalf, we worked together on issues directly related to those discussed below. These issues have become even more pressing than they were a decade ago.

To wit, this volume is appearing at a time when archaeology everywhere needs urgent attention. On the face of it, the discipline is prospering in most parts of the world with the ascent of the heritage management industry. A truly vast amount of archaeological work is undertaken every day, right around the planet, in the name of heritage preservation, employing orders of magnitude more archaeologists than all the universities and museums in the world combined. Driven both directly by development pressure and through 'trickle-down' by factors such as the growth in the World Heritage List (UNESCO 2022), the influence of the World Bank's heritage and Indigenous safeguards (World Bank Group 2022), and most recently, the G20's 2021 declaration on culture prioritising heritage protection (UNESCO 2021), the future for 'archaeology-as-heritage' looks bright.

Yet what is this supposed 'progress' doing to archaeology as a research discipline in historical science? Our claims to scholarly authority and professional integrity are under ever-increasing pressure from planners and developers, descendants and locals, looters and collectors, and politicians and diplomats, very few of whom have any real understanding of or sympathy for the basics, much less the nuances, of our craft. This pressure is being felt everywhere. What should we do in such circumstances? Should we just carry on as usual and hope for the best? Should we retreat into what an Australian anthropologist once dismissed as our 'cave-holes' (Tindale 1965, 162), trying to block out the noise, while focusing on the esoteric technical work that is so often our forte? Alternatively, should we abandon any pretence of scientific rigour and fall for the dubious attractions of 'New Age' spiritualism?

I have seen colleagues do all those things in different parts of the world, but I cannot agree with any of them. We certainly need to be politically savvy, strategically responsive to myriad circumstances, and open to a diversity of perspectives. At the same time, though, we need to embrace, value, and safeguard our own disciplinary heritage, so we have a clear vision of where we have come from and where that points us in the future. If we are unable to understand and appreciate our own history, how are we to convince others to value – or at least continue to pay for – what we do, rather than see it as an irrelevant waste of taxpayers' money, an impediment to progress, a colonialist imposition, or an instrument to advance or retard domestic and international political powerplays?

'Can he be serious?', some readers might ask. Is the history of the discipline not littered with examples of knowing or inadvertent complicity in the excesses of nationalism and colonialism? (*e.g.* Trigger 1984; Arnold 1990; Kohl & Fawcett 1996; Meskell 1998; Sommer 2017). Why would we want to embrace, value, and safeguard a heritage like that, when it is exactly those sorts of issues that continue to dog us in many parts of the world? Should we not make a clean break from our past and try to do things differently? In Australia, such matters are certainly 'up close and personal' for us. Although it was built on a platform of often-rigorous and insightful earlier work (Spriggs 2020), today's university-based form of archaeology only began here in the early 1960s, almost 200 years after initial European colonisation. Yet, for at least the last 40 years or so, or nearly all of my professional life, we have been intimately wrapped up in the often-painful politics of ongoing nation-building in a decolonising context.

I know many readers will have been entangled in these sorts of circumstances in various parts of the world. Yet, as someone who has worked all over the world as well as all over Australia, I can say that Australians often see themselves, and have frequently been held by others (*e.g.* Lilley 2009; Gfeller 2013; 2015), to be more involved than most at the 'pointy end' of postcolonial archaeology and heritage. Usually, and particularly from an Australian perspective of course, our involvement is seen as positive. We are out there with our Indigenous partners fearlessly changing the paradigm, and either being joined in that battle by similarly high-minded colleagues or dragging the less-enamoured along for their own good. That is the nice version. The not-so-nice version is that Australians are disproportionately responsible for visiting a postmodern, postcolonial nightmare upon the discipline. We are resented for eagerly prying open a Pandora's Box of issues that archaeology as a scientific enterprise really should not have to deal with, most obviously in connection with the heritage of Indigenous and other descendent communities.

There are numerous anecdotes that could be retold in this connection, but in a more scholarly vein, Canadian Robert McGhee's 2008 paper in *American Antiquity* explicitly implicated Australians in what he called 'aboriginalism and the problems of indigenous archaeology' in North America. Much of the concern expressed by McGhee and others like him (*e.g.* Stump 2013) is linked one way or another with the impact on 'research' or 'scientific' archaeology concerning questions of cultural heritage, notably regarding Indigenous and other descendent identity claims. Partnership with Indigenous and other descendent communities on questions of heritage, McGhee (2008, 591) and his ilk disparagingly believe, sees us 'sharing theoretical authority' in a way that will 'strip archaeology of the scientific attributes that make it a particularly powerful narrator of the past and accord it at most equal weight relative to Indigenous oral tradition and religious discourse', which he obviously does not rank highly. You may think this view would be an outlier these days, but in my experience, such perspectives are still surprisingly common around the world.

Ironically, although some archaeological colleagues may lay the blame for the discipline's current travails at the feet of Australian heritage practitioners, concern about the impact of postcolonial dynamics – and the subversive role played by Australians – is also voiced by sections of the international cultural heritage community. In print in 2009 and in person at the head of a group of like-minded delegates at the 2011 ICOMOS General Assembly in Paris, the late Michael Petzet, then the leading heritage figure in Germany, strongly asserted that what he glossed as 'Australian' approaches to community engagement are fatally undermining the management of cultural heritage around the globe. To paraphrase his words as I heard them firsthand, the terms 'community' and 'people' are not included in the title of ICOMOS, the International Council on Monuments and Sites, one of the three independent statutory Advisory Bodies to UNESCO under the World Heritage Convention. He held up his ICOMOS membership card, which features the winged horsed Pegasus, and asked the General Assembly whether we were going to change it to 'a flying kangaroo'! Even the late Willem Willems, who caustically criticised Petzet's views on this matter (*e.g.* Willems 2014, 120, note 2), regularly and emphatically reminded me 'Ian, this is not Australia!' when we were on international heritage missions (Lilley 2015). In this context, we should ask whether our discipline really has 'an Australian problem', as some people evidently believe. That requires a bit of history.

If any single person can be held accountable for what we might call the 'Australian turn' in archaeology and heritage at a global level, it is the late British archaeologist Peter Ucko. From 1972 to 1980, he was the principal of the then-Australian Institute of Aboriginal Studies, a Federal statutory authority now called the Australian Institute of Aboriginal and Torres Strait Islander Studies (or AIATSIS), what insiders just call 'the Institute'. I hesitate to quote Wikipedia in a scholarly context, but its entry on Ucko accords with my knowledge and experience of the man when it states that he was a 'controversial and divisive figure within archaeology... [whose] life's work focused on eroding western dominance by broadening archaeological participation to developing countries and indigenous communities'. It also notes that he 'sought to involve Indigenous Australians ..., hiring them in the [Institute] council and its committees' (Wikipedia 2022a). Two Aboriginal people, including Neville Bonner, the first Indigenous Australian elected to the country's Senate, had in fact been appointed to the Council just before Ucko took up his post at the Institute, but he greatly accelerated the trend. In 1976, Ephraim Bani became the first Torres Strait Islander appointed to Council, reflecting Ucko's insistence that both of Australia's Indigenous peoples be recognised, though the Institute's name did not formally show that until 1989.

What Wikipedia's Ucko entry does not say, though the entry for the Institute does, is that Ucko's much-needed moves to Indigenise the Institute were largely a response to the famous 'Eaglehawk and Crow Letter' from Aboriginal leaders demanding such changes (Wikipedia 2022b). Be that as it may, not long after he returned to the UK from Australia, Ucko became the driving force behind the creation of the World Archaeological Congress (WAC). WAC was created in 1986 as a break-away from the UNESCO-affiliated International Union of the Pre- and Proto-historic Sciences, of which Ucko was the UK Secretary, when the Union refused to sanction scholars from South Africa during the time of anti-Apartheid protests around the world. In addition to routine professional duties concerning 'the exchange of results from archaeological research...and the conservation of archaeological sites', WAC is dedicated to 'professional training and public education for disadvantaged nations, groups and communities; [and] the empowerment and support of Indigenous groups and First Nations people' (World Archaeological Congress 2022). WAC's 'One World Archaeology' book series (Routledge 2022) was central in establishing this scope and tone, and early volumes featured titles such as *Who Needs the Past? Indigenous Values and Archaeology* (Layton 1994a), *Conflict in the Archaeology of Living Traditions* (Layton 1994b), *The Politics of the Past* (Gathercole & Lowenthal 1994) and others in a similar vein.

The ethos of such works became and, in the background, remains a touchstone for archaeologists and heritage managers operating in turbulent postcolonial waters. Influential as Ucko and WAC have been, they did not appear out of thin air. Before WAC burst so dramatically onto the scene, pioneering Australian archaeologist Isabel McBryde published *Who Owns the Past?* in 1985, bringing together papers presented in a conference she organised through the Australian Academy of Humanities in 1983. McBryde's PhD was the first-ever awarded on the basis of Australian Indigenous archaeological fieldwork, and she was a highly influential figure in the same milieu that sensitised Ucko. Since that time, the themes first addressed in McBryde's volume have been continually revisited by scholars around the world. McBryde also had a profound impact on World Heritage policy and practice – and thus heritage practice at all levels around the world – by forcing the recognition of First Nations interests in cultural landscapes and intangible heritage based on her forward-looking partnerships with Aboriginal people. As Swiss researcher Aurelie Gfeller (2013, 497) put it:

> McBryde's concerns ... reflected the transformation of archaeology in the post-colonial context of Australia ... [and especially] growing indigenous political activism [which] prompted the rise of what would be termed 'indigenous archaeology', namely 'an expression of archaeological theory and practice in which the discipline intersects with Indigenous values, knowledge, practices, ethics, and sensibilities' and involves 'collaborative and community-orientated or -directed projects, and related cultural perspectives'.

'Australian' activism of the sort pursued by Ucko, McBryde, and others might seem quite radical in the long view of the discipline, but is it really? Has this intense focus on modern identity politics truly taken us off in a new and more ethical direction that substantively improves scholarly archaeology, as well as its relationships with the communities amongst which it works? It would seem hard to doubt, but how different is it really from the discipline's involvement in nationalist and colonialist projects over the last few centuries? We might be helping give voice to oppressed Indigenous minorities or other local communities, but in doing so are we not still interpreting the archaeological past in terms of essentialised and highly problematic ethnic or racial categories that archaeology cannot underpin empirically or that archaeology in fact empirically undermines? The 'Celts' is a well-known example of a term argued by some to flatten and diminish the rich diversity and dynamic histories of the varied Iron Age peoples to whom that label is attached (*e.g.* Dietler 1994; Megaw & Megaw 1996). So, too, with terms such as 'Aboriginal' in Australia. Even in its simplest original Latin meaning of 'from the beginning', the word 'Aboriginal' can imply that the descendants of Australia's first human occupants remained 'an unchanging people in an unchanging land' for tens of thousands of years, as Pulleine proclaimed in 1928. We know that is not true, just as we know the Celts were not a monolithic cultural bloc despite obvious commonalities, yet we have been using these sorts of questionable essentialising categories since the modern discipline emerged. How does continuing to do so in the context of modern identity politics improve things?

I am not suggesting that our attempts to work ethically with community stakeholders are doomed. Rather, I am suggesting that we look to the history of our discipline to see where our real strengths lie and, on that basis, think of how we can make use of those strengths in ways that benefit both the discipline as a scholarly endeavour and other parties with interests in the past.

Going back to the 19th century roots of modern scientific archaeology in Britain, and thus by extension Australia, we find that Sir John Lubbock spent a great deal of time bringing archaeology and heritage management together in mutually supportive ways. Lubbock was, amongst other things, a member of Parliament, as well as president of the Royal Anthropological Institute of Great Britain and Ireland and the Society of Antiquaries of London. In 1872, he began the very drawn-out process of establishing Britain's first Ancient Monument Protection Act (AMPA). His interest in doing so was mainly to ensure protection of sites for archaeological research, but he also used it as an opportunity to demonstrate the value of archaeology as a scientific discipline and advance the cause for its support by the public and funding by the government.

As historian of archaeology Tim Murray (2014, 15) put it, Lubbock's labours in connection with the first AMPA show how 'the exigencies of preservation have been an important context for the building of archaeological theory, and for the legitimation of claims to knowledge about the archaeological past'. In other words, Lubbock understood as clearly in 1872 as our Swedish colleague Kristian Kristiansen did in 2008 when he observed that 'Heritage is the dominant organizational/legislative framework for [all] archaeological practice, and it is where most of the money is spent' (Kristiansen 2008, 56).

Absolutely central to Lubbock's tactics was an understanding that the profession had to communicate effectively regarding the plausibility, reliability, and social value of archaeological findings with those whom today we would call 'community stakeholders', which in his time, like now, meant lawmakers and land developers, as well as archaeologists and others, including descendant communities with an interest in the human past. Even this cursory dip into archaeology's modern history shows that, far from the relationship between scientific or research archaeology and heritage management being some fashionable postmodern and postcolonial imposition from Australia or anywhere else, it has been there as a mutually beneficial interdependence from the very start of our modern discipline. That is something we should embrace, value, and safeguard as one of the discipline's foundation stones, as is Lubbock's emphasis on the necessity of effective stakeholder engagement to the success of this interdependence.

The problem for us today with Lubbock's work bringing archaeology and heritage together and engaging with stakeholder interests, and indeed the problem for us with archaeology in Lubbock's time more generally, was that it was closely aligned with nationalism and racialised theorising. As we know, this sort of thinking implicated archaeology in the devastation of the global colonialist project, as well as the horrors of Nazism. It is little wonder that many colleagues around the world, including most of the older generation of Australian archaeologists, shrink back from the discipline's recent re-engagement with racialised identity politics and the simplistic and archaeologically unsupportable essentialisms upon which they so often rest. Some, like the late John Mulvaney (1985; also Allen 1999), who lived through the Second World War and was a giant in modern Australian archaeology from its inception, have written quite explicitly about their fears in this regard. As we see on the nightly news, such concerns are looming large again now with the upswing in migrant and refugee movements around the globe. How were these issues countered by archaeologists in the lead-up to the Second World War, when such matters were trending towards their tragic conclusion in the Nazi death camps?

Another great archaeologist, Vere Gordon Childe, probably the most influential archaeologist of all time (Murray 1990), was literally a child when Lubbock was at the height of his powers. Childe was an Australian, who trained in archaeology at Oxford after studying classics at Sydney. On his return to Australia, he was banned from working in academia because of his socialist activism, so he went back to the UK and forged a momentous career there (Allen 1979; Green 1990). Childe's interpretive frameworks had an enormous impact on European and indeed world archaeology for a long time (and from my observation, still do in some quarters), and they would certainly have been influential when Mulvaney, McBryde, and the other Australian and British-born founders of modern Australian archaeology were training in Britain after the Second World War. So, can we attribute any so-called 'Australian problem' in today's community-oriented approaches to archaeology and heritage management to Childe himself or his deep influence on generations of scholars in Europe and the Antipodes?

Not in any direct sense (Murray 1990). While Childe was deeply involved in Australian Labor Party politics early in his life, there were no high-profile Indigenous identity politics of today's sort at that time, despite a degree of Aboriginal activism. In any case, Childe was preoccupied with other issues. In his work as an archaeologist, Childe was not completely untainted by the racial theorising of his era, but he was a card-carrying Marxist who was strongly critical of Nazi formulations of history and archaeology. In 1933, he wrote in the journal *Antiquity* that 'to admit as good only what is Celtic or Germanic or Indian, as exclusive nationalism would demand, is unscientific and unhistorical'. On this basis, I think one can assume that Childe would not have endorsed the more strident essentialising identity politics that so concern Australian colleagues such as Mulvaney (1985) and Allen (1999), cited earlier.

This common thread is not coincidental. After retiring from the directorship of the UCL Institute of Archaeology in 1956, Childe returned to Australia. Before falling (probably jumping) to his death off Govett's Leap in the Blue Mountains near where he was raised (Green 1990), he expressed some interest in the research prospects of Australian archaeology but publicly criticised academic racism towards Indigenous Australians. There was no Australian archaeology in today's sense then, so this activity at the very end of Childe's life could not have directly affected the way Australian practitioners thought about Indigenous issues and archaeology or heritage. Yet, there is a complication. It is highly likely that the influence of Childe's leftist politics and vocal anti-Nazi stance through his 20 books and 200 papers on archaeology did have an impact on the first generation of modern Australian archaeologists in training. Indeed, the entry on Childe in Australia's official *Dictionary of Biography* was written by the archaeologist Jim Allen (1979), my PhD supervisor and the person to whom Chris Gosden reported as a postdoctoral fellow. Along with Tim Murray, Allen was a central player in one of Australia's most difficult conflicts between Aboriginal people and archaeologists, the so-called Tasmanian Affair (Allen 1995; see also Allen 1999), largely because they pressed what might be characterised as a 'Childean' line on the essentialism underpinning Aboriginal identity politics, dismissing 'exclusive [Indigenous] nationalism' as 'unscientific and unhistorical'. Others value such essentialism as 'strategic', but that is a discussion for another paper.

The other thing that we should take from our disciplinary history, whether looking only as far back as Childe or all the way back to Lubbock, is the absolute centrality of 'good archaeology' in the technical scientific sense. Murray (2014, 15) notes that, when Lubbock was battling to have the AMPA passed, some opponents questioned whether archaeological knowledge was worth having, but even so 'the methodology of… archaeology was never questioned, especially its ability to produce rationally defensible knowledge about a past previously felt to be accessible only through speculation and the "tyranny of hypothesis"'. In other words, archaeological heritage preservation rests on doing good scientific archaeology, so we know what it is we are preserving and why. So, too, with Gordon Childe. His 1933 attack on Nazism (Childe 1933) underlined the need for professional competence, asserting that the archaeological heritage so grievously abused by Hitler and his henchmen 'can no more be profitably studied without laborious [scientific] preparatory training than can the movements of stars or the behaviour of electrons'.

Interestingly in this connection, after having a difficult relationship with the profession for at least a generation, First Nations Australians are now increasingly coming to archaeologists to have technically competent professional archaeology undertaken on their lands. Indeed, a handful have recently graduated with PhDs in archaeology, with more on the way. This shift is not just because First Nations communities have suddenly seen the light of archaeology. It is in large part because archaeologists have worked hard to communicate effectively with Indigenous people, bringing archaeology and heritage together in ways that John Lubbock would certainly have understood very well. As would have Childe, for that matter. As Allen (1979) wrote in the *Australian Dictionary of Biography*,

> Childe was concerned with archaeology's relevance (and therefore his own) in contemporary society. One method he used to demonstrate his own utility was to write a series of books designed to present the results of his research in a non-academic form for the general public. The most popular were *Man Makes Himself* (1936), and *What Happened in History* (1942), which by 1957 had sold over 300,000 copies. His works were also widely translated.

Childe, like Lubbock 60-odd years earlier, very clearly understood the need to bring people along with us on our journey of discovery. Which brings me to my conclusion. With continual and continuing strong prompting from First Nations people (*e.g.* Langford 1983; Dugay-Grist 2006), Australian archaeologists have made enormous strides throughout my professional lifetime in communicating effectively with Aboriginal and Torres Strait Islander communities about what we do and why, to the mutual benefit of all concerned. In doing so, we have had a profound influence on professional practice around the world, even in places such as North America, which have their own Indigenous politics and where similar postcolonial dynamics have also been unfolding (Lilley 2009). As my earlier quote from Kristiansen underlines, this has all been done in a context where 'Heritage is the dominant organizational/legislative framework for… [our] archaeological practice, and … where most of the money is spent'.

Despite positive contemporary evidence such as this regarding Indigenous archaeology and heritage, as well as the historical lessons provided by the likes of Lubbock and Childe, we have been much less effective in communicating our message to our other constituencies, the support of which we also need to secure the discipline's future (Lilley 2005). We really have not demonstrated any understanding that the intimate intertwining of archaeological and heritage value that we take for granted in our work with Indigenous and other local/traditional/descendent partners should extend to every kind of field

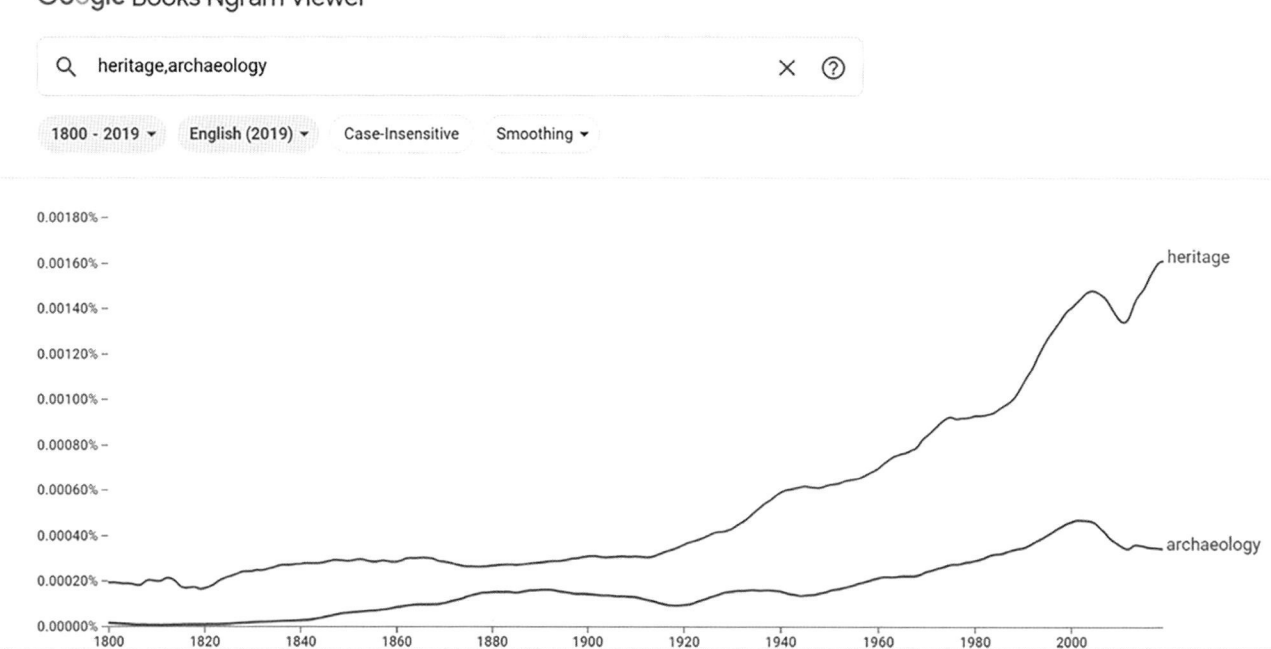

Figure 24.1 Google Ngram comparing use in English of the terms 'heritage' and 'archaeology' since 1800. Available through: Google Books Ngram Viewer (http://books.google.com/ngrams).

and laboratory archaeology in every part of the world if we are to continue to prosper as a scholarly enterprise. In this connection, around the time of writing this paper, I coincidentally received an email from Jeff Altschul, a now retired but still highly influential friend and colleague. He founded what became one of the largest heritage consultancies in the North America and was involved in my 2010–11 sabbatical project with Chris Gosden in Oxford, mentioned at the outset. Echoing what I asserted at the beginning of this chapter, he wrote that 'to someone who spent his whole career on the commercial side, it seems an odd time. Every… [heritage] firm I know is hiring, yet my academic colleagues tell me that all their programs are under threat' (see also Altschul & Klein 2022).

In Australia, this has not yet come to pass, owing primarily to the political weight attached to the strong connection between archaeology and Indigenous heritage. At a global 'whole of discipline' level, though, the writing seems to be well and truly on the wall. This assertion can be underlined by doing a quick Google Ngram search, contrasting use in English of the term 'archaeology' versus 'heritage' since 1800 (Fig. 24.1). I was shocked when I first saw these curves, but they have a blunt message. Public support and funding for archaeology as a worthwhile field of research scholarship will continue to wither if too many of our colleagues around the world hesitate to incorporate heritage-oriented perspectives centrally in their professional approach. If the discipline still has any so-called 'Australian problem', the n-gram suggests it is not the one we started out with. It is that we here in Australia have not taken our hard-won success with Indigenous communities to its logical conclusion, which is more effective communication with all stakeholders, and communicated that broadening of our effort with our colleagues globally. Archaeology as a discipline generating scholarly knowledge for its own sake is far from a universally recognised public good, while 'heritage' is demonstrably the framework in which non-archaeologists have always understood us and paid for us. That's why we need to embrace, value, and safeguard our own disciplinary heritage going back to the likes of Childe and Lubbock, and seamlessly integrate archaeology and heritage in all our professional thought and action.

Acknowledgements

I thank the editors of this volume for asking me to contribute. It has been a great privilege to know Chris Gosden and his partner Jane Kaye as good friends, as well as inspirational colleagues, for so many years. I would also like to thank Dr Robin Derricourt for suggesting the n-gram idea after he listened to an online precursor of this paper that I presented at the 2021 Annual Hall Lecture at the University of Queensland. Robin shared a view of the curve for 'heritage' and then I added 'archaeology' out of curiosity.

References

Allen, J. (1979) Childe, Vere Gordon (1892–1957), *Australian Dictionary of Biography*. National Centre of Biography,

Australian National University. [online] Available at: <https://adb.anu.edu.au/biography/childe-vere-gordon-5580/text9521> [accessed 12 March 2022].

Allen, J. (1995) A short history of the Tasmanian Affair. *Australian Archaeology* 41, 42–8.

Allen, J. (1999) Professorial inaugural address. The politics of the past. *Australian Archaeology* 49, 34–43.

Altschul, J. & Klein, T. (2022) Forecast for the US CRM industry and job market: 2022-2031. *Advances in Archaeological Practice, First View*, 1–16. doi:10.1017/aap.2022.18.

Arnold, B. (1990) The past as propaganda: Totalitarian archaeology in Nazi Germany. *Antiquity* 64, 464–78.

Childe, V. G. (1933) Is prehistory practical? *Antiquity* 7, 410–18.

Dietler, M. (1994) 'Our ancestors the Gauls': Archaeology, ethnic nationalism, and the manipulation of Celtic identity in modern Europe. *American Anthropologist* 96, 584–605.

Dugay-Grist, M. (2006) Shaking the pillars. In I. Lilley (ed.), *Archaeology of Oceania: Australia and the Pacific Islands*, 367–79. London, Blackwell.

Gathercole, P. & Lowenthal, D. (eds) (1994) *The Politics of the Past*. London, Routledge.

Gfeller, A. (2013) Negotiating the meaning of global heritage: 'Cultural landscapes' in the UNESCO World Heritage Convention, 1972–92. *Journal of Global History* 8, 483–503.

Gfeller, A. (2015) Anthropologizing and indigenizing heritage: The origins of the UNESCO Global Strategy for a Representative, Balanced and Credible World Heritage List. *Journal of Social Archaeology* 15, 366–86.

Green, S. (1990) V. Gordon Childe: A biographical sketch. *Australian Archaeology* 30, 18–25.

Kohl, P. & Fawcett, C. (eds) (1995) *Nationalism, Politics, and the Practice of Archaeology*. Cambridge, Cambridge University Press.

Kristiansen, K. (2008) The dialectic between global and local perspectives in archaeological theory, heritage and publications. *Archaeological Dialogues* 15, 56–69.

Langford, R. (1983) Our heritage – your playground. *Australian Archaeology* 16, 1–6.

Layton, R. (ed.) (1994a). *Who Needs the Past? Indigenous Values and Archaeology*. London, Routledge.

Layton, R. (ed.) (1994b). *Conflict in the Archaeology of Living Traditions*. London, Routledge.

Lilley, I. (2005) Archaeology and the politics of change in a decolonizing Australia. In J. Lydon & T. Ireland (eds), *Object Lessons*, 89–106. Melbourne, Australian Scholarly Publishing.

Lilley, I. (2009) Strangers and brothers? Heritage, human rights and a cosmopolitan archaeology. In L. Meskell (ed.), *Cosmopolitan Archaeologies*, 48–67. Durham NC, Duke University Press.

Lilley, I. (2015) This is not Australia! Willem Willems and international heritage management. In M. van den Dries, S. van der Linde & A. Strecker (eds), *Fernweh: Crossing Borders and Connecting People in Archaeological Heritage Management. Essays in Honour of Prof. Willem J.H. Willems*, 83–6. Leiden, Sidestone Press.

McBryde, I. (ed.) (1985) *Who Owns the Past? Papers from the Annual Symposium of the Australian Academy of the Humanities*. Melbourne, Oxford University Press.

McGhee, R. (2008) Aboriginalism and the problems of indigenous archaeology. *American Antiquity* 73, 579–97.

Megaw, J. V. S. & Megaw, R. (1996) Ancient Celts and modern ethnicity. *Antiquity* 70, 175–81.

Meskell, L. (ed.) (1998) *Archaeology Under Fire: Nationalism, Politics and Heritage in the Eastern Mediterranean and Middle East*. London, Routledge.

Mulvaney, D. J. (1985) A question of values. In McBryde 1985, 86–98.

Murray, T. (1990) Second Childhood? Gordon Childe and Australian archaeology. *Australian Archaeology* 30, 14–17.

Murray, T. (2014) *From Antiquarian to Archaeologist: The History and Philosophy of Archaeology*. Barnsley, Pen and Sword Archaeology.

Petzet, M. (2009) *International Principles of Preservation*. Berlin, Bäßler Verlag.

Pulleine, R. W. (1928) Presidential address. *Australian Association for the Advancement of Science* 19, 294–314.

Routledge (2022) *Book Series: One World Archaeology*. [online] Available at: <https://www.routledge.com/One-World-Archaeology/book-series/SE0145> [accessed 29 September 2022].

Sommer, U. (2017) Archaeology and nationalism. In G. Moshenska (ed.), *Key Concepts in Public Archaeology*, 167–86. London, UCL Press.

Spriggs, M. (2020) Everything you've been told about the history of Australian archaeology is wrong! *Bulletin of the History of Archaeology* 31, 1–16.

Stump, D. (2013) On applied archaeology, indigenous knowledge, and the usable past. *Current Anthropology* 54, 268–98.

Tindale, N. (1965) Stone implement making among the Nakako, Ngadadjara and Pitjandjara of the Great Western Desert. *Records of the South Australian Museum* 15, 131–64.

Trigger, B. (1984) Alternative archaeologies: Nationalist, colonialist, imperialist. *Man* 19, 355–70.

UNESCO (2021) G20 agrees first declaration on culture. *UNESCO*. [online] <https://en.unesco.org/news/g20-agrees-first-declaration-culture> [accessed 29 September 2022].

UNESCO (2022) *World Heritage List*. [online] Available at: <https://whc.unesco.org/en/list/> [accessed 29 September 2022].

University of Queensland (2021) *Archaeology, Heritage and the Heritage of Archaeology*. [online] Available at: <https://social-science.uq.edu.au/event/session/5723> [accessed 29 September 2022].

Wikipedia (2022a) Peter Ucko. *Wikipedia*. [online] Available at: <https://en.wikipedia.org/wiki/Peter_Ucko> [accessed 29 September 2022].

Wikipedia (2022b) Australian Institute of Aboriginal and Torres Strait Islander Studies. *Wikipedia*. [online] <https://en.wikipedia.org/wiki/ Australian_Institute_of_Aboriginal_and_Torres_Strait_Islander_Studies> [accessed 29 September 2022].

Willems, W. (2014) The future of world heritage and the emergence of transnational heritage regimes *Heritage and Society* 7, 105–20.

World Archaeological Congress (2022) *World Archaeological Congress*. [online] Available at: <https://worldarch.org/> [accessed 29 September 2022].

World Bank Group (2022) *Environmental and Social Framework*. [online] Available through: <https://www.worldbank.org/en/projects-operations/environmental-and-social-framework> [accessed 29 September 2022].

Selling Photographs: Collecting Archaeology

Elizabeth Edwards

This paper explores the absorption of commercially produced photographs into archaeological knowledge-making in the second half of the 19th century. I argue that these commercial photographers functioned as what has been termed 'knowledge workers'. They not only supplied archaeologists with informational images but drew on a shared repertoire of concepts of value and 'reliability'. Central to my account is an album, Our Ancient Monuments, *made by General Pitt Rivers between 1883 and 1887 in his role as first Inspector of Ancient Monuments, as he surveyed the prehistoric monuments named under the Ancient Monuments Protection Act of 1882. I consider the marketised flow of photographs across and within 19th-century knowledge spaces, as commercial producers of 'views' photographs intentionally positioned their images in order to sell into emerging scientific and disciplinary markets, including archaeology.*

Keywords: Photographs, Archaeology, Markets, Monuments, General Pitt Rivers

Introduction

From the 1850s on, commercial photographers produced a massive flow of images, which operated across great swathes of the knowledge landscape. Active in their pursuit of both learned and popular markets, they were hugely successful enterprises providing souvenir and antiquarian images to the middle classes. These same images could, however, serve as informational images within emergent disciplines, especially in archaeology. Further, a demonstrable number of photographers were part of the broader epistemic community of the historical sciences, in that they absorbed current debates in archaeology, anthropology, and geology, and drew on a shared repertoire of concepts of value and 'reliability' in articulating the past. Consequently, despite their disparate origins, these commercially produced photographs became 'recoded' to serve as what Lorraine Daston (2015) termed 'epistemic images'. That is, these images, collectively, contributed to understandings and perceptions that defined and stood for the core value systems of a body of knowledge, even if in practice their evidential force was less stable or certain.

I explore these entanglements between commercial photography and proto-modern archaeology through the prism of an album of watercolours, plans, and photographs, dating from the late 1880s, entitled *Our Ancient Monuments* (hereafter *OAM*). It was made by General Pitt Rivers in relation to his 1883–1885 tours of inspection as first Inspector of Ancient Monuments as he surveyed prehistoric monuments, most of which were named under the Ancient Monuments Protection Act of 1882.[1] While it has clear scientific intent in the meticulous surveys of henge monuments and circles, the album seems above all to be a souvenir and record, and was worked up after these tours – it is not in strict chronological order of inspections. Importantly, it contains 28 photographs of scheduled monuments taken by commercial photographers pasted into its pages and numerous loose interleaved photographs. Roughly contemporary with the making of the album (on technical and internal grounds), they create an assemblage of archaeological imaging and a density of information.

OAM exemplifies the presence and circulation of photographs of archaeological and historical monuments as active players in emergent historical imaginations at the period. It resonates with the shifting public sphere of the late 19th century. This saw not only the foundation of the Inspectorate but, for example, the foundation of the National Trust and the massive expansion of museums, notably local museums as vectors of a collective narrative. Even the album's name, *Our Ancient Monuments*, appears to stress growing ideals of ownership of 'national heritage'. Its engagement with the mass-provision of images of archaeological and historical interest points firmly in this direction and in Pitt Rivers' interests in education. Chris Gosden's work has been central to these interpretative developments in archaeology and museums, and the role of Pitt Rivers and his assistants within them. This paper addresses those particular historical, and indeed historiographical, interests as exemplified in his 2011 project with Jeremy Coote, 'Rethinking Pitt Rivers'.[2] While that project did not address Pitt Rivers' use of images in analytical terms, its address to the practices of collection have resonances here. This essay does not follow the well-trodden path of Pitt Rivers, as collector and archaeologist, and his methods nor that of practices of visual knowledge-making in 19th century archaeology (see for instance Smiles & Moser 2005; Bohrer 2011; Klamm 2021). Instead, it explores a largely overlooked element in these processes: the role of commercial photographers in the production and circulation of reliable visual knowledge, and the ways in which commercial photographers positioned themselves as makers and suppliers of reliable images that could feed into, and inform, the emerging disciplinary structures of the 19th century, including archaeology.

Noah Heringman (2013), writing of 18th-century antiquarianism and natural history, discussed as 'knowledge workers' the draughtsmen who worked with antiquarians and proto-archaeologists to provide reliable and credible visuals that serviced and absorbed the values of the emerging discipline. Commercial photographers often functioned as 19th-century knowledge workers, fulfilling a very similar role in the production and dissemination of reliable and credible graphic communication within an emergent disciplinary framework. The scientific/commercial interfaces also reflect the integral financial formations that attended the networks of discipline formation as 'scientific truths were produced and disseminated in a set of exploitative bureaucracies and forms of knowledge accumulation' (Flandreau 2017, 5). At a meta-level, these processes imbricate the *OAM* album.

The Album

The album comprises some 42 site plans, sections, and drawings surveyed by Pitt Rivers, largely paired with 33 atmospheric watercolours made by Pitt Rivers' surveyor/artist assistant William S. Tomkin, in which both Pitt Rivers and Tomkin – as a knowledge worker – appear in action (Fig. 25.1).[3] It opens with two anonymous commercial prints of Kit's Coty in Kent where the graffiti and wear had concerned Pitt Rivers (Thurley 2013, 42). There follow 14 pages of commercially produced photographs of Silbury Hill, Avebury, and Stonehenge, largely from the major topographical suppliers of Francis Frith, George Washington Wilson, and J. Valentine and Sons.[4] This strong photographic statement of longevity, decay, and preservation might be read as setting the agenda for the whole album. Stonehenge, among the key English sites scheduled in the 1882 Act, and long-established in the antiquarian, archaeological, and graphic imaginary (Barber 2014), sets the narrative of the album. It is the gold standard 'monument' for what follows,[5] and significantly, multiple views of Stonehenge were acquired from all three of these major commercial operators (Fig. 25.2).[6] The watercolours and plans follow. On occasion, these are reinforced visually and informationally with commercially produced photographs which complete and extend Pitt Rivers' and Tomkins' graphic record. Then towards the end of the album photographs reappear, clustered around incisions into more general archaeological interest such as the early Norman work of the round tower at Brechin, Forfarshire, or the stone cross at Sueno, Forres (Fig. 25.3).[7] In both cases a Tomkin watercolour is paired with photographs by Valentine & Sons.

Importantly, the album was also used as an assemblage space to file contemporary photographs of archaeological interest. Most of these comprise photographs from local and provincial commercial photographers, including photographs of Avebury, Silbury Hill, and Stonehenge, probably by C. W. Clarke of Devizes (Fig. 25.4), of the recently recognised Saxon church at Bradford on Avon by Richard Wilkinson of Trowbridge, and of Pembrokeshire cromlyns by H. Jackson of Fishguard: all local photographers who typify local quasi-archaeological production (Fig. 25.5). While the exact relationship with the pasted images is unclear – there are some duplicates of Stonehenge and Silbury Hill – all appear, on both technical and internal evidence, to date from around the time of the album and its compilation. It points to a steady flow of commercial photographs into Pitt Rivers' interpretative ambit.

It is unclear whether these photographs, both pasted and loose, were ordered by catalogue from photographers or acquired through the agents, the stationers, print sellers, and local shops which carried their stock. The stocks of such outlets tended to be very localised.[8] It is feasible to assume that those pasted in the album, including the Wilson and Frith views of Stonehenge and Valentine photographs of Celtic and Saxon crosses, were acquired during the inspectorate trips. Tomkin's journal for the Scottish tour of 1885 notes that on the 3rd of August they bought photographs in Oban,[9] spending eight shillings – this would have purchased between four and eight prints depending on size. Conversely, they might have acquired

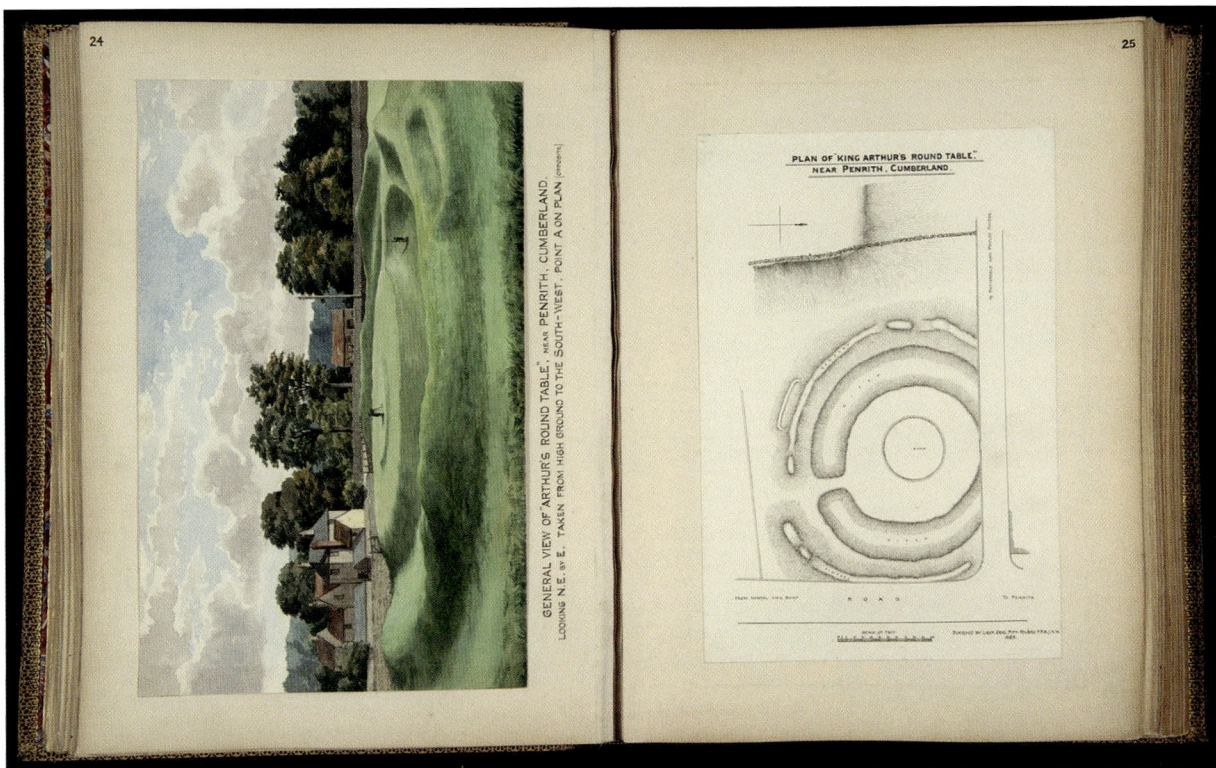

Figure 25.1 Our Ancient Monuments. *Arthur's Round Table, Penrith, Cumbria. Watercolour by W. S. Tomkin (left), plan by General Pitt Rivers (right). 1883. 2012.79.1.31/32 (© Pitt Rivers Museum University of Oxford).*

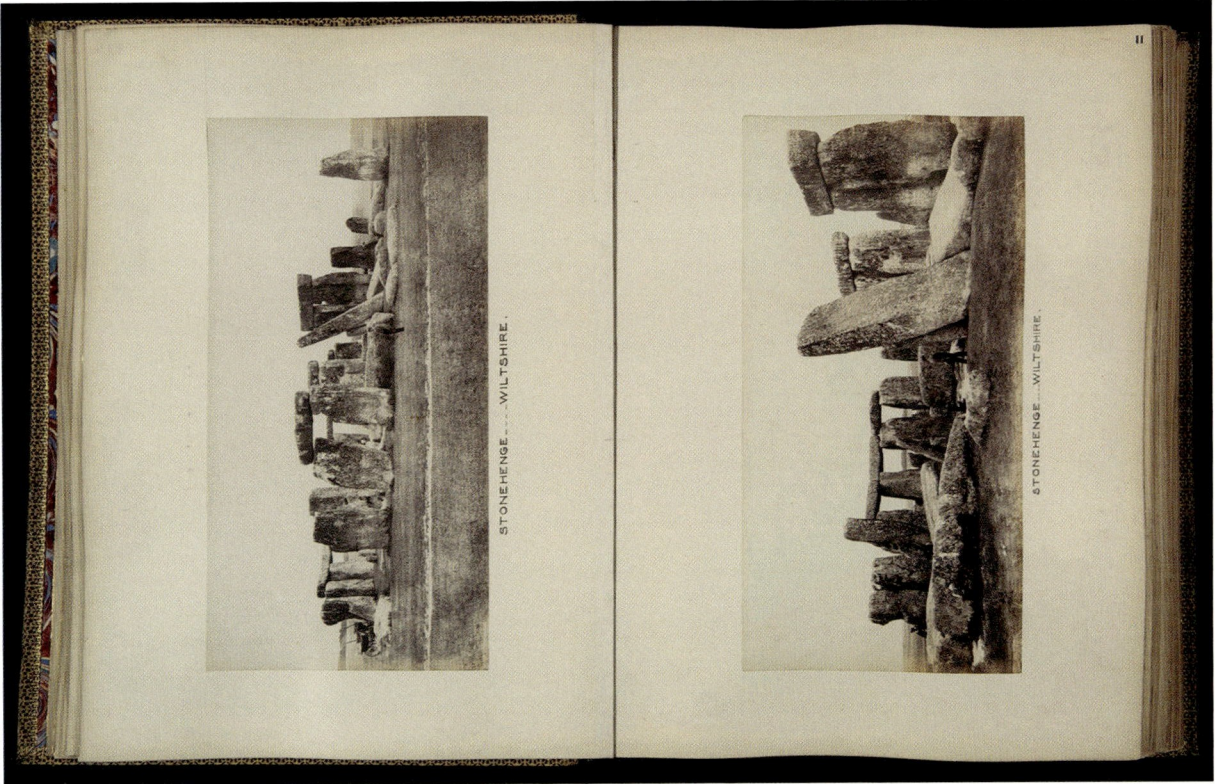

Figure 25.2 Our Ancient Monuments. *Stonehenge, photographed by George Washington Wilson?. Early 1880s. 2012.79.1.18/19 (© Pitt Rivers Museum University of Oxford).*

Figure 25.3 Our Ancient Monuments. *Sueno's Stone, Forres. Watercolour by W. S. Tomkin. Photographs by J. Valentine and Sons, Dundee. 1883. 2012.79.1.79/80/81*

Figure 25.4 Our Ancient Monuments. *Photographs of Avebury, Silbury Hill, Stonehenge, and Old Sarum. Photographers including C. W. Clarke, George Washington Wilson, and Francis Frith. Early 1880s. 2012.79.1.7-13.*

Figure 25.5 Loose albumen print. Cromlech, Trellys [Ffyst Samson], Pembrokeshire. Photograph by H. Jackson, Fishguard. 1888 (© Pitt Rivers Museum University of Oxford. 2012.79.170).

them by mail order while compiling *OAM*. Catalogues from the major firms were widely available, including through local agents.

Single, loose, commercially produced prints, which could be collected and inserted into albums, were known in the trade as 'photographic scraps'. Wilkinson's accounts reveal an extensive trade in 'scraps' being supplied to booksellers and stationers, where they were both displayed in shop windows and advertised locally. The important point is that, for Pitt Rivers and Tomkin, the possibility and existence of photographs to be had was assumed – that an integrated network of knowledge practices and of imaging practices was in place, linking the local to the national, and indeed international. They did not take photographic equipment with them on the early inspectorate tours (Morton 2014). Indeed, it appears that Tomkin did not properly add photographic skills to his technical repertoire as a knowledge worker until 1889, a couple of years after the likely compilation of the album. Pitt Rivers's accounts show that Tomkin received photographic tuition at the Polytechnic Institute, London (a leading provider of technical education at the time),[10] after which Tomkin's notebooks contain several pages of photographic chemical formulae details, lenses, and so forth.[11]

Thus, in the mid-1880s, at the beginning of their Inspectorate work, they were largely reliant on the networks of photographic production that they found locally, or that they could order through the big photographic dealers feeding into the cultural market.

These latter firms had, as noted, very wide distribution networks of local agents and sellers and could also be ordered by catalogue. The availability of photographs was advertised locally and nationally, with even quite modest suppliers carrying huge stocks to feed multiple markets. To give some sense of the scale of the business by 1886, approximately when Pitt Rivers and Tomkin were buying their photographs for *OAM*, Valentine's catalogue offered some 14,000 'views', many of archaeological and historical interest, including Stonehenge and Celtic and Saxon crosses. By 1890 it was 30,000. At the same time, Washington Wilson was advertising some 20,000 views. Importantly, these photographs were often 'packaged' by topic in ways that appealed to scientific and educational use – ancient abbeys or geological formations for instance.

Whatever the precise mode of acquisition, the work of these photographers filtered into many disciplinary uses. While specific archaeological photographic techniques were in place by this period, and photographs were part of the thickening inscription and interpretative practice of the time (Bohrer 2011; Klamm 2021), commercial photographs were used to extend the visualisation of the inspections and surveys. Such photographic sources are unsurprising, and probably a regular occurrence. Pitt Rivers and his assistants were familiar with marketised scientific evidence. Much of his collection, reframed as science, was garnered in the sale room or from dealers (Thompson 1977, 37–8, 71). This was standard practice of the time – Franks of the British Museum was engaged in equally commercial acquisitions, and photographs played an increasingly central part in creating that market and its networks (Edwards 2001, 49–79). Pitt Rivers made extensive acquisitions and repurposing of photographs through the late 1870s to support and demonstrate his archaeological and ethnographic theories (Morton 2014).

Epistemic Images and Reliable Forms

Central to the processes of image absorption is a sense of shared repertoire of concepts of value and 'reliability': photographs and their makers were part of the broader epistemic community, and archaeologists worked their networks hard to create that reliability. If the 19th century did not have uncritical acceptance of photographic evidence, there was a sense of inscriptive reliability and trustworthiness.

While commercial photographs carried what might be described as a 'corporate signature style' of certain patterns of visibility and management of attention that 'amply illustrat[ed] the monument in question, and minimizing any distracting idiosyncrasies that might inhibit the practical polysemy of the image' (Bear 2015, 113), they also implied a form of corporate authority in which could be translated into a discourse of reliability.

Wilkinson, who clearly supplied photographs to Pitt Rivers, described himself as a photographer of topography and architecture and photographic publisher (Fig. 25.6), while C. W. Clarke of Devizes, whose photographs Pitt Rivers also acquired, was also a bookseller and stationer, and described himself as a 'Photographer of architectural antiquities'.[12] Wilkinson's later postcard production makes this explicit. He often included short archaeological or historical texts superimposed on the image.[13] That is, they were aligning their work with values of reliability, accessibility and credibility.

The marketised flow of photographs across and within 19th century knowledge spaces was created as commercial producers of 'views' photographs intentionally positioned their images in order to sell into emerging scientific and disciplinary markets, including archaeology. It is likely that these photographers were in Pitt Rivers' and Tomkin's

Figure 25.6 Loose albumen print. Anglo Saxon chapel of St Laurence, Bradford upon Avon. Photograph by R. Wilkinson. Early 1880s (© Pitt Rivers Museum University of Oxford. 2012.79.163).

broad networks of knowledge and supply. Wilkinson, for instance, did business supplying photographs to the Larmer Grounds, Pitt Rivers' educational pleasure grounds on his Dorset estate (Bowden 1991, 144–9).[14] Wilkinson's surviving letters also show him to be linked to national networks of distribution of photographs of topography and antiquities, including the map and plan suppliers E. Stanford Ltd, London, who supplied Pitt Rivers[15] and who had a photo gallery on their top floor.

Photographers were broadly drawing on the same sets of values that informed an emergent archaeology and contributing to its visualising practices. However, like the knowledge workers Heringman discusses (who were excluded from membership of the Society of Antiquaries), photographers were seldom members of local antiquarian and archaeological societies. Reflecting their social status, neither Clarke nor Wilkinson is listed as a member of the Wiltshire Archaeological Society. This exclusion of technical skills is demonstrated by an incident when, in the 1890s, the Somerset Archaeological and Natural History Society's efforts to undertake a photographic survey of local antiquities foundered on lack of the requisite photographic skill within the membership. The suggestion that local photographers (that is, tradesmen) might be co-opted to the Society as associate members to help was strongly resisted. Their photographic survey quickly petered out. Although the commercial photographers' networks of image dissemination were extensive, their efforts were seen as supporting ancillary skills and providers of information for the emergent practices of archaeology.

Nonetheless, despite such exclusions and tensions, a steady flow of photographs moved across interactive spaces of archaeology, topography, antiquities, and popular consumption. Some photographers positioned themselves as active participants in the networks of knowledge which they wished to supply. Many had serious artistic and/or scientific interests themselves which enhanced their authority as providers of reliable and epistemic images and, at a local level, photographers were integrated in, and integral to, local civic society and its knowledge economy.

For instance, Gloucester photographer Arthur Pitcher was a keen antiquarian and took photographs at excavations of Roman remains in Gloucester, supplied photographs to the photographic survey of Gloucester, sold views of local antiquities, and contributed to local photographically illustrated guidebooks. His son Sydney, also a photographer, developed projects with the Bristol and Gloucester Archaeological Society and became an expert of medieval stained glass, selling prints to the V&A for use in their massive reference series on architecture and sculpture (Rosewell 2012).[16] While Pitcher senior was never a member of the local archaeological society, his son, in a combination of shifting social status and expertise, was elected in 1924. Conversely, Belfast photographer R.W. Welch was very active in the Belfast Natural History Society from the late 1880s onwards. As Ireland's leading commercial photographer, he produced numerous 'touristic' images of the essential Ireland – peat stacks, picturesque crofts, and slide carts – but also of stone circles, megalithic monuments, and geological formations. Significantly, his 'touristic' photographs were recoded as ethnographic or folkloric subjects which he sold into scientific markets, including to the Pitt Rivers Museum in Oxford and its Cambridge counterpart, having an extended correspondence with anthropologist A. C. Haddon.

Within this visual economy, proto-modern archaeologists themselves also sold photographs of their interests, excavations, and finds (Klamm 2021), the authority of which was based in their claims within a particular community of knowledge. An interesting example is John Parker, classical archaeologist and later Keeper of the Ashmolean Museum in Oxford. He was an ardent exponent of the value of photographs in archaeological and classical research to produce a history 'taught by the eye' (Parker 1870, 14). Parker collected photographs from the major photographers and dealers in Italy, especially in Rome, but, importantly, finding these too aesthetic in their corporate representational practices, he commissioned and supervised photographs made by a string of local Italian professional photographers. The mass of their names indicates the huge network of 'knowledge workers' underpinning archaeological and architectural studies by the third quarter of the 19th century. However, Parker also offered his photographs for sale to interested parties. The South Kensington Museum (V&A), for example, purchased over 1000 (Beattie 2018), whilst also buying some of the same archaeological and architectural images that I have been describing.

Information about 'reliable' photographers was also part of the flow of archaeological knowledge. Not only is this demonstrated by the overlaps in photographic holdings as archaeologists bought from the same sources – Franks' ethnological notebook, from his time in Berlin meeting luminaries of German anthropology and archaeology, has a scribbled list of addresses inside the front cover. They are the addresses of photographers, suggesting the exchange of information within disciplinary interests.[17] Pitt Rivers was part of this network of knowledge and the recoded sanctioning of commercially produced photographic material (Morton 2014, 172). An example here is a photograph of Kity Cot's House, interleaved in *OAM*, which was acquired from W. Mansell & Co in London (it bears their stamp). It is possible that the Kity Cot photographs which open *OAM* are from the same source. Importantly, for my networked argument, Mansell's acted as agents for a number of photographers, including Stephen Thompson's set of 929 photographs of British Museum collections produced in conjunction with the Anthropological Institute in the 1870s.[18] Pitt Rivers must certainly have known these images;

he was closely involved with the Anthropological Institute and its projects at this period, and thus known Mansell's as a source of reliable photographs.

Whatever the mode of acquisition, the work of these photographers filtered into many disciplinary uses, not only archaeology. There is evidence that this was intended. Photographs and lantern slides were packaged in ways that aligned them with the interests of the emergent historical sciences of archaeology, anthropology, and geology, and their evolutionary frames.[19] The large topographical firms, as noted, specialised in detailed and attentive recording which could be recoded as archaeology. For example, the 1913 report of the Inspectorate of Ancient Monuments (1913, 10), by Pitt Rivers's eventual successor, Sir Charles Peers, used a Frith photograph of St Botulph's church, Colchester, while as late as 1925, an annotated commercially produced picture postcard of Eggleston Abbey, County Durham, served as an impromptu Inspectorate site report.[20] One could keep piling up the examples, evidenced through marks, mount cards, photographer's stamps, and collections overlaps. However, I want to emphasise the symbiotic relationship between archaeology, collecting, and photography's commercial interests which reaches well into the 20th century, despite long-established techniques of archaeological field photography (see Baird 2017; Riggs 2018; McFayden & Hicks 2020). Yet simultaneously, archaeological imagery and its commercial absorptions were widespread, as *OAM* demonstrates so lucidly.

Conclusion

In the album, photographs work as 'boundary objects', that is, material objects and representations which form connections between different individuals and groups which might be interpreted differently within those networks with their different aims, but which are, at the same time, robust enough to maintain their own photographic and inscriptional identity across them (Davidson 2017, 10). What is important, for understanding larger overlapping networks of archaeological meaning-making at the period, are the dynamics between different actants – photographers, tradesmen, middlemen, and scientific consumers – as the recodability of photographs was translated, self-consciously, on the one hand as scientific data and on the other hand into business opportunity. As Flandreau (2017, 15) puts it: 'analyzing the very material forces at work in the production of knowledge structures can shed light on the meanings we attach to knowledge … styles of ownership of knowledge [and its enabling] [are] as important as understanding how it is produced.' The disciplinary consumption of images across the broad scientific field in the 19th century was not merely a matter of making epistemic images and meanings within those spheres of consumption, but also an active dialogic relationship with photographers and their commercial opportunism

and ambition, both as active knowledge workers and hard-headed business men and women. They represent different but inextricably networked styles of knowledge production and ownership which we can see as General Pitt Rivers and Tomkin gathered photographs for their album.

Notes

1. This album was acquired by the Pitt Rivers Museum, University of Oxford in 2012. See: https://excavatingpittrivers.blogspot.com/2012/12/our-ancient-monuments.html
2. https://web.prm.ox.ac.uk/rpr/index.php/progress-reports-index/11-projectreports/81-find-out-about-the-rethinking-pitt-rivers-project.html
3. For an account of Tomkin's archaeological contribution see: *Rethinking Pitt-Rivers*. https://web.prm.ox.ac.uk/rpr/index.php/articles-index/894-william-stephen-tomkin.html. Significantly, given my network argument, Tomkin was subsequently employed as a draughtsman for Waterlow & Sons, high quality lithographic printers who worked for both the British Museum and South Kensington Museum (V&A) (SSWM. Pitt Rivers Correspondence, L774 [Thompson 1976, 105]).
4. The latter are both Scottish firms. Not unlike whisky distilling, local water quality was seen as particularly efficacious in photographic processing.
5. Pitt Rivers himself, as Inspector, produced a report on the condition of Stonehenge in 1893 (TNA WORK 14/213).
6. All offered photographs of Stonehenge in their catalogues by the mid-1880s. In the 1890s a photographer from Shrewton was effectively in charge of access to Stonehenge and made his living selling photographs to visitors (Richards 2004, 23).
7. It is highly likely that these photographs, and others, fed into the making of models of Celtic crosses (now in Salisbury Museum) which were made for Pitt Rivers on his estate.
8. As demonstrated by Wilkinson's accounts ledger for 1884/5. WSHC. 1299/1.
9. TNA Works 39/16, f. 225.
10. SSWM. Pitt Rivers Accounts 1889 (Thompson 1976, 136).
11. TNA WORKS 39/10.
12. *Kelly's Directory of Wiltshire* (1889, 907 & 1037).
13. Examples in author's collection. By 1900, Wilkinson's business model, like that of Valentines, had turned towards the production of picture postcards, the new photographic mass-medium.
14. WSHC Wilkinson Business papers, 1299/1.
15. SSWM. Pitt Rivers Correspondence B496 (Thompson 1976, 117.)
16. Sydney Pitcher's photographs are now in the Historic England archives.
17. Franks Notebooks British Museum, DOAA, LS14. They are of photographers well represented in 19th century German archaeology and ethnology museums, such as Carl Dammann and Wilhelm Burger.
18. They were advertised as demonstrating cultural evolution (Edwards 2001, 69–73).
19. Wilson also recodes his Stonehenge photographs as geology, examples of Sarsen stones.

20. Inspectorate of Ancient Monument 1913, 10. HEA. Loose Photo files FL00724. This relates to a report on fallen masonry, TNA WORK 14/301.

Archives

Historic England Archives (HEA)
Pitt Rivers Museum Photograph Collections – 2012.79
The National Archives (TNA), WORK
Wiltshire and Swindon History Centre (WSHC), Wilkinson papers
Salisbury and South Wiltshire Museum (SSWM), Pitt Rivers Correspondence

References

Baird, J. A. (2017) Framing the past: Situating the archaeological in photographs. *Journal of Latin American Cultural Studies* 26(2), 165–86.

Barber, M. (2014) *Restoring Stonehenge 1881–1939*. London, English Heritage Research Report Series 6.

Bear, J. (2015) *Disillusioned: Victorian Photography and the Discerning Subject*. University Park, PA, Penn State University Press.

Beattie, S. (2018) Recording Rome: The John Henry Parker Collection [online]. Available at: <https://www.vam.ac.uk/blog/caring-for-our-collections/recording-rome-the-john-henry-parker-collection> [accessed 12 February 2022].

Bohrer, F. (2011) *Photography and Archaeology*. London, Reaktion.

Bowden, M. (1991) *Pitt Rivers: The Life and Archaeological Work of Lieutenant-General Augustus Henry Lane Fox Pitt Rivers DCL, FRS, FSA*. Cambridge, Cambridge University Press.

Daston, L. (2015) Epistemic Images. In A. Payne (ed.), *Vision and its Instruments: Art, Science and Technology in Early Modern Europe*, 13–35. College Station, PA, Penn State University Press.

Davidson, K. (2017) *Photography, Natural History and the Nineteenth Century Museum*. London, Routledge.

Edwards, E. (2001) *Raw Histories: Photographs, Anthropology and Museums*. Oxford, Berg.

Flandreau, M. (2017) *Anthropologists in the Stock Exchange: A Financial History of Victorian Science*. Chicago, Chicago University Press.

Heringman, N. (2013) *Sciences of Antiquity: Romantic Antiquarianism, Natural History and Knowledge Work*. Oxford, Oxford University Press.

Inspectorate of Ancient Monuments (1913). *Report of the Inspector of Ancient Monuments for the Year ending 31st March 1913*. London, HMSO.

Klamm, S. (2021) Refiguring the use of Photography in Archaeology. In S. Hillnhueter, S. Klamm & F. Tietjen (eds), *Hybrid Photography: Intermedial Practices in Science and Humanities*, 368–93. Abingdon, Routledge.

McFayden, L. & Hicks, D. (eds) (2020) *Archaeology and Photography: Time, Objectivity and Archive*. London, Bloomsbury.

Morton, C. (2014) The place of photographs in the collections, displays and other work of General Pitt-Rivers. *Museum History Journal* 7(2), 167–87.

Parker, J. H. 1870. *The Ashmolean Museum: Its History, Present State and Prospects*. Oxford, s.n.

Richards, J. (2004) *Stonehenge: A History in Photographs*. Swindon, English Heritage.

Riggs, C. (2018) *Photographing Tutankhamun: Archaeology, Ancient Egypt, and the Archive*. London, Bloomsbury.

Rosewell, R. (2012) The Life of Sydney Pitcher FRPS. *English Heritage Historical Review* 7(1), 111–24.

Smiles, S. & Moser, S. (eds) (2005) *Envisioning the Past: Archaeology and the Image*. Oxford, Blackwell.

Thompson, M. (1976) *Catalogue of the Correspondence and Papers of Augustus Henry Lane Fox ...in the Salisbury and South Wiltshire Museum*. For RCHM, typescript Pitt Rivers Museum, Oxford.

Thompson, M. (1977) *General Pitt-Rivers: Evolution and Archaeology in the Nineteenth Century*. Bradford upon Avon: Moonraker Press.

Thurley, S. (2013) *The Men from the Ministry: How Britain Saved its Heritage*. New Haven, CT/London: Yale University Press.

26

On the Origins of Khami: Evidence from the Henry Balfour Collection, Pitt Rivers Museum, Oxford

Innocent Pikirayi

Ceramics collected from the site of Khami in southwestern Zimbabwe in 1905 by Henry Balfour, curator of the Pitt Rivers Museum, represent some of the most decorated sherds recovered there and, arguably, in the Zimbabwe Tradition (AD 1100–1900). The decoration not only attests to the existence of a complex manufacturing tradition, but its replication on monumental architecture also suggests the value of pottery in understanding the development of socio-political complexity in the region. The ceramics were analysed using conventional typology, not only to characterise the assemblage, but also to assess the value of such approaches in understanding the role of ceramics in socio-political complexity. Henry Balfour's accompanying notes to the collection were also used to understand the assemblage, particularly on the role of lithics, often associated with earlier Stone Age contexts. The study discovered that, to understand the role of band-and-panel ware at Khami in the development of socio-political complexity, is important to bridge the gap between conventional typology and technology.

Keywords: Khami Pottery, Balfour Collection, Pitt Rivers Museum, Ceramics, Origins, Socio-political Complexity

Introduction

The following extract from Henry Balfour's diary entry of 19 September 1905, transcribed from the Archives of the Pitt Rivers Museum, provides the context of this chapter:

> Tues. 19th. Arrived at Bulawayo at 7.30 a.m. Put up at Grand Hotel… Col. Fielder + [and] I shared a trap [sic] to Khami (2 hours, 14 miles), Andrews riding with us. Explored the ruins at Khami + [and] found much pottery + [and] flint + [and] agate flakes. Andrews gave us a lot of wire work, spindle whorls, + [and] pottery sherds + [and] gave me an old Makalanga iron hoe blade found amongst, the debris [illustrated]. Took some photos of ruin walls. The debris heaps marvellously extensive and flint flakes very numerous. This is explainable by fact that of there being much pottery which has been incised (apparently with flint) <u>after baking</u>, in addition to the pottery decorated with incisions while still plastic, before baking. The same occurs at Dhlo Dhlo ruins, where flints are numerous, but not at Zimbabwe or Inyanga where flints are scarce, nor at Umtali where hardly any flints or quartz flakes have been found … Back to B'wayo on 2 hours, arriving at 5.30…

This chapter examines the pottery collected by Henry Balfour, the first curator of the Pitt Rivers Museum, from the site of Khami on the Zimbabwe plateau, following a visit there in 1905 (Fig. 26.1).

Khami is the second largest group of monuments on the Zimbabwe plateau after Great Zimbabwe, located on the banks of the Khami River, after which it is named (Robinson 1959) (Fig. 26.2). In an environment dominated by granite and dolerite, *Colophospermum mopane* and *Vachellia karroo* trees, and varieties of grasses suitable for cattle grazing (*e.g. Themeda triandra, Heteropogon contortus,* and *Cynodon dactylon*), Khami occupied an area more than 0.4 km in extent. The archaeology shows 10 residential clusters, which probably housed members of the ruling class. The ruler resided in the biggest cluster, the Hill Complex. There, a secret room was discovered in 1947 by archaeologist Keith Robinson (1959) at the top of the passage, where royal regalia was found wrapped in cloth and hidden in a basket. The basket contained iron axes (one with a wooden handle covered in copper sheets), iron spears, two copper ingots

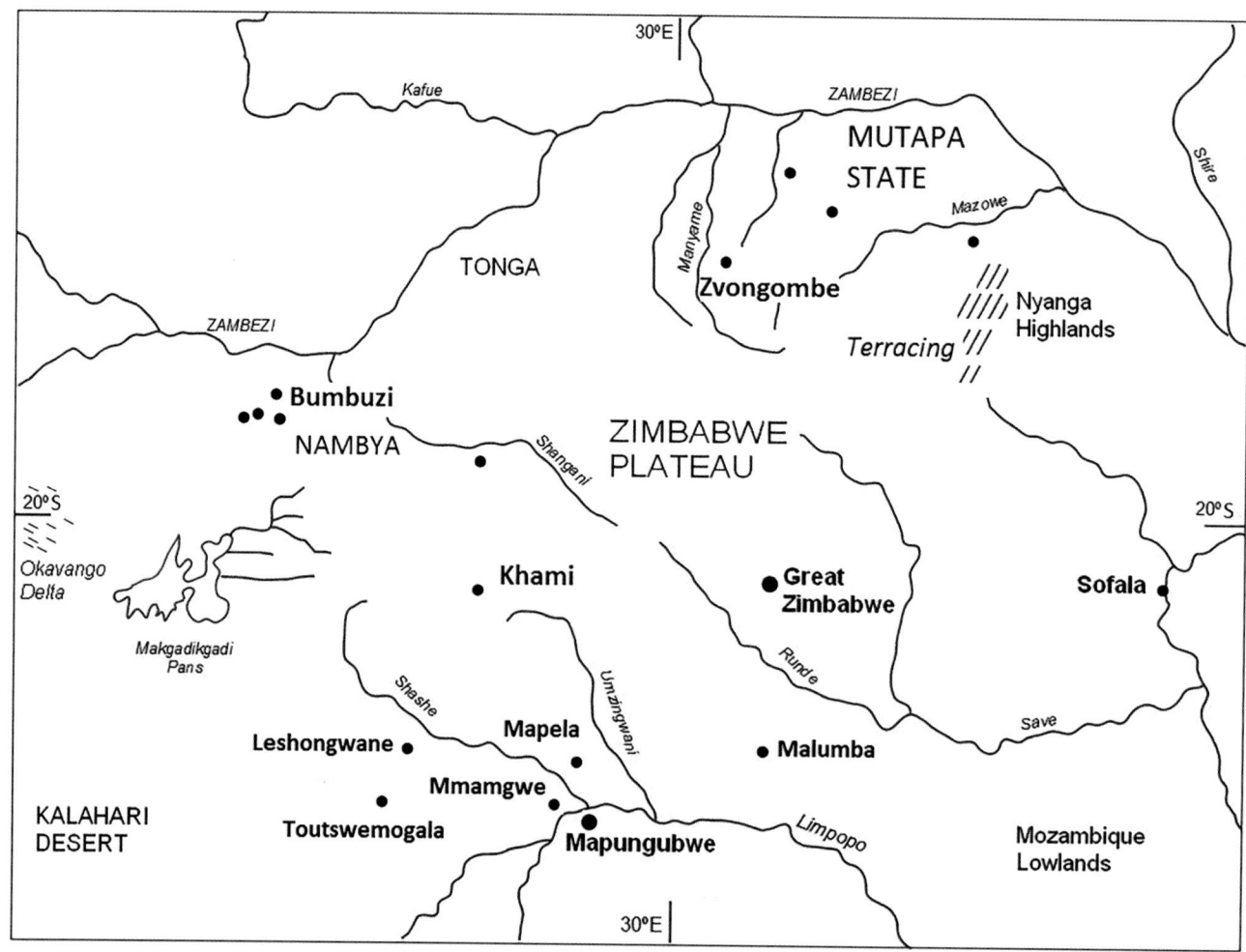

Figure 26.1 The location of the site of Khami on the southwestern Zimbabwe plateau.

(each of a different design), small ivory carvings of lions and leopards for the tops of ceremonial staffs, a set of carved ivory divining dice, and drinking pots decorated with red and black patterns. According to oral traditions and written accounts, Khami was the capital of the Torwa State (AD 1490–1650), which rose in the southwest regions of the Zimbabwe plateau and, following a dispute, was destroyed by the Portuguese, led by Sismundo Dias Bayao, a *prazero* (estate holder and warlord), around 1644 (Beach 1980; Van Waarden 2012).

The stone architecture of Khami, for example, is a combination of the building tradition seen at Great Zimbabwe and the local Leopard's Kopje Tradition. This includes terraced stone platforms and elaborately decorated stone walling (see Fig. 26.2). Most of the stone walls are retaining rather than freestanding, with houses erected on top of stone platforms or on small hills. There are also roofed passages plastered with clay (*dhaka*). The pottery recovered at Khami is decorated in band and panel motifs (Robinson 1959), and the origins of this ceramic tradition continues to be the subject of archaeological enquiry, especially in understanding the origins of socio-political complexity in the region.

Research History

Like Great Zimbabwe, Khami attracted the attention of late 19th and early 20th century European explorers, antiquarians, fortune seekers, and early professional archaeologists (see *e.g.* White 1901; Randall-MacIver 1906; R. N. Hall 1909; Caton-Thompson 1931). The most extensive excavations at Khami were conducted by Keith Robinson (1959) on the Hill Complex. Huffman (1984; 1986; 1996) identified the central area of the site with the elite and the periphery with commoners. Carolyn Thorp (1984; 1995) dug some middens within and around the Hill Complex, recovering faunal data that provided information on the inhabitants' subsistence patterns. Gwilym Hughes (1997) excavated the northwest peripheral area of Khami, recording complex architectural developments and cultural activities. He demonstrated complementarity between the peripheral buildings and those in the central area of the site, the latter of which are more monumental in scale.

During the late 1990s, University of Zimbabwe archaeologists conducted research at Khami to assess the infrastructure, visitor facilities, publicity, and conservation status

Figure 26.2 The Hill Complex at Khami (Image reproduced with the permission of Senzeni Khumalo).

of the site. I conducted a desktop study on research, both published and unpublished, undertaken at Khami. This resulted in a detailed historiography of the site from the 19th century onwards. During the exercise, I realised that Henry Balfour had collected some artefacts from Khami in 1905. Records also indicated that some material from Khami was loaned to the British Museum for exhibition during the 1930s. A follow-up was made with the Pitt Rivers Museum in Oxford in July 1999. Curator Jeremy Coote allowed me access to the material, but due to limited time, I only managed to examine 25 elaborately decorated pieces, all of them from display. A fellowship grant from the Commonwealth Commission made it possible to examine the material in more detail between November 2000 and April 2001.

Between early 2000 and the end of the first decade of the 21st century, not much archaeological research has been conducted at Khami that ties in with the aforementioned studies of the collections. Focus has been on the conservation of stone-walled structures. Fieldwork by Catrien van Waarden (2012) in adjacent eastern Botswana, a region with numerous Khami type sites, and work on other Zimbabwe culture-type sites in southern and south-central Zimbabwe (Chirikure *et al.* 2013a; 2013b), however, triggered a revisit of the site of Khami. Research by Mukwende (2016) is the most definitive in this regard. Chirikure and various colleagues (2001; 2012; 2013a; 2013b; 2016; 2018) and Mukwende (2016) have not only attested to a much earlier antiquity of Khami but also suggest a multidirectional evolution of socio-political complexity in southern Africa. Thus, Khami may not have developed directly from Great Zimbabwe (see *e.g.* Garlake 1972; Pikirayi 2001) but may have developed as a primary state. Although both Great Zimbabwe and Khami presided over state systems characterised by monumental stone architecture, there are innovations that are typical of Khami and its contemporaries in southwestern Zimbabwe.

Methodology

A total of 145 sherds from the site of Khami in the Balfour collection at the Pitt Rivers Museum were examined and subjected to basic typological analysis. Attributes measured to understand vessel manufacture include the lip-form, vessel thickness, surface treatment, the degree of firing, and decoration. Using published oral and other accounts, a discussion on the origins and identity of the makers of the 'band and panel ware' (Khami) was attempted. Terms such as pot/bowl were employed in a non-functional sense. Robinson

(1959) identifies zoomorphic and special features such as lids, lugs, and handles. Balfour (1906) did not collect these from Khami, although some sherds had special features, such as bosses and mouldings on the neck-shoulder region.

My approach synchronised categories defined from Balfour's collection with those established by previous investigators. The complexity of decorative motifs on Khami pottery presents difficulties with class definition, such that a class may only be represented by a sherd or vessel. Thorp's (1995) classification, for example, sometimes differentiates sub-classes based on the absence of one attribute/attribute state (see *e.g.* Class 3B3 & 3B4), while the status of some sub-classes (*e.g.* 3C, 6 & 7) remains unclear. The problem could be sample size itself, which, if fragmented, obscures decoration layout. This calls for broader definitions of vessel classes/shape groupings. Also, with Khami pottery, no single researcher, including Mukwende (2016), has demarcated the full range of vessels available. Despite extensive excavations, Robinson (1959) still consulted Randall-MacIver (1906) on certain shapes not identified from his assemblages. Instead of using rim-shape morphology in defining vessel class, I link shape with decoration motif. This allows for broad vessel groupings sharing the same attributes to be compared with other assemblages.

There has been a dearth in typological approaches to archaeological ceramics in southern Africa since 2000. I have previously critiqued how ceramic typology poorly informs us about past forms of socialisation, including the construction of human group identities (Pikirayi 2007; see also M. Hall 1984; Huffman 1974; 1989; 2007). Pottery manufacture and use should inform us about social meaning, values, and other human activities. Ceramics reflect multiple identities and meanings, which may be expressed in graphic terms, including decoration. Similar expressions appear on other material objects, *e.g.* drums, grain silos, headrests, etc. (Aschwanden 1987; Huffman 1989). Recent approaches on the Zimbabwe plateau have interrogated ethnography using decolonial approaches to better understand these archaeological contexts (see *e.g.* Pikirayi & Lindahl 2013; Nyamushosho & Chirikure 2020); however, none of these studies directly relate to the material recovered from Khami.

Results

Balfour collected at least eight shape categories from Khami (Table 26.1), which I compare to typologies from previous works (Robinson 1959; Thorpe 1995; Hughes 1997) and also discuss in light of recent work (Mukwende 2016). Though Mukwende (2016) has conducted the most comprehensive analysis of pottery from the site, details are not replicated here due to length constraints.

The characterisation of the pottery collected by Balfour is further presented in Figures 26.3–4. The material collected by Henry Balfour is dominated by sherds with well-finished rims, necks, and shoulders, which fall broadly into shape-forms I and III (see Table 26.1). Rim morphology (Fig. 26.3) was determined from a sample of 40 vessels. Rims are essentially rounded, with a large proportion being everted.

Most vessel walls are thin (6–9 mm). With the exception of two sherds, Balfour collected sherds that were elaborately decorated with complex geometric designs, comprising lines, triangles, bands, diamonds, and zones. These shapes were filled in with graphite burnish, red ochre, or brown buff polish, as well as diagonal, vertical, or horizontal engraved incised lines, and these patterns were made on the rim-neck and shoulder body regions. A high percentage of the material is well-fired brown to reddish brown, but some sherds have brownish-black or blackened exterior surfaces, suggesting functions related to cooking or exposure to fire after discard. Some sherds show variegated firing, suggesting poor firing conditions. Due to pre-treatment, it was difficult to quantify the soot coating accurately. Only five sherds had clear evidence of soot coatings on the exterior.

Only a sample of decorated sherds is presented here. Figure 26.5 shows motifs primarily from bowls (see Robinson 1959; Mukwende 2016). The complex geometric designs on the rim and neck regions are incised. Inside are bands of graphite (GB) and red ochre (RO). The majority of sherds in Figure 26.6 derive from constricted bowls. One sherd comes from a globular vessel with a short in-sloping neck and a vertical rounded lip. Decoration is on the rim/neck/shoulder region. The decorative motifs are triangular designs etched by incision. The triangles are sometimes separated by oblique or chevron bands filled in by graphite burnishing or red ochre. Black polish appears on some sherds. Figures 26.7 and 27.8 show globular vessels with tall, vertical necks incised with complex triangular designs, filled in with red ochre or graphite burnish. These shapes may carry complex geometric designs on their shoulder/body regions. These designs are also depicted on bowls. Mukwende's (2016) motifs 15a–15e occur on the bodies of deep, straight-sided or open bowls (see Mukwende 2016, 94, shape profiles D2–D4). Figure 26.9 illustrates sherds incised with herringbone designs on the shoulder/body region. For some sherds, the transition from shoulder to rim is enhanced by thickening the vessel body, which appears on the exterior as a 'ribbed' neck. Some sherds carry horizontal bands of crosshatching on the shoulder. The entire vessel body was polished with either graphite or red ochre. Mukwende (2016) identified herringbone motifs on constricted jars (shapes C1, C2, motifs 10a–10d), and crosshatching on vessels with tall necks, thickened neck regions, and fluted rims. Figure 26.10 depicts vessels with tall necks and fluted rims. One body sherd carries a bobble. A globular vessel with an in-sloping rim has brown polish or red ochre. Most sherds are decorated with multiple horizontal and sometimes alternating bands of graphite and red ochre.

Table 26.1 Vessel shapes from the Balfour collection compared with other assemblages excavated at Khami.

Shape Form	Robinson (1959)	Thorp (1995)	Hughes (1997)	Total
I: Globular Pot	Form A: Spherical vessels without necks; thick rolled rims	Class 5: Constricted vessels with rolled rims decorated with graphite and graphite bands inside the rims Undecorated vessels with short, concave necks and everted or flattened rims Vessels with rolled rims decorated with graphite with a band of incised triangles, some interlocking, separated by graphite bands on the shoulders Vessels with plain, rolled rims and incised interlocking triangles on the shoulder	Form III: Constricted vessels with spherical or spheroidal bodies with little or no neck	13
II: Globular Pot	Form B: Spherical vessels with short, concave necks, rolled rims and undercut. Rarely decorated	Class 1: Shouldered jars with everted rims and straight necks. Some vessels have an incised line on the shoulder/neck region Class 2: Shouldered jars with everted rims and straight necks, covered with a band of graphite on the inside rim and an incised line on the neck/shoulder region	Form I: Shouldered jars with short necks, v(i): vertical necks with plain rims v(ii): vertical or slightly concave necks with everted rims	3
III: Globular Pot	Form C: Spheroidal vessels with comparatively high necks, either vertical, concave or convex; conical bowl-shaped or funnel-shaped. Fluted forms common. Decoration also below the neck	Class 4: Shouldered jars with tall, straight necks, everted rims, decorated with graphite and graphite bands on the inside rim, graphite on the rim, and panels of graphite, red ochre, and incised herringbone on the neck, graphite on the rim and inside rim with red triangles separated by a graphite band on the neck, graphite on the rim and a band of alternating red and incised triangles on the neck Class 6: Plain jars with slightly in-sloping necks, tapered rims Class 7: Shouldered jars with flared necks and tapered rims, with a row of pendant triangles filled on the inside with bangle impressions	Form II: Shouldered jars with short necks v(i): vertical concave or funnel-shaped necks with everted rims v(ii): fluted necks with everted rims	40
IV: Globular Pot	Gourd-shaped vessels	Class 3: Constricted pots decorated with graphite or a band of black graphite on the rim, a plain or red band and black or graphite band on the shoulder separated by incised lines, a band of graphite on the rim above a row of incised, interlocking triangles separated by coloured bands on the shoulder, a red band on the rim above a red bordered band of oblique incisions and incised triangles on the shoulder with graphite on the body, a graphite band on the rim above bordered bands of incised cross-hatch on the shoulder with a graphite body, a plain rim and interlocking incised triangles on the shoulder		5
V: Pot	Zoomorphic objects			-
VI: Hemispherical Bowls	Bowls			0
VII: Deep, straight-sided bowls	Beaker Bowls			1
VIII: Features like handles, lugs, lids, etc.	Special Features			4
Uncategorised				79
			Total	**145**

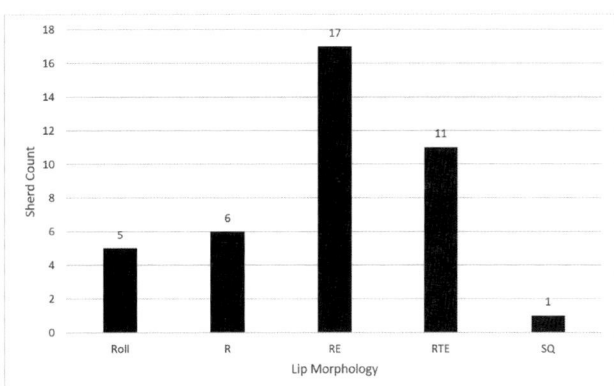

Figure 26.3 Lip morphology. Roll = rolled; R = rounded; RE = rounded and everted; RT = rounded, tapered, and everted; SQ = square lip-form.

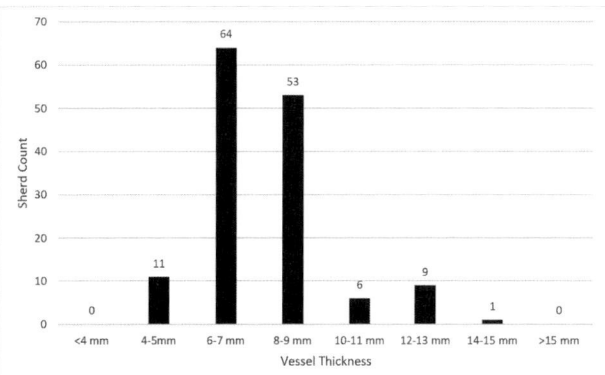

Figure 26.4 Distribution of vessel thickness in the assemblage, 145 sherds total.

Discussion

Ceramic Production

Some of the data presented here relates to aspects of ceramic production and use. Potting clay was most probably sourced locally. Based on some of the stone tools recovered by Balfour (1906), clay preparation and mixing were done within the site of Khami itself. The clay was ground to fine powder and tempered with fine (>0.5 mm) grains of sand-quartz to obtain the required potting formula. Balfour (1906, 18) observed that the 'pottery is very full of sand, apparently derived from granite denudation, in which quartz crystals predominate'. The vessels were moulded and then both the exterior and interior surfaces were polished smooth. Figure 26.4 shows that potters made generally thin and thick-walled pots and bowls, and potters seem to have so expertly mastered knowledge of the clay that they knew how it would react during open firing. Most shoulder-body sherds are fractured along the neck or zone of transition, further suggesting how these vessels were formed. A tall-necked globular pot might have been produced in two stages: first, a constricted bowl and hollow cylindrical shape were produced; second, the cylinder was attached to the bowl's neck/shoulder region. Some interior surfaces of these sherds show additional clay was used to strengthen the curvature. While this may appear as 'decoration' on some vessels, it could have been integral to the production technique to reduce thermal stress and breakage. Balfour's observations that '... much of the pottery had evidently been engraved while yet in a soft and plastic state' but also that 'a considerable quantity ... [was] evidently engraved after passing through the firing process' supports this view.

Balfour (1906) reports how he managed to relate sherds from the middens at Khami with stone flakes found in association with them and concluded that the pottery was 'stone-carved'. Comparing the lines of incision etched on vessels that are still partly plastic with those made after the vessels had completely dried, he observed that the patterns produced on the latter were ragged and irregular, lacking the blurred edges of the former. This was due to the numerous quartzes or chalcedony flakes found associated with the pottery. Balfour demonstrated this claim by engraving a plain sherd with a stone flake. This area was also inhabited by hunter-gatherer groups from Middle Pleistocene times (Cooke 1957), which accounts for the stone tools, including microliths. Khami potters may have reused these tools, as well as the numerous other flakes. Caton-Thompson (1931, 127) observed similar contexts at Chiwona, in eastern Zimbabwe:

> At the base of the kopje, particularly on the south-west side and up its slope, a considerable number of quartz flakes are noticeable. Veins of white quartz abound in the granite, and the materials has evidently been used in the manufacture of implements. None that I could find, however, showed secondary trimming: they were simple flakes with striking platform and bulb, but their human origins is unquestionable. As the pottery at this site is certainly not incised with a stone point, but by an impression on the wet clay, I am unable to suggest the purpose to which these quartz flakes were put and I have no reason to believe them contemporary with the ruins. Our finds at Dhlo Dhlo ... hint at the occupation of kopjes by Stone Age man long before stone walls were erected upon them.

Lithics, it appears, were an integral part of ceramic manufacture at Zimbabwe culture sites.

The pots attest to investment in high-quality production. Balfour (1906, 17) remarked that:

> A very large proportion of the pottery is decorated with incised patterns, in addition to a general coating of plumbago relieved by patterns partly of the natural grey, red, or yellowish colour of the baked clay, and partly of a bright red colour painted on to the surface, apparently before the pots were baked, in rough but effective style, the colour effects being combined with the incised lines in producing decorative designs. Raised bands and bosses also occur.

Balfour (1906, 18) was also concerned with incised designs on vessels, describing them as:

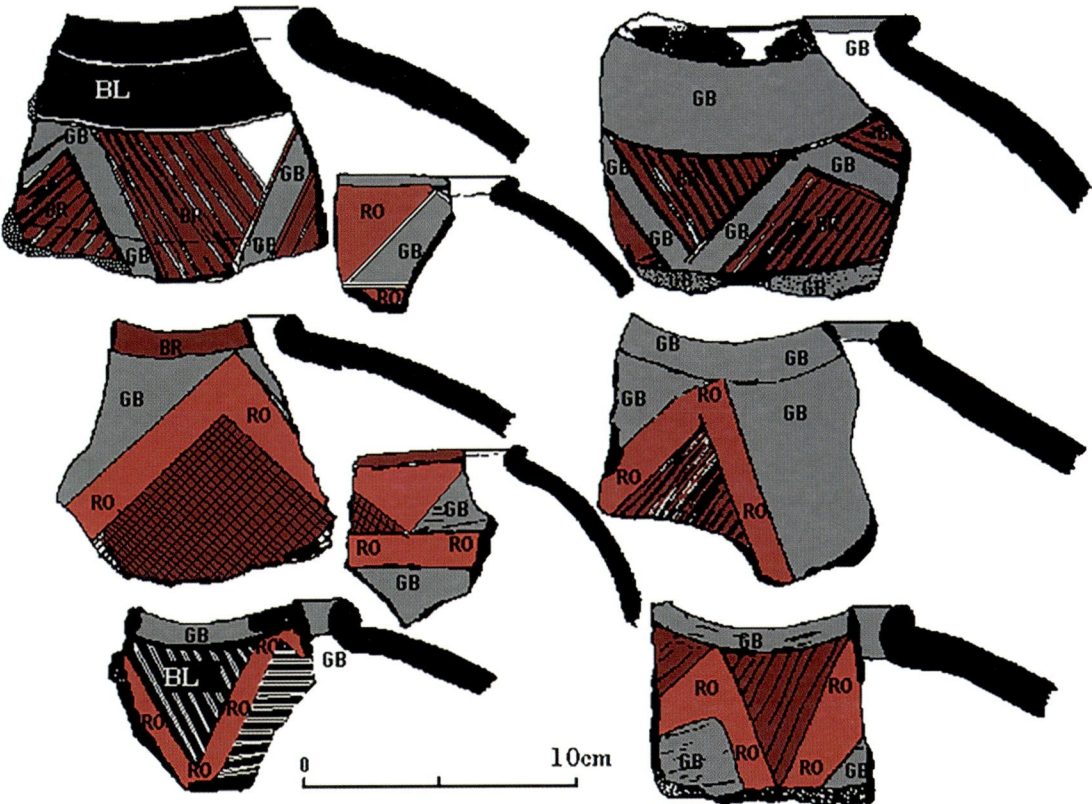

Figure 26.5 Decorated bowls from Khami (Illustration: J. Chikumbirike).

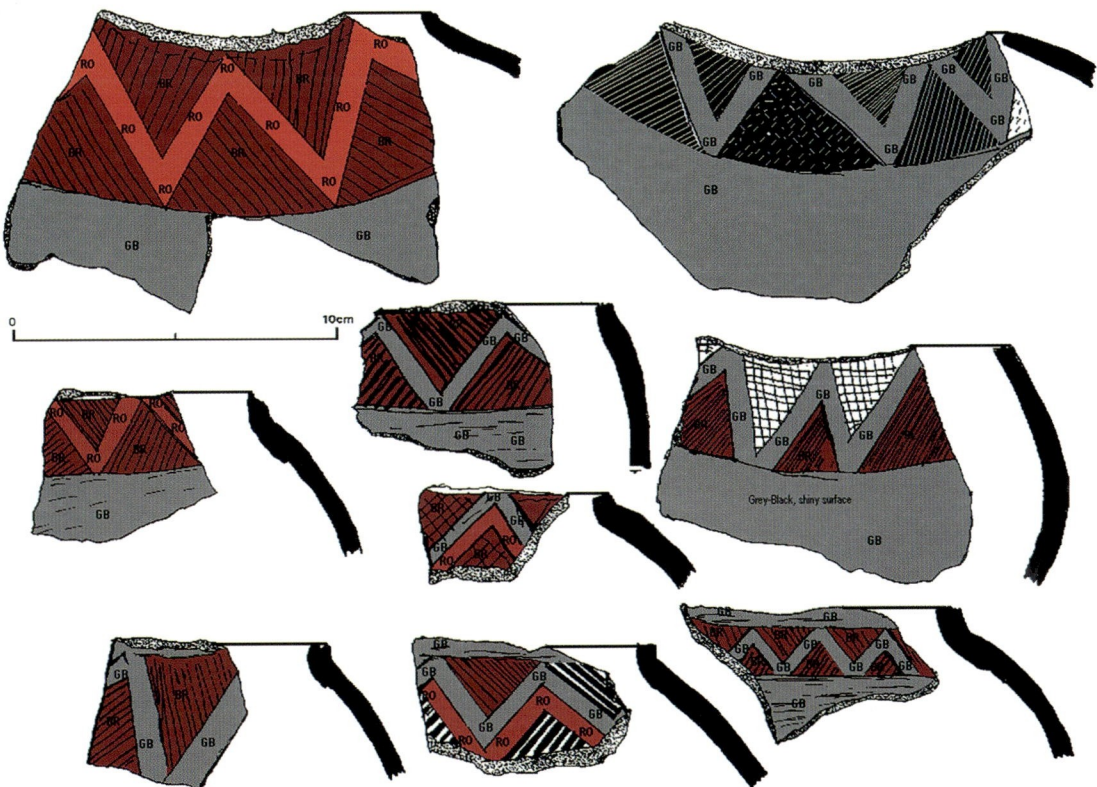

Figure 26.6 Sherds from constricted bowls or globular vessels with in-sloping necks/rims (Illustration: J. Chikumbirike).

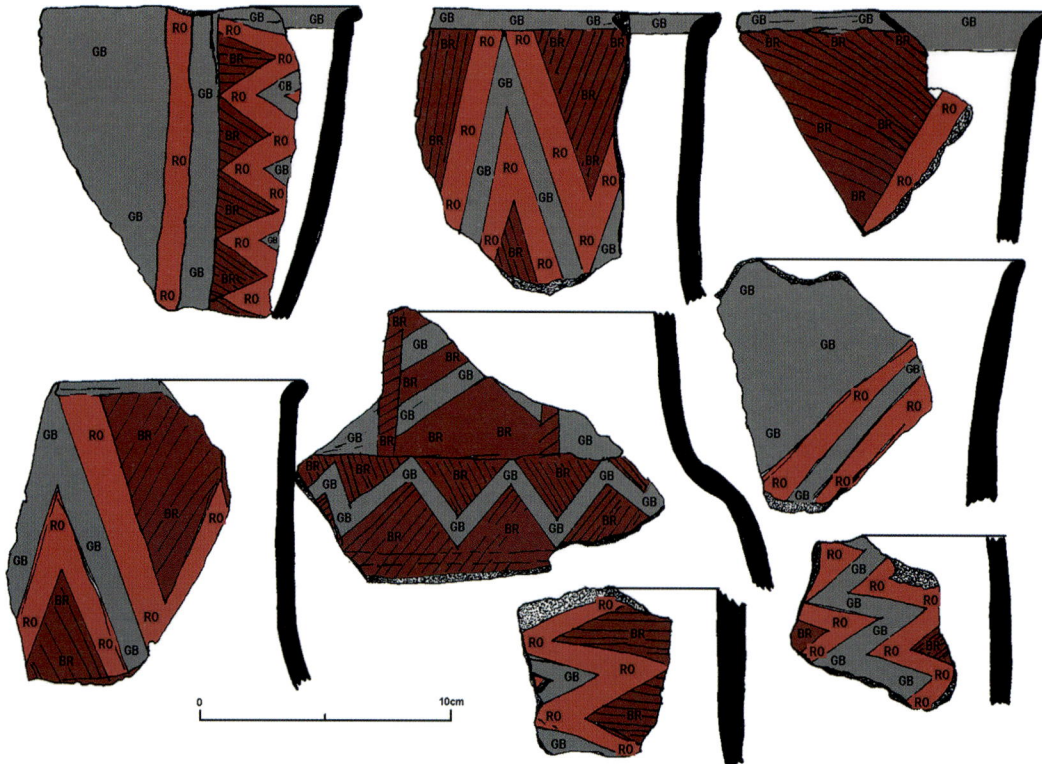

Figure 26.7 Vessels with tall necks decorated with complex incised triangular designs painted with alternative zones of red ochre and graphite burnish (Illustration: J. Chikumbirike).

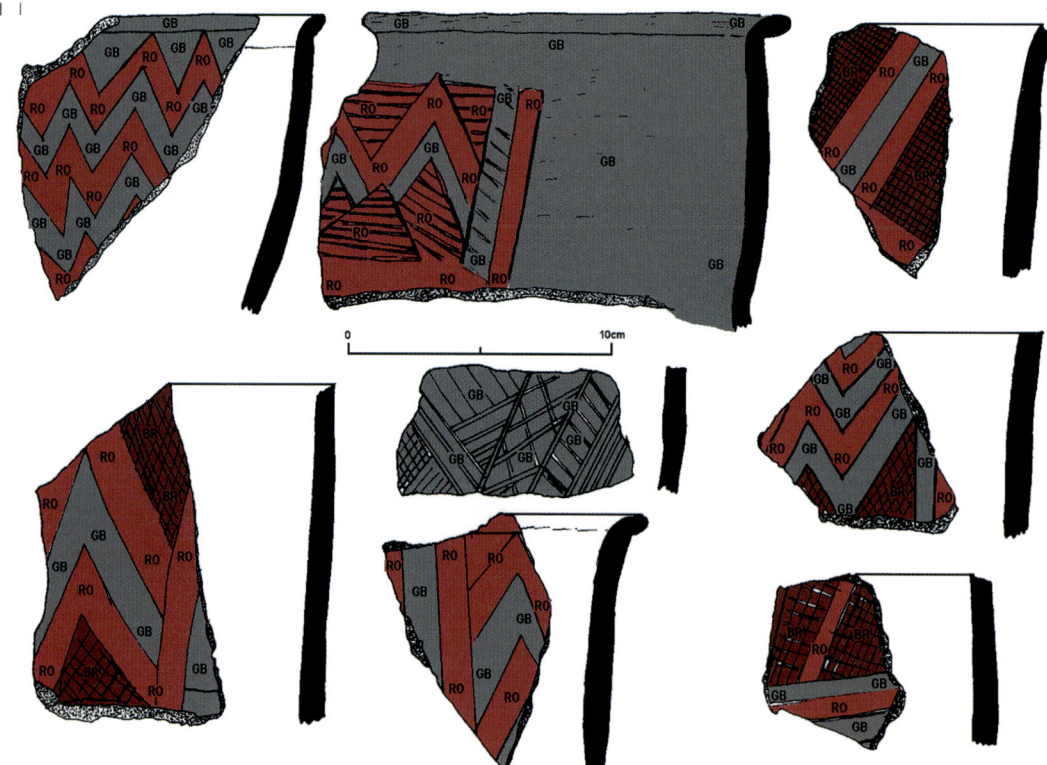

Figure 26.8 Vessels with tall necks decorated with complex incised triangular designs painted with alternative zones of red ochre and graphite burnish (Illustration: J. Chikumbirike).

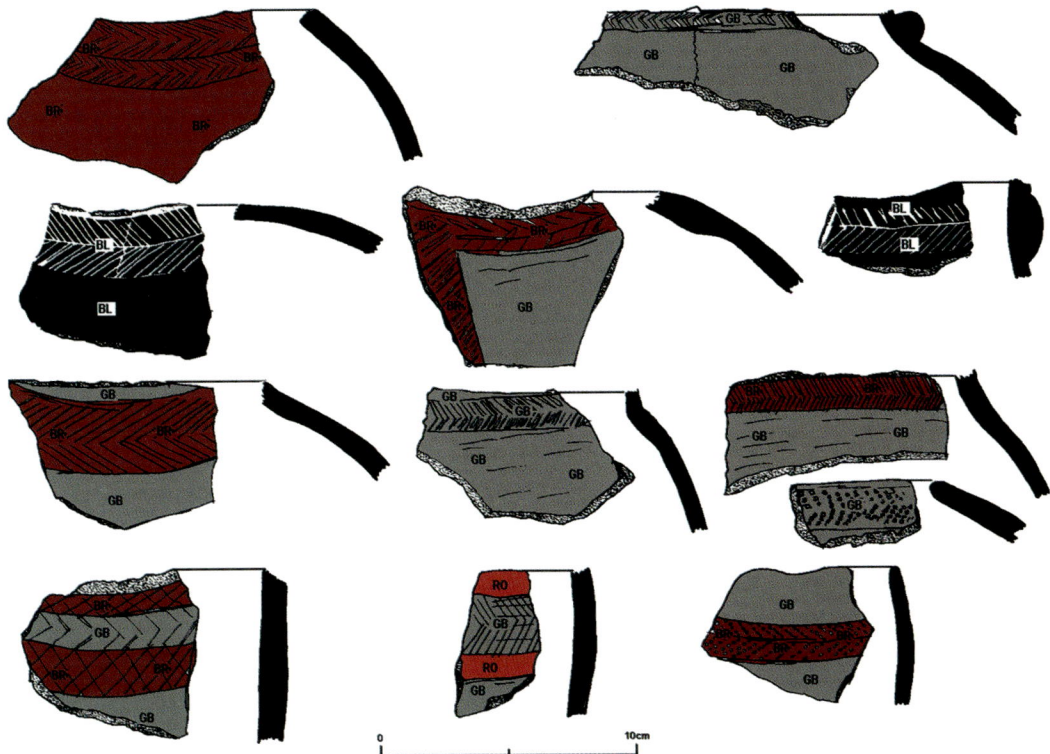

Figure 26.9 Vessels decorated with incised patterns of herringbone on the neck-shoulder region (Illustration by J. Chikumbirike).

groups of parallel lines, cross-hatched lines, chevrons, herring-bone, zig-zag and chequer patterns, triangular, lozenge-shaped, and irregular quadrilateral figures, filled with parallel or cross-hatched lines or with impressed dots, in varying combinations. Curved lines are relatively rare, as also are arrow designs. The incised lines serve also for outlining the coloured bands and zig-zags.

These descriptions provided a baseline for subsequent work. Schofield (1948, 113) described Khami pottery as 'thick tough pottery with a black, red or reddish brown surface', and Robinson (1959, 130) as 'pottery with a polychrome band and panel style of decoration'.

Origins of 'Band and Panel' Ware

The 'band and panel' ware was apparently well developed by the time it was introduced at Khami, suggesting the tradition may have developed elsewhere. Oral traditions cited by Robinson (1959) speak of the Hambe and Lilima 'tribe' specialists, who made this pottery on behalf of residents at Khami. Robinson also associates the more elaborate pottery with ceremonial functions, suggesting Sotho influence. Caton-Thompson (1931) identified the band and panel ware as a later development at Great Zimbabwe. Plate XVIII.2 shows 10 sherds (Caton-Thompson 1931, 50–51), with sherds 1–3 decorated with bands of interlocking triangles of incision on the rim/neck/shoulder region. She categorised these sherds as Class B2. Sherds 48 are undecorated, slightly rolled rims, which she categorised as Class B. Other vessels in this category are subsequently described as having a tall, vertical or slightly in-sloping neck/rim zone leading to the shoulder/body zone. Sherds 9 and 10 carry complicated triangular panels placed on either the body or shoulder of the vessels. The triangles, as well as the short bands, are filled in with cross lines. This is her Class D, regarded as 'geometrically incised', and attributes it to the Rozvi. Her Class C is defined as brown polished 'decidedly thicker and coarser than B'. This is the pottery with heavily rolled rims leading directly to the neck region, which is short, and the body (Caton-Thompson 1931, plate LXX.2, figs 50–71).

Caton-Thompson also described the pottery from other Zimbabwe Culture sites in south-central and eastern Zimbabwe. Ceramic sherds carrying a herringbone band decoration were recovered at Great Zimbabwe, Matendere, and Chiwona. Sherds with cross-hatching on the neck were recovered at Muchuchu and Chiwona, and Caton-Thompson's (1931) Plate XXXI.3 shows complete specimens, as well as Matendere. The site of Gombe yielded Class D pottery, while Chiwona had Class B sherds mixed with Class C. Caton-Thompson (1931, 147) remarked on the complete absence of Class B ware at Muchuchu, where she only recovered 'coarse brown or grey rough ware similar to that at Matendere'.

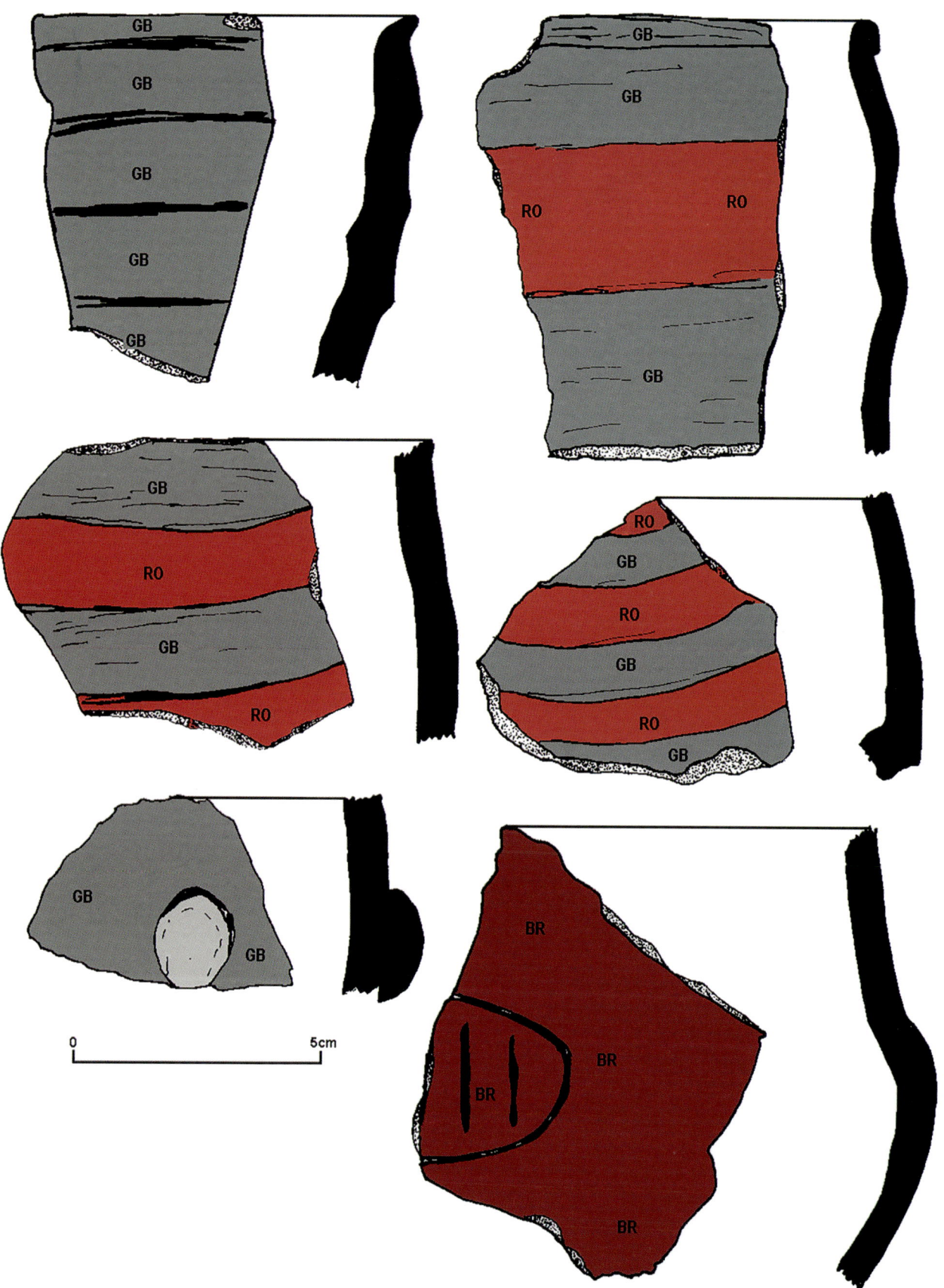

Figure 26.10 Vessels with tall necks and fluted rims (Illustration: J. Chikumbirike).

Caton-Thompson was apparently attempting to differentiate a single ceramic tradition but encountered challenges when examining sites beyond Great Zimbabwe. John Schofield (1948) suggested that some of the polychrome pottery could be identified with other assemblages from Khami and Vukwe in Botswana. Schofield also suggested that some of Caton-Thompson's Class C actually belonged to her Class D. Traditions collected by Robinson (1959, 159–62) during the late 1940s to early 1950s, which were primarily aimed at establishing the identity of the builders of Khami, suggest that Khami was built by the Kalanga or Rozvi. One tradition links Khami with the 'Pasipasi' (literally meaning 'much further down') or south of the Limpopo River, while another names the Nguni as the people who eventually destroyed it. Nguni conquest, according to the tradition, resulted in Rozvi dispersal. However, according to Robinson (1959), these traditions are unreliable, given their inability to cope with events more than a century old. They only refer to the 19th century and do not cover much earlier events. It was these accounts, as well as Schofield's (1948) ethnographic accounts on the distribution of Sotho-Tswana pottery, that shaped Robinson's conclusions on the origins of 'band and panel ware' from Khami. Robinson (1959) further suggested that the pottery at Khami was made by a select group of Kalanga women.

Robinson (1966) excavated Leopard's Kopje sites in the vicinity of Khami and made detailed descriptions of their pottery. Some of his illustrations suggest that ceramics from Khami were inspired by some shape-forms such as beaker bowls used by the Leopard's Kopje people, while the red ochre and graphite burnished finish derived from Great Zimbabwe and related sites. Huffman (1974) suggested that the Leopard's Kopje culture continued to exist in the area and can be identified with contemporary Kalanga speakers. If this is the case, then Khami was established on a pre-existing population. However, Leopard's Kopje decoration motifs are not replicated in Khami pottery, which may suggest the specialist nature of the latter. Robinson (1959, 113) states that:

> The presence on the bedrock of a sherd probably belonging to the Leopard's Kopje Culture, in hut Cb1, is in keeping with the evidence obtained from the other excavations in the vicinity of the Hill Ruin. Platform B also produced a small sherd which may have a similar origin; this was also on bedrock. They can be no doubt that, before the builders of the dressed stone walls appeared on the scene, the hill was occupied by people with an entirely different material culture, probably related to the early inhabitants of both Zimbabwe and the Limpopo valley sites.

The pottery from Khami requires an understanding of how socio-political complexity must be read from material culture and how traditions of manufacture are adapted to serve specialist production contexts.

Ceramics and Socio-political Complexity

Given that Khami pottery is highly decorated, carrying complex patterns, a discussion on the relationship between material culture objects, such as ceramics, and socio-political complexity is required. There is need to read messages conveyed by ceramics beyond ascribing them to mere traditions of manufacture and human group identities, which remain the cornerstone of typology in southern Africa. The 2014 Society for American Archaeology (SAA) meeting in Austin, Texas, session titled 'Characterization of Andean Ceramics' (Druc 2015) highlighted ideas relevant beyond South American contexts, linking typology with manufacturing technology. In the Tarapacá Atacama Desert, ceramic typology and technology have been used to assess the development of social complexity in northern Chile for the period 900 BC to AD 900. Petrographic and typological analyses attest to a diversity of compositional groups that infer past technological innovation. Ceramics show considerable societal diversity in later periods, which is different from that of earlier formative periods. In north-coastal Peru, the Moche (AD 300–900) were not politically organised as a monolithic state but rather as autonomous polities that shared a common culture. This is attested to by their elaborate iconography as well as their monumental architecture. Moche ceramic styles speak to both internal technological and external typological connections between regions. It is apparent that exterior style was ideological, expressing a worldview. Andean ceramics also express gendered, multi-scalar political expressions in some contexts. From 500 BC to AD 600 (Formative), AD 600–1000 (Middle Horizon), and AD 1000–1437 (Later Intermediate), ceramics play an increasing role in political centralization, showing connections between craft production and elite control. Socio-political dynamics are inferred in areas such as the Moquegua Valley, where the socio-political environment may have impacted ceramic production, trade, and exchange. Homogenisation in ceramics attests to local responses to production contexts.

Khami ceramics recovered from the archaeology, as well as those in the Balfour collection, attest to the development of socio-political complexity not just in terms of craft specialisation, but also in terms of how ceramics were produced and exchanged in the region. At this level, there is need to determine the level of elite control in the production process (Mukwende *et al.* 2018). This provides an opportunity to link typology with technical processes as described by Balfour (1906) to understand both the internal and external connections, including societal worldview. The highest expression of Khami worldview was undoubtedly its monumental architecture, where some decoration on the stone walls is also found on the pottery.

Conclusion

This study adds to ongoing archaeological work in southwestern Zimbabwe on Khami-type sites, focusing on material culture. It was initially drafted at a time when I started questioning the value of ceramic typologies in archaeology, especially in addressing human origins and identities. I shelved the project altogether after leaving Oxford. If the premise of typology is accepted, the origins of Khami ceramics lie in the preceding Leopard's Kopje culture, rather than Great Zimbabwe, as originally thought. This claim further supports recent claims on the origins of state societies on the Zimbabwe plateau and adjacent regions, which suggest multidirectional evolution of socio-political complexity in southern Africa (Chirikure et al. 2016). Ceramics should be part of the conversation on the origins of socio-political complexity. Khami and Great Zimbabwe ceramics present themselves for this discussion, since they speak to elite and non-elite material remains, as well as the potential meanings of the contexts in which they were recovered. How we read socio-political complexity through ceramics is a research gap in the southern African Iron Age that needs addressing.

Acknowledgements

This study was funded by a grant from the Commonwealth Commission administered by the British Council in 2000. I am grateful to Jeremy Coote, then Curator of the Pitt Rivers Museum, for access to the collection. Chris Gosden, Claire Harris, and Laura Peers inspired me through their research and teaching. Haas Ezzett rendered computing support. Norman Weller and Susan Brooks handled my administrative needs. Outside the Pitt Rivers, Drs John Sutton and Nick Allen arranged my affiliation with Wolfson College, whose membership I cherished during my tenure at Oxford. At the University of Zimbabwe, Joseph Chikumbirike prepared the illustrations of ceramics.

References

Aschwanden, H. (1982) *Symbols of Life: An Analysis of the Consciousness of the Karanga*. Gweru, Mambo Press.

Balfour, H. (1905) *Diaries of Henry Balfour (1863–1939), anthropologist and museum curator: South Africa 1905*. [manuscript] Oxford, Archives of the Pitt Rivers Museum.

Balfour, H. (1906) Flint-engraved pottery from the ruins at Khami and Dhlo Dhlo, Rhodesia. *Man* 6, 17–19.

Beach, D. N. (1980) The *Shona and Zimbabwe, 900–1850: An Outline of Shona History*. Gweru, Mambo Press.

Caton-Thompson, G. (1931) *The Zimbabwe Culture: Ruins and Reactions*. Oxford, Clarendon Press.

Chirikure, S., Pikirayi, I. & Pwiti, G. (2001) A comparative study of Khami pottery, Zimbabwe. In F. Chami & G. Pwiti (eds), *People, Contacts and the Environments in the African Past*, 106–31. Studies in the African Past 1. Dar es Salaam, Dar es Salaam University Press.

Chirikure, S., Manyanga, M. & Pollard, M. (2012) When science alone is not enough: Radiocarbon timescales, history, ethnography and elite settlements in southern Africa. *Journal of Social Archaeology* 12(3), 356–79.

Chirikure, S., Manyanga, M., Pikirayi, I. & Pollard, M. (2013a) New pathways of socio-political complexity in southern Africa. *African Archaeological Review* 30, 339–66.

Chirikure, S., Pollard, M., Manyanga, M & Bandama, F. (2013b) A Bayesian chronology for Great Zimbabwe: Re-threading the sequence of a vandalised monument. *Antiquity* 87(337), 854–72.

Chirikure, S., Bandama, F., House, M., Moffett, A., Mukwende, T. & Pollard, M. (2016) Decisive evidence for multidirectional evolution of sociopolitical complexity in Southern Africa. *African Archaeological Review* 33(1), 75–95.

Chirikure, S., Mukwende, T., Moffett, A., Nyamushosho, R., Bandama, F. & House, M. (2018) No big brother here: Heterarchy, Shona political succession and the relationship between Great Zimbabwe and Khami, Southern Africa. *Cambridge Archaeological Journal* 28(1), 45–66.

Cooke, C. K. (1957) The waterworks site at Khami, Southern Rhodesia: Stone Age and Proto-historic. *Occasional Papers of the National Museums of Southern Rhodesia* 3(21A), 1–43.

Druc, I. C. (2015) Linking technology and society – An introduction to the volume. In I. C. Druc (ed.), *Ceramic Analysis in the Andes*, 7–14. Blue Mounds, WI, Deep University Press.

Garlake, P. S. (1972) *Great Zimbabwe*. London, Thames and Hudson.

Hall, M. (1984) Pots and politics: Ceramic interpretations in southern Africa. *World Archaeology* 15(3), 262–73.

Hall, R. N. (1909) *Prehistoric Rhodesia*. London, F. T. Unwin.

Huffman, T. N. (1974) *The Leopard's Kopje Tradition*. Museum Memoir 6. Salisbury, National Museums and Monuments of Rhodesia.

Huffman, T. N. (1984) Expressive space in the Zimbabwe Culture. *Man* 19(4), 593–612.

Huffman, T. N. (1986) Iron Age settlement patterns and the origin of class distinction in southern Africa. *Advances in World Archaeology* 5, 291–338.

Huffman, T. N. (1989) Ceramics, settlements and Late Iron Age migrations. *African Archaeological Review* 7, 155–82.

Huffman, T. N. (1996) *Snakes and Crocodiles: Power and Symbolism in Ancient Zimbabwe*. Johannesburg, Witwatersrand University Press.

Huffman, T. N. (2007) *Handbook to the Iron Age: The Archaeology of Pre-colonial Farming Societies in Southern Africa*. Pietermaritzburg, University of KwaZulu-Natal Press.

Hughes, G. (1997) Excavations within the peripheral area of settlement at Khami, 1989. *Zimbabwea* 5, 3–21.

Mukwende, T. (2016) *An archaeological study of the Zimbabwe Culture capital of Khami, south-western Zimbabwe*. Unpublished PhD thesis, University of Cape Town.

Mukwende, T., Bandama, F., Chirikure, S. & Nyamushosho, R. T. (2018) The chronology, craft production and economy of the Butua capital of Khami, southwestern Zimbabwe. *Azania: Archaeological Research in Africa* 53, 477–506

Nyamushosho, R. T. & Chirikure, S. (2020) Archaeological implications of ethnographically grounded functional study of pottery from Nyanga, Zimbabwe. *Quaternary International* 555, 150–64.

Pikirayi, I. (2001) *The Zimbabwe Culture: Origins and Decline of Southern Zimbabwe States*. Walnut Creek, CA, Altamira Press.

Pikirayi, I. (2007) Ceramics and group identities: Towards a social archaeology in southern African Iron Age ceramic studies. *Journal of Social Archaeology* 7(3), 286–301.

Pikirayi, I. & Lindahl, A. (2013) Ceramics, ethnohistory, and ethnography: Locating meaning in southern African Iron Age ceramic assemblages. *African Archaeological Review* 30, 455–73.

Randall-MacIver, D. (1906) *Medieval Rhodesia*. London, Macmillan.

Robinson, K. R. (1959) *Khami Ruins*. Cambridge, Cambridge University Press.

Robinson, K. R. (1966) The Leopard's Kopje Culture: Its position in the Iron Age of Southern Rhodesia. *The South African Archaeological Bulletin* 21 (81), 5–51.

Schofield, J. F. (1948) *Primitive Pottery: An Introduction to South African Ceramics, Prehistoric and Protohistoric*. The South African Archaeological Society Handbook Series No. 3. Cape Town, The South African Archaeological Society.

Thorp, C. (1984) A cultural interpretation of the faunal remains from Khami Hill ruin. In M. Hall, G. Avery, D. M. Avery, M. L. Wilson & A. J. B. Humphreys (eds), *Frontiers: Southern African Archaeology Today*. Oxford, British Archaeological Reports, 266–76.

Thorp, C. (1995) *Kings, Commoners and Cattle at Great Zimbabwe Tradition Sites*. Harare, National Museums and Monuments of Zimbabwe.

Van Waarden, C. (2012) *Butua and the End of an Era: The Effect of the Collapse of the Kalanga State on Ordinary Citizens: An Analysis of Behaviour Under Stress*. BAR International series 2420. Oxford, British Archaeological Reports.

White, F. (1901) On the ruins of Dhlo-Dhlo in Rhodesia. *Journal of the Anthropological Institute* 31, 31–28.

27

In Dreams the Heart: Impermanence at the Museum

Chantal Knowles

This chapter addresses the transformative qualities of a collection object. In Dreams the Heart, *a sculpture in the form of a shield, by Danie Mellor was acquired by the National Museums Scotland, Edinburgh, in 2013. Made from the lid of a travelling trunk, the metal has been cut and reworked into the form of a rainforest shield from Far North Queensland. This paper examines the tension of a transformed 'thing' from European trunk to Aboriginal shield and utilitarian object to sculpture, which, suffering from weeping iron corrosion, has the potential to affect other objects stored or displayed around it. Thinking through the different forms or identities of the object – trunk, sculpture, shield, museum object, European, Aboriginal – I explore what permanence, agency, and effect mean through the consideration of one object in a museum collection. The shield's inherent instability is used to explore the role of museums to corral, contain, and preserve, as well as the potential for objects to chart their own journeys and interactions with other objects and people.*

Keywords: Museums, Material Culture, Agency, Effect

Objects and Museums

In 2012, I contacted artist Danie Mellor to acquire an item for the collections. I was keen to acquire a work on paper, in a gilded frame, that depicted the landscape of the Queensland rainforest in blue and white, with fauna, people, and artefacts highlighted in colour. The work had connections with a range of objects already in the encyclopaedic collections of the National Museums Scotland. The blue and white palette echoed the fashionable chinaware produced in English potteries that travelled with settlers to the colonies. In discussing the acquisition of *The Story Place (a History of Two Worlds)* 2008, Mellor offered a second work – *In Dreams the Heart* (Fig. 27.1). This sculpture, in the form of a shield, was acquired with less curatorial intention; instead, in some sense, through the artist's intervention. When the works arrived at the museum, each was unwrapped separately, the large work on paper in a gilt frame was opened in the large object store, and staff immediately responded and were drawn to view it. The metal shield provoked less fanfare; it was more rough and ready, shedding rust fragments, and was swiftly relocated to the conservation lab.

Once this object became a museum object, alongside its other identities as a sculpture, a shield, *balan bigin*, a memorial, and an artwork, it gained a number (V.2013.27), a physical location, was ascribed the name 'sculpture', and given a Collections Management System entry that 'presenced' it online (National Museums Scotland n.d.). As a thing, it has a lineage, a past, and a future. It is the sum of its parts: the material, its making, its use, and decay. It creates networks of relations between the maker and his genealogy, between museum staff and visitors. Currently on display in the museum, it relates to the objects around it and the people who view it, but it also connects to places in Australia, and objects and people from the past, from the present, and across several generations. But what are the consequences and meanings of these multiple, interconnecting relationships? How independent of these people, places, and other things is this object's future? What is the material agency of *In Dreams the Heart* in the museum setting?

For decades, the discussion on object biographies and agency has circulated, providing new ways of exploring objects as evidence, objects as interactions, and objects

Figure 27.1 In Dreams the Heart *by Danie Mellor (© Danie Mellor, Image: © National Museums Scotland).*

as actors (see *e.g.* Gell 1998; Gosden & Marshall 1999; Hoskins 2006; Ingold 2007; Joy 2009; Bennett 2010; Byrne *et al.* 2011; Knowles 2013). In Australia, Aboriginal and Torres Strait Islander peoples may describe an object as an *ancestor*, putting it in a genealogical relationship with themselves. Academic papers on object agency may acknowledge, but frequently exclude, cultural approaches from their theoretical debates, instead locating culture firmly in relation to the disciplinary approaches of the Western academe. This creates two separate realms or cultural lenses for considering agency, and in the museum, these worlds unavoidably collide. The object becomes representative of and interpreted as a cultural *representation*, placed within the theoretical and classificatory world of the museum, frequently setting aside the Indigenous relationship to artefacts, relegating it to an attribute rather than a state of being.

As a museum curator with a quarter of a century of practice, I accept objects are agents; in my experience, they interact and react to the objects with which they sit on the storage shelf or in the display case. They create unforeseen or unlooked-for connections and networks of relationships with other objects and people. Though neither sentient nor conscious actors, an object's form, history, and locale of interaction have the potential to shape the actions and reactions of others (human and thing) around them. The assemblages created in the museum store or display are created as groups or juxtapositions intended to speak to scholarship and provide scientific meaning; assembled to engage with audiences and academics in research and education. Embedded in objects are the histories of their past, the before of the material, the process of making, the resultant form, and then its 'use-life'; these experiences are written on the object, through morphology and decoration, and wear and tear (Gosden 2005, 197). Objects often accrue layers of these physical manifestations of use or neglect. Use may wear away parts or build up an object, layering it with accretions. Objects physically manifest these interactions: on display, they may become damaged by light; in storage, subject to the teeth marks of a rat, the holes from the larvae of moths, or mould adhering to its surface from humidity, depending on its purpose, at any moment it may be damaged or repaired.

In Dreams the Heart is one such object – a mutable object, a recycled object (Fig. 27.1). The chosen material attests to its journey and transformation. Once the lid of a tin travelling trunk, it has been physically reworked by Mellor into a sculpture. Its chosen form, a shield, echoes its cultural lineage in Queensland's rainforest across time and space.

The maker creates an identity for the object in dialogue with the material, but at the moment of entry into a museum, all objects are given multiple new identities in relation to the museum space. Entries and descriptions provide data for the records and place objects in multiple categories: Class, Name, Material, Description, Date, etc., these fields are usually constrained by thesauri, which limit language and choice. At National Museums Scotland, *In Dreams the Heart* is identified primarily through its number and then by its object name 'sculpture'. The properties, size, material, region, and type are most often used to locate the object in relation to other collection items and any one or several of these attributes may determine where it is located in the store. In the museum store, it could be placed amongst an assemblage of wooden shields from the same region or amongst a group of 'strangers' – objects related by museum criteria, such as material – and not necessarily related through culture or use.

In Dreams the Heart is difficult to categorise and gathers around it a complex set of relationships, presenting itself as an object worthy of deeper exploration. As a sculpture, Danie Mellor's series of shields crafted from old, metal travelling trunks have been referenced, published, discussed, and displayed more often in art galleries than museums.[1] To date, each label or catalogue entry foregrounds Mellor's motivation for exploring the material, the use of metal travelling trunks, to create *balan bigin*, and how the works embody his relationship to Country and his maternal ancestors. The works explore his own Indigenous and settler identities in relation to the experiences of his Aboriginal ancestors (see Sculthorpe 2021, 139–41) and his own experiences reconnecting to Country through his art, as well as the guidance and mentorship of Aboriginal Elder Ernie Grant. Into these narratives, Mellor often weaves the museum's institutional displays of material culture by displaying his work in ways reminiscent of typological displays and dioramas (Fig. 27.2).

Mellor has shared how the shield series references the imposition of the settler culture on Indigenous Country and the forcible integration of Aboriginal people as they were pushed from their lands to the margins of towns. The shields, whether displayed singly or in groups, evoke the collecting tropes, classification of culture, and the othering by imperial processes and European science. The shield connects personal stories and forms a link between two worlds: Danie's relationship with his maternal great-grandmother May Kelly and great-great-grandmother Ellen Kelly and the renewal of connection to Country through walking on Country with Ernie Grant.

Material Ancestors

For Mellor the choice of sculptural form, the shield, was determined by its matrilineal connections to his Aboriginal identity, as the word for shield, *balan bigin*, is feminine. Although shields were made and used by men, referencing gender links Mellor as maker to his reflections on the experiences of his female ancestors. Shields figure in large numbers in museum collections and are often arrayed together both in storage and on display. Each shield provides a canvas for cultural designs or expression as well as having a utilitarian purpose as a defensive weapon.

Like Mellor, the material used for the sculpture has its own ancestors. *In Dreams the Heart* began as iron ore, extracted by smelting, an alloy created and then pressed into a metal travelling trunk for carrying personal possessions across long distances. In 1850, the introduction of the Bessemer process of making steel meant trunks made of wood, board, paper, and leather could be manufactured more cheaply from sheet metal. This innovation coincided with the transition from sail to steam and more affordable travel and migration overseas. Metal trunks became synonymous with one-way, long-distance travel. On arrival in a new country, they could provide good protection for

Figure 27.2 A grouped display of shields in 'Danie Mellor: Exotic Lies Sacred Ties', a University of Queensland Art Museum 10-year touring survey curated by Maudie Palmer AO, 2014. Installation view, Tarrawarra Museum of Art, Victoria (Image: Andrew Curtis 2015).

personal treasures and, although prone to rust, they were an improvement on wooden chests, which were susceptible to mould and termites. Metal was lighter, more hard-wearing and potentially more watertight. Metal was often pressed to mimic or reflect the patterns or embellishments of the more expensive wooden trunks, which were fitted with leather straps and fixed with brass nails.

The metal shield's industrial ancestor, the tin trunk, is reformed, echoing its Aboriginal ancestor, the shield of the rainforest region. Wooden, decorated with red and yellow earth pigments, with a raised central boss aligned with the hollowed-out recess for the handle on the shield's interior face. *Balan bigin* are carved out from the buttress roots of the rainforest fig tree, the flanged roots leave the trunk and reach into the earth. The shape of the buttress root determines the shape of the shield and must be chosen carefully. The metal trunks from which Mellor made his shields were no less carefully chosen. Trunks fascinated Mellor, and he had built up a personal collection, their utility as receptacles or surfaces within the home allowing him to acquire several. As the trunks accumulated, he started to imagine their transformation, and gradually the shields series began. In working the lids of metal trunks, the lids' forms and malleability, just like the flange root of the tree, shaped Mellor's work, creating constraints around scale and proportion; in each case, the physical and conceptual were engaged and reformed. These two shields, one a defensive weapon, the other a sculpture, never fully shed their previous identities, tree and travelling trunk.

Once his trunk collection was exhausted, Mellor had to hunt out others. Like the roots, there were qualities Mellor valued in selecting a trunk: wear, often manifesting as rust, dents, or even perforations, attested to use and activities and were sought after by Mellor. Tin trunks with mouldings were preferred. The matrilineal language of *balan bigin* connected him through his maternal grandmothers to shields. The pressed imitation brass buttons echoed the buttoned-up Sunday best worn by his Granny May in her studio portrait as a child (Fig. 27.3; Mellor 2022).

At the very moment that the British were able to move more quickly, comfortably, and cheaply overseas to North America, Australia, and New Zealand, the Indigenous populations' lives were restricted further, with increasing numbers of settlers and new laws enacted that cemented the unequal relationship between settler and Indigenous, further disenfranchising Aboriginal men and women.[2] Settlers violently unsettled those they encountered, and they were forced to relocate and move. For Mellor, using the trunk as raw material speaks strongly to these histories, the tensions in his family history, and the tension glimpsed in the pose of his Great-Gran in her Sunday best – the tension of Aboriginal people forced to move with their possessions as they were dispossessed of their land.

To work the trunk into the shield, Mellor used industrial tools, an oxy-acetylene torch to cut the maximum length from the lid of the tin trunk, applying heat and the hammer to bash out the upturned sides of the lid and then gently curve the metal to its new form. He then welded 'handles' or crosspieces on the reverse side to suspend the work. The friable surface of the metal was smoothed with an application of beeswax, bringing a delicate sheen to the work and further highlighting wear on the surfaces. Mellor's male ancestors would have cut with adze blades and applied heat to strengthen, shape, or cure the wood before applying pigments to create surface designs.

The form and materiality of the sculpture echo into the past across both worlds, Aboriginal and European, speaking of tensions and genealogies of material things and cultural lives. The rainforest shield is a defensive weapon, a feminine object that was made, owned, and used exclusively by initiated men. The shield has been transformed by the marks of spears, surface damage that has flaked the surface pigments, termite holes from long-term storage and fading of the yellow and red ochres; these are physical manifestations of the shields' social lives. In the same way, the ochre rust that flakes from the sculpture's surface attests to the object's journey through varied climates. A trunk for treasured possessions absorbed the salty moisture from the marine environment, the bangs from mishandling, and the changing climates from temperate to tropical. Just like the protective shield, its protective purpose leaves marks on the surface and protects the treasures within.

Agency and Effect

When an object enters a museum, it is no longer free to circulate as it might in other contexts. The object may move between store, lab, or display; may enter a series of different contexts in exhibitions; may temporarily move beyond the museum walls to an offsite store or on loan to another institution, but these circulations are always tethered to its identity as a collection item, legally framed, where activities are circumscribed, locations tracked, and condition monitored. These conditions frame the context in which the object can act or cause effect. For example, museums dull sensory engagement; in their halls, sight is privileged, and smell, touch, taste, and sound are muted. Yet the act of bringing an object into a museum singles it out as worthy of preservation; here it is ascribed *significance* that is affirmed by its special treatment.

Although the object's material agency is transformed and potentially constrained within the museum at the same time, all museum agencies are in consort to actively manage the item preserving it beyond its expected 'use life' so that it remains a marker of a moment in time, evidence of a present, which will become the past. How can material have agency in this constructed space? According to Van Oyen (2018, 2)

Figure 27.3 Granny May Kelly photographed by Alfred Atkinson ca. *1908, Gimuy/Cairns, Queensland (Image © Danie Mellor 2022, courtesy Mellor/Hamill family archive).*

'objects only act under certain circumstances, as part of particular assemblages, in specific settings'. It is the placement of *In Dreams the Heart* inside the museum that creates a specific locale that constrains some agentive a/effects and enhances others. Once placed amongst the collections, it is immersed in a series of relational assemblages with other collections, items, and museum departments. This can be generative, creating a series of conditions or circumstances in which the object's agentive qualities are changed.

The trunk's decay demanded Mellor's attention, it was this quality that drove him to acquire multiple examples, as well as their utility, scale, and aesthetic. However, once transformed into a sculpture and brought into the museum, these qualities became a challenge. The conservator, Jill Plitkinas, highlighted the issues of adhered dirt and weeping corrosion. Beneath the painted interior face of the shield and the beeswax applied to the obverse lay chloride-containing salts, dormant, and under the right conditions these would act. A small rise in humidity would cause the salts to dissolve into water, and droplets had the potential to lift the surface paint and wax, separating it from the sculpture and exposing it to further loss.

Despite being transformed by the artist, it is the potential for decay that drives the conservator's actions, monitoring, isolating, testing, and documenting; and it is the decay or potential change to the object and perhaps those around it that has caught and held my curatorial attention. There are of course many objects where a rise in humidity or a sharp temperature change will trigger change, including deterioration that can be arrested but may never be recovered. But where the artist embraces these qualities, the shield creates a certain expectation of an uncertain future.

In Dreams the Heart provides an opportunity to explore relations differently, to think about actions and effects and the networks or re-action that coalesce or circulate around and with the work. It lends itself to the exploration of Bennett's (2010, 6) idea of '*Thing-Power*: the curious ability of inanimate things to animate, to act, to produce effects dramatic and subtle'. We agreed to meet, artist, curator, and conservator in the museum's conservation lab to discuss the issues. For me, the meeting was revelatory: I had not considered the impact that the state of the sculpture's raw material would have on the conservator. For her, the work needed to be actively managed, and display in a case posed risk to other items; if it started to off-gas, it could harm and discolour other works. The very qualities that had attracted Danie to the trunk were the qualities that meant that the object once in the store would need to be separately contained so that it was buffered against a rise in humidity and other objects were buffered from it. On display, a 'leaky' case or open display was favoured to limit potential damage to other items, so that if the corrosion reactivated as acidic vapours, these would be absorbed and neutralised in the larger ambient atmosphere of the museum.

The genealogy of the artefact gave it a history and use that impacted the materials that it was made from and in turn was impacting our approach to the sculpture and our response, professional and emotional, to it. The effect of the thing was not just emotional or semiotic but rather these responses, combined with the shield's 'materiality in motion', was moving us artist, conservator, and curator to act based on our roles, knowledge, experience, and history. With all of the skills and experience we brought to that moment in the lab, *In Dreams the Heart* embodied an unpredictability that the artist embraced (and had perhaps sought?), I was engaged, and the conservator was frustrated and concerned.

The potential for *In Dreams the Heart* to act or react provoked me to reflect on the museum as holding space, catching objects in a 'moment in time', stalling them. The sculpture in the museum, connected across the collections, moved it beyond an artwork with all the meanings and symbolism that Mellor had imbued it, and across disciplines. Each transformation of the object and materials – raw material extraction, material processing, travelling trunk, furniture and store, raw material, sculpture, museum object – collected relationships and connections but also new adherences, such as the salt, no doubt a testament to a long-forgotten journey by sea.

Impermanence at the Museum

Museums are suggestive of permanency, where objects are in general assumed to be held *in perpetuity*.[3] For curators, acknowledging the agency and emotional effect of objects is moot; ask any curator or collection manager and they will tell you which objects are troubling them. This will be because they are not yet nested, located, or documented as they should be, or because they believe they are of significance and that significance is not understood by others. Perhaps they are uncertain about the object's placement within the museum, and its presence (or potential absence) plays on their mind.

A visitor to a collection will come for a myriad of reasons: personal connection or research, and they too will be affected, not least by the way the object is handled, wrapped, nested, or retrieved by the collections manager, always held with two hands, placed on a clean surface, supported. Rarely does a museum staff member place an object down and not step back to review it – a moment of appraisal, appreciation, connection, and respect. These performances play out continuously behind the scenes at the museum, and agency and effect are embedded in these relations and performative acts.

All of these processes that are inherent in museum practice – significance, documentation, provenance, and relationships (to other artefacts, other individuals, and collections) – take the object out of one realm of circulation, the quotidian, into another, the specialist, to become an example or exemplar of its kind in relation to others. In that placement, the museum takes on the responsibility to maintain the item *in perpetuity* for an infinite time that will transcend beyond the individual lifetimes of the maker, user, curator, and collections manager, outliving each of them and potentially the building, the store, and its location. The museum object is slowed but not static, its relationships are transient but 'sticky' individuals' stories and biographies can become part of the object's history. In the case of *In Dreams the Heart*, the disease, dormant below the surface, gives the potentiality that it may exist only fleetingly in museum time scales. Jill's frustration at the acquisition of an active object, actively deteriorating and with the potential to do harm and increase the deterioration of other objects, was understandable given her role as a conservator. Yet this has caused me to further reflect on how permanence and impermanence are played out in the museum. Museums sit outside of the general passage of time but are inextricably linked to it, external factors can accelerate unforeseen change, changing fashions and personnel can prompt reviews and reassessment, and objects may cease to exist in their museum form, become lost to the public but may yet change and transform, re-emerge. In many ways, Mellor's work preserves the lives of the cheap, accessible metal travelling trunks, often invisible in museum collections, mundane, yet inspiring, reflective, and corrosive. It may be that, by being transformed into a sculpture, the trunk will have a longer life than in its original form.

The future of *In Dreams the Heart*? She depends on relationships, the relationship to the building and the case around her, the relationship with climate and whether it can continue to be controlled. There are the relationships that tie her to the past from Granny May, and to the future, as new relationships are formed. *In Dreams the Heart* is an actor in these evolving relationships between nature and culture, the raw materials from the earth, industrially transformed into a utilitarian trunk pressed into service to carry important items overseas, and store and protect them for the future. Just like the abrupt unsettlement brought about by colonialism over decades, later the trunk's reworking was redolent with upheaval and change; the transformation from the trunk lid to shield was quick, brutal, and unsettling. The echo of a past life is presented for the future, entangling multiple histories and physically embodying them. It is these 'emergent and relational phenomenon' (Van Oyen 2018, 4) that give *In Dreams the Heart* its material agency, both affective and effective.

Perhaps it is the transformation, the mutability of the trunk-sculpture-museum object, that is the object's power, to affect other objects and people that it draws into its orbit ensuring relational interactions that form long-standing ties that resonate and create echoes between past, present, and future. This mutability means that at different moments the thing or the materials that we encounter in the museum have echoes of previous pasts and existence that transcends nature (the minerals and raw materials) and culture, and changes the object's form. The mutable qualities of metal may see the object take a new form in the future (see Sainsbury *et al.* 2020 for a discussion on mutable objects). The artist's intent and knowledge, deeply connected to their journey, are imbued in the artwork. The audience brings their journey, the 'meeting' is generative, and new stories emerge enmeshed in these relational moments.

When Ingold writes about the artist/maker being guided or influenced by the material's resistance or supplication to being formed into something new, I reflect on Mellor's material choices and each trunk's properties and how this limited or enhanced his actions. Once *In Dreams the Heart* was placed within a museum, not in a gallery or a home, the museum itself, the stores and gallery spaces, documentation and record-keeping, and the staff roles created a specific set of circumstances that is generative of particular agency and effects.

Set in an assemblage of genealogy in the context of the Queensland shields found in museum collections, it embraces a longer lineage and interacts with those other shields, causing the curator/s and conservator/s to interact differently. Categorised, its identities multiply, shifting beyond cultural knowledges (Aboriginal, Australian, and European) embodied by Mellor's heritage to the systems of academic and scientific knowledges, both Aboriginal and European, that are present within the museum. There, European systems are privileged over Aboriginal knowledge that is preserved but in reserve, set behind the scenes subject to western principles.

One museum object created levity, frustration, engagement, and inspiration, and located and linked itself to the lives of others. At one moment in a conservation lab in a museum, three professional lives converged, and our emotions engaged as we responded to the sculpture's materiality and effect. The object's instability is unknown to the visitor, but behind the scenes this means that it will always require a little more attention, regular checking in and monitoring, commanding space and attention even when in isolation from the myriad of other objects and specimens held in the museum.

Acknowledgements

This paper is written in memory of Brenda McGoff, Assistant Curator at National Museums Scotland. Brenda spent her whole working life as a member of staff at NMS, and after she passed away in 2010 her widower, Gordon McGoff, donated a sum of money to acquire an object in Brenda's memory. NMS matched the donation, and we were able to purchase two works from Danie Mellor. Danie and I have stayed in touch and worked together again intermittently over the years. Danie is an inspiring artist and generous colleague, and this paper could not have been written without him. I have been honoured during my time at Queensland Museum to visit his mentor Ernie Grant, Elder and educator, who shared stories about his relationships with objects and Country. I wish to thank Michael Aird for introducing me to Ernie in 2016. I also thank Jill Plitkinas, whose initial response to *In Dreams the Heart* revealed a complex, dynamic object and lodged it in my conscious for continuous reflection. Finally, I wish to thank Chris Gosden, who connected me with Pacific collections and whose confidence in my museum work and research supported my career development. My current PhD supervisor, Professor Lynette Russell, has guided my thinking throughout the development of this research. I am lucky to have worked alongside so many talented individuals and generous scholars.

Notes

1. Shields were clustered together for display at 'Culture Warriors', the inaugural Indigenous Art Triennial at the National Gallery of Australia, October 2007 to February 2008, and in the major survey exhibition of a decade of Mellor's practice 'Exotic Lies Sacred Ties' at the University of Queensland Art Museum, January to April 2014.
2. In Queensland, this was the *Aboriginals Protection and Restriction of the Sale of Opium Act* 1897.
3. For example, the *National Heritage (Scotland) Act* 1985 makes provision for a Board of Trustees, whose first responsibility is to '(a) care for, preserve and add to the objects in their collections'. Section 8 of the Act gives limited grounds for disposal of items based on concepts of utility, duplication, and degradation.

References

Bennett, J. (2010) *Vibrant matter: A Political Ecology of Things*. Durham, Duke University Press.

Byrne, S., Clarke, A., Harrison, R. & Torrence, R. (2011) *Unpacking the Collection*. New York NY, Springer New York.

Gell, A. (1998) *Art and Agency: An Anthropological Theory*. Oxford, Clarendon Press.

Gosden, C. (2005) What do objects want? *Journal of Archaeological Method and Theory* 12(3), 193–211. doi:10.1007/s10816-005-6928-x.

Gosden, C. & Marshall, Y. (1999) The cultural biography of objects. *World Archaeology* 31(2), 169–78. doi:10.1080/00438243.1999.9980439.

Hoskins, J. (2006) Agency, biography and objects. In C. Tilley, W. Keane, S. Kuechler-Fogden, M. Rowlands & P. Spyer (eds), *Handbook of Material Culture*, 74–84. London, SAGE Publications.

Ingold, T. (2007). Materials against materiality. *Archaeological Dialogues* 14(1), 1–16. doi:10.1017/S1380203807002127.

Joy, J. (2009) Reinvigorating object biography: Reproducing the drama of object lives. *World Archaeology* 41(4), 540–56. doi:10.1080/00438240903345530.

Knowles, C. (2013) Artefacts in waiting: Altered agency of museum objects. In R. Harrison, S. Byrne & A. Clarke (eds), *Reassembling the Collection: Ethnographic Museums and Indigenous Agency*. School for Advanced Research Advanced Seminar Series, 229–58. Santa Fe, School for Advanced Research Press.

Mellor, D. (2022) *Interview*. Interviewed by Chantal Knowles. Online, 22 March 2022.

National Museums Scotland (n.d.) In Dreams the Heart. *National Museums Scotland*. [online] Available at: <https://www.nms.ac.uk/explore-our-collections/collection-search-results/?item_id=713013> [accessed 05 October 2022].

Sainsbury, V.A., Bray, P., Gosden, C. & Pollard, A. M (2021) Mutable objects, places and chronologies. *Antiquity* 95(379), 215–27. doi:10.15184/aqy.2020.240.

Sculthorpe, G. (2021) 'Silent testimonials': Shields from Queensland frontiers. In G. Sculthorpe, M. Nugent & H. Morphy (eds), *Ancestors, Artefacts, Empire: Indigenous Australia in British and Irish Museums*, Chapter 13. London, The British Museum Press.

Van Oyen, A. (2018) Material Agency. In S. L. López Varela (ed.), *The Encyclopedia of Archaeological Sciences*, 1–5. Hoboken, NJ, John Wiley & Sons, Inc. doi:10.1002/9781119188230.saseas0363.

28

A Civil Servant Walks onto a Neolithic Barrow…: Sir Lindsay Scott and the Whiteleaf Oval Barrow

Gill Hey

This article examines changing perspectives on a Neolithic oval barrow in Buckinghamshire beginning with Lindsay Scott, the civil servant who excavated it in the 1930s, through its publication by Gordon Childe and Isobel Smith in the 1950s and then to those of the Oxford Archaeology staff who re-excavated and reconstructed it in the 2000s.

Keywords: Lindsay Scott, Gordon Childe, Isobel Smith, Neolithic, Buckinghamshire, Barrows

Introduction

In July 1934, Warwick Lindsay Scott, a civil servant in the Air Ministry, working and living in London and with no formal archaeological training, began to excavate a Neolithic oval barrow on the crest of the Chilterns Ridge above Princes Risborough, Buckinghamshire (Fig. 28.1). He had planned this work from at least 1931 when the Commissioners of HM Office of Works granted his request to dig on the site, then part of the Whiteleaf Cross Scheduled Monument (letter from Edward Muir to Scott, 26 March 1931; LIN02). Work continued intermittently until the outbreak of the Second World War, when the exposed chalk surface of the excavations was hurriedly covered with some soil from the adjacent spoil heap, and Scott became immersed in his role as Second Secretary at the newly formed Ministry of Aircraft Production, a role for which he was knighted in 1942. After the war he neither had an opportunity to resume the excavations as he intended, beyond digging a section across the barrow ditch to the west, nor to prepare material for publication before his untimely death in 1952 (Childe & Smith 1954, 212). It was Vere Gordon Childe who accomplished this task with Isobel Smith in 1954, with Smith analysing and contextualising the important Neolithic pottery assemblage.

There matters rested, and Whiteleaf, lying distant from the major centres of later 20th-century archaeological endeavour in the Thames Valley or the western end of the Ridgeway, faded into the intellectual background. On the ground, however, the monument was taking a battering as a result of increasing visitor footfall and, particularly, the attentions of off-road motorbikers and rabbits. Nobody could determine any more what was surviving Neolithic barrow and what was spoil heap. As part of a bid to the very modern concept of the Heritage Lottery and the Onyx Environmental Trust, Buckinghamshire County Council successfully gained funding to enhance the landscape setting of the hill and improve public understanding and appreciation of its ancient monuments. Oxford Archaeology was commissioned to undertake the surveys and excavations as a community archaeology project, including investigating the Neolithic barrow remains; removing Scott's backfill; examining, sampling, and re-interpreting the surviving prehistoric deposits; and reinstating the barrow to its early 1930s' form (Hey *et al.* 2007). I was fortunate enough to be chosen to project manage this work.

Scott and the Neolithic Barrow

Scott was born and brought up in London and his only link to the Whiteleaf barrow seems to be that he had a country cottage near to the site (letter from Scott to Ralegh Radford, March 1931; Ministry of Works file AA06175/1). This was not Scott's first experience of archaeology, however, for he had already excavated sites in the Hebrides: at Kraiknish

Figure 28.1 Whiteleaf Hill from the air, taken by Major Allen in August 1934 just as work was beginning. The exposed chalk beneath part of the barrow can be seen just above and to the right of the Cross (© Allen Collection, Ashmolean Museum, Allen 0957/AA0603).

and Rudh 'an Dunain on Skye in 1929 and 1931 respectively (Scott 1929; 1932), and at Clettraval (North Uist) in 1930 (UCL Childe Notebooks 23 & 50; Scott 1935a). The reason for Scott's involvement in archaeology is something of a mystery, but he seems to have had family connections in Scotland (Brian Smith pers. comm.). He may have been drawn to the Hebrides because of his love of sailing and the prehistory he saw there piqued his interest. His yacht apparently served as his excavation base (Sharples 2015, 3). He was also excavating at Pant-y-Saer on Anglesey in October 1930 and April 1932 (Scott 1933), and it was this connection that led to his query to Ralegh Radford (then Inspector of Ancient Monuments for Wales) about permission to investigate Whiteleaf; his friendship with W. J. Hemp, who was digging nearby, provided the lever to gaining it. Over this period he was writing and publishing reports on all this work (see dates above) in addition to holding down a full-time job: a remarkable feat demonstrating his passion for archaeology.

Scott's 1932 report on the Neolithic cairn at Rudh 'an Dunain, which he dug with his wife in September 1931, is an accomplished, confident, and well-illustrated piece that demonstrates three enduring features of his work:

1. The importance he gave to methodology and his use of careful excavation techniques. Allied to this was his desire to employ up-to-the moment techniques of analysis: he sought out and developed relationships with a wide range of specialists, in this case for human bone (including chemical analysis), animal bone, and geological advice about stone finds, in addition to persuading Stuart Piggott to draw the pottery and the British Museum to restore the Beaker. He was to use some of these specialists for the Whiteleaf material.

2. His keen interest in the physical positioning of monuments and sites in relation to their surrounding landscape/seascape, and a desire to reconstruct the

environment in the past (*e.g.* the extent of woodland and of peat development, and what this might reveal in terms of population density).

3. The importance he attached to explaining the broader context of what he had found. He had evidently read extensively, anthropology as well as archaeology, and had sought the views of other archaeologists in order to understand the parallels for the site and the social practices represented. The continental origins of megalithic tombs feature in his summary for, as Sharples has noted (2015, 3), his views were shaped by the diffusionist views of the time, amongst them, those of Gordon Childe.

It was said, 'Like all good archaeologists, he often relied on instinct, but having once started on an excavation or other piece of research he was painstaking in testing with scientific accuracy every link in a complicated chain of evidence' (Obituary in *The Times*, June 1952).

Leaving St Paul's School, London in 1910, he had studied mathematics at Clare College, Cambridge and, on graduating in 1914, immediately joined the navy where he was employed in minesweeping in the war. This extremely dangerous work (for which he was awarded the Distinguished Service Cross) required 'nerve, skill and unremitting watchfulness' (East Riding Museums, Heritage Learning & Normanby Hall Country Park n.d.) and would undoubtedly have honed Scott's practical skills and the application of precise surveying techniques. One senses also an excitement, of archaeology as adventure.

Scott and Childe I

In the Rudh 'an Dunain report Scott acknowledges Professor V. G. Childe for his 'very valuable assistance ... in the preparation of this paper' (Scott 1932, 211), but this was not the first time that the two had come into contact. Childe's notebooks suggest that they had met sometime between August 1930 and October 1931, either at the site of Clettraval, North Uist, or to talk about it and the pottery finds, some of which Childe sketched (UCL 0093332, Notebook 50). That this was a new contact is suggested by the fact that in his notebook Childe initially mis-writes 'Rob W Lindsay Scott' followed by his London address. Scott cites a number of Childe's publications in Rudh 'an Dunain and thanks him for a pre-publication view of his paper, 'The continental affinities of British Neolithic pottery' (Childe 1931). At the time, Childe was Abercromby Professor of Archaeology at Edinburgh and, as Scott became a Fellow of the Society of Antiquaries of Scotland in 1931 and read papers at the Society, it is probable they would have met in the city too.

Enter Oxford Archaeology

Walking out along the Ridgeway National Trail from the public car park in 2002, my first encounter with the Neolithic barrow was confusing. Its very lumpy outline (Fig. 28.2a) did not resemble the contour plan created by Scott and reproduced (with slight amendments) in the publication (Childe & Smith 1954, fig. 1); it was much more dispersed and the prominent dual lobes described were not obvious. Bike tracks were very visible and the fence constructed to deter them did not add to its allure. The attempt to backfill the excavations seemed to have been half-hearted at best and, indeed, upon excavation it became evident that, not only was the 'mound' mostly spoil heap offset to the south and southeast of the original barrow profile, but that the excavations had been more extensive than published. It was hard to dispel the prejudice (assisted by the discovery of a gin bottle in the backfill at the start of our work) that Scott was something of a dilettante, sitting and watching workmen dig in the tradition of many Victorian barrow digging expeditions. Nothing could be further from the case.

The 1930s' Excavations Begin

Scott approached the Whiteleaf barrow in the same spirit of scientific endeavour that he applied in Scotland. In a letter to R. E. Simms at HM Office of Works in July 1934, when he explains that he needs to fence the excavation area because, having a full-time job, he cannot work quickly and meticulous work was needed (LIN02, correspondence). He would use 'labour to fill in, but workmen with picks and shovels would not be good enough to excavate, even if there was money to pay them'. He states his intention to carry through the job 'as my scientific conscience demands'.

The Whiteleaf barrow posed an entirely different excavation challenge to the monuments he had been digging on the western seaboard. There he had dug in layers but the size of the areas within which he worked were relatively small; at Whiteleaf the mound was *c.* 22 m × 16 m with no obvious physical divisions. Beginning with a contour survey at 6-inch intervals and working from the interned east side, he dug the mound in 2 ft vertical slices, drawing the sections every 2 feet on both the x and y axes in gridded notebooks (LIN02). This was not especially unusual at the time: it was, for example, how General Pitt Rivers had excavated Wor Barrow (Pitt Rivers 1898, 65 & plate 249; Bowden 1991, 132), but was also employed at a number of sites in the 1920s, for example by Keiller at Windmill Hill (Oswald *et al.* 2001, 22). More than 20 photographs were taken, and every find was recorded three-dimensionally according to each 1 ft square: 'in no previous excavation has every scrap ... been collected and registered with such exhaustive thoroughness', reported Childe and Smith (1954, 217), perhaps a little ruefully. Our work indicated that, as at Rudh 'an Dunain, Scott had sieved the soil for finds; precious few were discovered when we sieved the backfill (Hey *et al.* 2007, 27) and the sections through it revealed finely-sorted chalky soil interspersed with lenses of brown humic material which we interpreted as the topsoil from each season (Fig. 28.2b).

Figure 28.2 (a) The Neolithic 'barrow' as work began in 2003; (b) Cleaning the northwest quadrant of Scott's excavation area, showing his spoil heap in section. A site tour is in progress (© Oxford Archaeology).

Scott was quick to consult specialists from the outset, for example Miss M. L. Tildesley of the Royal College of Surgeons about the human remains and Wilfred Jackson of the Manchester Museum about animal bone. In order to reconstruct the Neolithic environment, A. S. Kennard at the University of Cambridge was approached about the land snails, an early use of this technique (letter Kennard to Scott, 18 August 1936; LIN02), and its Department of Geology consulted about the soils. The appendix to the publication includes brief reports from them (Childe & Smith 1954, 219–20, 228–30).

Scott's Interpretation

We know less about how Scott interpreted what he found and how his perceptions changed as digging continued. There are only three short interim summaries published in the 'Notes on Excavations' sections of the *Proceedings of the Prehistoric Society* for 1935, 1936, and 1937, and the only notes he seems to have made on site were annotations to his notebook sketches (Scott 1935b; 1936; 1937).

An adult male skeleton was found in the first season. The articulated left foot, skull fragments, and a tooth were lying within what Scott thought to be a wooden chamber, the postholes of which were found to either side, and the remaining bones were scattered about on the adjacent 'forecourt' to the east. This prompted a great deal of (unpublished) correspondence between Scott, Miss Tildesley and A. J. E. Cave (both of the Royal College of Surgeons), and C. W. Phillips (then Honorary Secretary of the Prehistoric Society [LIN02]) about whether this was evidence of a successive burial rite (there seems to have been a misunderstanding at first about how many people were represented). In his initial (1931) letter to Ralegh Radford, Scott states that the mound 'at some distant period has been partially excavated' but by 1934 he was adamant that there had been no later intrusion into the mound and that its 'kidney' shape was an original feature. This line was followed in the publication despite the presence of photographs that appear to show very clear intrusion cutting at an angle into the mound from the east (animal activity rather than human). A comment by Miss Tildesley that the bones had evidently been broken when they were dry and brittle, after they had been buried for more than a few decades, is printed in the publication appendix, but attracts no further notice (Childe & Smith 1954, 220). The chamber had been covered by an inner earth mound which incorporated burnt layers and many finds (animal bone, pottery, and flint flakes), and this in turn was capped by chalk.

In subsequent seasons, as he progressed steadily west, north, and south, Scott claimed that the inner mound was retained on three sides by large tree trunks laid horizontally open on east to crescentic forecourt. Childe and Smith could find no trace of these in the records; we also failed.

He found some features hard to understand and struggled with the unfamiliarity of the terrain, in particular the evident presence of solution hollows in the chalk on the edge of the ridge, producing distinctive 'pits' and 'postholes' and even (in 1937) a 'peristalith trench'. And so he resorted to the familiar: the barrow 'shows increasing analogy to megalithic chambered tombs' (Scott 1936), and he decided its distinctive interned east side and the peristalith trench were original features which he compared to those at Pant-y-Saer (Scott 1933) and Bryn yr Hen Bobl on Anglesey (Hemp 1936).

The *Buckinghamshire Herald* did, however, print an article in June 1939 'with caution and under correction' about the excavations on Whiteleaf and also on the adjacent Bledlow Cop. There are some curiosities within this account but under the title 'Invaders from the South' it states that the barrow, which belonged to a very ancient period, may have been 'erected by an invading people pushing from the west along the Icknield Way'. These 'long-headed people' who did not practise cremation at that early time may be 'non-Aryan'. Although this account chimes with diffusionist views of the time, it sounds rather incautious and the extent to which it might reflect Scott's views is very uncertain. More reliably, there is an account in the Records of Buckinghamshire (1941–6, 298) of a talk he gave to the Buckinghamshire Archaeological Society in 1942, in which it is stated that 'The pottery is of Neolithic A2 type, linking with the Upper Thames Valley and with the Cambridge district'. Relationships with the wider region were obviously being considered.

The War and After

Even when at the Ministry of Aircraft Production during the Second World War, with the Blitz raging and a key role of resolving difficulties between the irascible Minister, Lord Beaverbrook, and his professional civil servants, Scott still found time to write a short article in *Antiquity* for 1942. Revealingly, he wrote to its editor, O. G. S. Crawford, that, knowing Scott's commitments, Childe had offered to fill in the missing references (MS Crawford 25: 23/4/42).

Scott retired after the war and, by 1946, when the Prehistoric Society recommended recording members' addresses, he had moved to Princes Risborough permanently. Nevertheless, as Childe relates, he only had time to excavate one more section through the ditch before he died in 1952. The records suggest that little post-excavation work had been done.

By this time Scott, the amateur, had become part of the archaeological establishment. A keen and effective networker, he was very involved in the (London) archaeological social scene: he was President of the Prehistoric Society from 1946–49; Vice-Chair of the Council for British Archaeology; on the Management Committee of the London

Institute of Archaeology; and on the council of the Society of Antiquaries of London. He was writing reviews for *Antiquity* at the behest of Crawford and a chapter on pottery for the Oxford *History of Technology* (Scott 1954), as proposed by Childe (Brian Smith pers. comm.). He was also spending time on his Scottish work: undertaking further research and excavations there, publishing his results, and writing three long papers for the *Proceedings of the Prehistoric Society* on the Neolithic and Iron Age settlement of Scotland in the culture-historical style.

Scott and Childe II

In contrast to the paucity of written information about Scott, the man who brought Whiteleaf to publication must have had more ink devoted to him than any other archaeologist (Díaz-Andreu 2009). And it is hard to think of two more dissimilar characters: Scott, a decorated serviceman and civil servant, 'a man of fine physique and striking presence' (Obituary in *The Times*, June 1952), a natural London club member who obviously loved the physical process of excavation; Childe, who expressed strong views against the Great War, was a Marxist, was said to be socially awkward and was not by natural instinct a field archaeologist (Harris 2009, 132), although he excavated and published important sites. Nevertheless, however unlikely, they struck up a strong bond of mutual respect and friendship from the 1930s when Childe was in Edinburgh (Green 1981, 59). Scott sought the views of Childe on his work from this time, Childe wrote to Scott at the beginning of WW2, asking him to support the settlement of the Bersus in the UK (Green 1981, 87), and it was Childe who wrote a glowing obituary of Scott when he died:

> a series of brilliant articles – of which 'Gallo-British Colonies' published in 1948 might be cited as outstanding for its originality and thoroughness – constitute masterly contributions to the prehistory not only of the British Isles but of Europe as a whole; for they are distinguished both by freshness and humanity of approach and also by an erudition which professional colleagues might envy (Obituary in *The Times*, June 1952).

Childe was a complex character and struck up collaborations with a number of senior Scottish businessmen during his time at Edinburgh (Ralston 2009, 59–61), was himself a London club member and, in any case, was in continuous discussion about new findings with all those who were creating the data (László 2009). Archaeology was a passion they shared.

Publishing the Site

There is no indication that Childe ever visited the Whiteleaf excavations himself and it is evident that he and Isobel Smith struggled with the records, in particular the lack of a complete plan of the site. The sections are also difficult to interpret, even with the photographs. The extent to which the many features Scott excavated beneath the mound were cut or natural was uncertain, and Childe notes that even Scott was doubtful. He omitted those features he thought that Scott would have excluded, including the peristalith trench.

Reconstructing the distribution of finds would have been an arduous process by hand, but the significance of the assemblage from the inner mound was obvious and factors such as the discovery of sherds from the same vessel in different areas of the mound and the freshness of the broken edges were noted. Feasting on the site; the strewing of the mound with artefacts as part of the burial rite; or the use of refuse and hearths from a nearby encampment? Childe poses these questions but leaves the answer open. Indeed, Childe seems to have added little of his own interpretation to what is a concise account of the excavation results. Was this the result of modesty and an 'unwillingness to impose himself and his ideas too strongly on other people' (Trigger 1980, 18); did the site not fit into any obvious category; or was it insufficiently interesting or difficult to interpret? Perhaps he was fulfilling an obligation to a friend in order that at least the important pottery assemblage was reported? Or was it to allow one of his most able students to shine?

It was left to Isobel Smith, Childe's PhD student and assistant at the Institute of Archaeology, to draw conclusions in her excellent analysis of the pottery assemblage, which formed part of her PhD thesis (Smith 1956). 'Diversity in decoration and form illustrates the complex nature of the relationships which obtained among the Neolithic communities of southern England and the fact that the Chilterns were, as their geographical position suggests, a focal area for both the reception and transmission of influences' (Smith 1954, 228); a mingling of traditions. These views are understandably influenced by those of Childe, but judging by the articles that Scott himself wrote in the late 1940s/early 1950s, it is probable that he would have taken a very similar, if not a less restrained, approach. Had Scott completed the publication himself, he would certainly have made much more of the landscape position of the barrow and the environment in which it was set, and the contributions of the other specialists on the project.

Something should be said about the other women in this story, for they are too often relegated to the footnotes of early 20th century publications. It is evident on the large plans in the Historic England archive that Childe made some of the annotations, but the hand of someone else other than Scott was involved. Some of these were undoubtedly the hand of Isobel Smith, but an intriguing possibility is that Mrs Scott (Fig. 28.3) might have been involved. According to Clark, in his obituary of Scott: 'with the help of his wife he investigated three chambered cairns among other sites' (Clark 1952, 234). Scott (1932, 187) did acknowledge his wife in the publication of Rudh 'an Dunain: 'With the kind

Figure 28.3 Lindsay and Winifred Scott sailing in the 1920s.

consent of Macleod of Macleod, excavations were carried out by my wife and myself in September 1931', and he described the obviously very arduous work to excavate the chamber with no other labour available, including lifting capstones etc. (Scott 1932, 187–8), though she does not feature in his writings again. Just possibly, might the sheer volume of archaeological work he undertook whilst also working full-time in part be due to her contribution?

Other women who worked with Scott included Mrs Peggy Piggott, who assisted on Scott's 1947 excavations at Clettraval and again with its 1948 publication, and Alison Young, who was Scott's assistant on excavation of a wheelhouse at Tighe Talamhanta, Barra, a site she brought to publication (1952); she became an important figure in the archaeology of the Western Isles (Sharples 2015, 3–5; Brian Smith pers. comm.).

The HLF Project

> Childe ... stands halfway between the heroic age of later nineteenth-century prehistory and the anonymous professionalism of the present-day discipline (Sherratt 1989, 184).

At the turn of the 21st century, we were particularly concerned with the significance of Whiteleaf as a place and the experience of the past users of the site, a relevance that has continued to the present day. We were hampered by there being so little of the original barrow remaining and the desire to preserve as much as possible of what remained, but it was possible to take samples from the surviving sections and to dig new cuttings through the surrounding ditch in order to undertake further analysis on the environment. To what extent could the monument be seen from afar and what could be seen from it? Kennard's conclusion that the barrow was situated in damp woodland was borne out by our work, though we were able to track subtle changes through time as the barrow ditch filled (Stafford 2007). Scott would have approved.

We had the undoubted advantage of a wealth of new data that had accumulated over the intervening 50 years, including many examples of complex barrow development, with the form and apparent meaning of monuments changing over a period of generations. We also had parallels for internal wooden chambers and for the use of a pair of split tree trunks to contain a burial (Wayland's Smithy farther west along the Ridgeway being a good example; Whittle 1991). The interpretation of finds in the inner mound being placed from an adjacent feasting or settlement area or being part of ritual practice chimed with current thinking.

Childe would have been envious of our ability to deploy radiocarbon dating (see, for example, his *Valediction*; Childe 1959). We were thus able to propose that the man buried in the barrow had probably died in the second quarter of the 4th millennium cal BC and between 45 and 150 years before at least some parts of the inner mound were raised. An additional capping of chalk took place between the 34th and 32nd centuries cal BC (Bayliss & Healy 2007). We discussed the social status of the burial given the relatively few early Neolithic individual burials that were known, and perhaps a widespread assumption that tombs of the time were for communal burial. We did not see pots as representing the identities of particular communities, and it would not have crossed our minds to speculate on any distant origins of Whiteleaf man.

On Reflection

Twenty years on, were we to embark on our post-excavation work afresh, we would submit samples for a range of isotope analyses and aDNA, techniques that, in the first decade of the 21st century, were in their infancy and rarely deployed. If successful, it is entirely likely that we would be discussing where Whiteleaf man had been brought up and had been living since, and the European origins of his ancestors. Warwick Lindsay Scott (and V. G. Childe) would have felt himself on familiar territory.

Acknowledgements

I would like to acknowledge the help of Brian Smith of Shetland Museum and Archives, David Wilkinson of Historic England and the staff of Historic England Archive and Library; UCL Special Collections; and Bodleian Library Special Collections in the preparation of this paper. I am also very grateful to the staff of Oxford Archaeology and members of the Princes Risborough Countryside Group who worked on the site, Buckinghamshire County Council for their support, and Richard Bradley for his unstinting advice. The project won the Institute for Archaeologists Award for the best archaeological project undertaken by a professional/voluntary partnership at the 2006 British Archaeological Awards.

Chris Gosden became Chairman of the Trustees of Oxford Archaeology in 2008, after the Whiteleaf project had finished. He has been a long-standing champion of collaborative

Figure 28.4 Chris Gosden with staff and students on tea break at Dorchester (Photo: Paul Booth).

community/professional partnerships, however, and already, in 2006, we had (with Helena Hamerow & Paul Booth) been involved in setting up an OA/Oxford University and local community project at Dorchester-on-Thames. The success of Whiteleaf was certainly a spur on OA's part, but Chris's enthusiasm and energy was a huge boost which saw us through eight field seasons there. As Chairman of the Trustees over the eight-and-a-half years that I was CEO of OA, his support, encouragement, and helpful advice was hugely appreciated; always calm and ready to listen.

Primary sources

LIN02: Historic England Archive and Library, Swindon
Ministry of Works files held by Historic England: AA0610571/1 for Whiteleaf Cross Scheduled Monument
MS Crawford: Bodleian Library Special Collections
The Times (20th June 1952)
University College London (UCL) Special Collections: Childe Collection, including Notebooks. (UCL 0093325-009339)

References

Bayliss, A. & Healy, F. (2007) Radiocarbon dates. In Hey *et al.* 2007, 68–70.
Bowden, M. (1991) *Pitt Rivers: The Life and Archaeological Work of Lieutenant-General Augustus Henry Lane Fox Pitt Rivers, DCL, FRS, FSA.* Cambridge, Cambridge University Press.
Childe, V. G. (1931) The continental affinities of British Neolithic pottery. *Archaeological Journal* 88, 37–66.
Childe, V. G. (1959) Valediction. *University of London Institute of Archaeology Bulletin* 1, 1–8.
Childe, V. G. & Smith, I. (1954) Excavation of a Neolithic barrow on Whiteleaf Hill, Bucks. *Proceedings of the Prehistoric Society* 8, 212–30.
Clark, J. G. D. (1952) Sir Lindsay Scott K.C.B., D.S.C., F.S.A.. *Proceedings of the Prehistoric Society* 18, 234.
Díaz-Andreu, M. (2009) Introduction. *European Journal of Archaeology* 12(1–3), 7–9.
East Riding Museums, Heritage Learning & Normanby Hall Country Park (n.d.) Dangerous work for fishing trawlers used as minesweepers. Webpage [online]. Available at: <https://www.mylearning.org/stories/minesweeping-during-the-first-world-war/785>
Green, S. (1981) *Prehistorian: A Biography of V Gordon Childe.* Bradford-on-Avon, Moonraker Press.
Harris, D. (2009) A new professor of a somewhat obscure subject: V. Gordon Childe at the London Institute of Archaeology, 1946–1956. *European Journal of Archaeology* 12(1–3), 123–44.
Hemp, W. J. (1936) The chambered cairn known as Bryn yr Hen Bobl near Plas Newydd, Anglesey. *Archaeologia* 85, 252–92.

Hey, G., Dennis, C. & Mayes, A. (2007) Archaeological investigations on Whiteleaf Hill, Princes Risborough, Buckinghamshire, 2002–5. *Records of Buckinghamshire* 47(2), 1–80.

László, A. (2009) The young Gordon Childe and Transylvanian archaeology: The archaeological correspondence between Childe and Frerenc László. *European Journal of Archaeology* 12(1–3), 35–46.

Oswald, A., Dyer, C. & Barber, M. (2001) *The Creation of Monuments: Neolithic Causewayed Enclosures in the British Isles*. Swindon, English Heritage.

Pitt Rivers, A. H. (1898) *Excavations in Cranborne Chase 1893-1896, Vol. IV*. Privately Published.

Ralston, I. (2009) Gordon Childe and Scottish archaeology: The Edinburgh years 1927–1946. *European Journal of Archaeology* 12(1–3), 47–90.

Records of Buckinghamshire, 1941–6. *Journal of the Architectural and Archaeological Society for the County of Buckingham* 14.

Scott, W. L. (1929) Discovery of Beakers in a cairn at Kraiknish, Loch Eynort, Isle of Skye. *Man* 29, 165–6.

Scott, W. L. (1932) Rudh 'an Dunain chambered cairn, Skye. *Proceedings of the Society of Antiquaries of Scotland* 66, 183–213.

Scott, W. L. (1933) Chambered tomb at Pant-y-Saer, Anglesey. *Archaeologia Cambrensis*, 88, 185–228.

Scott, W. L. (1935a) The chambered cairn of Clettraval, North Uist. *Proceedings of the Society of Antiquaries of Scotland* 69, 480–536.

Scott, W. L. (1935b) Whiteleaf Barrow, Monks Risborough, Buckinghamshire, *Proceedings of the Prehistoric Society* 1, 132.

Scott, W. L. (1936) Whiteleaf Barrow, Monks Risborough, Buckinghamshire, *Proceedings of the Prehistoric Society* 2, 213.

Scott, W. L. (1937) Whiteleaf Barrow, Monks Risborough, Buckinghamshire, *Proceedings of the Prehistoric Society* 3, 441.

Scott, W. L. (1942) Neolithic culture of the Hebrides. *Antiquity* 16, 301–6.

Scott, W. L. (1954) Pottery. In C. Singer, E. J. Holmyard, A. R. Hall & T. I. Williams (eds), *A History of Technology, Vol. 1*, 376–412. Oxford, Oxford University Press.

Sharples, N. (2015) A short history of archaeology in the Uists, Outer Hebrides. *Journal of the North Atlantic*, Special Volume 9, 1–15.

Sherratt, A. (1989) V. Gordon Childe: Archaeology and intellectual history. *Past & Present* 125, 151–85.

Smith, I. F. (1956) *The decorative art of Neolithic ceramics in south-eastern England and its relations*. Unpublished PhD thesis, University of London Faculty of Arts.

Stafford, E. C. (2007) Land mollusca. In Hey *et al.* 2007, 61–8.

Trigger, B. G. (1980) *Gordon Childe: Revolutions in Archaeology*. London, Thames and Hudson.

Whittle, A. (1991) Wayland's Smithy, Oxfordshire: Excavations at the Neolithic tomb in 1962-63 by R. J. C. Atkinson and S. Piggott. *Proceedings of the Prehistoric Society* 57(2), 61–101.

Young, A. (1952) An aisled farmhouse at Allasdale, Isle of Barra. *Proceedings of the Society of Antiquaries of Scotland* 87, 80–105.

29

Redirecting the Field – Total Archaeologies, Flagships, and Sample Design

Christopher Evans

With the Joneses' Mucking and David Clarke's Great Wilbraham, excavations serving as case-studies, aspirations to, and diverse concepts of Total Archaeology are explored. Particularly relevant is the impact of New/Processual Archaeology upon British fieldwork during the 1970s and, with it, issues of appropriate sample design. The paper raises questions pertinent to today's developer-led practices and concerning the idea of 'excavation as experiment'.

Keywords: Total Archaeology, Sample Design, New Archaeology, Developer-Funding, Excavation as Experiment, Fieldwork Historiography

Introduction

Stately sailing over the erratic waters of the day's general-standard sites, self-acknowledged 'flagship' excavations often come down to their resourcing and sponsorship. The Society of Antiquaries' campaigns at Silchester of 1889–1909, Wheeler's Maiden Castle (1934–36), and Carver's Sutton Hoo (1983–2005) provide obvious instances (see *e.g.* Fulford 2007), and the Prehistoric Society's 'orchestration' of Bersu's Little Woodbury could equally be cited (1938–39). More recent examples include Framework's Heathrow Terminal 5 and Hodder's Catalhöyük. Their accompanying manifesto-like pronouncements are intriguing (Hodder 1997; Andrews *et al.* 2000), carrying a promise that, somehow, this time things will be done right and, thereby, redirect 'the field' at large.

Invariably, the degree that such grand projects succeed in their ambitions, and the extent of their influence, is variable, with David Clarke's (1968, 3) appraisal of the ascent (and passing) of sites still appropriate:

> Every decade produces one or more sites of outstanding importance and impact, that linger on in the literature or sparkle briefly on the glossy pages of ephemeral publications. *The archaeologists come and go, new names and sites outshine the old ...* [emphasis added].

Related, proclamations of 'total archaeology' have long been a chimera of fieldwork.[1] The term was first widely applied by Christopher Taylor (1974), referring to the full range of sources (from excavation and earthwork recording to historical maps and folklore) that should be brought to bear within area surveys. Here, two 'total' initiatives will be reviewed: the Joneses' Mucking and Clarke's Great Wilbraham. Both had lofty aspirations and, for a variety of reasons, arguably failed. Neither were published in their time – at least fully in the case of Mucking – and only much later appeared as Cambridge Archaeological Unit (CAU) *Historiography and Fieldwork* studies (Evans *et al.* 2006; 2009; 2016a; Lucy & Evans 2016). The series' abiding premise is that we now 'dig after origins', and as much in relationship to what has been written before as anything in the ground. Accordingly, it beholds us to appreciate the influence of any region's early-day sources and their conceptual underpinning.

The account here is punctuated by disputes. They tell of the day's practices at a time when things were not finalised and set in 'professional stone'; like failures, they have been/are a major means of how we learn. To these two site ventures will be added Pryor's Maxey/Welland Valley (Pryor & French 1985). Attempting a redirection of field practice, it represents a commendable attempt to do things differently. Like Clarke's Wilbraham, it was very much a

New/Processual Archaeology exercise and, as such, reflects upon the impact – and various aftermaths – of 'The New' in British Archaeology (see *e.g.* Champion 1991). The sites discussed here moreover raise issues pertinent to today's development-led fieldwork and the arising massive increase in site-data: research prioritisation, information redundancy, appropriate site sample-design, and 'the challenge of numbers' generally.

Two Totalities

The 'totalities' espoused at Mucking and Great Wilbraham could not have been more different. Determined by an immediate threat of quarrying, Mucking's excavations progressed across *ca.* 18 ha over 13 years (1965–77; Fig. 29.1). Involving in total some 5000 participants, it was a case of rescue archaeology *par excellence* (Evans *et al.* 2016a). The intensity to which it was excavated – the target was to dig 100% of all features but upwards of a quarter of the area either went unexcavated or its features were dug less intensively – was essentially to maximise find retrieval. This largely stemmed from that, in the early days of rescue, the archaeological 'past' was thought to be quite rare and that there were relatively few sites left to dig (Evans *et al.* forthcoming). For Margaret Jones, the project's director and driving force, *total destruction warranted total excavation*.

With an estimated 44,000 features, including more than 400 structures and 1145 burials, the site's bald figures are staggering. Apart from Neolithic (4000–2400 BC) pits and a rich Beaker burial, it encompassed Bronze Age (2400–800 BC) barrows and a field system, a concentric-plan late Bronze Age ringwork and extensive settlement – as there was during the Iron Age (800 BC–AD 43; including a number of enclosure-compounds) – a Roman estate supply-centre and 'industrial' settlement complex (with five cemeteries), and two main Saxon settlement clusters and their cemetery. Its sequence essentially 'took off' in the late Bronze Age, with high settlement levels maintained for more than a millennium thereafter. It was, in short, a completely unparalleled site. Much can be attributed to its specific location perched on a riverside gravel terrace at the Thames' last downstream bend, with its eastern viewshed effectively opening to France.

Through its many interim reports and summary notices, we were able to piece together much of what would likely have been the Joneses' interpretative framework. Unsurprisingly, given the time and their findings, it was predominantly historicist. Evoking folk movements/migration and invasion, things have now gone full circle, with recent aDNA studies beginning to provide a measure of validity to such 'dynamic pasts' (*e.g.* Patterson *et al.* 2022). As was widely practiced at the time by its director's regular checking of what finds were forthcoming and what initial post-excavation study was conducted, they had a firm grasp of the site's broad development. Producing a series of period-phase plans (Evans *et al.* 2016a, fig. 1.18), their understanding of most of its sequence was largely 'right', if falling short of formal proof (see *e.g.* papers in Chapman & Wylie 2015); their only significant omission being the recognition of its late Iron Age barrow-flanked 'plaza space' (Evans *et al.* 2016a, 336–49, fig. 4.81).

As an excavation, Mucking grew organically, with its eventual scale never envisaged from the outset. Starting in the days before single-context recording, its basic documentation ran to 360 notebooks. Without concurrent computing (nor even readily available photocopying), there was an incredible audacity in what they attempted. While arguably failing as a flagship, Mucking was undoubtedly a 'great' excavation (Barford 2011) and, given the circumstances, what they achieved was truly heroic.

The fraught history of the site's various post-excavation initiatives and its partial publication are fully outlined in the 2016 volume. Working with the archives, it was evident that the Joneses really did not want 'everything'. Instead, it was *a selective type-based totality* they were after. What was emphasised and separately gazetteer-numbered were main feature-types: – graves, Saxon *Grubenhäuser*, Roman kilns and wells, as well later prehistoric roundhouses – with little recorded heed paid to ditch-segment fills or building postholes. Equally, while bulk pottery coarse wares were recorded, it was a period's specific type-wares that commanded attention (*e.g.* Samian). In other words, Mucking's totality involved recognisable entities (*i.e.* building-block 'types') and things that would readily 'speak'.

They were, nonetheless, aware of the value of per-period finds distributions and exemplars were run when the data was eventually computerised in the early 1980s (*e.g.* Evans *et al.* 2016a, figs 5.20 & 5.21). Yet, it is telling that they – as others at the time – seem to have had little appreciation of the possibilities of plotting residual finds, with their occurrence potentially serving as, in effect, a proxy sampling grid of surface-deposit material. For our work, the flintwork within later features allowed for the reconstruction of earlier occupation/use 'clusters' (*e.g.* early Neolithic, Grooved Ware, and Beaker; Fig. 29.2), something that even today can only rarely be convincingly documented on major site exposures.

Crucially, the day's publication goals did not extend to 'completeness', but rather what was representative of a site's sequence. What scuppered their efforts was the Frere Report of 1975. Standardising and codifying post-excavation procedures into staged levels, to obtain funding required a degree of archival presentation that it is unlikely they could ever have achieved. Nor do our site publications amount to a 'totality'. While hopefully doing justice to the Joneses' sustained efforts and the site's extraordinary sequence, Mucking's vast archives still have much untapped potential.

Strictly focused upon a small, *ca.* 2 ha double-circuit circular Neolithic causewayed enclosure 8 km east of

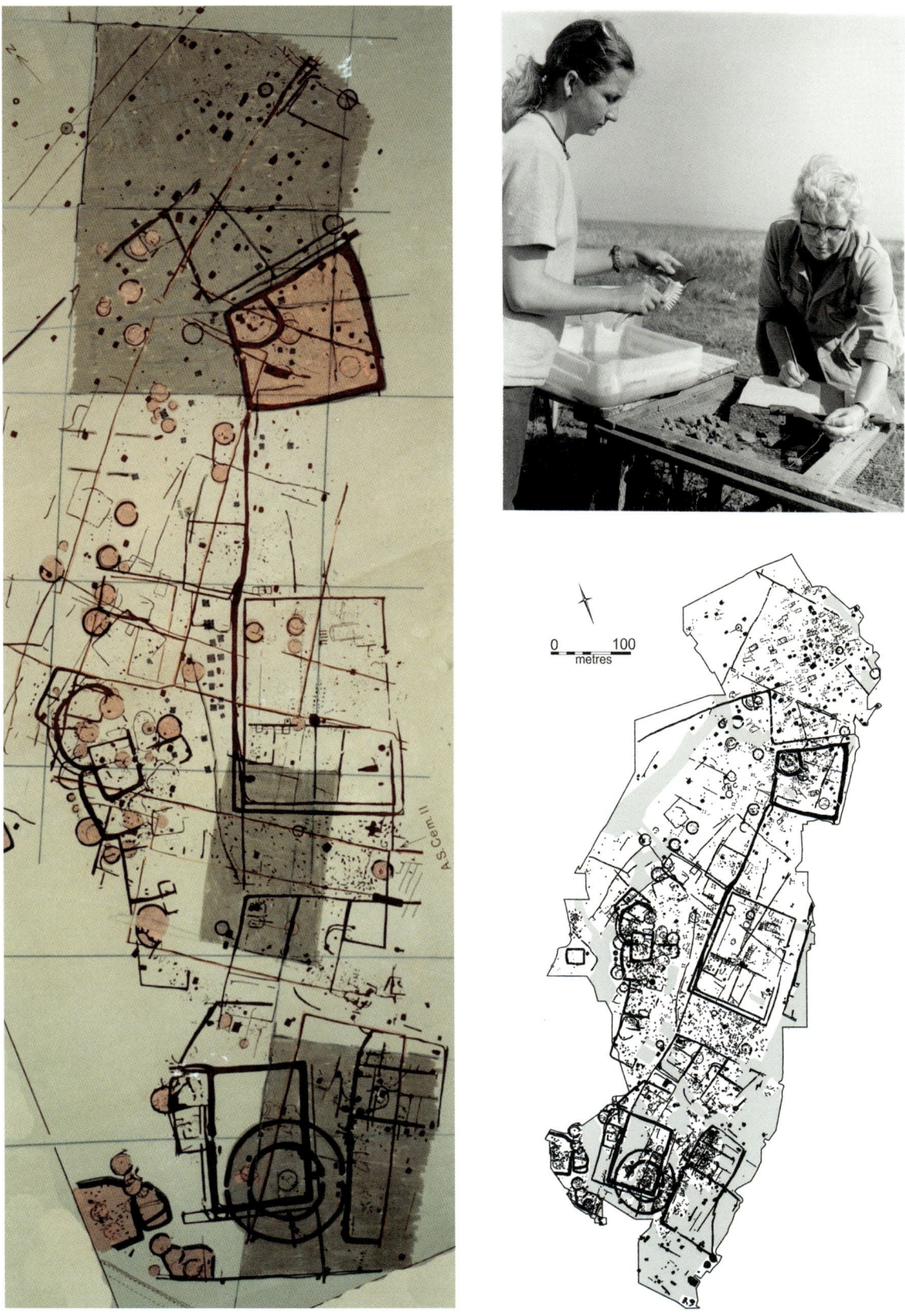

Figure 29.1 Mucking (I): (left) archival plan with Iron Age compounds and roundhouses highlighted; (top right) Margaret Jones (1916–2001); (bottom right) base-plan with swathes of incomplete feature-recovery in grey-tone (Evans et al. 2016a, figs 1.6, 2.1 & 3.36).

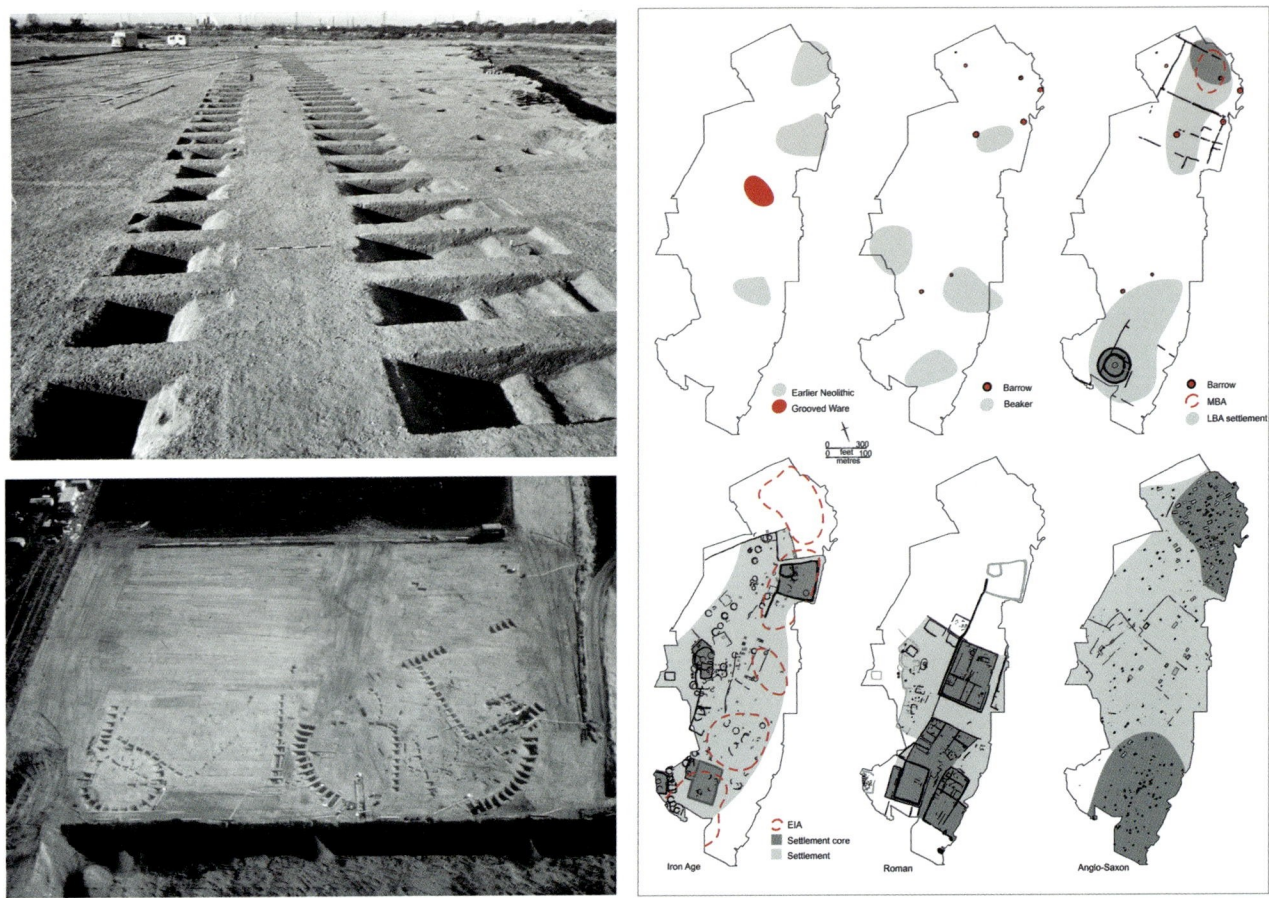

Figure 29.2 Mucking (II): (left) photographs showing excavation intensity (the intervening baulks would later be reduced); (right) CAU period phase-plans (Evans et al. 2016a, figs 1.5 & 6.20).

Cambridge and there, situated between the chalklands and the fens, David Clarke's Great Wilbraham was to be an exercise in total archaeology on two accounts (Evans *et al.* 2006). First, that the intention was to excavate the enclosure's interior and ditches in their entirety. Second, by the breadth of its on-going analytical programme and intense sampling procedures, it was to be highly systematic.

As to the first aspiration, the fieldwork was planned for five years, with an application made to the British Museum for £2000 funding *per annum*. In the end, involving some 30 Cambridge students at any one time, just two preliminary seasons were conducted in 1975 and 1976, with only three trenches ever dug (Fig. 29.3c). Clarke's co-director during the first season was John Alexander, with Ian Kinnes replacing Clarke in the second upon his untimely death.

Clarke primarily wanted a finds-rich Neolithic site, and Wilbraham was in many respects ideal. Having high finds densities, that it lay so close to the University's facilities, its laboratories and computers, was stressed, with such a 'home-base' infrastructure held essential for the provision of both interdisciplinary 'science' and on-going information feedback. Yet, given the available funding, the directors'

experience (especially with machining), and lack of support-infrastructure – the on-site facilities amounting to a single garden shed – it is honestly difficult to see that, had Clarke lived, they could have ever succeeded in their 'total' aims. In the end, no immediate publication of the work was forthcoming; it being left to us to assemble its archival fragments 30 years later.[2]

We were able, though, to achieve some understanding of what Clarke's intentions were from his various application documents and brief first-year summary. He clearly considered the enclosure a settlement, probably a winter basecamp. The archives have a checklist of causewayed enclosure attributes (Evans *et al.* 2006, fig. 3); while ritual is included amongst its entries, nothing was forthcoming from the first season to alter his interpretative slant.

With fieldwalking and magnetometer surveys conducted beforehand (Fig. 29.3), the project would certainly have been pioneering. It was to follow a 'cascade system' providing oversight of the ongoing findings and ensuring that deposits were intensively sieved, processed, and analysed. Hand-recovered finds were closely plotted and, with the preserved ground surfaces spit-dug by metre-square, it

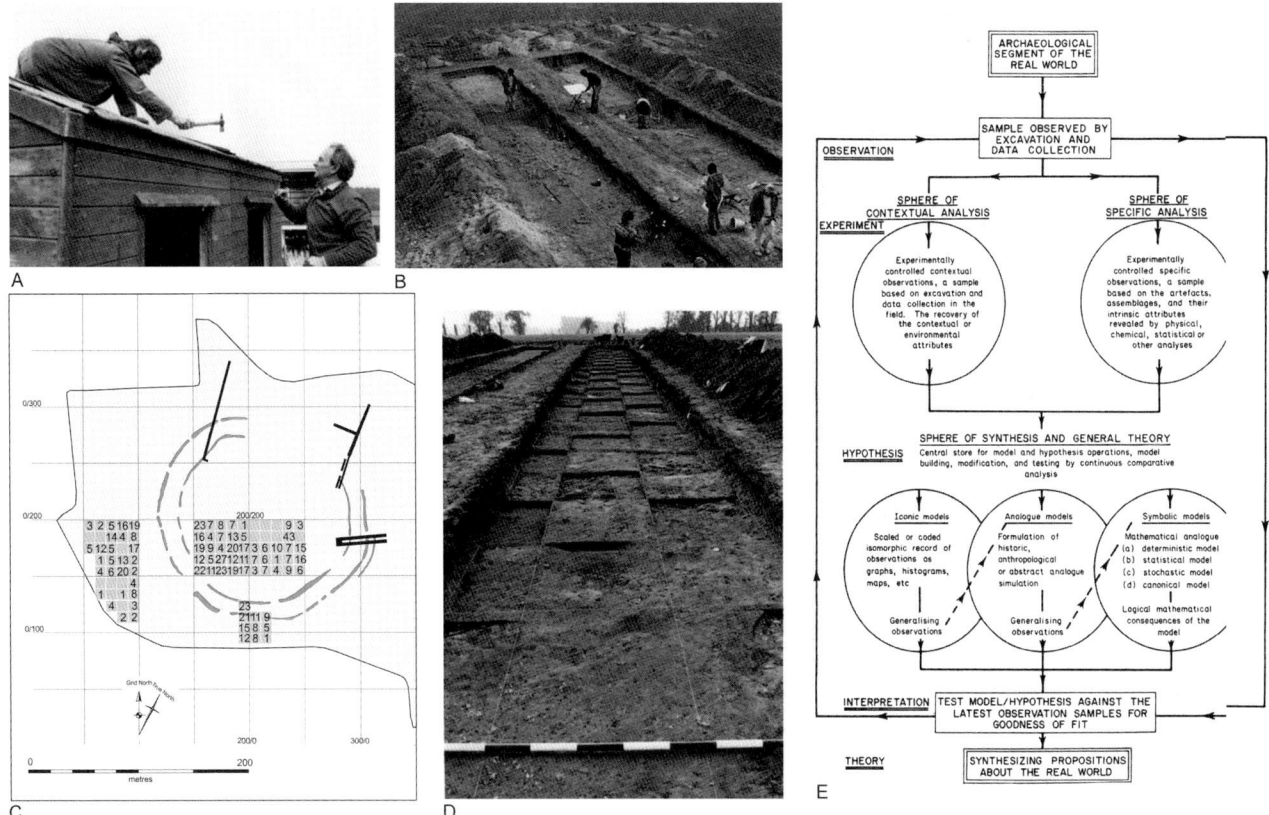

Figure 29.3 Great Wilbraham: (a) Clarke & Alexander re-roofing the site office during the 1975 season; (b) 1976 trenching, with Ian Kinnes in the background on the left; (c) enclosure plan showing trench locations and worked flint values from fieldwalking; (d) metre-square excavation of old ground surface deposits; (e) Clarke's 'General Model for Archaeological Procedure' (1968, fig. 2).

would have made for detailed distributional studies had this policy been continued throughout.

Telling here is that this exercise in total archaeology was considered an *experiment*, and that its cascade system would have facilitated a formal feedback process, one surely reminiscent of *Analytical Archaeology*'s general (looping) model of archaeological procedures as a whole (Clarke 1968, fig. 2; see Fig. 29.3e). While excavation is only accorded the briefest of mentions within that volume (Clarke 1968, 14 & 34–5), in the 'Archaeology: The Loss of Innocence' paper's General Theory (Clarke 1973, 16), it was very much framed – on the grounds of loss/preservation and fieldwork procedures – as amounting to 'a sample of a sample of a …'.

In the project documentation, Clarke stressed the enclosure's 'betwixt' environmental situation. Presuming that he would have undertaken its landscape modelling in the manner of Glastonbury's (Clarke 1972), in our publication both Site Catchment Analysis and Thessian Polygon 'enclosure territories' were applied, the latter based on what few causewayed enclosures were then known in the area (with a number discovered shortly thereafter; Evans *et al.* 2006, 153–7, figs 22 & 24).[3] While Clarke may have gone on to develop a more nuanced appreciation of its landscape, there is something inherently mechanical in such approaches. It is as if, parachuting into a spot and focusing on just one period, the site's Neolithic (approximately half the recovered pottery was later), effectively time-sliced its land-use history/sequence. Certainly, it reflects nothing like the commitment of the Joneses at Mucking, with the title *Lives in Land …* of our volume (Evans *et al.* 2016a) both referencing the longevity of theirs' within its landscape and the many lived therein over time. But then, Clarke's strict time-slicing focus can, in effect, be seen as setting the parameters of the 'site as experiment'.

Contestations: The Impact of 'the New'

During the 1970s, many sites were still intensively dug; in their entirety (100%) and without, for example, interval sample-slotting of their linear features. Portraying the day's 'proper excavation-standard', *Current Archaeology* of 1979 still abounded with exquisite photographs of completely excavated small- and medium-sized sites. But, amidst these, there is the opinion section on 'SAMPLING' by Cherry, Gamble, and Shennan (1979). Introduced by

Colin Renfrew's 'the days of the innumerate are numbered' (Shennan 1988, 7), it was basically a 'young bloods' call-to-arms announcing the publication of their edited volume, *Sampling in Contemporary British Archaeology* (1978). With American New Archaeology sources prominent (*e.g.* Binford 1964), at some 400 pages the book was massive, and amongst its 24 contributions were papers by Foley, Haselgrove, Orton, and Champion. The latter's concerned how Chalton's entirely dug settlement could have been variously sample-investigated. Champion (1978, 208) critiqued 'the recovery of repetitive and redundant information' arising from most sites and, citing Mucking, disputed claims of their total excavation:

> a consideration of such total excavations is most revealing. The strategy usually involves as a first stage the mechanical stripping of the topsoil, thus ignoring and destroying all the information contained therein, such as the spatial distribution of artefacts and their relationship to subsoil features. Already part of the total information has been lost ... *Total excavation, in fact, all too often involves nothing more than the total recovery of such traditionally determined features as houses and such artefacts as pottery* [emphasis added, citations omitted].

Both the damage wrought to sites by ploughing and the potential of ploughsoil finds were themes then receiving wide attention (see *e.g.* papers in Hinchliffe & Schadla-Hall 1980). Although attitudes towards sampling, research orientation and theory generally were beginning to turn – in part, cost-determined (*e.g.* Wainwright 1978; 2000; see also Renfrew 1978) – this flew in the face of rescue archaeology's 'dig everything' ethos, with many opposed (*e.g.* Barker 1977; Fowler 1977). Matters were further exasperated with the publication of Pryor's (1980a) *Survey Excavation* 'manifesto' for his Maxey/Welland Valley Project in *Rescue News*:

> I'm afraid, however, that enormous set-piece excavations such as Fengate, are a thing of the past; for a start, they cost too much and post-excavation work ... is time consuming, dull and needlessly repetitive. *Vast quantities of data are produced, but there is often precious little information; in short, we are learning more and more about less and less* [emphasis added].

As evinced in their letter-exchanges, this was an approach with which Margaret Jones took umbrage (11/02/80; Evans *et al*. 2009, 12–14 & 250–2).

Aside from generic references to the value of ploughsoil data (e.g. Binford et al. 1970), concerning 'fallacious' total archaeology strategies, Haselgrove's 1978 paper was cited in the Welland Valley volume (Pryor & French 1985, 34). He distinguished two classes of (rural) site survival. The first where surface strata survived (Class I) that clearly deserved total excavation if threatened (Glastonbury being an obvious example). The other, Class II, being those whose surface deposits were disturbed, largely through plough-damage.

> However, most archaeologists would agree that we should also attempt to obtain the maximum possible amount of useful and representative information from Class II sites using a combination of judgement and probabilistic sampling procedures, and it is here that more experiment is urgently needed.

The reason for this is that one form of excavation procedure for Class II sites has achieved an unhealthy monopoly in British archaeology, *i.e.* the removal of disturbed topsoil over large areas using earth-moving machines, followed by the excavation of most or all the undisturbed sub-soil features and levels revealed. As Renfrew (1974, 39) observes:

> 'Many archaeologists feel today that we now know how to excavate a settlement site, that it can be done adequately following recognised procedures. *The truth is the opposite* [Haselgrove's emphasis]. *Any archaeological site contains so much potential information that any kind of excavation is simply a sampling procedure and a very partial one at that.*'

> *In this strategy, often misguidedly known as total excavation, the conscious decision is made to ignore the information potential of the surface stratum by discarding it unscreened*, on the grounds that it is more important to explore the plan of structures and features exposed and to obtain cultural debris from reliable contexts [Haselgrove 1978, 169; emphasis added].

The original source of Pryor's 'New' influences were apparently his American colleagues while working at the Royal Ontario Museum (Pryor pers. comm.). These evidently underpinned his Second Fengate Report concerned with the Storey's Bar Road sub-site (Pryor 1978), whose organisation and logic were entirely 'foreign' (*i.e.* 'New'). Indeed, based on premises of labour expediency, this led to the mis-attribution of its Bronze Age ring-ditch/barrow and field system to the site's later Neolithic Grooved Ware occupation, whose traces otherwise occurred in pits or hollows and postholes, with its presence in the Bronze Age features residual (Evans & Pollard 2001; Evans *et al*. 2009, 89, fig. 3.20). As was the case at Mucking, there was little appreciation of residual-context material.

Pryor's Maxey/Welland Valley programme involved two facets. There was Maxey Quarry's 'survey excavation'. Conducted over 3.75 ha with the aim of 'digging for questions', it was far more explicit in its 'New' goals than Fengate's Storey's Bar Road. It included significant portions of a major cursus and a large henge, an oval barrow, Iron Age square barrows and a Late Iron Age/Roman settlement-enclosure system, with the latter having a small square shrine setting (Fig. 29.4). Upon site stripping, the target sampling level was a 20% minimum of all features. With linear features generally tested through two-metre segment interventions, this also involved considerable judgmental flexibility. Its structural components – namely roundhouse eaves-gullies – were more intensively dug, to at least a level of 40% and often up to 80%, with the henge's outer circuit almost completely emptied. The excavation procedures also involved a complicated sieving regime (Pryor & French 1985, 24–9). Prior to all this was the ploughsoil's intense interrogation,

Figure 29.4 Maxey: (top) site base-plan with sample locations indicated; (bottom) East Field base-plan showing excavation interventions (Pryor & French 1985, figs 151 & 40).

with a barrage of techniques applied. In addition to metre-grid fieldwalking collection were sedimentological, chemical, botanical, geophysical, and metal-detecting surveys. In the volume, 24 pages are devoted to their detailed analyses, including modelling of the ploughsoil finds depositional behaviour (Fig. 29.5; Pryor & French 1985, 34–58; see also Crowther 1983). Amounting to a genuine attempt to do something different in the field, little in British archaeology would then compare to Maxey's excavation strategy. Oxford's Mount Farm, Dorchester-on-Thames, could be cited (1977–78; Jones 1978), but when finally published there was limited integration of its ploughsoil and cut-feature

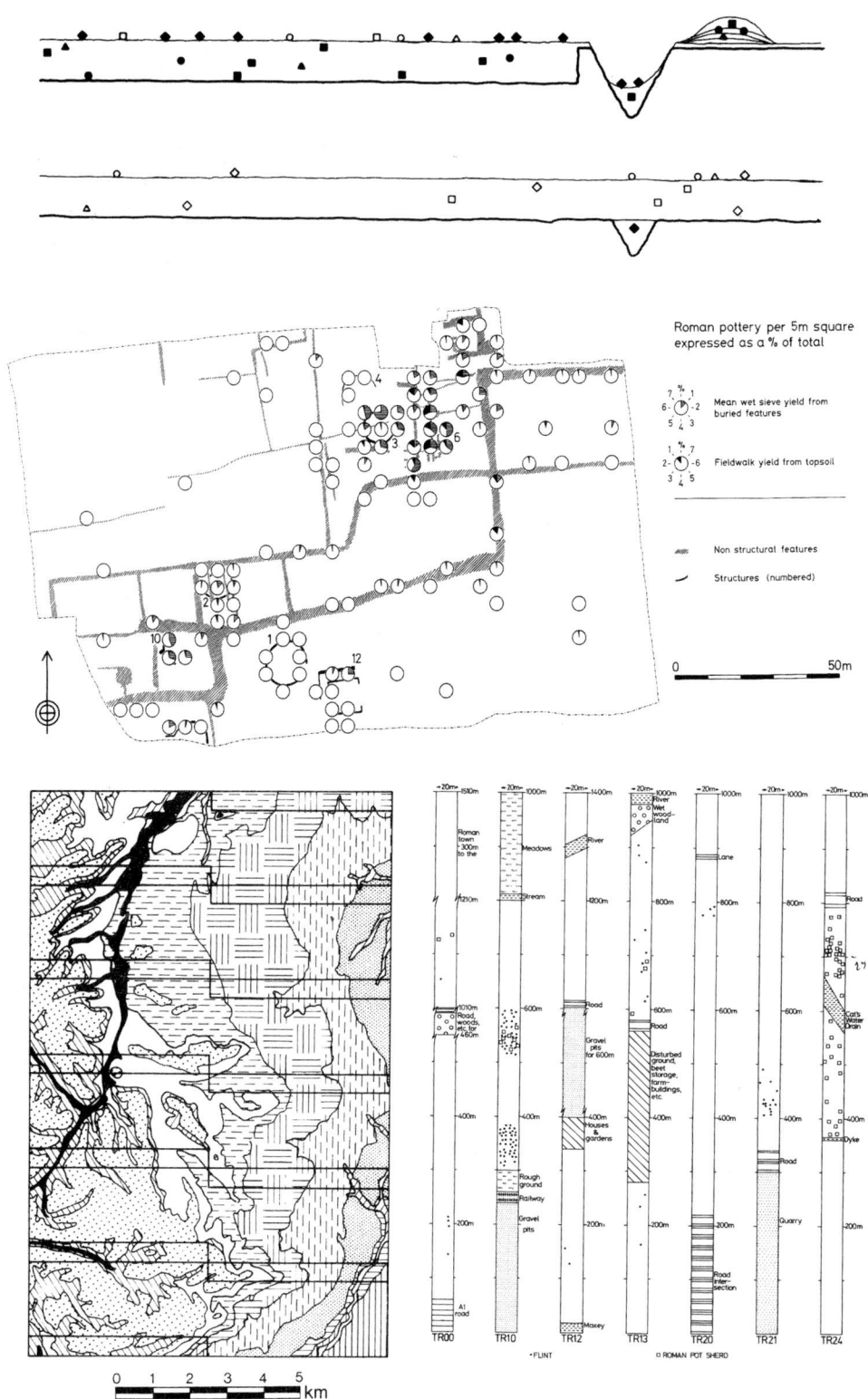

Figure 29.5 Surface Deposit Investigations: (top) Maxey's artefact depositional model (Proposal/Stages d & g only; Pryor & French 1985, fig. 34); (middle) Maxey East Field, showing Roman pottery fieldwalking values in relationship to wet-sieve sample values from cut features (Pryor & French 1985, fig. 31); (bottom left) Shennan's East Hampshire fieldwalking transects (Shennan 1980, fig. 45); (bottom right) Welland Valley's fieldwalking transects (Pryor & French 1985, fig. 10). The kind of ploughsoil artefact depositional modelling conducted at Maxey was just one of a number at the time, including Haselgrove's (1985, 14–20 & fig. 1.3) and Reynolds and Schadla-Hall's (1980) 'ideal' earthwork and barrow models, with the most sophisticated being Foley's (1981) generic artefact distributional taphonomic studies.

data (Lambrick 2010; see also *e.g.* Miles 1983 concerning Claydon Pike's procedures and fig. 57 for its Site Catchment Analysis).[4]

In many respects, the main directive of the day's sampling agenda was not so much site-focused as relating to macro-scale fieldwalking surveys. In part, this was prompted on pragmatic grounds: to provide a firmer basis of a region or county's number of sites, largely to assess their threat through plough-damage (see papers in Fowler 1972; Hinchliffe & Schadla-Hall 1980). Yet, it also arose through theoretical concerns, and a desire to move beyond single type-site foci, interrelating them in socio-cultural networks and their landscape context (Cherry & Shennan 1978). Foley's 'off-site archaeology' was central. Revolving around the idea that archaeologists should not be constrained by strictly bounded site definitions and, instead, consider group activity within broader territorial ranges (Foley 1978; 1981, 163–6), it has had long-lasting impact, especially for prehistory.

Concurrent with the region's broader Fenland Survey (Hall & Coles 1995), the Welland programme's other main component accordingly involved a series of fieldwalking transects targeting a 30 km stretch of the river valley. Although curtailed by access and cover, advised by Clive Orton (and referring to the Cherry *et al.* 1978 volume's papers), this was 'formal' in its design. Taking into account landscape and environmental variability, its sampled portions also involved a 'random' component. In the end, some 23 ha were surveyed as a pilot study, but just 53 Roman sherds and 160 flints were forthcoming (Fig. 29.5; Pryor & French 1985, 15–23 & 310–12). While not directed towards site-prospection as such, the exercise would have been far better suited to an arid Mediterranean environment (*e.g.* Cherry 1982) or Wessex's downland (Shennan 1985), and not an alluvial lowland landscape that demands more in-depth procedures, such as dyke survey (Crowther *et al.* 1985) or deep test pitting and trenching (Evans 2000; Evans *et al.* 2014).

With the Welland volume appearing in 1985 and four years after the completion of Maxey's fieldwork, it included few references to its New Archaeology inspiration. These were paraded in an earlier paper titled 'Will it all come out in the Wash' (Pryor 1980b) that both further announced its programme and reviewed aspects of Fengate's. Alluding to Schiffer's C- and N-transforms, Binford's Hatchery West, and Middle Range Theory, it widely cited American sources and, though acknowledging shortcomings of the 'processual school', argued (Pryor 1980b, 486):

> it is difficult to see why this school has been received in such a lukewarm fashion by the vast majority of full-time field archaeologists in Britain, *for it has, in effect, been completely rejected*. The initial adverse comments that greeted its appearance still seem to stifle discussion. We have thrown an often imaginative baby out with some jargon-ridden bathwater and a few trivial 'Micky Mouse' laws [emphasis added, citations omitted].

Maxey/Welland did not fully realise its goals. Due to logistical factors (and funding), many of the features along the western third of its East Field evidently went untested (Fig. 29.4). Also, too few of its Iron Age and Roman features were dug and too little material of the periods retrieved to fully articulate the settlement's dynamics, with the result that its excavation and ploughsoil components do not really mesh. Arguably then, another 'failure', but that is beside the point: we learn from such efforts, and it only remains a brave attempt to redirect things.

One is hard-pressed to find other sites of the day that were excavated according to explicitly New/Processual agendas. With its specific remit of obtaining dating evidence, Renfrew's (1979) 1972–74 keyhole excavations of Orkney's monuments could be cited. Although having few 'New' references (*e.g.* Binford and Schiffer), it included a simulation of chambered cairn locations and, remarkably enough, quantitative artefact analyses by John Cherry and Robin Torrence (Renfrew 1979, Appendix H). Dug in 1973 and '74 anticipating motorway construction, a stronger case can be made for Hodder's Iron Age settlement and Roman villa at Wendens Ambo, Essex. With its analyses conducted as a case-study in Behavioural Archaeology (Halstead *et al.* 1978), there is some irony that its final publication appeared in the same year as *Symbols in Action* and *Symbolic and Structural Archaeology* (Hodder 1982a; 1982b), that effectively announced Post-Processualism (see, though, also Hodder 1981). On the whole, Wendens' format now seems familiar. Yet, having a strong palaeo-economic emphasis, spatial/distributional and depositional analyses (with *e.g.* per cubic-metre densities), and with citations to Clarke's 'Glastonbury' and 'Loss of Innocence' (Clarke 1972; 1973) – including, also, (quasi-) Site Catchment Analysis (Hodder 1982c, figs 8 & 65) – it had a distinctly Processual/Behavioural orientation.

A reason why New/Processual Archaeology seemingly had such a limited impact in Britain is that its formal procedures were so very 'foreign'. Equally, it was a time of rapid paradigm shift, with some sites dug under one set of influences only published (if at all) much later, when the interpretative ground had alerted and, unless issuing 'manifesto' interims/retrospects, their original impetus went undocumented. An off-shoot of the attention then paid to research designs generally (*e.g.* Binford 1964; Schiffer *et al.* 1978), these circumstances gave rise to a specific type of 'what I should/could have done' publication, including Champion's aforementioned 1978 contribution, Woodman (1982; 1985) on Mount Sandel, and even Fasham and Whinney (1991, 149–53) for the M3's fieldwork.

Aftermaths and Partial Pasts

New/Processual Archaeology should not be accredited as having invented site sampling, just its far more rigorous and widespread application. Admittedly, prior to the 1960s, most sites were investigated though hand-dug trenches subsequently expanded into larger exposures. Yet, there were also instances of more systematic, and occasionally elegant, grid-based sample-detection and expansion procedures; for example, the 1955 Linford Site (Evans *et al.* 2016a, fig. 1.8) or the Hawkeses' Longbridge Deverill Cow Down (1956 and 1958–60; Hawkes & Hawkes 2012).

With the greater availability of earthmoving machinery, during the later 1960s and 70s, many medial-scale sites (*ca.* 1–5 ha; Evans *et al.* 2016a, table 1.1) were excavated with the aim of establishing 'type' period-exemplars. While there began to be larger quarry-related exposures during that era's latter years, nothing really matched Mucking. Only with Heathrow Terminal 5's fieldwork of 1996–2007 was there a comparable rescue/development-led excavation (*e.g.* Framework Archaeology 2010). Yet it lacked Mucking's long-term intensity/feature-density and, if parading the Joneses' site as a 'rescue-flagship', it is difficult to accredit it as having much methodological impact. While a touchstone for the day's other sites in the region and the decades that followed (*e.g.* Shoebury North, Springfield Lyons, and Orsett Cock), by not seeing full publication in its time, Mucking's findings failed to accrue their due impact.

The British Museum's funding of Great Wilbraham (£100 for each of its two years) was on account of its potential for Neolithic waterlogged finds. The funding dissipated during the second season; funding was withdrawn, and fieldwork ceased. Although it is therefore difficult to accredit it as having any significant influence, it did later resonant with other Fenland-area projects.

Also, in part British Museum resourced, Pryor's 1982–87 rescue excavation of Etton's causewayed enclosure can be considered a successor of sorts. With its circuit locally waterlogged, the entirety of the enclosure's ditches that were threatened were dug, its interior buried soil cover was intensively sampled, and the available ditch circuit completely excavated (Pryor 1998). This strategy was entirely appropriate given its impending quarry destruction and the nature of the enclosure's deposits. Etton's finds densities were, though, nothing like Wilbraham's (see Evans *et al.* 2006, table 12) and, unlike it, was interpreted as a ritual foci within an over-arching Post-Processual framework.

Although by no means directly ancestral, undertaken in a Post-Processual agenda, the University's Haddenham Project of 1981–87 also had affinities with Great Wilbraham. Primarily directed towards the excavation of the Upper Delphs causewayed enclosure (*ca.* 8.75 ha), located on the fen-edge north of Cambridge, it too had rigorous sampling methodologies, with on-going site feedback an aim (Evans & Hodder 2006). Like Clarke's site, it similarly failed in the latter aspiration due essentially to limited funding and infrastructure, particularly the restrictions of the day's main-frame computing. Where it succeeded was in its landscape-scale buried soil sampling, enabling widespread 'in-depth' distributions of artefact and chemical traces (Evans & Hodder 2006, 212–15, figs 4.4 & 4.8–4.10). Crucially, despite the project's initial directives, interests were pulled beyond its enclosure focus, with a nearby long barrow, plus Bronze/Iron Age and even Roman sites dug. If concerned with landscape and its contextual long-term history, then the kind of strict 'time-slicing' focus as announced for Wilbraham seems both inadequate and insensitive.

It is here telling that rapid data-feedback was an aspiration of both New/Processual and Post-Processual projects. As far as can be established, this was first regularly achieved at Catalhöyük and Heathrow Terminal 5. If ignoring their interpretative 'paraphernalia' – for instance, the former's Site Catchment Analysis and Thessian Polygons, and the latter's multi-vocality and reflexivity (*e.g.* Hodder 1997; Tilley 1989; 1994) – then certain procedural directives have been common to both. This raises the question to what degree basic field methods are actually theoretically determined, as opposed to technologically and/or resource-based. Just like modes of finds plotting, sieving, or environmental sampling, they became naturalised as 'best practices'. In other words, *process after Processualism* and no longer directly tied to any original 'school'.

Case studies supported the inception of both New and Post-Processual Archaeology. These were generally of 'pristine' single-phase sites or else ethno-archaeological exemplars (Fig. 29.6). Yet, at least in much of Europe, sites and their locales are rarely so straightforward. Often seeing multiple phases of use, involving both open surface-scatters and successive cut-feature sequences, this is why the limited appreciation of residual finds has been highlighted here. Accordingly, what is required are the means to address the complexity and mess of diverse multi-phase site-use – as well as estimate total artefact populations – and here ploughsoil/overburden sampling provides crucial insights.

While (contra Champion 1978) some limited testing of Mucking's horizontal deposits occurred, it unfortunately proved impossible for us to detail. Nevertheless, by the intensity of its excavation, the representation of its distinct pottery wares provides some gauge of horizontal find 'loss'. One is its La Tène-style pottery. Having 133 sherds attesting to at least 72 vessels, if applied uniformly throughout, approximately only a tenth of the site's finds may have ended up in features (Evans *et al.* 2016a, 440–3, fig. 5.25). Seemingly a very low ratio, at Great Wilbraham, just 22% of the early Neolithic wares occurred within the enclosure's circuits. With the site's material – including all its later pottery – otherwise retrieved from its horizontal contexts (*i.e.* ploughsoil, 'old ground surface'/buried soil horizons),

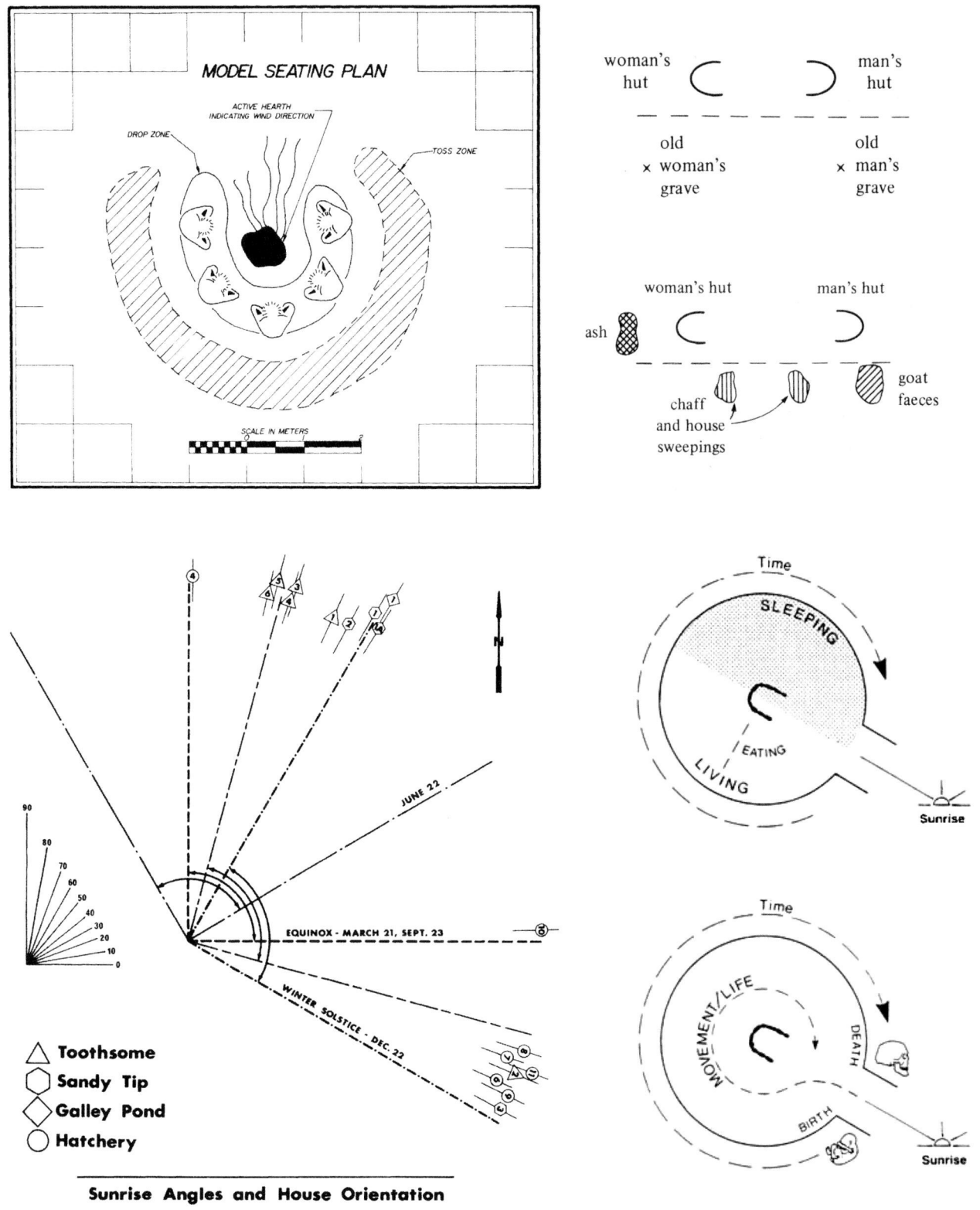

Figure 29.6 'School' Affinities: (top left) Ethnoarchaeology – Binford's (1978, fig. 4) Nunamiut Eskimo, Mask Site hunting stand seating-plan model; (top right) Moore's (1982, figs 3 & 4) rendering of East African Marakwet house-related deposition and burial; (bottom left) Excavation – Hatchery West's and related sites' house orientations in relationship to sunrise (Binford et al. 1970, fig. 23); (bottom right) Parker Pearson's roundhouse sunwise-orientation model (Parker Pearson & Sharples 1999, fig. 1.6).

surface-deposit sampling has the capacity to distinguish episodes of use or occupation without any feature-register.

Ultimately, what is an appropriate sampling ratio comes down to the goals of an excavation and the extent its percentage is actually representative of 'the whole' to be elucidated and what are a site's overarching aims. In this capacity, it could be argued that most sites in Britain are still today dug within a (vaguely) Post-Processualist framework, albeit informed by a high level of environmental science, with an emphasis upon ritual or placed deposits on the basis that they more clearly 'speak'. This can seem at odds to such low-intensity excavation sample ratios that are widely implemented. On the basis of expediency, most developer-funded sites – of the 'Class II' category – are dug at a level comparable to that of Maxey but usually with limited, if any, ploughsoil sampling. While, in theory, you can sample for 'ritual', arguably it is contradictory to the aims of long-term history and contextual 'specificity'. It essentially amounts to digging according to a Post-Processualist perspective using a New/Processual methodological framework.

Building upon Haddenham's area-wide buried soil sampling, now after nearly 30 years, Barleycroft Farm/Over's flint-density plots extend over some 600 ha (Evans *et al.* 2014; 2016b). A key aim throughout has been the inter-meshing of its landscape-scale sampling with excavation procedures, with intensive metre-grid sampling of many of its sites variously conducted. These consistently demonstrate phases of use having no cut-feature register. Whereas the independent occurrence of ploughsoil lithic scatters without accompanying sub-soil features has long been recognised (*e.g.* McInnes 1971), by the depth of the area's floodplain deposits, pottery also survives within its buried soils, allowing for rare detailing of site sequences.

The project's most ambitious sampling exercise occurred along the mid-stream Godwin Ridge (5.4 ha; Evans *et al.* 2016a, chapter 3). With continuous period-representation from the late Mesolithic (*ca.* 10,000–4000 BC) through to early Roman times, 460 flints and 3027 pottery sherds were recovered from its features. These figures pale in relationship to the *ca.* 20,800 flints and 7340 sherds forthcoming from the *ca.* 700 hand-dug metre-square buried soil test pits and collected from that horizon's surface (Evans *et al.* 2016a, fig. 3.4 & table 3.45). With 98 and 71% of these find categories therefore deriving from the buried soil, it is estimated that, in total, that ridge would have had in excess of a million surface finds. What is crucial is of the 35 separate occupation events/sites identified; more than half had no cut-feature register whatsoever.

Undertaking such scaled sampling made one aware that, in effect, this just amounts to the tip of a depositional iceberg. If the sample-grid were intensified, still further discrete use-events/sites would have been evident. This certainly does not amount to any 'total archaeology', but it does allow you to gauge 'the missing'. If such buried/ploughsoil sampling is not conducted, then what is left are only *truncated sequences*, with much of a locale's use-history undetected. It is only by such intense surface-deposit sampling do we approach making 'time' and sequences work with any conviction (Evans *et al.* 2014).

Not all locales will have seen such an intense sequence as that mid-stream ridge. Equally, the fiscal constraints of most development-led fieldwork dictates that such an intensity of surface-sampling cannot usually be achieved. It is, though, imperative that there is awareness – as there was in the 1970s – of just what a loss of data there is if buried/ploughsoil horizons are not adequately sampled. The conditions and means of field practice may have radically changed, but many of the issues remain the same.

Failing Better

Further to Clarke's ascent/passage of sites, a directive of fieldwork-related historiography must be to ensure that they do not just 'go' altogether. Given the degree and pace of paradigm shift to Post-Processualism in the 1980s and 90s, this has largely been the case with the earlier impact of New Archaeology upon British field practice. Yet, apart from a pleasurable read, Flannery's (1976) *The Early Mesoamerican Village*, for example, remains *the* sampling primer; just as, in their basics, Middle Range Theory and Behavioural Archaeology are still useful operational concepts.

In part a product of their short, edited volume-format, rereading papers in Hodder's (1982a) 'brown book' and others in Cambridge's New Directions series that followed in rapid succession, it can be difficult to see that they would engender such sea change. Requiring acknowledgement is that the uptake of Post-Processualism was not solely positive but, as its name implies, reactive to the extremes that New Archaeology was taken by some, with the over-elaboration of systematic methodologies – and 'laws' – touted as if for their own sake and without intent of understanding the past. In this capacity, 'hard' and 'soft' New Archaeologies can be distinguished. The former – almost exclusively American – was explicitly logical-positivist in its approach and in its concern with nature of proof and scientific procedures; the latter 'home' version coalescing around matters of methodology, a greater use of statistics, and a more systematic archaeology generally. Accordingly, the impetus of the 'school' was almost entirely one-sided, with British developments having little or no impact in America (C. Renfrew, pers. comm.; also, Renfrew 1983).

Charting of the subject's near-tectonic theoretical shift and its relationship to fieldwork warrants far more comprehensive review than has been possible here (for comparative disciplinary-change see *e.g.* Le Grand 1988, also O'Brien *et al.* 2007). Its scope should be sufficient to encompass the impact of American 'New' advocates' visits to Britain (*e.g.* Binford, Schiffer, and Plog), the influence of 'New' geography on archaeology and its spatial analyses (see *e.g.* Clarke 1977; Foley 1981, 180), and the background role of Cambridge's Palaeoeconomic School, as well as nuance the respective directives of 'New' and Processual approaches and explore the interconnections of practitioners working both aboard and in British archaeology (*e.g.* G. Barker, Hodder, and Cherry).

By way of conclusion, with many pasts and lives evidently caught up in our sites, for reasons of preservation/survival, the archaeological past is invariably partial. Apropos Renfrew's (1974, 39) aforementioned quote that any excavation amounts to a sample, aspiring to 'totality' you will invariably fail. Although a compelling funding hook, to seriously hold that total archaeology is a realistic possibility is as to evoke Borges' Cartographers Guild and their one-to-one mapping of the world in his 'On Exactitude in Science' of 1946.[5]

On the grounds that we do not control their variables, sites cannot be considered laboratories. Barker (1977, 12), of course, famously held that the act of excavation does not amount to an experiment as it is inherently destructive and unrepeatable (see also Piggott 1959, 14) and, rather, compared it to surgery, as Binford *et al.* (1970, 1) applied 'dissection' to describe excavation. That is certainly true of 'one-off' sites, such as Barker's Wroxeter. Yet now, in awareness of how densely settled large tracts of lowland Britain were in later prehistory and Roman times (see *e.g.* Smith *et al.* 2016; Gosden *et al.* 2021), many of its components are fundamentally *repetitive* (Evans *et al.* forthcoming). Raising a 'challenge of numbers' and highlighting the need for more statistically comparative approaches, once having established such sites' base-line characteristics, if through current development-led 'process' their excavation is to continue, then there seems little justification to just conduct this according to set procedures. Therefore, further to Clarke above (see also *e.g.* Foley 1981, 157), it seems entirely appropriate to undertake the excavation of, at least, some site-type categories as *applied experiments* (Evans 2012). Not just having fieldwork straightjacketed by the day's 'professional standards', but actively push things by the application of innovative techniques and diverse approaches, further mobilising site data to make more insightful and robust statements about the past in the hope of 'failing better'.

Acknowledgements

Uniquely bridging 'the field' and university lecture rooms, it has only been an inspiration to work – and, on occasion, spar in his 'Grey Literature' workshops – with Chris over decades now. In the winter of 2022, we together hosted a 'How Do We Learn' seminar at the Society of Antiquaries on behalf of Historic England and the British Academy.

Compiling this account, I am indebted to the many colleagues who contributed to Mucking and Great Wilbraham's publications. Beyond this, Oscar Aldred, John Cherry, Rob Foley, Charly French, Colin Haselgrove, Gill Hey, Ian Hodder, Martin Jones, George Lambrick, John Lewis, Tim Murray, Francis Pryor, Colin Renfrew, Niall Sharples, Simon Schaffer, Michael Schiffer, Steve Shennan, Marie Louise Stig Sørensen, and Alison Wylie have all variously provided pertinent insights into the day's fieldwork and, otherwise, issues raised here.

Notes

1. See also, for example, Lucas (2001a, 19–20) on Pitt Rivers' 'total archaeological record' and Carver (2009, 26–7) for comparison between his and Petrie's approaches. Barker (1977, 39–40) did not actually use the term 'total …' but rather 'ideal excavation' and, admitting that it was unattainable, recognised the necessity of some sampling procedures; this receiving greater emphasis in the book's later editions. See Olsen (1980) and his exchange with Barker concerning the more prosaic aspects of total excavation (vs. partial preservation) and, too, Reynolds and Barber's 'Analytical Excavation' (1984).
2. Elements of the site's archives had evidently gone missing over the years, their loss and/or whereabouts being a source of dispute between the two surviving co-directors. After our publication and upon Ian Kinnes' passing, more was unearthed in his garage. Now deposited with the British Museum, these have no significant implications for the 2006 paper's account.
3. This being further informed by Tilley's 1979 *Post-Glacial Communities* volume. Later disavowed by him, as the publication of his undergraduate dissertation (supervised by Clarke), it was very much a student work and a Processual/'New' exercise concerned with home-base and off-/away-camp site dynamics. Site Catchment Analysis was conducted on a number of the region's Mesolithic/Neolithic sites, with most uninformed by any relevant in-depth environmental analysis (see also Barker & Webley 1978).
4. Occurring prior to the regular implementation of area-wide evaluation trenching, and with sites during the 1970s and '80s still largely identified on the basis of aerial photography, such programmes were essentially conducted to provide insights as to what could be anticipated upon their stripping. See Barker 1982 (Fig. 19) for the 1972–73 Beckford on Avon quarry excavations. Its large-scale machined exposure was augmented by hand-stripping across a 40×50 m portion, wherein the recovery of minor features was much greater.
5. See also, *e.g.* Aldred and Lucas (2019). Lucas (2001b, 43–6) discusses the pitfalls of conceiving the archive as a (site-) representational 'copy' – and any desire for its 'totalisation' – and also on the 'repetition' possible through both archival re-analysis and the further excavation of previously investigated sites (see *e.g.* Bradley 2015).

References

Aldred, C. & Lucas, G. (2019) The map as assemblage: Landscape archaeology and mapwork. In M. Gillings, P. Haciguzeller & G. Lock (eds), *Re-Mapping Archaeology: Critical Perspectives, Alternative Mapping*, 29–46. London, Routledge.

Andrews, G., Barrett, J.C. & Lewis, J. S. C. (2000) Interpretation not record: The practice of archaeology. *Antiquity* 74, 52–30.

Barford, P. M. (2011) Mucking: Real heritage heroism or heroic failure? In J. Schofield (ed.), *Great Excavations: Shaping the Archaeological Profession*, 212–30. Oxford, Oxbow Books.

Barker, G. & Webley, D. (1978) Causewayed camps and early Neolithic economies in central southern England. *Proceedings of the Prehistoric Society* 44, 161–86.

Barker, P. (1977) *Techniques of Archaeological Excavation*. London, Batsford.

Binford, L. (1964) A consideration of archaeological research designs. *American Antiquity* 29, 455–41.

Binford, L. (1978) Dimensional analysis of behavior and site structure: Learning from an Eskimo hunting stand. *American Antiquity* 43, 330–61.

Binford, L. R., Binford, S. R., Whallon, R. & Hardin, M. A. (1970) Archaeology at Hatchery West. *Memoirs of the Society for American Archeology* 24, i–vii & 1–96.

Bradley, R. (2015) Repeating the unrepeatable experiment. In R. Chapman & A. Wylie (eds), *Material Evidence: Learning from Archaeological Practice*, 23–41. London, Routledge.

Carver, M. (2009) *Archaeological Investigation*. London, Routledge.

Champion, T. (1978) Strategies for sampling a Saxon settlement: A retrospective view of Chalton. In Cherry *et al.* 1978, 207–25.

Champion, T. (1991) Theoretical archaeology in Britain. In I. Hodder (ed.), *Archaeological Theory in Europe: The Last Three Decades*, 129–60. London, Routledge.

Chapman, R. & Wylie, A. (eds) (2015) *Material Evidence: Learning from Archaeological Practice*. London, Routledge.

Cherry, J. F. (1982) A preliminary definition of site distribution on Melos. In C. Renfrew & M. Wagstaff (eds), *An Island Polity: The Archaeology of Exploitation in Melos*, 10–23. Cambridge, Cambridge University Press.

Cherry, J. F. & Shennan, C. (1978) Sampling cultural systems: Some perspectives on the application of probabilistic regional survey in Britain. In Cherry *et al.* (1978), 17–48.

Cherry, J. F., Gamble, C. & Shennan, S. (eds) (1978) *Sampling in Contemporary British Archaeology*. British Series 50. Oxford, BAR.

Cherry, J. F, Gamble, C. & Shennan, S. (1979). Opinion. *Current Archaeology* 66, 223.

Clarke, D. L. (1968) *Analytical Archaeology*. London, Methuen.

Clarke, D. L. (1972) A provisional model of an Iron Age society and its settlement system. In D. L. Clarke (ed.), *Models in Archaeology*, 801–85. London, Methuen.

Clarke, D. L. (1973) Archaeology: The loss of innocence. *Antiquity* 47, 6–18.

Clarke, D. L. (ed.) (1977) *Spatial Archaeology*. London, Academic Press.

Crowther, D. R. (1983) Old land surfaces and modern ploughsoil: Implications of recent work at Maxey. *Scottish Archaeological Review* 2, 31–44,

Crowther, D., French, C. & Pryor, F. (1985) Approaching the fens the flexible way. In C. Haselgrove, M. Millett & I. Smith (eds), *Archaeology from the Ploughsoil: Studies in the Collection and Interpretation of Field Survey Data*, 59–76. Sheffield, Dept. of Archaeology and Prehistory.

Evans, C. (2000) Testing the ground - Sampling strategies. In A. Crowson, T. Lane & J. Reeve (eds), *Fenland Management Project Excavations 1991–1995*. Lincolnshire Archaeology and Heritage Reports Series 3, 15–21. Heckington, Heritage Trust of Lincolnshire.

Evans, C. (2012) Archaeology and the repeatable experiment: A comparative agenda. In A. Jones & J. Pollard (eds), *Image, Memory and Monumentality: Archaeological Engagements with the Material World*, 295–306. Oxford, Oxbow Books.

Evans, C. & Hodder, I. (2006) *A Woodland Archaeology: Neolithic Sites at Haddenham*. The Haddenham Project, Vol. I. Cambridge, McDonald Institute for Archaeological Research.

Evans, C. & Pollard, J. (2001) The dating of the Storey's Bar Road fields reconsidered. In F. Pryor (ed.), *The Flag Fen Basin: Archaeology and Environment of a Fenland Landscape*, 25–7. Swindon, English Heritage.

Evans, C., Edmonds, M. & Boreham, S. (2006) 'Total archaeology' and model landscapes: Excavation of the Great Wilbraham causewayed enclosure, Cambridgeshire, 1975–76. *Proceedings of the Prehistoric Society* 72, 113–62.

Evans, C., Beadsmoore, E., Brudenell, M. & Lucas, G. (2009) *Fengate Revisited: Further Fen-edge Excavations, Bronze Age Fieldsystems/Settlement and the Wyman Abbott/Leeds Archives*. Cambridge Archaeological Unit, Historiography and Fieldwork Series 1. Cambridge, Cambridge Archaeological Unit.

Evans, C., Tabor, J. & Vander Linden, M. (2014) 'Making time work': Sampling floodplain artefact frequencies and populations, *Antiquity* 88, 241–58.

Evans, C., Appleby, G. & Lucy, S. (2016a) *Lives in Land – Mucking Excavations by Margaret and Tom Jones, 1965–1978: Prehistory, Context and Summary*. Cambridge Archaeological Unit, Historiography and Fieldwork Series 2. Oxford, Oxbow Books/Historic England.

Evans, C., Tabor, J. & Vander Linden, M. (2016b) *Twice-crossed River: Prehistoric and Palaeoenvironmental Investigations at Barleycroft Farm/Over, Cambridgeshire*. The Archaeology of the Lower Ouse Valley, Vol. III. Cambridge, McDonald Institute for Archaeology.

Evans, C., Aldred, O. & Cooper, A. (in review) Landscape after Fox's ... *Cambridge Region* (2023): Settlement Densities.

Fasham, P. J. & Whinney, R. J. B. (1991) *Archaeology and the M3: The Watching Brief, the Anglo-Saxon Settlement at Abbots Worthy and Retrospective Section*. Hampshire Field Club and Archaeological Society Monograph 7. Stroud, Hampshire Field Club/Trust for Wessex Archaeology.

Flannery, K.V. (ed.) (1976) *The Early Mesoamerican Village*. London, Academic Press.

Foley, R. (1978) Incorporating sampling into initial research designs: Some aspects of spatial archaeology. In Cherry *et al.* (1978), 49–65.

Foley, R. (1981) Off-site archaeology: An alternative approach for the short-sited. In I. Hodder, G. Isaac & N. Hammond (eds), *Pattern in the Past: Essays in Memory of David Clarke*, 157–84. Cambridge, Cambridge University Press.

Fowler, E. (ed.) (1972) *Field Survey in British Archaeology*. London, Council for British Archaeology.

Fowler, P. (1977) *Approaches to Archaeology*. London, Adam and Charles Black.

Framework Archaeology (2010) *Landscape Evolution in the Middle Thames Valley: Heathrow Terminal 5 Excavations, Volume 2*. Salisbury, Wessex Archaeology.

Fulford, M. (2007) The grand excavation projects of the twentieth century. In S. Pearce (ed.), *Visions of Antiquity: The Society of Antiquaries of London 1707–2007*. Archaeologia 111, 353–81. London, Society of Antiquaries of London.

Gosden, C., Green, C., Cooper, A., Creswell, M., Donnelly, V., Franconi, T., Glyde, R., Kamash, Z., Mallet, S., Morley, L., Stansbie, D. & ten Harkel, L. (2021) *English Landscapes and Identities: Investigating Landscape Change from 1500 BC to AD 1086*. Oxford, Oxford University Press.

Hall, D. N. & Coles, J. (1995) *The Fenland Survey: An Essay in Landscape and Persistence*. London, English Heritage.

Halstead, P., Hodder, I. & Jones, G. (1978) Behavioural archaeology and refuse patterns: A case study. *Norwegian Archaeological Review* 11, 118–31.

Haselgrove, C. (1978) Spatial pattern and settlement archaeology: Some reflections on sampling design. In Cherry *et al*. (1978), 159–75.

Haselgrove, C. (1985) Inference from ploughsoil artefact samples. In C. Haselgrove, M. Millett & I. Smith (eds), *Archaeology from the Ploughsoil: Studies in the Collection and Interpretation of Field Survey Data*, 59–76. Sheffield, Dept. of Archaeology and Prehistory.

Hawkes, S. C. & Hawkes, C. F. C (2012) *Longbridge Deverill Cow Down: An Early Iron Age Settlement in West Wiltshire*. Oxford University School of Archaeology Monograph 76. Edited by L. Brown. Oxford, School of Archaeology.

Hinchliffe, J. & Schadla-Hall, T. (eds) (1980) *The Past under the Plough*. Department of Environment Occasional Paper 3, 114–24. London, Department of the Environment.

Hodder, I. (1981) Introduction: Towards a mature archaeology. In I. Hodder, G. Isaac & N. Hammond (eds), *Pattern in the Past: Essays in Memory of David Clarke*, 1–13. Cambridge, Cambridge University Press.

Hodder, I. (ed.) (1982a) *Symbolic and Structural Archaeology*. Cambridge, Cambridge University Press.

Hodder, I. (1982b) *Symbols in Action*. Cambridge, Cambridge University Press.

Hodder, I. (1982c) *Wendens Ambo: The Excavation of an Iron Age and Romano-British Settlement. The Archaeology of the M11*. Passmore Edwards Museum Monograph Series 2. London, Passmore Edwards Museum.

Hodder, I. (1997) 'Always momentary, fluid and flexible': Towards a reflexive excavation methodology. *Antiquity* 71, 691–700.

Jones, M. (1978) Sampling in a rescue context: A case study in Oxfordshire. In Cherry *et al*. 1978, 191–205.

Lambrick, G. (2010) *Neolithic to Saxon Social and Environmental Change at Mount Farm, Berinsfield, Dorchester-on-Thames*. Oxford Archaeology Occasional Paper 19. Oxford, Oxford Archaeology Ltd.

Le Grand, H. E. (1988) *Drifting Continents and Shifting Theories: The Modern Revolution in Geology and Scientific Change*. Cambridge, Cambridge University Press.

Lucas, G. (2001a) *Critical Approaches to Fieldwork: Contemporary and Historical Archaeological Practice*. London, Routledge.

Lucas, G. (2001b) Destruction and the rhetoric of excavation. *Norwegian Archaeological Review* 34, 35–46.

Lucy, S. & Evans, C. (2016) *Romano-British Settlement and Cemeteries: Mucking Excavations by Margaret and Tom Jones, 1965–1978*. Cambridge Archaeological Unit, Historiography and Fieldwork Series 3. Oxford, Oxbow Books/Historic England.

McInnes, I. J. (1971) Settlement in later Neolithic Britain. In D. D. A. Simpson (ed.), *Economy and Settlement in Neolithic and Early Bronze Age Britain and Europe*, 113–30. Leicester, Leicester University Press.

Miles, D. (1983) An integrated approach to the study of ancient landscapes. In G. S. Maxwell (ed.), *The Impact of Aerial Reconnaissance on Archaeology*. CBA Research Report No. 49, 74–84. London Council for British Archaeology.

Moore, H. L. (1982) The interpretation of spatial patterning in settlement residues. In Hodder 1982a, 74–9.

O'Brien, M. J., Layman, R. L. & Schiffer, M. B. (2007) *Archaeology as Process: Processualism and Its Progeny*. Salt Lake City, University of Utah Press.

Olsen, O. (1980) Rabies archaeologorom. *Antiquity* 54(210), 15–20.

Parker Pearson, M. & Sharples, N. (1999) *Between Land and Sea: Excavations at Dun Vulan, South Uist*. Sheffield, Sheffield Academic Press.

Patterson, N. *et al*. (2022) Large-scale migration into Britain during the middle to late Bronze Age. *Nature* 601, 588–94.

Piggott, S. (1959) *Approach to Archaeology*. Harmondsworth, Penguin.

Pryor, F. (1978) *Excavation at Fengate, Peterborough, England: The Second Report*. Royal Ontario Museum of Archaeology Monograph 5. Toronto, Royal Ontario Museum.

Pryor, F. (1980a) Survey excavation. *Rescue News* 21, 6.

Pryor, F. (1980b) Will it all come out in the Wash? Reflections at the end of eight years' digging. In J. Barrett & R. Bradley (eds.), *Settlement and Society in the British Later Bronze Age*. British Series 83, 483–99. Oxford, BAR.

Pryor, F. (1983) Talking heads. *Scottish Archaeological Review* 2, 98–100.

Pryor, F. (1998) *Etton: Excavations at a Neolithic Causewayed Enclosure near Maxey Cambridgeshire, 1982–87*. Archaeological Report 18. London, English Heritage.

Pryor, F. & French, C. (1985) *The Fenland Project, No. 1: Archaeology and Environment in the Lower Welland Valley*. East Anglian Archaeology 27. Cambridge, Cambridgeshire Archaeological Committee.

Renfrew, C. (1974) British prehistory: Changing configurations. In C. Renfrew (ed.), *British Prehistory*, 1–40. London, Duckworth.

Renfrew, C. (1978) Archaeology and society in 1978. In T. C. Darvill, M. Parker Pearson, R. W. Smith & R. M. Thomas (eds), *New Approaches to Our Past: An Archaeological Forum*, 157–78. Southampton, Department of Archaeology, University of Southampton.

Renfrew, C. (1979) *Investigations in Orkney*. Reports of the Research Committee of the Society of Antiquaries of London No. 38. London, Society of Antiquaries of London.

Renfrew, C. (1983) Divided we stand: Aspects of archaeology and information. *American Antiquity* 48, 3–16.

Reynolds, P. J. & Schadla-Hall, T. (1980) Measurement of plough damage and the effects of ploughing on archaeological monuments. In Hinchliffe & Schadla-Hall 1980, 114–24.

Reynolds, N. & Barber, J. (1984) Analytical excavation. *Antiquity* 58, 95–102.

Schiffer, M. B, Sullivan, A. P & Klinger, T. C. (1978) The design of archaeological surveys. *World Archaeology* 10, 1–28.

Shennan, S. J. (1980) Meeting the plough damage problem: A sampling approach to area-intensive fieldwork. In Hinchliffe & Schadla-Hall 1980, 125–33.

Shennan, S. J. (1985) *Experiments in the Collection and Analysis of Archaeological Survey Data: The East Hampshire Survey*. Sheffield, Dept. of Archaeology and Prehistory.

Shennan, S. J. (1988) *Quantifying Archaeology*. Edinburgh, Edinburgh University Press.

Smith, A., Allen, M., Brindle, T. & Fulford, M. (2016) *New Visions of the Countryside of Roman Britain. Volume 1: The Rural Settlement of Roman Britain*. Britannia Monograph Series 29. London, The Society for the Promotion of Roman Studies.

Taylor, C. (1974) Total archaeology or studies in the history of the landscape. In A. Rogers & R. T. Rowley (eds), *Landscapes and Documents*, 15–26. London, Bedford Square Press.

Tilley, C. (1979) *Post-glacial Communities in the Cambridge Region*. British Series 66. Oxford, BAR.

Tilley, C. (1989) Excavation as theatre. *Antiquity* 63, 275–90.

Tilley, C. (1994) *A Phenomenology of Landscape*. Oxford, Berg.

Wainwright, G. (1978) Theory and practice in field archaeology. In T. C. Darvill, M. Parker Pearson, R. W. Smith & R. M. Thomas (eds), *New Approaches to Our Past: An Archaeological Forum*, 11–27. Southampton, Depart of Archaeology, University of Southampton.

Wainwright, G. (2000) Time please. *Antiquity* 74, 909–43.

Woodman, P. C. (1982) Sampling strategies and problems of archaeological visibility. *Ulster Journal of Archaeology* 44/45, 179–84.

Woodman, P. C. (1985) *Excavations at Mount Sandel 1973-77*. Northern Ireland Archaeological Monographs 2. Belfast, Her Majesty's Stationary Office.

30

Oxford Intelligence

Lynn Meskell

This contribution examines a formative historical moment when four major archaeologists from Oxford University (David Hogarth, Gertrude Bell, T. E. Lawrence, and Leonard Woolley) found themselves at the crossroads of empire and espionage during the First World War. Recruited specifically for their archaeological training and fieldwork in the Middle East, they occupied the frontline of military intelligence at the Arab Bureau, advancing British interests and fending off competing powers. Not only were these archaeologists preoccupied with mapping the Middle East and its ancient heritage, but they also devised military strategy, classified peoples, established new borders, and helped forge new subject nations. This blurring of military and academic expertise has long characterised our discipline. I argue that Hogarth and his proteges were high-profile participants in an emergent military-industrial-academic complex that has shaped the development of modern archaeology.

Keywords: Middle East, Espionage, Military, British Colonialism, Archaeological Adventurism

Introduction

It seems only fitting to have the words 'Oxford' and 'intelligence' in a volume honouring Chris Gosden and his remarkable career. During time spent at Oxford over the years, Chris inspired me to follow new lines of enquiry, such as working on UNESCO, that led to all kinds of unexpected fieldwork and archival analysis which have taken me around the globe. As always, his instincts were right on the mark, and that prompted me to examine world-making projects on a grand scale from the League of Nations to the imperial ventures of British archaeologists in the Middle East, themselves inextricably bound up in those organisations. But what Chris always encouraged me to do is capture the specifically archaeological aspect of these broader global histories, the disciplinary formations and key individuals that are connected to, and indeed constitute, the story of archaeology.

Archaeology has been deeply entwined with colonial rivalries and struggles for self-determination that continue to have lasting legacies across the Middle East. Neo-imperial ambitions and conflict over territory, religion, and antiquities have similarly been accompanied by heritage claims and the rhetoric of high cultural humanism. Throughout the 20th century, foreign occupation and military adventurism galvanised archaeological elites, including David G. Hogarth, Frederick Kenyon, T. E. Lawrence, C. Leonard Woolley, and Gertrude Bell, to use their particular archaeological skills and sensibilities for the furtherance of British interests. Whether situated in London or Jerusalem, Western archaeologists were instrumental in the colonial carve-up that extended to historical sites, concessions, collections, and even the creation of new states, often working through international agencies such as the League of Nations under the banner of salvage and uplift. The connected histories of archaeology, espionage, and the end of empire have cast a long shadow, prompting deeper analysis of our role as participants and beneficiaries.

Employing archival materials, some newly released, we can trace how imperatives to govern the material vestiges of the past in order to shape the future were effectively materialised through archaeology. Competition and control, contestation and conflict over archaeological access and excavations afforded moments of redrawing, re-inscribing, and reinforcing territorial and historical mastery in the

Middle East. And with that came a reinvigorated imperialism, paternalism, and opportunism. Acquiring archaeological concessions in the Middle East has often been as much about securing supplementary resources, whether oil or antiquities, as it has been about claims to religious or civilisational inheritance (Meskell 2020). Genuflections to internationalism and collaboration have often served as expert cover for extending imperial claims, whilst keeping exactly the same measures of empire in place.

Archaeological Adventurism

At the close of Ottoman empire, the Oxford archaeologists Hogarth, Lawrence, Bell, and Woolley were all involved in covert military activities (Barr 2011). In 1913, the British War Office had engaged the Palestine Exploration Fund (PEF) to map large unrecorded areas of the Wadi Araba and the al Naqab Desert, the first surveys since the 1880s. Founded by the Archbishop of York in 1865, the PEF infamously linked English Protestantism, patriotism, science, and England's role in a global colonial competition with the archbishop's oft-quoted remark that: 'This country of Palestine belongs to you and me. It is essentially ours'.[1] The map of Palestine created by the PEF surveyors from the War Office essentially became the map of Britain's Mandate for Palestine after the First World War and defined its border with the French Mandate in Lebanon. Its legacy formed the basis of proposals to divide the land between Palestine's Arab and Jewish populations; it ultimately defined the borders of the state of Israel, and thus at least one border for several modern Middle Eastern states.

The imputed purpose of the PEF survey was to 'find any evidence of the wanderings of the Israelites during the Exodus period' (Moscrop 2000, 207). Ostensibly an illegal mapping exercise, archaeology provided a convenient 'whitewash' for espionage. Leonard Woolley and T. E. Lawrence (1936) were dispatched to the desert for their now-famous Wilderness of Zin Survey from their excavations at the Bronze Age site of Carchemish, a Hittite capital recorded in the Bible, that today straddles the frontier between Turkey and Syria. They had been recommended for the job by the British Museum's director, Sir Frederick Kenyon (Winstone 1990, 36).

Woolley read theology at New College, Oxford. In 1905, he was appointed assistant to Sir Arthur Evans, then Keeper of the Ashmolean Museum, before moving into field archaeology with expeditions in Nubia and Italy (1907–1911). In 1912, he was chosen to succeed Dr R. Campbell-Thompson as leader of the British Museum expedition to Carchemish.[2] Lawrence, the younger of the two, graduated from Jesus College in 1910 before receiving a scholarship to Magdalen College (1911–14), then later becoming a Fellow of All Souls (1919–26). So much has been written about Lawrence that he hardly requires introduction; suffice to say that he was an internationally celebrated archaeologist, army officer, diplomat, and writer, known for his role in the Arab Revolt and the Sinai and Palestine Campaign against the Ottoman Empire. Lawrence famously said of his survey work in Sinai, 'we are obviously only meant as red herrings, to give an archaeological color to a political job' (Brown 1994, 56). Moreover, it was 'his connection with Middle East archaeology and with the Sinai survey which led him to finding suitable war work' (Brown 1994, 62). Whereas Woolley was given a commission in the Royal Artillery, Lawrence went to work for Military Intelligence, his principal function being map officer in the Arab Bureau, using his archaeological skills and experience of the sites and landscapes acquired through years of survey and excavation.

While Lawrence is undoubtedly the most famous of the four archaeologists, it was David Hogarth who was perhaps the real mastermind of the Arab Bureau (1915–1919), and indeed minder for the likes of Woolley, Bell, and Lawrence. Hogarth was an archaeologist, writer, Naval Intelligence officer, and a member of Magdalen College. Between 1887 and 1914 he worked as an archaeologist in Greece, Turkey, Egypt, and Syria (see Fig. 30.1). He served as Director of the British School at Athens (1897–1900) and was Keeper of the Ashmolean Museum (1909–1927).[3] As his Oxford correspondence clearly shows, Hogarth was a close friend and mentor to Lawrence, giving him his first job as an archaeologist. 'T.E.L is in the middle of the most serious adventure he has yet tried,' Hogarth confided to his wife Laura; 'Don't tell his mother any more than that he was safe and sound'.[4] Ever thankful to Hogarth, Lawrence wrote, he

> got me on to the Carchemish staff, when I was happier than I'd ever been since in my life. Then when war broke out he got me into his geographical section at the War Office and afterwards to Egypt ... till I went up country into Arabia and he took charge of my political roots in G.H.Q. I owe him for all of that ... for he was the only perfectly finished and four-square human being of all the thousands I have met in my chequered life.[5]

After the infamous Wilderness of Zin survey, Woolley and Lawrence returned to Carchemish. Woolley secured the Carchemish concession from the Ottoman Kaymakam (district governor) at gunpoint: 'Taking my revolver out of its holster I got up and walking to the side of his chair put the muzzle against his left ear. "On the contrary," I said, "I shall shoot you here and now unless you give me permission to start work to-morrow"' (Woolley 1920, 156). Carchemish was originally Hogarth's dig, and Woolley and Lawrence were sent there to excavate and gather intelligence on the Germans, in what Lord Kitchener referred to as 'archaeological whitewash' (Winstone 1990, 60); Kitchener himself had taken part in an earlier British survey of Palestine in the 1880s. The Carchemish excavations and dig house were in close proximity to the Berlin-Baghdad railway. Funded by Deutsche Bank and constructed by German engineers, the railway would reach the oil fields and Persian Gulf,

Figure 30.1 D. G. Hogarth in the Middle East, Hogarth Papers, P452/PER/3/2. By kind permission of the President and Fellows, Magdalen College Oxford.

thereby bypassing the Suez Canal, controlled by British and French interests.

Excavation had started in 1911 and in August that year Lawrence wrote to Hogarth, 'I hear there is a good deal of stuff in the village, at present afraid to come out. Thompson cut the bakshish after you went: I fear false economy! Still if I can recover the booty for English hands, so much gained'.[6] Similarly, the next year Lawrence was hunting for seals in Aleppo, dealing with 'one villainous looking dealer after another … I have a dozen or fifteen (my best year)'. International competition was rife, he wrote, 'for I'm not going to let that go to America or Berlin. When Kenyon's money comes I'm going in for it'. In that same correspondence Lawrence describes the Berlin-Baghdad railway construction at Jerablus that he was tasked to spy upon. Lawrence wrote extensively in his letters to Hogarth about the German engineers, their plans for the railway and bridge, their surveyors, and, with some disdain, their archaeologists. Local villagers regaled him about the Germans: 'They are ignorant of antikas – not recognizing a Hittite inscription they know no language … they do no work with their hands, but sit in the tents'.[7]

Tensions were running high and various hostilities broke out between the British excavators, their workmen, and the German engineers (Moscrop 2000, 208). Competition over archaeological expertise and endeavours, especially in light of German superiority in Biblical archaeology, ran in tandem with larger colonial ambitions. Mapping and excavation played a 'critical role linking the study of the past with concrete military and political realities' (Richter 2008, 216) and to territorial claims and economic advantage; there was also a core religious imperative behind occupation and colonial rule. What is significant, and what I hope to convey here, is the porous borders of archaeological work, military intelligence, and the furtherance of British interests in both spheres. The recently released correspondence between Lawrence and Hogarth exemplifies these intercalations, so much so that both men viewed their expeditions and expertise as part of the same mission. Moreover, the acquisitive nature of the archaeological enterprise, namely securing materials for British institutions, neatly conjoined with larger British imperial interests. Thus, archaeological access and adventurism were sutured together.

After Carchemish, Woolley was posted to the Port Said intelligence office, tasked with tracking spy planes, reconnaissance, and interrogating prisoners of war. In August 1916, Woolley took Lord Roseberry's yacht on an intelligence mission headed for the Bay of Alexandretta. The boat struck a mine and sank, and Woolley was rescued by a Turkish vessel and conveyed to prison, where he spent the next two years with the survivors of the devastating Siege of Kut (Winstone 1990, 80–2). As for Lawrence, by 1916 after the Arab Revolt, he had left his paper war in Cairo for the field. As Richter argues (2008, 235),

> Archaeology not only served as a useful cover for espionage on multiple occasions, but was also a source of skills useful in both archaeology and intelligence, such as surveying, languages, photography, making detailed descriptions or sketches, report writing, personal contacts, and an intimate and usually personal knowledge of geography. Archaeology and espionage could be carried out side-by-side since the skills and techniques used could easily be utilized by both, and therefore one activity informed the other.

With Oxford archaeologists Hogarth, Lawrence, and Woolley, we see the initial stirrings of an early military-industrial-academic complex.

Camouflage for Politics

It was the British Arabist Colonel Mark Sykes who proposed the establishment of the Arab Bureau in Cairo (1916–1920), as part of British military intelligence, to collect and disseminate information and propaganda. Archaeologist David Hogarth, Director of the British School in Athens and Keeper of Oxford's Ashmolean Museum, led the Bureau; the archaeologists he recruited included Lawrence and Bell (see Fig. 30.2). Archaeological expeditions and intelligence gathering were rendered seamless. For example, Hogarth (1927, 1) wrote that Bell's journey to Hayil in the winter of 1913–1914 resulted in a 'mass of information that she accumulated about the tribal elements ranging between the Hejaz Railway on the one flank and the Sirhan and Nefud on the other, particularly about the Howeitat group, of which Lawrence, relying on her reports, made signal use in the Arab campaigns of 1917 and 1918'. Archaeology and patriotic duty were one and the same before, during, and after the war.

From his correspondence with Gilbert Clayton, British Army intelligence officer and colonial administrator, it is clear that Hogarth worked very closely with Mark Sykes (along with François Georges-Picot), planning for a Middle East carve-up. With reference to Sykes, Hogarth boasted that 'M.S. says that last Tuesday he had "almost" got Arabia for us minus a neutralized Hijaz and plus dual control of the Arab Legion … I have never been able to get out of M.S. what and whom P. (Picot) represents. (I know it is the Catholics, but who else?)'.[8] Mixing with the peers, politicians, and royalty, Hogarth's correspondence reveals information on everything from British aspirations to control oil reserves in Fersan (Yemen) to Churchill's political woes at home, to Russian ideology and territorial interests, and even to Italian designs on Arabia. Of American interests, he wrote, 'I am assured the Yankees are in for real business and mobilizing enormous forces of all sorts and kinds. What will be in the latter and thereof. An armed U.S.A running Europe?'.[9] What is revealing about educated elites such as Hogarth is that these deliberations were not simply about territorial claims to empire: archaeology supplied the imputed ideological and moral justification for national ambitions in the Holy Land. These were, after all, religious wars. For the French, capturing the Middle East was about Catholicism and the Crusades. For the British, places like Jerusalem were also about reliving the Crusades and capturing the Holy Lands they thought rightfully theirs, but through redemptive Protestantism. While Hogarth was stationed in Jerusalem even espionage, the 'eldest profession,' was shrouded in Biblical metaphor, referring to a female spy as 'a daughter of Rahab herself in Jericho'.[10] For the Americans, the relative latecomers, their presence was tethered to the Old Testament and their development of Biblical archaeology (Corbett 2015). Hogarth dismissed them as the 'quaint "Second Coming" folk'.[11]

In January 1918, Hogarth, then Commander and head of the Arab Bureau, was dispatched to Jeddah with a letter penned by Mark Sykes on behalf of the British Government. Hogarth was tasked with explaining the Balfour Declaration to King Hussein, the Sharif of Mecca. Known today as the Hogarth Message, this much-debated historical moment pivots on whether Hogarth pledged that Britain would respect not only the economic but also the political freedom of the Arabs in Palestine. When Britain subsequently refused to recognise Arab independence, Hussein accused Britain of betrayal.

In his role as advisor to the British delegation at the 1919 Paris Peace Conference, Hogarth was part of the Big Four (Britain, France, Italy, the US): the nations that decided the fate of others, presiding over the redistribution of German and Ottoman overseas possessions as mandates overseen by British and French authorities. He was literally at the centre of a new world government. Hogarth observed that 'the problem of dealing with the Ottoman carcass seems to get no nearer solution. The commission itself, promised to Allenby, hangs fire, the French, the Jews, and our own Mespot. people all protesting'.[12] Hogarth was well aware of the embittered struggle for the Middle East, in which he played no small part. 'Personally I cant hold on much longer', he told Clayton; 'I must resign and go back to Oxford, sick at heart at all this fiasco … Heavens! what a business it is to make "Peace"! So much easier to make war!'.[13] A paternalistic ethos came to infuse the Conference and the League of Nations: the promises of assistance for the Middle East and its heritage were merely imperial ploys and proxies in the context of post-war rebuilding.

At the Peace Conference archaeological matters were continually inserted into the terms of peace, finding their way into the Treaty of Versailles. While vested interests remained the same, both archaeologists and politicians realised that stewardship over archaeological places and objects could no longer appear overtly tethered to empire, as was the case before the war. Lawrence and Bell were

Figure 30.2 T. E. Lawrence, D. G. Hogarth and Lt. Col. Dawnay. By kind permission of the President and Fellows, Magdalen College Oxford.

there to deliver those arguments and to speak for the Arab cause. Britain obtained mandates over Palestine and Mesopotamia and inserted archaeological clauses into the Treaty of Peace with Turkey.[14] A committee was established with one archaeologist from each of the Big Four nations. One British Foreign office memo summarised the political centrality and opportunism archaeology bestowed upon the empire:

> Archaeologists are notoriously touchy and quarrelsome, even for men of science, and there will probably be great difficulties in reaching any agreed form of organisation and, still more, of policy, but it might be well to consult Sir Arthur Evans, Mr. Hogarth and one or two other leading British authorities (a) on the provisions to be inserted in the various mandates and (b) on what international organisation, if any, should be recognised under the League of Nations. The matter is of some importance both because archaeology proper is a fruitful cause of jealousy and friction and because archaeology is an extremely useful camouflage for politics.[15]

More than an extremely useful camouflage, archaeology had become the very matter of politics, suturing past and future for the furtherance of empire. At war's end, through the imputed processes of internationalism and co-operation, governing archaeology and access to antiquities took on an even more public role in the bargaining between nations. As head of the British Academy, Frederick Kenyon wrote to the Foreign Office, suggesting that the government should consider the future (archaeological) mastery of the region. Installing this 'future machinery'[16] would entail appointing archaeologists to the army in Palestine and starting negotiations with the French. The British Foreign Secretary, Arthur Balfour, acknowledged Kenyon's letter and a month later the War Office informed Kenyon that there were plans to safeguard the antiquities of Palestine and Mesopotamia. A Joint Organising Committee was recorded by the Foreign Office in late 1918, while other bodies raised the issue of instituting an Archaeological Intelligence Division. Competition was rife, demonstrated by Kenyon's 1919 veiled threat against monopolies, clearly directed at the French: 'any nation whose law on antiquities imposes harsh conditions on foreigners or which maintains a selfish monopoly of archaeological privileges in its sphere of control should modify its methods if it desires to enjoy rights of research outside its own provinces'.[17]

Both war and archaeology are global mobilities that share an entwined history. By the close of the First World War, the British had mapped, excavated, and imagined Palestine and Transjordan into a new reality. Significantly, the British Mandate period continues to be regarded as the formative 'Golden Age' of archaeology in Palestine (Gibson 1999, 115). By 1920, General Allenby and the High Commissioner for Palestine, Herbert Samuel, were elected Vice-Presidents of the British School of Archaeology in Jerusalem. Imperial ambition generated new forms of governance, uniting archaeology and politics. As Secretary of State for War, Winston Churchill assured the House of Commons that restoration and preservation of monuments in Palestine were to be given full attention. Ultimately, the British devised policies regarding antiquities that ultimately forged an enduring separation between 'living' and 'archaeological' heritage that was as unnatural as the new borders they drafted (Corbett 2015, 16). Transjordan was negatively defined in regard to Palestine, being ostensibly *not* the land of the biblical Israelites, but of their adversaries. Created as a colonial afterthought, archaeology could offer no evidential account that bound this territory together.

Oversight

T. E. Lawrence wrote that the British had entered the Middle East like sphinxes. Throughout this 'covert empire' of intelligence gathering, cultural representations mattered: in both the strategies the British state employed and the varying standards of the empire's 'geographical morality' (Satia 2011, 23). This synergy was exemplified by the developments in archaeology and aerial photography during the First World War, when together they were used for military reconnaissance and strategy, and where archaeologists both in the field and at home were commissioned to read and interpret the data gathered. In Iraq, the Royal Air Force (RAF) deployed archaeological air photography with the technics of surveillance and policing from the air. British archaeologists like O. G. S. Crawford ostensibly appropriated military techniques and introduced new methods of air photography (Hauser 2007; Melman 2019, 136). Such technologies of development and security share common military-industrial and cultural roots (Satia 2011, 5).

One need only see the stunning aerial images of Middle Eastern sites taken by the Royal Flying Corps, advised by Lawrence and Bell, both to survey and later to bomb villages. Along with Churchill, Lawrence was convinced that 'aircraft could rule the desert' in what was called 'pacification'. Lawrence argued that for the Arabs 'it is not punishment, but a misfortune from heaven striking the community'.[18] It is noteworthy that after the war, Lawrence enlisted in the air force (with a brief stint in the army), serving from 1922 to 1936. Taking full control of Iraq in 1922, the RAF patrolled the country from a network of bases, bombarding villages and tribes as needed to put down unrest and subversive activities (Satia 2011, 39). Despite the horrific violence inflicted, British moralising and justification was directed towards saving and restoring the 'cradle of civilization' and the land of the Bible (Satia 2011, 34). British authorities argued that it was redemption, not conquest, that stirred the British establishment and its archaeological elite and thus extended to governing archaeology itself.

By the end of the Great War, archaeology and archaeologists had become firmly entrenched in Middle Eastern state-making. In the case of Iraq, Gertrude Bell was instrumental in the carving up of territories in her role as Oriental Secretary. Bell studied at Lady Margaret Hall, Oxford, and later earned fame as a traveller, author, archaeologist, and intelligence officer, playing a critical role in the establishment of the states of Iraq and Jordan. At the invitation of David Hogarth, Bell joined the military intelligence department in Cairo in autumn 1915 on the basis of her extensive expert knowledge of pre-war Arabia. In 1916, Bell was called to India and tasked by Lord Hardinge to travel to Basrah as the viceroy's personal envoy in order to assess the impact of the Arab Bureau's schemes, whose approach differed from the India Office's imperial policy. The Bureau was keen to exploit rising Arab nationalism to perpetuate Britain's presence in the area. Annexed to the military intelligence department at the headquarters of the Mesopotamian Expeditionary Force, Bell was then charged with intelligence gathering on Bedouin movements in central Arabia and the Sinai Peninsula. She contributed to Hogarth's *Arabian Report* and the famous *Arab Bulletin*, perhaps the best source of information on wartime events in the Middle East.[19]

Often described as a king-maker, Bell, historians have recently argued, shaped policy much more than previously recognised. One lasting legacy is her assertion that the Kurds were not deserving of an independent Kurdistan and would always be reliant on Iraq (Bell 2015, 110). She confessed her role to be 'rather like the Creator'.[20] She mediated between the Arab government and British officials, famously remarking that 'I'll never engage in creating kings again; it's too great a strain'. As Honorary Director of Antiquities, she decided on the share of archaeological finds apportioned to foreign expeditions for their museums. And as confidante to King Faisal, she supervised the selection of appointees for his cabinet and other leadership posts in the new government.

As the foregoing illustrates, archaeologists and archaeological expertise provided expert cover for espionage and military operations. At the same time, that expertise extended imperial ambitions, whether territorial, cultural, or religious, and invariably incorporated them, in a way that no other discipline afforded. This nexus of military and academic opportunism, underwritten by technologies of rule like surveillance and mapping, was typically shrouded in the redemptive claims of restoring civilisation. Regrettably, those moments are not relegated to history but have continued to influence the fate of the Middle East for more than a century. Hogarth, Lawrence, Bell, and Woolley may have been exceptional individuals, yet their aspirations and entanglements have had lasting consequences across the region. It is thus timely that we uncover how archaeologists from many nations have played a role in the 'future machinery' of espionage and empire.

Acknowledgements

I extend my thanks to the President and Fellows, Magdalen College Oxford, for their kind permission to use their archives and images. Charlotte Berry and Lucy Smith deserve special mention for their assistance. The visit was made possible by my colleagues Jas Elsner and Constanze Guthenke at Corpus Christi College. From the School of Archaeology, I thank Amy Bogaard for her support during my time in Oxford. Felicity Cobbing at the Palestine Exploration Fund has been incredibly generous with her time and expertise, as has Phillip Freeman at Liverpool University.

Notes

1. Speech by the Archbishop of York, Palestine Exploration Fund, *Report of the Proceedings at a Public Meeting, London, June 22, 1865*, 4.
2. https://www.penn.museum/sites/expedition/sir-leonard-woolley/
3. https://archive-cat.magd.ox.ac.uk/records/P452
4. April 22, 1918. Letter from D. G. Hogarth to Laura Hogarth, P452/PER/2/4/5/17, Hogarth Letters, Magdalen College Oxford.
5. January 1927. A. C. L. Shaw (a.k.a T. E. Lawrence) to Laura Hogarth, Hogarth Letters REL/2/1, Magdalen College Oxford.
6. August 6, 1911. Letter from T. E. Lawrence to D. G. Hogarth, Hogarth Letters REL/2/1, Magdalen College Oxford.
7. December 16, 1911(?). Letter from T. E. Lawrence to D. G. Hogarth, P452/REL/2/1/ii, Magdalen College Oxford.
8. August 17, 1916. Letter from D. G. Hogarth to G. Clayton, Hogarth Letters P452/MIL/2, Magdalen College Oxford.
9. July 11, 1917. Letter from D. G. Hogarth to G. Clayton, Hogarth Letters P452/MIL/2, Magdalen College Oxford.
10. April 29, 1918. D. G. Hogarth to Laura Hogarth from Jerusalem, Hogarth Letters P452/PER/2/5/18 (1–2), Magdalen College Oxford.
11. January 30, 1918, D. G. Hogarth to Laura Hogarth from Jerusalem, Hogarth Letters P452/PER/2/4/4, Magdalen College Oxford.
12. March 30, 1919. D. G. Hogarth to G. Clayton, Hogarth Letters P452/MIL/2, Magdalen College Oxford.
13. May 19, 1919. Letter from D. G. Hogarth to G. Clayton, Hogarth Letters P452/MIL/2, Magdalen College Oxford.
14. FO 608/240, http://discovery.nationalarchives.gov.uk/details/r/C2058238. Their justification was 'no security that the representatives on the spot of each mandatory power will have any sympathy with antiquarian matters'.
15. FO 608/116/6 Foreign Office Minutes, March 24, 1919 (see Corbett 2015, 88).
16. FO 371/3398/57. I thank Phil Freeman for this point and alerting me to these documents, and for sharing his knowledge on the topic. See also FO 608/82, FO 608/116, FO 371/3398/58, FO 371/3398/60, and FO 371/3398/122.
17. See the National Archives FO 608/116, Article 422 of the Treaty of Sèvres, August 10, 1920 (National Archives FO 93/110/81).
18. T. E. Lawrence, June 1930, quoted in Hart 1933, 159. For full discussion see Satia 2011, 40.
19. https://www.oxforddnb.com/view/10.1093/ref:odnb/9780198614128.001.0001/odnb-9780198614128-e-30686
20. December 5, 1918. Letter from Gertrude Bell to Florence Bell (Burgoyne 1958–61, ii, 10). See also Satia 2011, 173.

References

Barr, J. (2011) *A Line in the Sand: Britain, France and the Struggle that Shaped the Middle East*. New York, Simon and Schuster.

Bell, G. (2015) *A Woman in Arabia: The Writings of the Queen of the Desert*. London, Penguin.

Brown, M. (ed.) (1994) *T. E. Lawrence: The Selected Letters*. New York, Paragon House.

Burgoyne, G. (ed.) (1958–61) *Gertrude Bell: From Her Personal Papers*. 2 vols. London, Ernest Benn Ltd.

Corbett, E. (2015) *Competitive Archaeology in Jordan: Narrating Identity from the Ottomans to the Hashemites*. Austin, University of Texas Press.

Gibson, S. (1999) British archaeological institutions in Mandatory Palestine, 1917–1948. *Palestine Exploration Quarterly* 131, 115–43.

Hart, B. H. L. (1933) *The British Way in Warfare*. New York, Macmillan.

Hauser, K. (2007) *Shadow Sites: Photography, Archaeology, and the British Landscape 1927–1955*. Oxford, Oxford University Press.

Hogarth, D. G. (1927) Gertrude Bell's journey to Hayil. *The Geographical Journal* LXX, 1–17.

Melman, B. (2019) Ur: Empire, modernity, and the visualization of antiquity between the two World Wars. *Representations* 145, 129–151.

Meskell, L. M. (2020) Imperialism, internationalism, and archaeology in the un/making of the Middle East. *American Anthropologist* 122, 554–67.

Moscrop, J. J. (2000) *Measuring Jerusalem: The Palestine Exploration Fund and British Interests in the Holy Land*. Leicester & New York, Leicester University Press.

Richter, T. (2008) Espionage and Near Eastern archaeology: A historiographical survey. *Public Archaeology* 7, 212–40.

Satia, P. (2011) A rebellion of technology. Development, policing, and the British Arabian imaginary. In D. K. Davis & E. Burke (eds), *Environmental Imaginaries of the Middle East and North Africa*, 23–59. Athens, Ohio University Press.

Winstone, H. V. F. (1990) *Woolley of Ur: The Life of Sir Leonard Woolley*. London, Harvill Secker.

Woolley, C. L. (1920) *Dead Towns and Living Men: Being Pages from an Antiquary's Notebook*. London, Milford.

Woolley, C. L & Lawrence, T. E. (1936) *The Wilderness of Zin*. London, Cape.